CEREBRAL CORTEX

Volume 4

Association and Auditory Cortices

CEREBRAL CORTEX

Edited by Edward G. Jones and Alan Peters

Volume 1: Cellular Components of the Cerebral Cortex

Volume 2: Functional Properties of Cortical Cells

Volume 3: Visual Cortex

Volume 4: Association and Auditory Cortices

Volume 5: Sensory-Motor Areas and Aspects of Cortical Connectivity

CEREBRAL CORTEX

Volume 4
Association and Auditory Cortices

Edited by

ALAN PETERS
Boston University School of Medicine
Boston, Massachusetts

and

EDWARD G. JONES
California College of Medicine
University of California, Irvine
Irvine, California

Plenum Press • New York and London

Library of Congress Cataloging in Publication Data

Main entry under title:

Cerebral cortex.

Vol. 2, edited by Edward G. Jones and Alan Peters.
Includes bibliographies and indexes.
Contents: v. 1. Cellular components of the cerebral cortex—v. 2. Functional properties of cortial cells—[etc.]—v. 4. Association and auditory cortices.
1. Cerebral cortex—Collected works. I. Peters, Alan, 1929- . II. Jones, Edward G., 1939- . [DNLM: 1. Cerebral Cortex—anatomy and histology. 2. Cerebral Cortex—physiology. WL 307 C4136]
QP383.C45 1984 612′.825 84-1982
ISBN 0-306-42040-6

A Division of Plenum Publishing Corporation
233 Spring Street, New York, N.Y. 10013

Printed in the United States of America

This volume is dedicated to the memory of a dear friend, Dr. Norman Geschwind (1926–1984), who at the time of his death was James Jackson Putnam Professor of Neurology at Harvard Medical School. We were lucky enough to have Norman agree to serve on the Advisory Committee for this treatise, and he served us well, for he could always be relied upon to recommend the names of potential authors and to give an account of what he considered to be their strengths and weaknesses. This reflected one of Norman's greatest strengths: his ability to analyze and to consider broad issues. This ability placed him in a special category of neuroscientists, for it allowed him to synthesize information which changed many views about the functional architecture of the brain, the specialized talents of the two cerebral hemispheres, and the anatomical substrates for language and its dysfunction. It is fitting that four of the authors in the present volume—Drs. Pandya, Vogt, Mesulam, and Mufson—have at some time worked directly with Norman Geschwind and have benefited from his insight into cortical functions, and that the chapter by Seldon makes reference to the work of Norman Geschwind on the asymmetry of the planum temporale.

Contributors

John F. Brugge Department of Neurophysiology and Waisman Center on Mental Retardation and Human Development, University of Wisconsin, Madison, Wisconsin 53705

Irving T. Diamond Departments of Psychology and Anatomy, Duke University, Durham, North Carolina 27706

David Fitzpatrick Departments of Psychology and Anatomy, Duke University, Durham, North Carolina 27706

Joaquin M. Fuster Department of Psychiatry and Brain Research Institute, University of California School of Medicine, Los Angeles, California 90024

M.-Marsel Mesulam Bullard and Denny-Brown Laboratories and Behavioral Neurology Section, Neurology Department, Harvard University, and Charles A. Dana Research Institute, Beth Israel Hospital, Boston, Massachusetts 02215

Elliott J. Mufson Bullard and Denny-Brown Laboratories and Behavioral Neurology Section, Neurology Department, Harvard University, and Charles A. Dana Research Institute, Beth Israel Hospital, Boston, Massachusetts 02215

Deepak N. Pandya Edith Nourse Rogers Memorial Veterans Hospital, Bedford, Massachusetts 01730; and Departments of Anatomy and Neurology, Boston University School of Medicine and Harvard Neurological Unit, Beth Israel Hospital, Boston, Massachusetts 02215

Richard A. Reale Department of Neurophysiology and Waisman Center on Mental Retardation and Human Development, University of Wisconsin, Madison, Wisconsin 53705

H. Lee Seldon Physiology Institute, Free University of Berlin, D-1000 Berlin 33, Federal Republic of Germany

James M. Sprague Department of Anatomy, University of Pennsylvania, Philadelphia, Pennsylvania 19104

Brent Alan Vogt Departments of Anatomy and Physiology, Boston University School of Medicine, Boston, Massachusetts 02118

I. C. Whitfield Neurocommunications Research Unit, University of Birmingham, Birmingham B15 2TJ, England. *Present address:* Department of Electronic and Electrical Engineering, University of Birmingham, Birmingham B15 2TT, England

Edward H. Yeterian Department of Psychology, Colby College, Waterville, Maine 04901

Preface

This volume deals with some of the association areas of the cerebral cortex and with the auditory cortex. In the first chapter, by Deepak Pandya and Edward Yeterian, the general architectural features and connections of cortical association areas are considered; as these authors point out, in primates the association areas take up a considerable portion of the total cortical surface. Indeed, it is the development of the association areas that accounts for the greatest differences between the brains of primate and nonprimate species, and these areas have long been viewed as crucial in the formation of higher cognitive and behavioral functions. In the following chapter, Irving Diamond, David Fitzpatrick, and James Sprague consider the question of whether the functions of the association areas depend on projections from the sensory areas of the cortex. They use the visual cortex to examine this question and show that there is a great deal of difference between species in the amount of dependence, the differences being paralleled by variations in the manner in which the geniculate and pulvinar nuclei of the thalamus project to the striate and extrastriate cortical areas. One of the more interesting and perhaps least understood of the association areas is the cingulate cortex, discussed by Brent Vogt. Cingulate cortex has been linked with emotion and with affective responses to pain, and in his chapter Vogt gives an account of its cytoarchitecture, connections, and functions. The chapter on prefrontal cortex by Joaquin Fuster considers an association area in a different light and puts forward the hypothesis that the principal and most characteristic function of the prefrontal cortex is the temporal organization of behavior. In the next contribution, Marsel Mesulam and Elliott Mufson describe the extensive connections of the insula with other portions of the cerebral hemisphere and with the thalamus, and they discuss how these connections might account for some of the possible functions of the insula.

Of the chapters dealing with auditory cortex, the first one is by John Brugge and Richard Reale, and it deals with the disposition of the various auditory cortical fields in cat and monkey. These authors describe the response properties of single auditory neurons, before going on to discuss the functions of the auditory fields and their connections. Next is an account of the neuronal structure and connections of human auditory cortex given by H. Lee Seldon, who considers the role of this cortex in speech and the differences between right and left temporal fields. In the last chapter, I. C. Whitfield gives an analysis of the role of auditory cortex in behavior and discrimination.

We wish to thank the contributors to this volume for their cooperation and the time they have spent to produce such scholarly contributions. The staff of Plenum Press has clearly made every effort to ensure that their contributions are presented as well as possible, and we are grateful to them for maintaining such high standards in the production of this treatise. In particular we thank Kirk Jensen, Senior Editor for Life Sciences, and Peter Strupp, Assistant Managing Editor.

The development of this work would not of course have been possible without advice from our editorial board members, to one of whom, Norman Geschwind, this volume is dedicated. After his untimely death, Norman is sorely missed.

Alan Peters
Edward G. Jones

Boston and Irvine

Contents

I. Association Cortices

Chapter 1

Architecture and Connections of Cortical Association Areas

Deepak N. Pandya and Edward H. Yeterian

Chapter 2

The Extrastriate Visual Cortex: A Historical Approach to the Relation between the "Visuo-Sensory" and "Visuo-Psychic" Areas

Irving T. Diamond, David Fitzpatrick, and James M. Sprague

Chapter 3

Cingulate Cortex

Brent Alan Vogt

Chapter 7

The Anatomy of Speech Perception: Human Auditory Cortex

H. Lee Seldon

Chapter 8

The Role of Auditory Cortex in Behavior

I. C. Whitfield

I

Association Cortices

1

Architecture and Connections of Cortical Association Areas

DEEPAK N. PANDYA and EDWARD H. YETERIAN

1. Introduction

The execution of behavior depends on the integration of activity at various levels of the central nervous system. Nevertheless, different regions of the nervous system have been shown to participate in specific functions. Whereas, at the cortical level, the primary and secondary sensory areas are involved in the elementary analysis of incoming information, the other regions of the cerebral cortex contribute to a variety of complex processes (Figs. 1 and 2). These latter regions are designated as association areas and may be conceptualized as being interposed between the external and the internal environments (Figs. 1 and 3). These areas are highly developed in primates and have been subdivided on the basis of architecture, their cortical and subcortical connections, and function, into several different types. For example, cortical areas subjacent to the primary sensory regions are termed first-order parasensory areas (Fig. 3), while those beyond the first-order parasensory areas are designated as second-order or third-order association areas (Jones and Powell, 1970a; Chavis and Pandya, 1976). Other cortical regions, which are situated at the junctions of modality-specific

DEEPAK N. PANDYA • Edith Nourse Rogers Memorial Veterans Hospital, Bedford, Massachusetts 01730; and Departments of Anatomy and Neurology, Boston University School of Medicine and Harvard Neurological Unit, Beth Israel Hospital, Boston, Massachusetts 02215. EDWARD H. YETERIAN • Department of Psychology, Colby College, Waterville, Maine 04901.

regions in parietal, occipital, and temporal cortices (Fig. 4), are termed multimodal areas (Jones and Powell, 1970a; Seltzer and Pandya, 1976, 1978, 1980). Within the frontal lobe, beyond the motor cortex, a similar categorization of association areas has been identified (Jones and Powell, 1970a; Chavis and Pandya, 1976). Finally, the medial and ventral portions of the cerebral cortex contain groups of specialized regions (Fig. 5) which are termed paralimbic regions (Locke *et al.,* 1964; Yakovlev *et al.,* 1966; Pandya and Seltzer, 1982b).

It is thought that for information reaching the cerebral cortex to be further elaborated upon and ultimately lead to the execution of behavior, it must pass from the primary sensory areas through the different cortical association areas. This process of advancing information within the cerebral cortex must also involve the interlinking of cerebral connections among different cortical and subcortical regions. With this in mind, several investigators have attempted to study the long and short cortical connections of these regions in primates using different techniques. As more refined and precise neuroanatomical methods became available, more exact and detailed connectivity of these regions has been described. Thus, in recent years, it has been shown that within a given modality, related regions are interconnected in a stepwise fashion, beginning in the primary sensory area and progressing sequentially through post-Rolandic parasensory and multimodal areas and finally reaching frontal association and paralimbic regions (Pandya and Kuypers, 1969; Jones and Powell, 1970a; Van Hoesen *et*

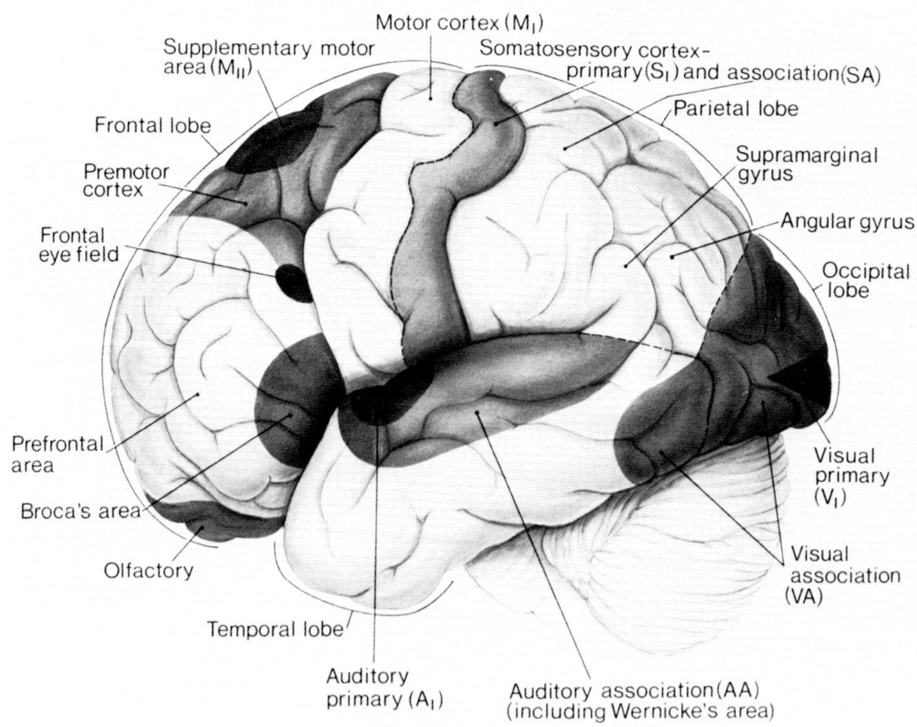

Figure 1. Map of human cerebral cortex showing major functional areas. From Lassen *et al.* (1978) with slight modification.

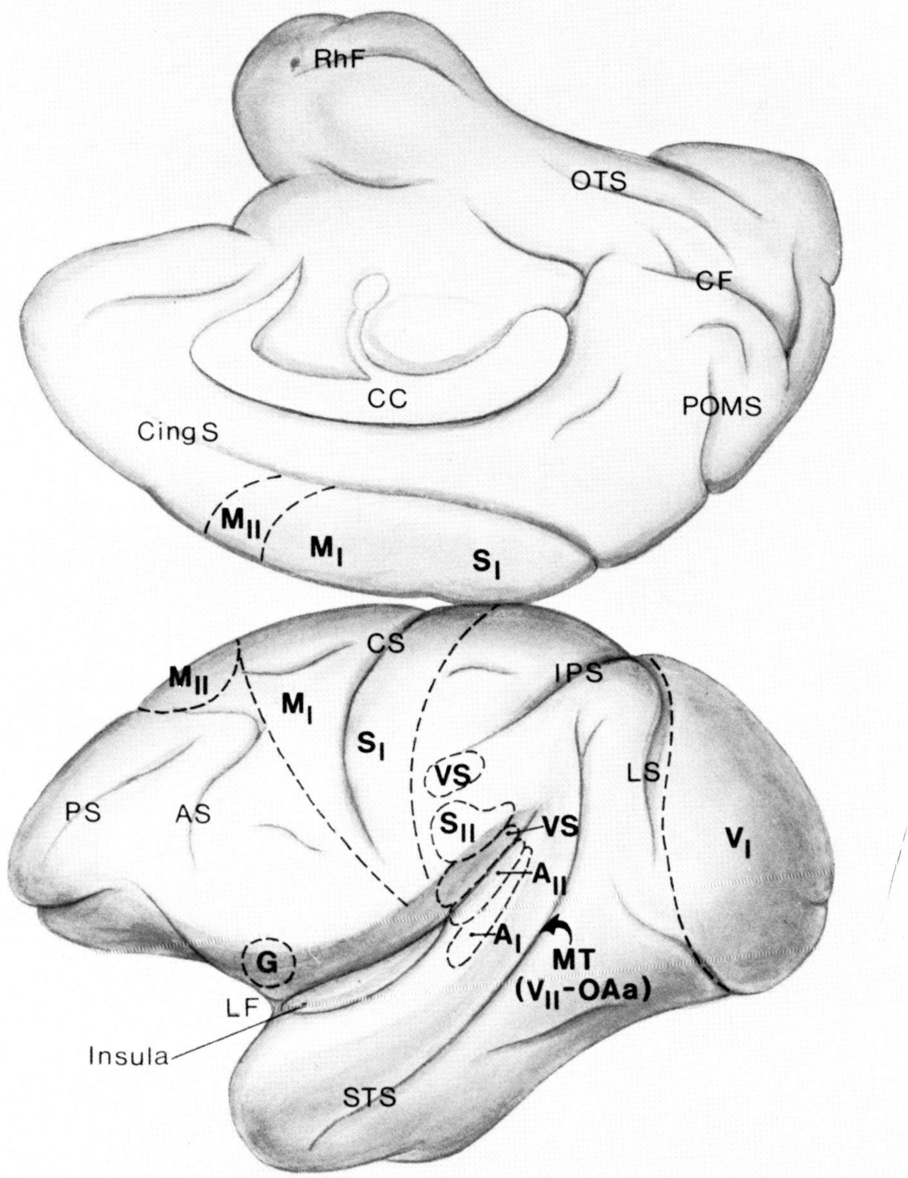

Figure 2. Diagrams of the lateral and medial surfaces of the cerebral hemisphere of the rhesus monkey showing primary and secondary sensory areas as well as primary motor and supplementary motor areas. Abbreviations in this and subsequent figures: AS, arcuate sulcus; CC, corpus callosum; CF, calcarine fissure; CING S, cingulate sulcus; CS, central sulcus; ENTO, entorhinal cortex; G, gustatory area; HIPPO, hippocampus; IOS, inferior occipital sulcus; IPS, intraparietal sulcus; LF, lateral (Sylvian) fissure; LS, lunate sulcus; OLF, olfactory bulb and tract; OS, orbital sulcus; OT, olfactory tubercle; OTS, occipitotemporal sulcus; POMS, medial parieto-occipital sulcus; PS, principal sulcus; RhF, rhinal fissure; Rspl, retrosplenial cortex; Sub-hp, subiculum-hippocampus.

al., 1972; Pandya and Seltzer, 1982b). The sequential nature of the flow of information through the long connections of association areas has been viewed as an important substrate for attention, sensory integration, learning, memory, as well as the performance of skilled behavior (e.g., Geschwind, 1965a,b; Heilman *et al.*, 1970; Mishkin, 1972; Watson *et al.*, 1973; Petrides and Iversen, 1978; Ungerleider and Mishkin, 1982; Mesulam, 1983; Mishkin *et al.*, 1983). While studies of the long association connections revealed parallel organization for post-Rolandic sensory systems in terms of their relations with the multimodal areas, frontal lobe, and paralimbic regions, other investigations have addressed

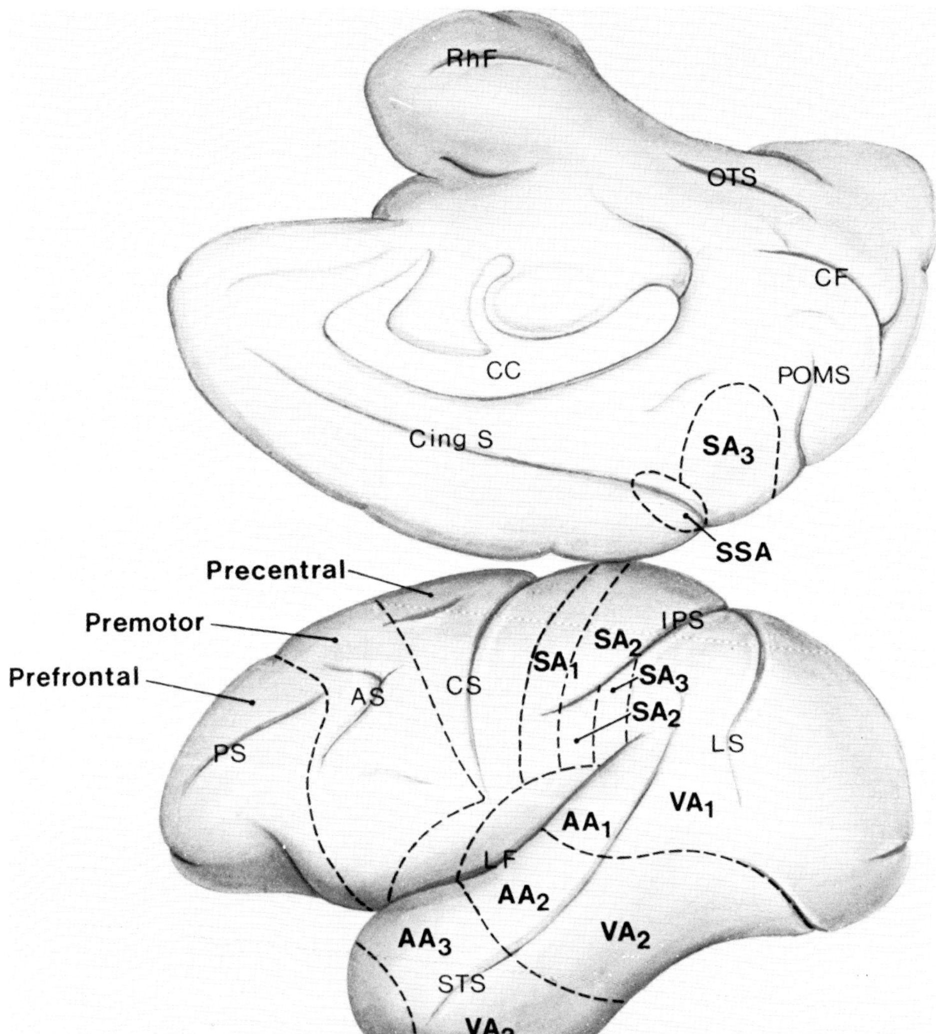

Figure 3. Diagrams showing post-Rolandic parasensory association areas as well as frontal lobe association areas.

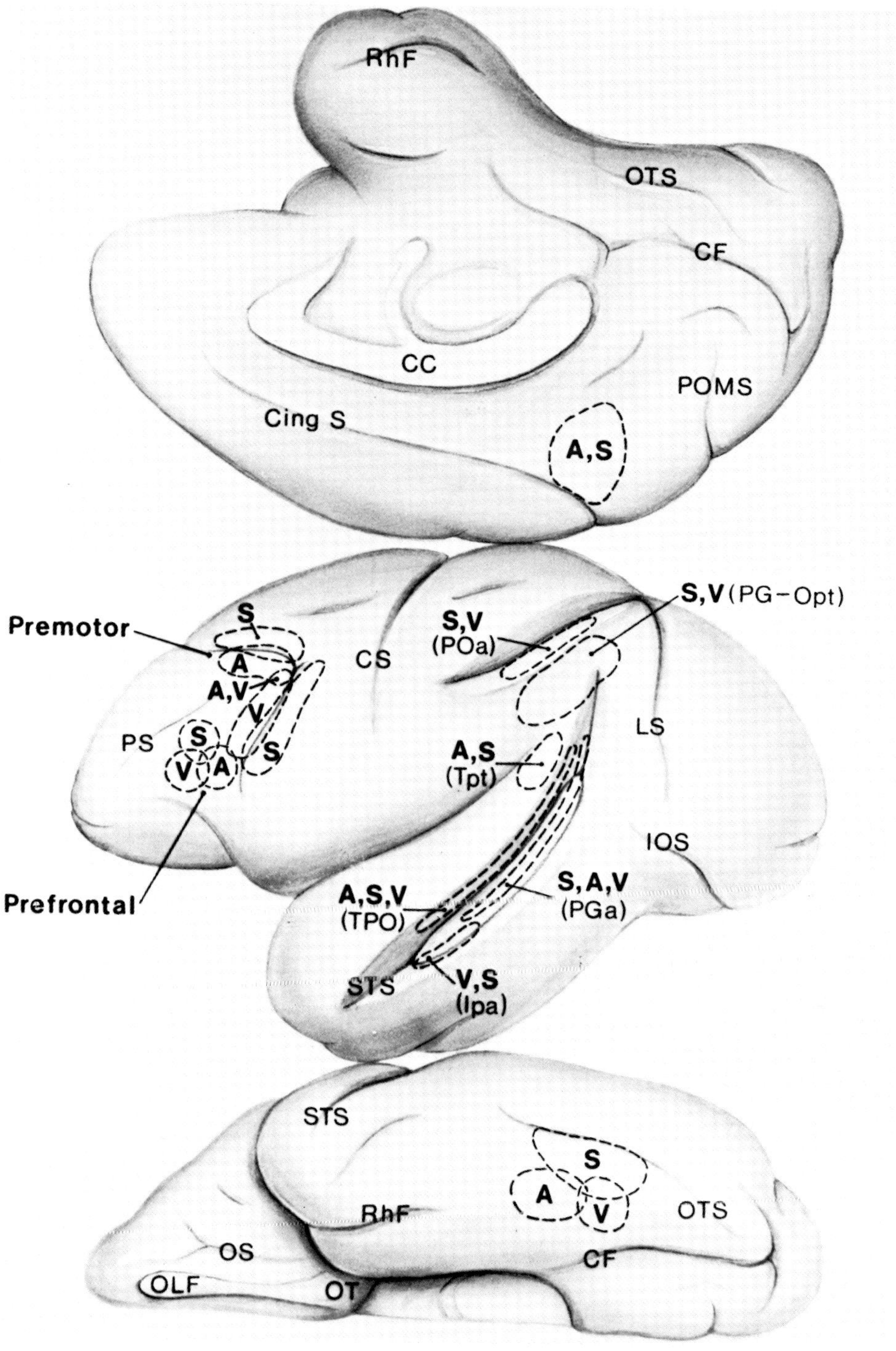

Figure 4. Diagrams showing the locations of multimodal sensory convergence areas in pre- and post-Rolandic regions on the medial, lateral, and ventral surfaces.

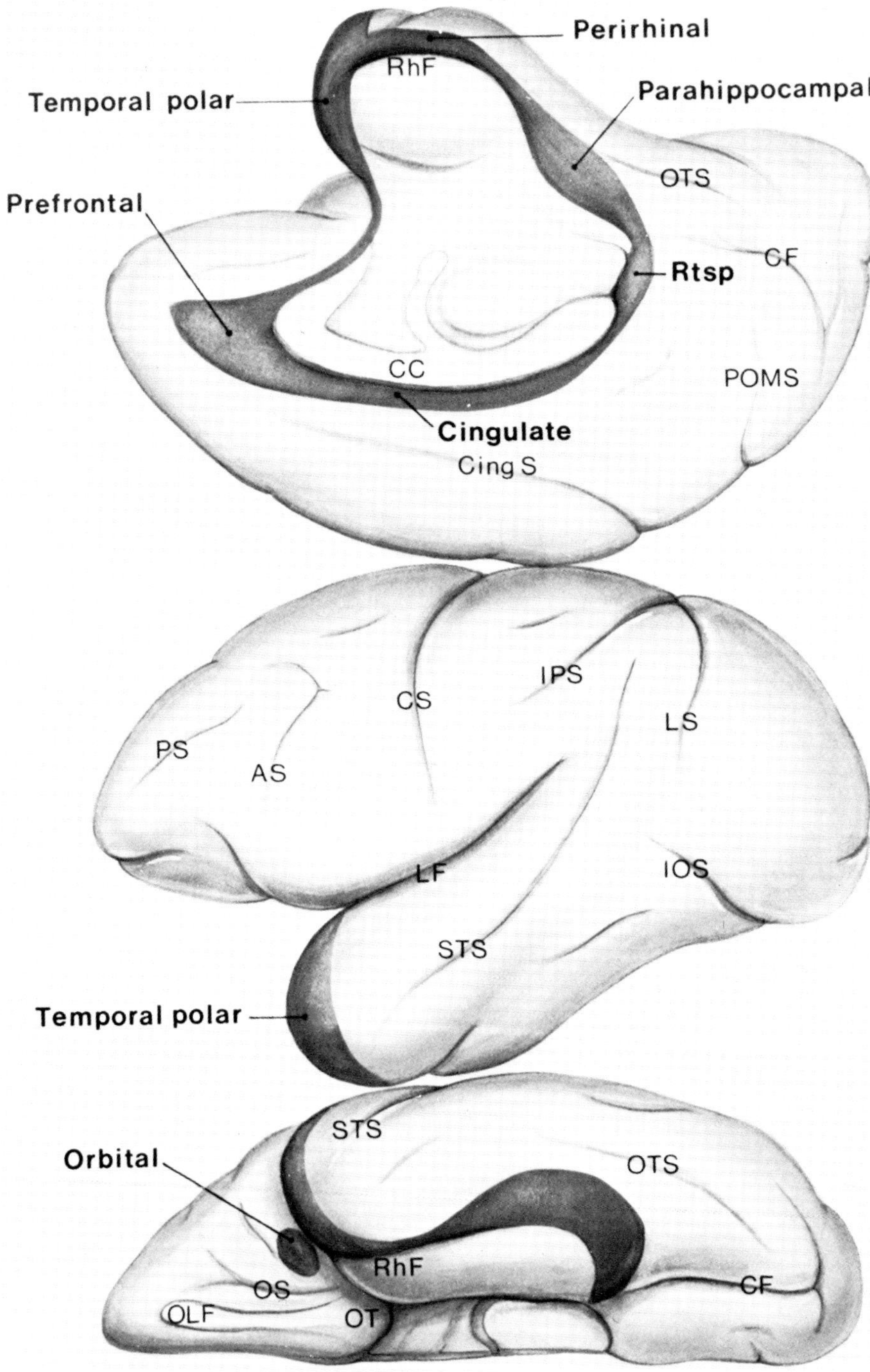

Figure 5. Diagrams showing the extent of paralimbic association areas (dark).

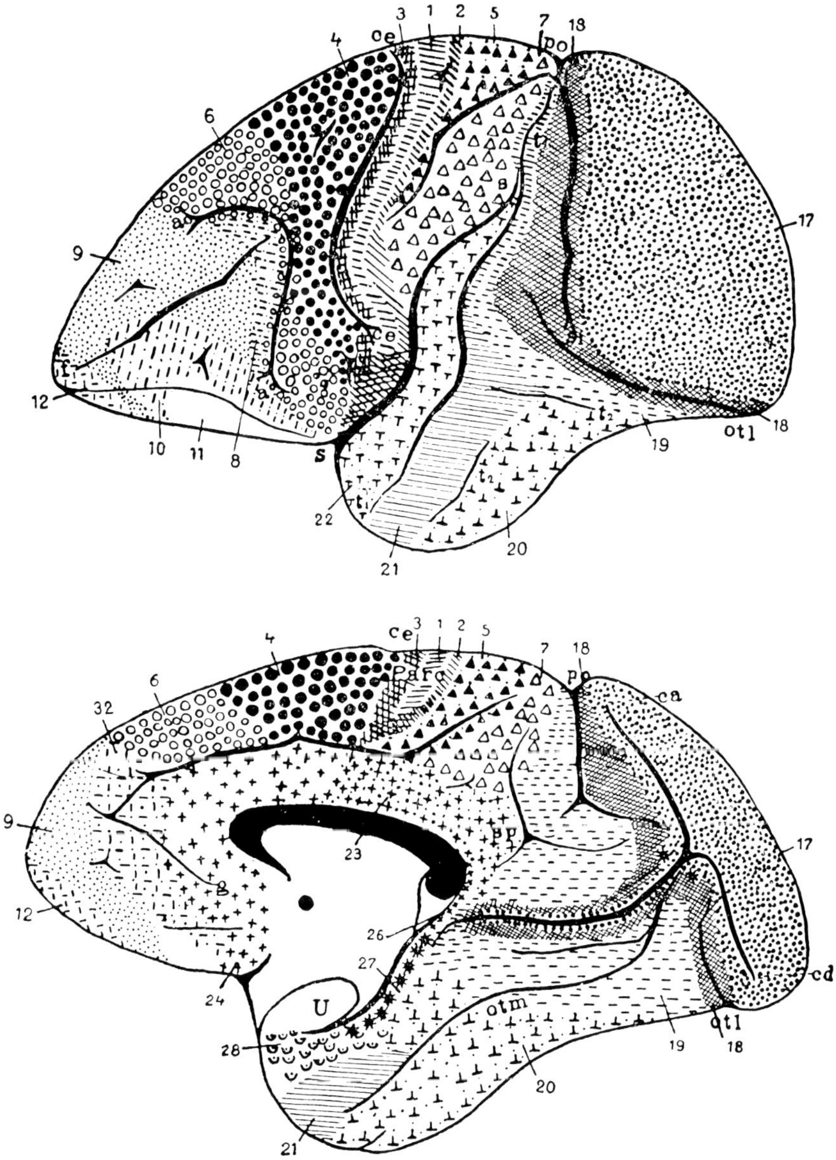

Figure 6. Architectonic map of the cerebral cortex of *Cercopithecus* according to Brodmann (1909).

the issue of the intrinsic connections* of cortical areas within one specific sensory modality (Kuypers *et al.*, 1965; Jones and Powell, 1969; Jones and Wise, 1977; Tigges *et al.*, 1977; Vogt and Pandya, 1978; Zeki, 1978; Rockland and Pandya, 1979, 1981; Fitzpatrick and Imig, 1980; Pandya and Seltzer, 1982a; Kaas, 1983; Van Essen and Maunsell, 1983; Galaburda and Pandya, 1983). These latter studies have demonstrated common organizational features for the adjacent areas in terms of cells of origin and laminar terminations of the intrinsic connections.

It is equally important to understand the nature of cortical neuronal organization since neurons give rise to the intrinsic as well as long cortical fiber systems. Thus, the cytoarchitecture of the cerebral cortex in primates has been delineated since the turn of the century by a number of investigators (e.g., Campbell, 1905; Brodmann, 1909; Vogt and Vogt, 1919; von Economo and Koskinas, 1925; von Bonin and Bailey, 1947). These studies have divided the cerebral cortex into different regions according to cellular makeup (Fig. 6). It is only within the past 25 years that a coherent principle of cortical architecture has been appreciated. From comparative architectonic studies in primitive insectivores, prosimians, and other primates, Sanides (1969, 1972) has proposed the dual nature of development of the cerebral cortex. According to Sanides, the isocortex evolves from two prime moieties (Fig. 7), namely archicortical (hippocampal) and paleocortical (olfactory or insular). Both archi- and paleocortices give rise to nearly six-layered cortex, proisocortex, which in turn gives rise to isocortex (true six-layered cortex). Further steps in the progression of isocortex are from two structures, namely the paralimbic (cingulate; see Vogt, this volume) and parinsular (insula; see Mesulam and Mufson, this volume) areas which are related to archi- and paleocortical moieties, respectively. Paralimbic areas show predominant emphasis on pyramidal cells and give rise to the supplementary motor (MII) and sensory (SSA) areas located on the medial surface of the primate cerebral cortex (Fig. 8). From this stage, the primary motor (MI) and somatosensory (SI) cortices are proposed to have evolved. The parinsular trend shows progressive emphasis on granular cells as it differentiates into second somatosensory (SII), second auditory (AII), and second visual (VII or MT in the superior temporal sulcus) areas. Further development from these regions is thought to lead to a stage in each sensory modality from which primary sensory areas originate. From both the pyramidal and granular trends also evolve the so-called post-Rolandic sensory association areas as well as frontal association areas. Thus, this architectonic approach integrates the structure and the phylogenetic development of cortical areas, rather than merely parcellating the cortex according to cellular physiognomy.

In this chapter our primary aim is to present a summary of the corticocortical connections of association areas as described by a number of investigators in recent years. In doing so, we shall try to correlate connections with the architectonic trends described by Sanides and others. Since progressive evolution of

* For the purpose of this review, intrinsic connections are defined as those which exist between adjacent architectonic areas within each lobe of the cerebral cortex. The long association connections refer to those projections which are directed to cortical areas distant to the region where they originate, i.e., connections between nonadjacent areas within a given lobe and those interrelating the various lobes of the cerebral cortex.

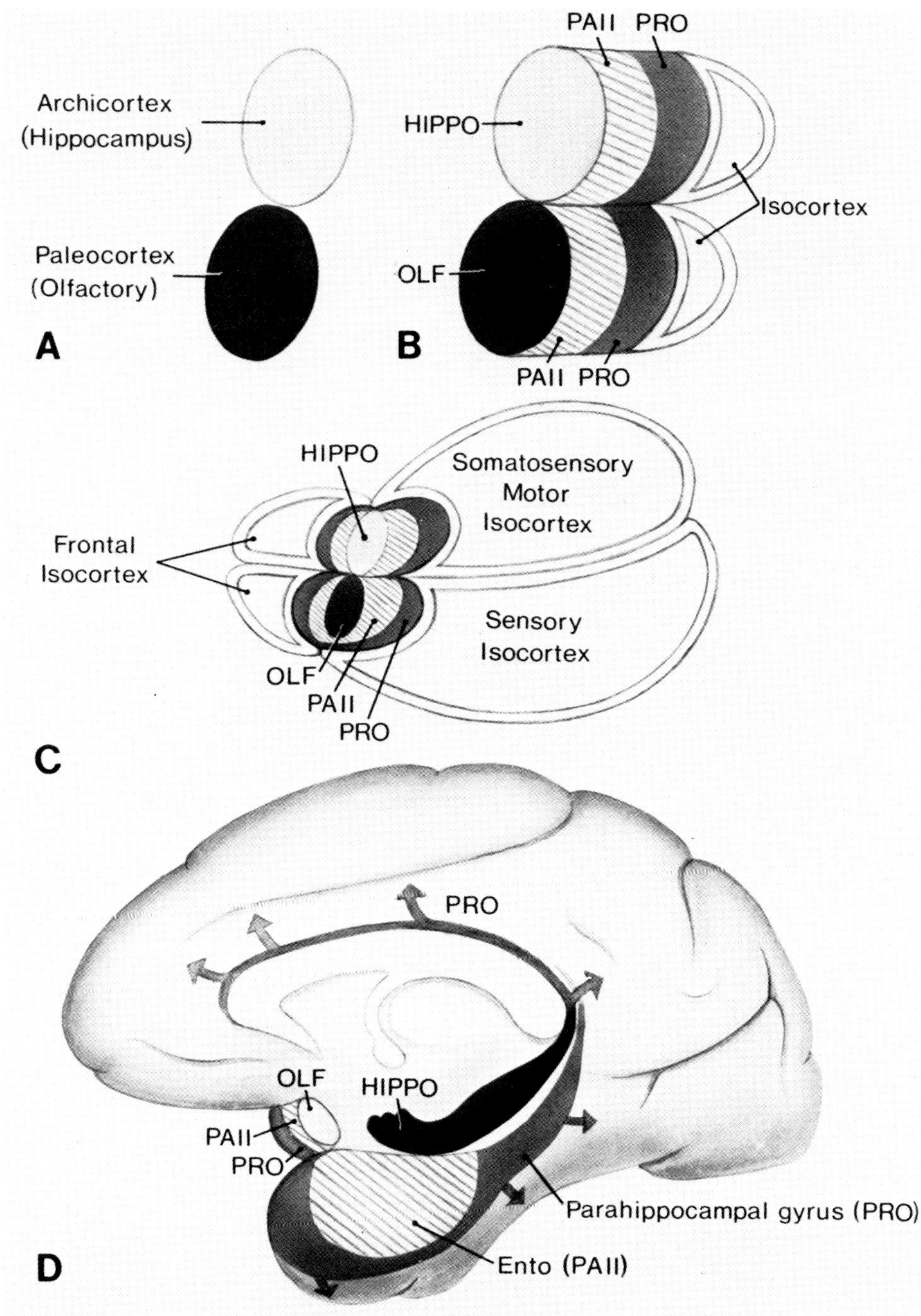

Figure 7. Diagrammatic representations (A–D) depicting the evolutionary development of cortical areas from two moieties, the archicortical (hippocampal) and the paleocortical (olfactory).

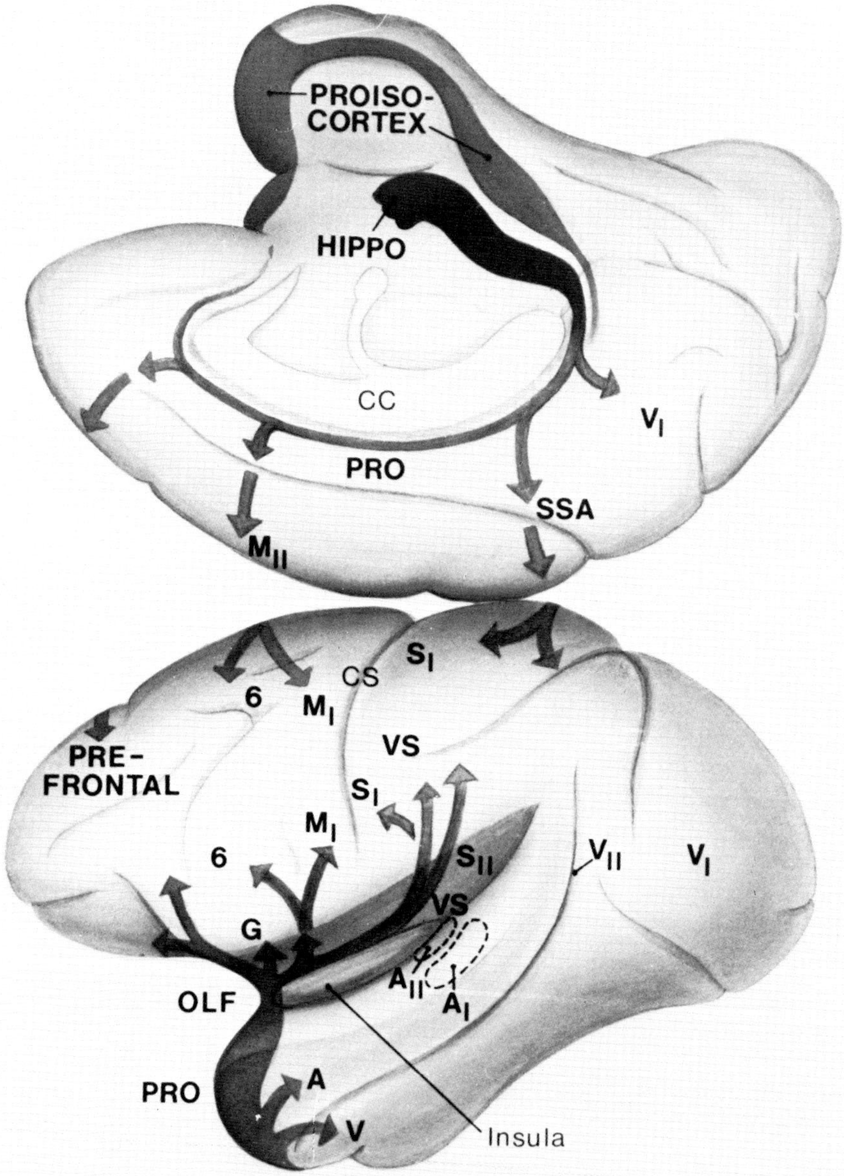

Figure 8. Diagrammatic representation of further progression of the two evolutionary trends shown in Fig. 7, on the lateral and medial surfaces, which culminate in the pre- and post-Rolandic cortical areas.

sensory and motor regions and their connections reflects increasing functional complexity, it is important to interrelate these two morphological entities. It would be most desirable to be able to do this for the human brain. However, whereas architectonic parcellations of human brains have been carried out (e.g., Brodmann, 1908), precise information regarding the connections of the human brain is not available. For this reason, we will discuss connections and architecture in rhesus monkeys, which have a basically similar brain structure and relatively well-developed cognitive and behavioral capacities. Moreover, since the early part of this century a substantial amount of information has been gathered in this primate species not only in terms of morphology but also in regard to neurophysiology and behavior. We hope that this review will provide further insight into cortical morphological organization and hence a better understanding of cortical function and a framework for future investigations.

In presenting the connectional and architectonic data, we will first describe the post-Rolandic sensory and association areas (Figs. 2 and 3), then the frontal areas (Fig. 3), and finally the multimodal (Fig. 4) and paralimbic regions (Fig. 5). For the sensory association areas, first the auditory regions of the temporal lobe will be described, since architectural and connectional analyses have been carried out in greater detail than for other cortical systems. This will be followed by the somatosensory region of the parietal lobe and the visual areas of the occipitotemporal cortices.

2. Auditory Association Areas

2.1. Architectonics

The cortical auditory system has long been known to be within the superior temporal region (STR) in both human and nonhuman primates (Ferrier, 1876; von Economo and Horn, 1930; Walker, 1938). In general, the primary auditory area in rhesus monkey has been shown to be in the supratemporal plane (STP) by its architectonic specialization, i.e., markedly granular cortex (koniocortex), its thalamic connections from a specific nucleus, the medial geniculate body (MGB), and its electrophysiological properties (Walker, 1938; Ades and Felder, 1942; Walzl and Woolsey, 1943; Kennedy, 1955; Merzenich and Brugge, 1973; Mesulam and Pandya, 1973; Pandya and Sanides, 1973). The primary auditory area is flanked by the auditory association area in the superior temporal gyrus (STG) and by the second auditory area (AII) medially around the circular sulcus (Fig. 9A; Woolsey and Fairman, 1946). The auditory association area has distinct architecture and has been designated area TA by von Bonin and Bailey and area 22 by Brodmann (Fig. 6). The thalamic connections of these regions are related more to the pulvinar nucleus than to the MGB (Walker, 1938; Mesulam and Pandya, 1973; Trojanowski and Jacobson, 1975). Behavioral studies of the primary auditory area have implicated it in elementary auditory processing such as the analysis of frequency and amplitude (Walzl and Woolsey, 1943; Merzenich and Brugge, 1973), while the association regions are considered to be involved in more integrative functions such as auditory pattern recognition and sound localization (Weiskrantz and Mishkin, 1958; Wegener, 1969; Leinonen *et al.*, 1980; Hyvärinen, 1982). Despite the morphological and functional observations, the exact limits of the primary auditory area and the auditory association regions are not fully understood. Merzenich and Brugge (1973) have shown the existence of multiple auditory representations in the superior temporal plane and superior temporal gyrus (Fig. 9B). Likewise, the thalamic projections from the MGB are shown to extend quite rostrally in the superior temporal region (Mesulam and Pandya, 1973; Wegener, 1976).

More recent architectonic analyses of the superior temporal region have revealed a beltlike organization for the auditory areas, with a koniocortex core

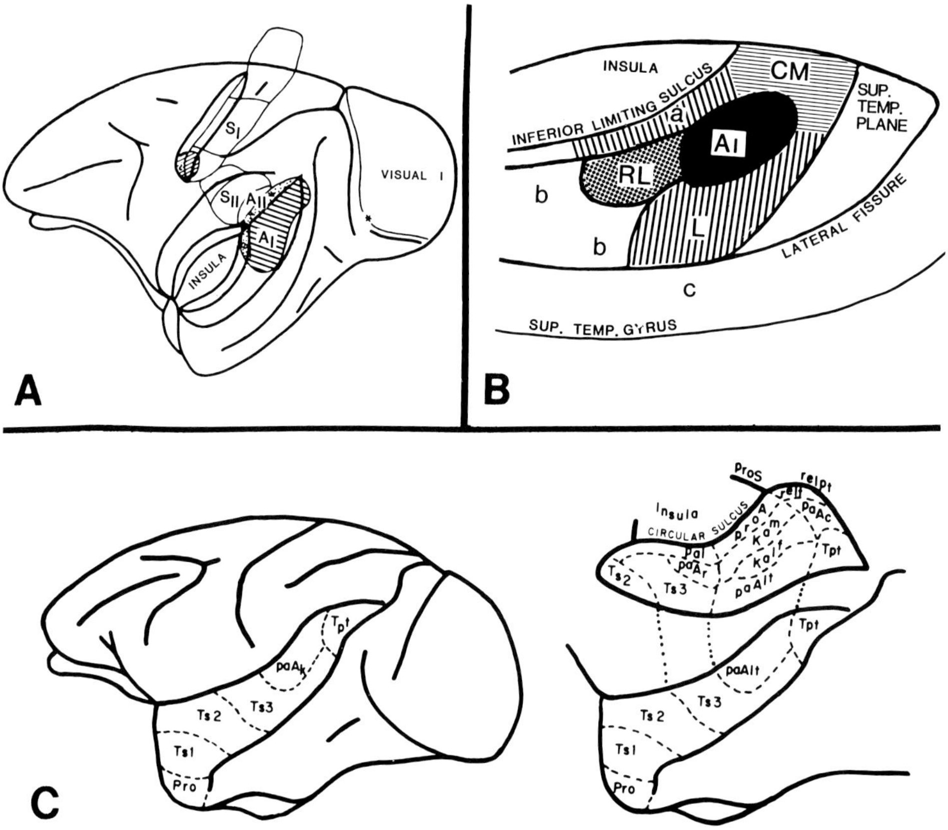

Figure 9. (A) Diagram of the lateral surface of the cerebral hemisphere of the rhesus monkey showing locations of primary and secondary auditory (AI and AII), somatosensory (SI and SII), and visual (VI) areas as described by Woolsey and Fairman (1946). Note that the sylvian fissure is opened to expose the depths. (B) Schematic drawing of the supratemporal plane and superior temporal gyrus showing major subdivisions of auditory areas based upon the cytoarchitectonic and physiological observations of Merzenich and Brugge (1973). AI, primary auditory field; RL, rostrolateral field; L, lateral field; CM, caudomedial field; a, field bordering AI medially and rostrally, buried in the circular sulcus; b, area of auditory cortex rostral and lateral to other fields; c, lateral surface of superior temporal gyrus with additional auditory fields. (C) Architectonic parcellation of the superior temporal gyrus and supratemporal plane as described by Pandya and Sanides (1973).

(corresponding to area AI), surrounded by belt areas (Pandya and Sanides, 1973; Jones and Burton, 1976). As shown in Fig. 9C, the belt is composed of one prokoniocortical area (proA, corresponding to AII) in a parinsular location, and of a caudal (paAc), a lateral (paAlt), and a rostral (paAr) parakoniocortical area. Besides outlining these opercular areas, Pandya and Sanides have demarcated additional areas in the superior temporal gyrus (areas Tpt, Ts3, Ts2, Ts1 and the proisocortical region, Pro, in the temporal pole) and in the circular sulcus (Pro, paI, proA, and reit). The studies of Jones and Burton have revealed a somewhat similar parcellation of the superior temporal region, albeit using a different nomenclature.

How all these areas relate to the auditory system has been the subject of more recent anatomical investigations. These studies have extended and elaborated upon the evolutionary architectonic concept of a dual pattern of cortical development. As indicated above, all the sensory systems have evolved as granular trends progressing outward from paleocortical moieties (Figs. 7 and 8). According to this concept, the next step from the paleocortex is the proisocortex from which development of sensory areas with progressive architecture may have proceeded. For the auditory system, the temporal polar cortex is the proisocortex which has a bilaminated appearance and a marked emphasis on infragranular neurons. Galaburda and Pandya (1983) have suggested that three distinct architectonic lines can be traced from the proisocortex of the temporal pole caudally through the superior temporal region (Fig. 10A,B). One of these lines lies along the circular sulcus in a peri-insular location. It consists of four distinct areas, namely Pro (proisocortex); paI (parinsular region); proA (prokoniocortex); and reit (retroinsular temporal cortex). These areas tend to be relatively hypocellular and retain an emphasis on infragranular layers. Progressive specialization in this line occurs in a caudal direction and consists mainly of a stepwise decrease of the relative prominence of deeper layers and an increase in cellularity in layers III and IV, although the layer V cells still remain prominent in each of these areas. Since these peri-insular areas continue to maintain primitive features as compared to the other two lines (see below), this line is termed the *root line*. Another line of progressive differentiation is found along the superior temporal gyrus, and is termed the *belt line,* because it is located at the lateral border of the auditory regions. This line is represented most rostrally by area Tsl, followed caudally by Ts3, paAlt (lateral temporal parakoniocortex), and finally area Tpt (temporoparietal region). In this belt line the most striking architectonic feature is a stepwise progressive differentiation of the third layer, which occurs along with a deemphasis of the infragranular layers. Interposed between the root and belt lines, a *core line* of differentiation can be demonstrated to lie in the supratemporal plane. It is termed core line because its fields lie between the root and belt lines. In this core line, the rostralmost field is area Ts2, followed caudally by paAr (rostral temporal parakoniocortical area), then KA (auditory koniocortex) and paAc (caudal parakoniocortical area). Progressive rostrocaudal differentiation in this line is manifested primarily by a stepwise increase in the number of small neurons in all layers of the cortex and by a successive widening of layer IV. In this line, area KA, so-called "koniocortex" or the classical primary auditory area, achieves the highest degree of granule cell proliferation of the entire superior temporal region (Fig. 10B).

Galaburda and Pandya have further suggested that these superior temporal regions may be grouped into four stages oriented in a mediolateral direction. Each stage contains a core field flanked by a medial root field and a lateral belt field (Fig. 10C). Thus, the rostralmost stage (Stage 1) contains root area Pro, area Ts2 from the core line, and Ts1 from the belt line. Stage 2 is made up of areas paI, paAr, and Ts3 as root, core, and belt areas, respectively. In Stage 3, areas proA, KA, and paAlt represent these three lines, while Stage 4 is comprised of areas reit, paAc, and Tpt in order of root, core, and belt representation. Thus, according to these architectonic observations, the auditory areas are confined not only to so-called areas AI and AII and the immediately adjacent association areas of the superior temporal gyrus but also extend in the form of stages to the temporal pole rostrally and the parietotemporal junction caudally.

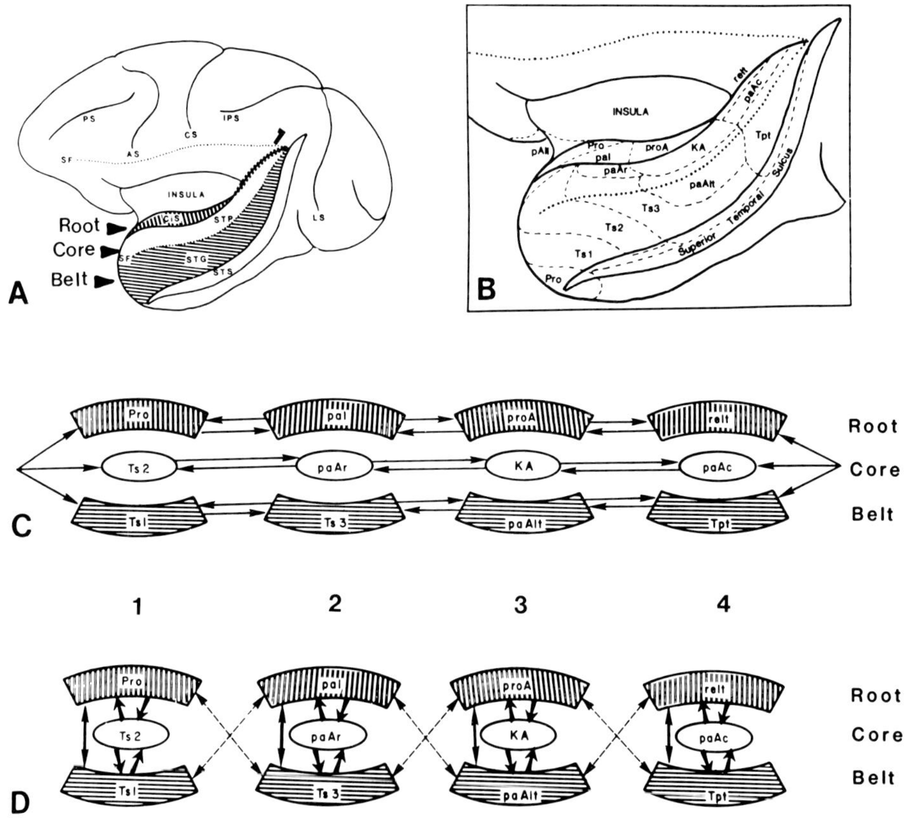

Figure 10. (A) Diagram of the lateral surface of the cerebral hemisphere showing three architectonic trends in the superior temporal region. Note the root trend in the circular sulcus, the core trend in the supratemporal plane, and the belt trend in the superior temporal gyrus. (B) Distribution of the architectonic areas in the three trends shown in (A). (C, D) Four architectonic stages within the superior temporal region, and the intrinsic connections between (C) and within (D) these stages (shown by arrows).

This architectonic notion that the entire superior temporal region is involved in the cortical auditory system receives further support from the intrinsic connectional studies carried out with antero- and retrograde tracer techniques (Fitzpatrick and Imig, 1980; Galaburda and Pandya, 1983). It has been observed that intrinsic connections within each architectonic stage as well as between adjacent stages show specific and consistent patterns (Fig. 10C,D). Thus, the core area within each stage is interconnected with its root and belt areas. Root and belt areas in each stage are also interconnected. Additionally, adjacent areas within each line are interconnected, in both rostral and caudal directions. These connections also indicate a similar pattern in terms of cells of origin and laminar terminations (Fig. 11). Thus, paAlt in the belt line receives connections from area Tpt which originate from third layer neurons in Tpt and terminate in the supragranular layers in area paAlt in a columnar fashion (Fig. 11B). In turn, area paAlt projects back to layer I of area Tpt, and these projections originate in layers V and VI (Fig. 11A). This type of reciprocal connectivity is shown between areas paAlt and Ts3 and between Ts3 and Ts1, as well as Ts1 and Pro. Laminar specificity is also seen in the connections of adjacent fields belonging to the different lines. Thus, e.g., area KA projects to its belt area paAlt where the terminations are distributed in a columnar manner in supragranular layers. Area KA also projects to its root area proA. These projections are distributed to all layers. Projections from area paAlt to its core area KA are distributed to layer I, while those to its root area are given to all cortical layers. Similar patterns of intrinsic connectivity are evident in each stage. Thus, the intrinsic connections of the STR indicate that the entire STR is related to the auditory system. Since

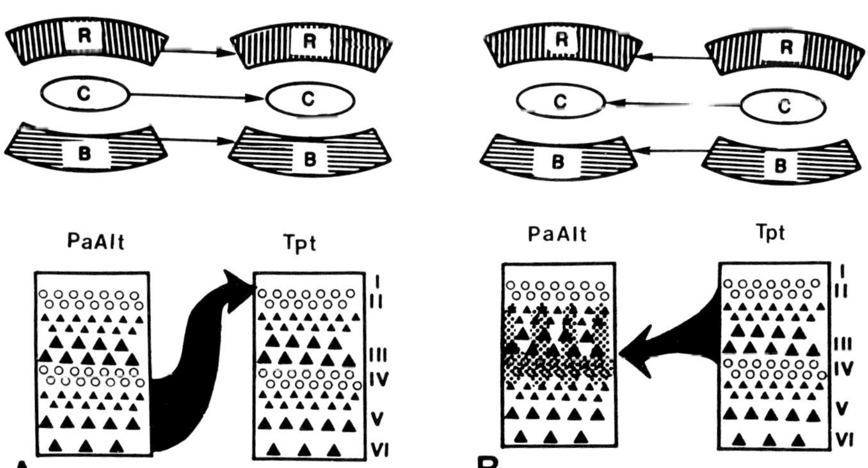

Figure 11. Common patterns of cells of origin and laminar terminations of intrinsic connections between two architectonic areas in the STR. (A) Caudally directed connections; (B) rostrally directed connections.

the main objective of this chapter is to describe the connections of association areas, we will now examine the long connections of the belt areas of the STG, which are considered to be the association areas of the auditory system on the basis of both architecture and intrinsic connections.

2.3. Long Association Connections

Many investigators have studied the intracortical connections of the auditory association cortices in the rhesus monkey (Whitlock and Nauta, 1956; Hurst, 1959; Myers, 1967; Pandya and Kuypers, 1969; Pandya *et al.*, 1969, 1971; Jones and Powell, 1970a; Trojanowski and Jacobson, 1975; Chavis and Pandya, 1976; Seltzer and Pandya, 1976, 1978; Jacobson and Trojanowski, 1977; Barbas and Mesulam, 1981; Amaral *et al.*, 1983). On the basis of long association connections, the belt areas of the STG are divided into three major sectors in a caudorostral

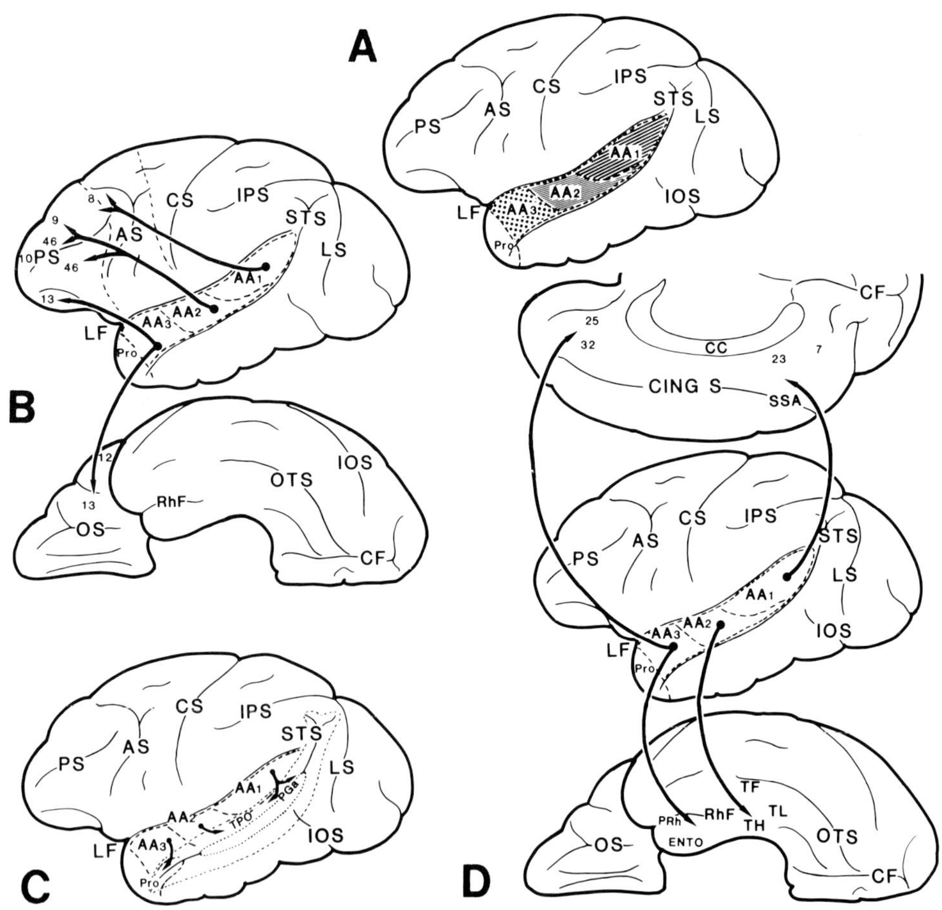

Figure 12. Diagrams showing three subdivisions of auditory association areas of the STR (A), and their frontal (B), superior temporal sulcus (C), and paralimbic (D) connections.

direction (Fig. 12A), the first, second, and third auditory association areas, AA1, AA2, and AA3, respectively (Chavis and Pandya, 1976). The *first* auditory association area (AA1) consists of architectonic regions Tpt, paAlt, and the caudal portion of Ts3. The *second* auditory association area (AA2) is comprised of the rostral portion of Ts3, and area Ts2, while the *third* auditory association area (AA3) includes the Ts1 and Pro regions. Each of these auditory association regions has a distinctive pattern of long association projections to the frontal lobe, parietotemporal region, and the paralimbic area (Fig. 12–D).Thus, area AA1 projects mainly to the dorsal periarcuate cortex in the frontal lobe, in particular dorsal area 8 in the concavity of the arcuate sulcus. The frontal lobe projections from AA2 are directed primarily to the prearcuate region, area 46 below the principal sulcus, and to the dorsal prefrontal cortex, areas 9 and 10. The third auditory association area, AA3, in the temporal polar region projects mainly to medial prefrontal and orbitofrontal cortices, i.e., areas 12, 13, 25, and 32 (Fig. 12B). Each of these auditory areas projects to a specific portion of the parietotemporal cortex, within the superior temporal sulcus (STS) (Pandya *et al.*, 1969; Jones and Powell, 1970a; Seltzer and Pandya, 1978). Thus, AA1 sends connections to areas TPO and PGa in the STS caudally, whereas AA2 projects to area TPO rostrally. Area AA3 projects most rostrally within the STS, to the proisocortical portion (Fig. 12C). Finally, the paralimbic connections of each of these areas are also quite distinctive (Pandya *et al.*, 1969, 1981; Jones and Powell, 1970a; Seltzer and Pandya, 1976; Van Hoesen, 1982). Area AA1 projects to the caudal portion of the cingulate gyrus (area 23 and medial area 7—SSA). In contrast, areas AA2 and AA3 both project to the ventral temporal region, AA2 to area TH of the parahippocampal gyrus and AA3 to the perirhinal cortex, area 35 (Fig. 12D). It should be pointed out that most of these connections between the auditory association areas and the frontal, parietotemporal, and paralimbic regions are in fact known to be reciprocal (Figs. 25, 27, 29, and 30).

2.4. Overview

Each of the auditory association areas has specific connectional relationships with other association areas in the frontal, parietotemporal, and paralimbic regions. Thus, area AA1 is connected with the periarcuate region (area 8), parietotemporal region (caudal area TPO), and paralimbic region (area 23). Area AA2, on the other hand, projects to the prefrontal region (area 46), rostral area TPO in the STS, and the paralimbic region (area TH). Finally, area AA3 projects to the orbitofrontal and medial prefrontal regions (areas 12, 13, 25, and 32), the proisocortex in the rostral STS (Pro), and the perirhinal cortex (area 35). As will be discussed, the distant projections of the auditory association areas to the frontal, parietotemporal, and paralimbic cortices are directed to specific regions, some of which are recipient of afferents from other sensory modalities. Even more striking is the fact that each of the auditory association regions projects specifically to those other association areas, in the frontal, parietotemporal, and paralimbic regions, which appear to have analogous architectonic features and appear to occupy comparable stages within their own architectonic lines. (See Section 5 for further discussion.)

The analysis of the intrinsic connections of the auditory association areas reveals that the main flow of cortical connections occurs in two directions. Thus, a forward flow of connections originates from the third layer neurons and terminates in and around layer IV of the next region. This connectivity might comprise a caudorostral system within which the successive elaboration of thalamic input may take place. Equally important are the reciprocal intrinsic connections which originate from layers V and VI and terminate in layer I of the caudally adjacent association area. These connections, which have as their cortical source the temporal proisocortex, might serve as a pathway through which limbic influences are integrated with incoming auditory information. The principles outlined above for the architecture and intrinsic and long connections of the auditory system are also evident to some extent in the somatosensory system.

3. Somatosensory Association Areas

The somatosensory system has been shown to be located in the parietal lobe. The cortex of the postcentral gyrus (i.e., architectonic areas 3, 1, and 2) has a distinct architecture (Fig. 13B) and receives thalamic input from the ventroposterior thalamic nuclei and is known as a primary sensory area, SI (Powell and Mountcastle, 1959; Jones and Powell, 1970b). The physiological studies of Woolsey (1958) have shown a topographic somatic representation within the postcentral gyrus (in areas 3, 1, and 2) in the rhesus monkey (Fig. 13A). More recent physiological investigations have suggested that there are multiple somatic representations in the postcentral gyrus (Paul *et al.*, 1975; Kaas *et al.*, 1981; Merzenich *et al.*, 1981). In terms of function, SI is involved in basic processing of somatic sensation, e.g., texture and angularity (Randolph and Semmes, 1974). In contrast, the somatosensory association areas are considered to occupy most of the posterior parietal cortex (areas 5 and 7 of Brodmann, and Vogt and Vogt; areas PE, PEm, PF, and PG of von Bonin and Bailey), and are thought to be involved in more complex and integrative functions in the somatosensory sphere (Duffy and Burchfiel, 1971; Mountcastle *et al.*, 1975; Sakata, 1975; Robinson and Goldberg, 1978; Lynch, 1980; Hyvärinen, 1982). Unlike the primary somatosensory areas, these association areas maintain connections with the associative thalamic nuclei, predominantly the lateral posterior and pulvinar nuclei, and also intralaminar and reticular nuclei (Trojanowski and Jacobson, 1975; Jones *et al.*, 1979; Yeterian and Pandya, 1982; Weber and Yin, 1984).

3.1. Architectonics

An architectonic analysis in terms of root, core, and belt areas as described for the auditory system in the STR, has not yet been carried out for the somatosensory system. Nevertheless, it appears that unlike the auditory system which originates from a single moiety, paleocortex, somatosensory areas have progressed from two different moieties, archicortical as well as paleocortical (Sanides, 1972). Thus, the superior parietal region is thought to have progressed from the hippocampal moiety, passing through the paralimbic-proisocortical stages (within the cingulate region) in successive steps to culminate in the trunk, tail, and extremity representations in the postcentral gyrus on one hand and the association areas in the superior and medial parietal lobule on the other (Fig. 13C). Likewise, the face representation of the postcentral gyrus and its association areas in the inferior parietal lobule are thought to be derived from the paleocortical moieties in successive steps passing through the parinsular proisocortical stages in the frontal and pericentral opercula. These two trends seem to merge in the depths of the intraparietal sulcus (Fig. 13C).

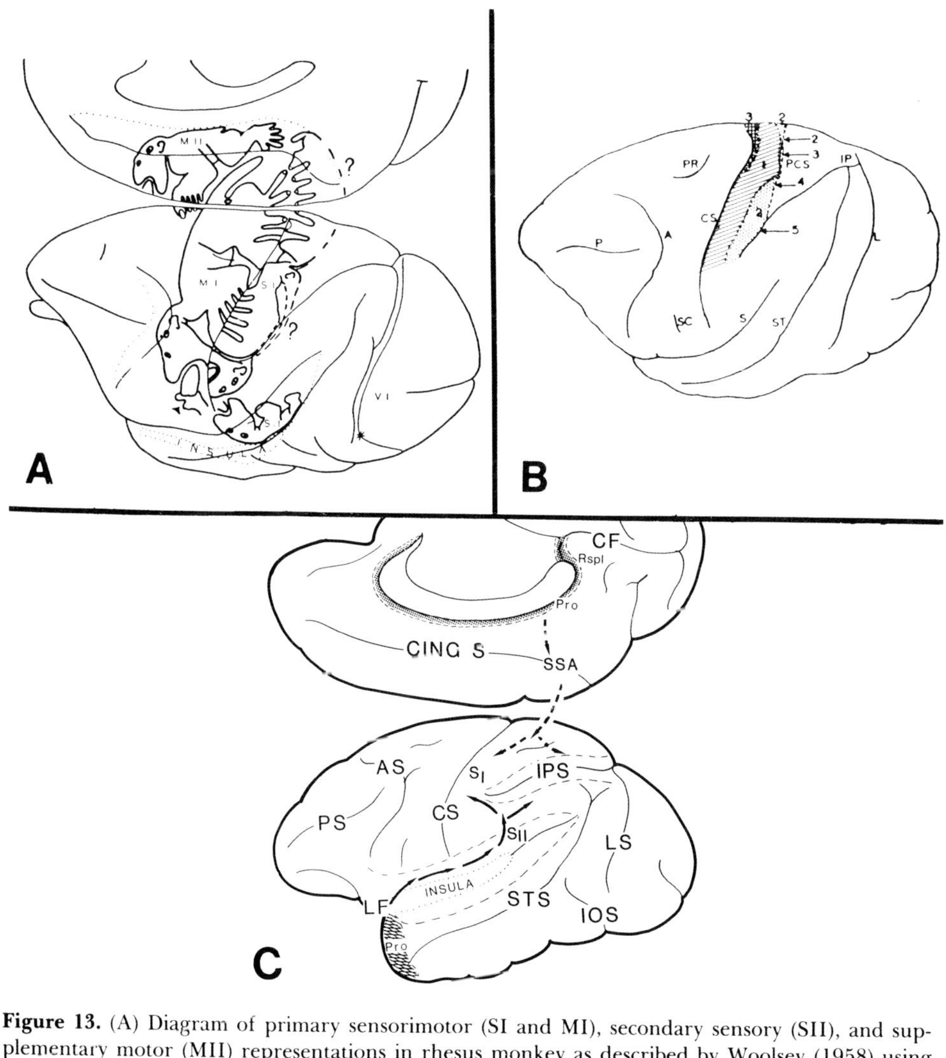

Figure 13. (A) Diagram of primary sensorimotor (SI and MI), secondary sensory (SII), and supplementary motor (MII) representations in rhesus monkey as described by Woolsey (1958) using physiological methods. (B) Architectonic boundaries of area SI of the postcentral gyrus as described by Powell and Mountcastle (1959). (C) Two progressive architectonic trends leading to the primary somatosensory and association areas of the parietal lobe. The dorsal trend (archicortical) is shown by dashed lines and the ventral trend (paleocortical) by solid lines.

Recent architectonic analyses (Fig. 14) of the posterior parietal cortices have supported the notion of the superior and inferior parietal regions as comprising dual architectonic trends (Pandya and Seltzer, 1982a). Thus, the superior parietal lobule (SPL) is occupied by one basic architectonic type, termed area PE, with two regional variants (Pandya and Seltzer, 1982a). The first, area PEc, is located at the apex of the SPL, while the second area, PEa, is in the dorsal bank of the intraparietal sulcus (IPS). Areas PE and PEc also extend onto the adjoining medial surface of the hemisphere where a third variant of PE-type cortex, PEci, is situated in the caudal portion of the cingulate sulcus. Areas PEci and PEc

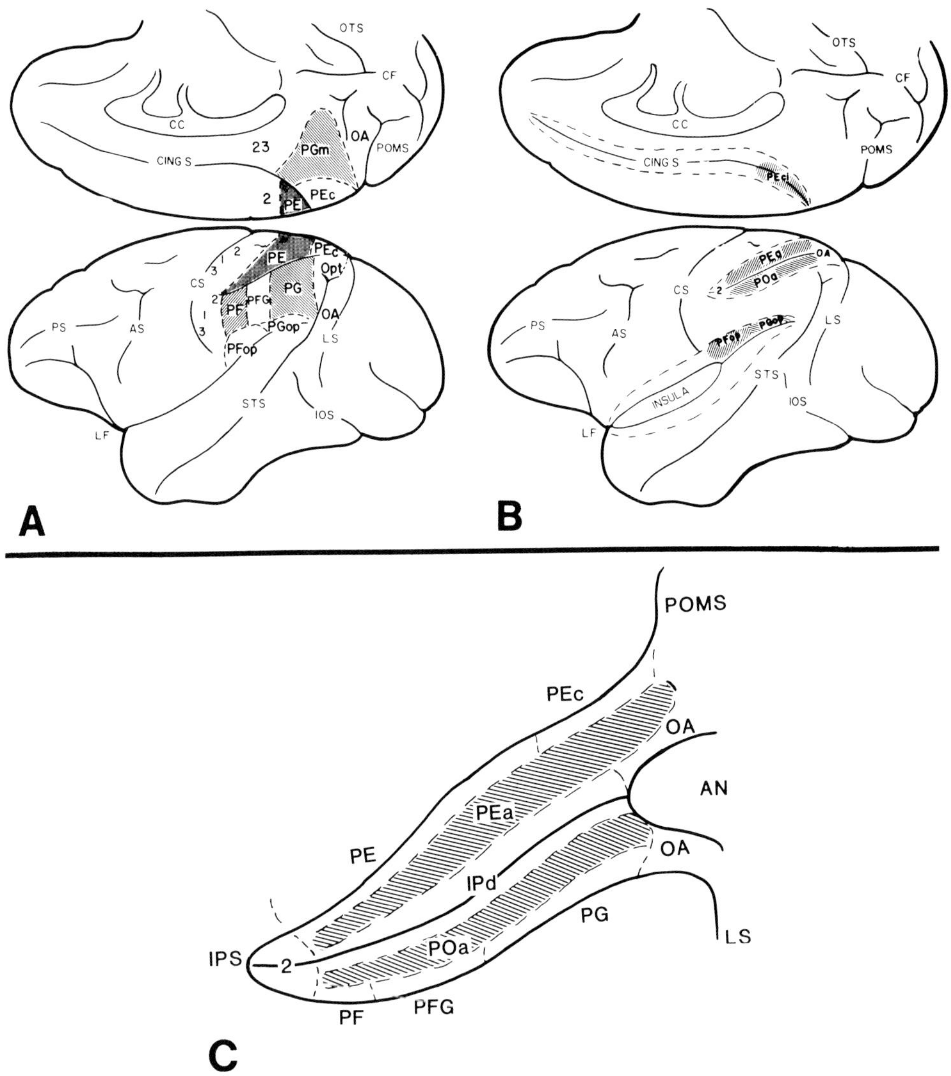

Figure 14. Diagrams showing the architectonic parcellation of the posterior parietal cortex. (A) Architectonic areas on the exposed lateral and medial surface. (B) Architectonic areas within the cingulate, and intraparietal sulci and the caudal Sylvian fissure. (C) A magnified view of the IPS showing different regions within the sulcus; note area IPd in the depth of the sulcus.

together correspond to the supplementary sensory area (SSA) as described by Murray and Coulter (1981). More ventrally on the medial surface is another distinct architectonic zone, area PGm. Next to area PGm is a caudal cingulate region, area 23. These areas thus comprise the upper architectonic trend within the parietal association cortex (Fig. 14A,B).

Within the inferior parietal lobule (IPL) three main architectonic areas can be discerned, namely areas PF, PFG, and PG in a rostrocaudal direction. In the caudalmost IPL, bordering on area OA of the occipital lobe, there is an additional architectonic sector, area Opt. Another architectonically distinct area lies within the ventral bank of the IPS and is termed area POa (Seltzer and Pandya, 1980). Finally, the parietal opercular region contains area PFop rostrally (corresponding in part to area SII of Woolsey) and area PGop caudally. Thus, these areas of the IPL belong to the lower architectonic trend. In a more recent study, another region at the junction of the dorsal and ventral trends has been identified in the depths of the IPS (Fig. 14C), and is designated as area IPd (Seltzer and Pandya, 1983).

3.2. Intrinsic Connections

The intrinsic connections of the parietal lobe follow the architectonic trends as described above (Jones and Powell, 1969; Pandya and Seltzer, 1982a). The dorsal half of area 2, which receives input from areas 3 and 1 of the postcentral gyrus, projects to caudally adjacent area PE (Jones *et al.*, 1978; Vogt and Pandya, 1978; Fig. 15). Area PE in turn projects to a more posterior region of the SPL, area PEc. Both areas PE and PEc also send connections to area PEa in the upper bank of the intraparietal sulcus, and to area PEci in the caudal cingulate sulcus. The principal flow of connections from areas PE and PEc proceeds caudally and medially to area PGm on the medial surface of the parietal lobe (Fig. 16A).

With regard to the IPL, the rostral area (PF) receives connections from areas 3, 1, and 2 of the postcentral gyrus (Fig. 15) and projects to the adjoining area PFG. Areas PF and PFG also project to the cortex in the lower bank of the IPS (area POa). The main sequence of connections continues, however, with a projection from the rostral to the caudal portion of the IPL. Areas PF and PFG send projections to two distinct sectors, one located dorsally along the lip of the IPS (area PG) and the other situated ventrally in the parietal operculum (area PGop). These two sectors project further caudally to the rostral portion of area PG (Fig. 16B).

Areas PGm and rostral PG therefore represent the final destinations of two different but parallel rostral-to-caudal sequences of connections. One leads through the SPL from the dorsal postcentral gyrus to the medial surface of the parietal

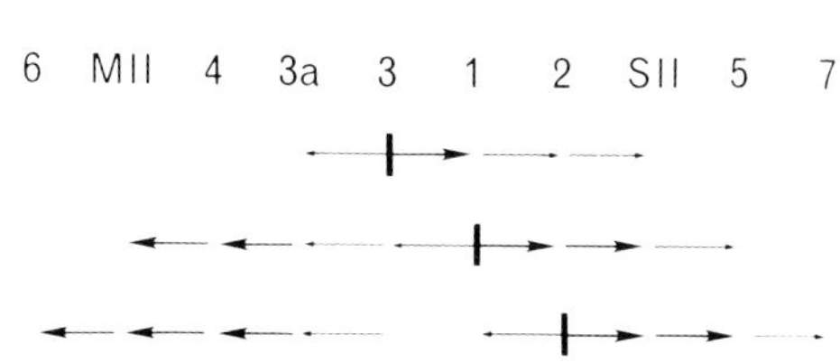

Figure 15. Corticocortical connections (arrows) of areas 3, 1, and 2 (indicated by thick vertical bars).

lobe and area PGm. The other passes in successive stages from the ventral postcentral gyrus through the IPL to rostral area PG. In addition to sharing certain similar architectonic features, areas PGm and rostral PG also have similar patterns of cortical connectivity (see Section 3.3). Thus, each stands at the end of a chain of connections beginning in the primary somatic sensory cortex. Both project to the intraparietal sulcus, area PGm to the upper bank, area PEa, and rostral area PG to the lower lip of the IPS. Both areas PGm and PG also project to area IPd in the depth of the IPS. Furthermore, areas PGm and rostral PG are reciprocally interconnected. Rostral area PG, however, has the additional feature of projecting to caudal area PG and area Opt (Fig. 16A,B).

While the rostrocaudal sequence of projections is a major feature within both SPL and IPL, there exist interconnections between these two areas as well (Pandya and Seltzer, 1982a). Thus, areas PE, PEc, PEa, and PEci of the SPL have topographic connections with the parietal opercular cortex (PFop—area SII) in the IPL, while area PGm projects to rostral area PG and the caudalmost portion of the parietal operculum (PGop) of the IPL. The connections from the

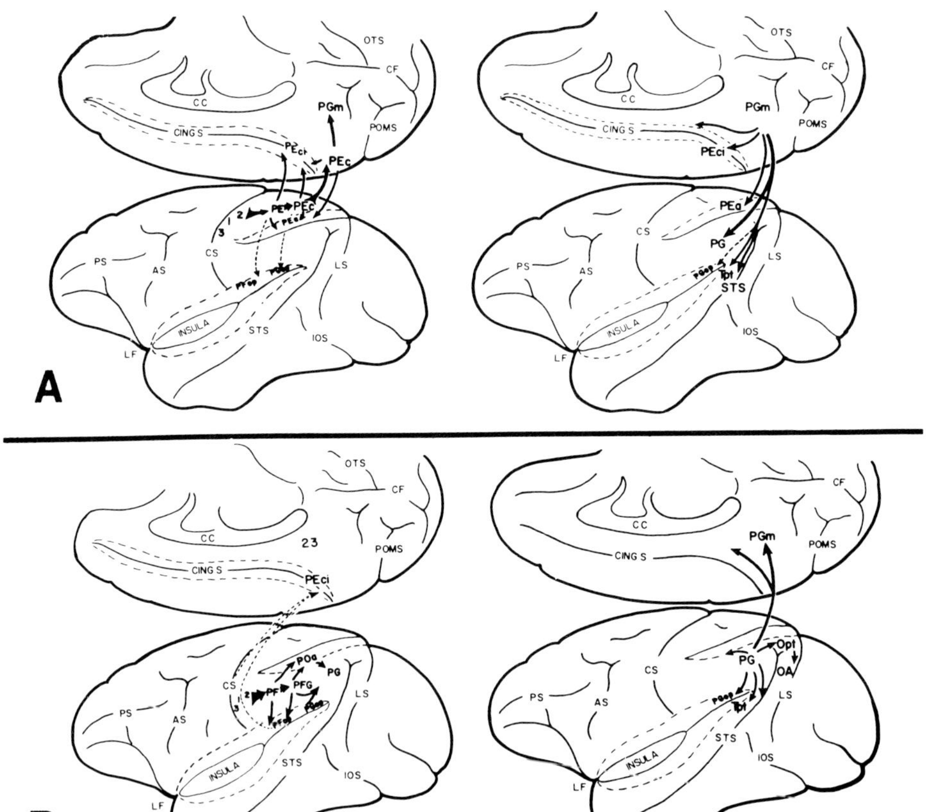

Figure 16. Diagrams of intrinsic connections of posterior parietal cortex. (A) Connections in the superior parietal lobule and medial surface. (B) Connections in the inferior parietal lobule.

IPL (areas PFG and PG) to the SPL are mainly to area PGm, while the parietal opercular cortex also has connections with areas PE and PEci (Fig. 16).

On the basis of the above-described connections, the association areas of the SPL and IPL can be divided into two broad categories. Thus, the rostral areas of the SPL (PE, PEa, PEci, and PEc) and IPL (PF, PFG, PFop, and PGop) are closely related to the primary and secondary somatosensory areas. In contrast, the caudal regions of the SPL (PGm) and IPL (PG) are only indirectly related to the primary and secondary sensory areas. Overall, this differential connectivity is consistent with the suggestion that rostral parietal association areas are involved in relatively more basic somatosensory processing, whereas the caudal areas may subserve more complex functions (see Hyvärinen, 1982, for discussion).

Although the data regarding the cells of origin and termination patterns of the intrinsic connections of the posterior parietal region are limited, a consistent and specific laminar pattern is evident. Thus, the rostrally directed intrinsic connections from the parietal regions terminate in layer I, whereas the caudally directed projections are distributed in a columnar fashion (Bowker and Coulter, 1981; Pandya and Seltzer, 1982a). This laminar pattern of connections, in part, resembles that of the auditory areas of the STR.

3.3. Long Association Connections

As in the auditory system, the association areas of the parietal lobe may be grouped into three sectors (SA1, SA2, and SA3) on the basis of their long connections (Chavis and Pandya, 1976; Fig. 17A). The long association connections of these sectors are directed toward frontal, parietotemporal, and paralimbic regions. Since the connections to each of these regions are quite complex, we will describe them separately.

3.3.1. Frontal Lobe Connections

It has been well established that the postcentral gyrus (area SI) is connected with the motor cortex, MI, as well as the supplementary motor cortex, MII (Fig. 15), whereas the association areas of the parietal lobe are connected with the premotor, supplementary motor, and prefrontal regions (Pandya and Kuypers, 1969; Jones and Powell, 1969, 1970a; Chavis and Pandya, 1976; Barbas and Mesulam, 1981; Petrides and Pandya, 1983). The postcentral gyrus is flanked caudally by the association areas which have been designated as first-order sensory association regions (SA1). The SA1 of the SPL consists of areas PE and PEa, and projects to rostral area 4, to dorsal premotor cortex, area 6, and to the supplementary motor cortex. Likewise, area PF, representing the SA1 for the IPL, projects to ventral premotor area 6 and the pericentral and frontal opercula (Fig. 17B). The next regions in the parietal lobe are designated SA2 or second somatosensory association regions. In the SPL, area SA2 consists of area PEc and projects to dorsal area 6 as well as MII. However, the area 6 projections are considerably more rostral than those of area SA1. Area SA2 of the IPL consists

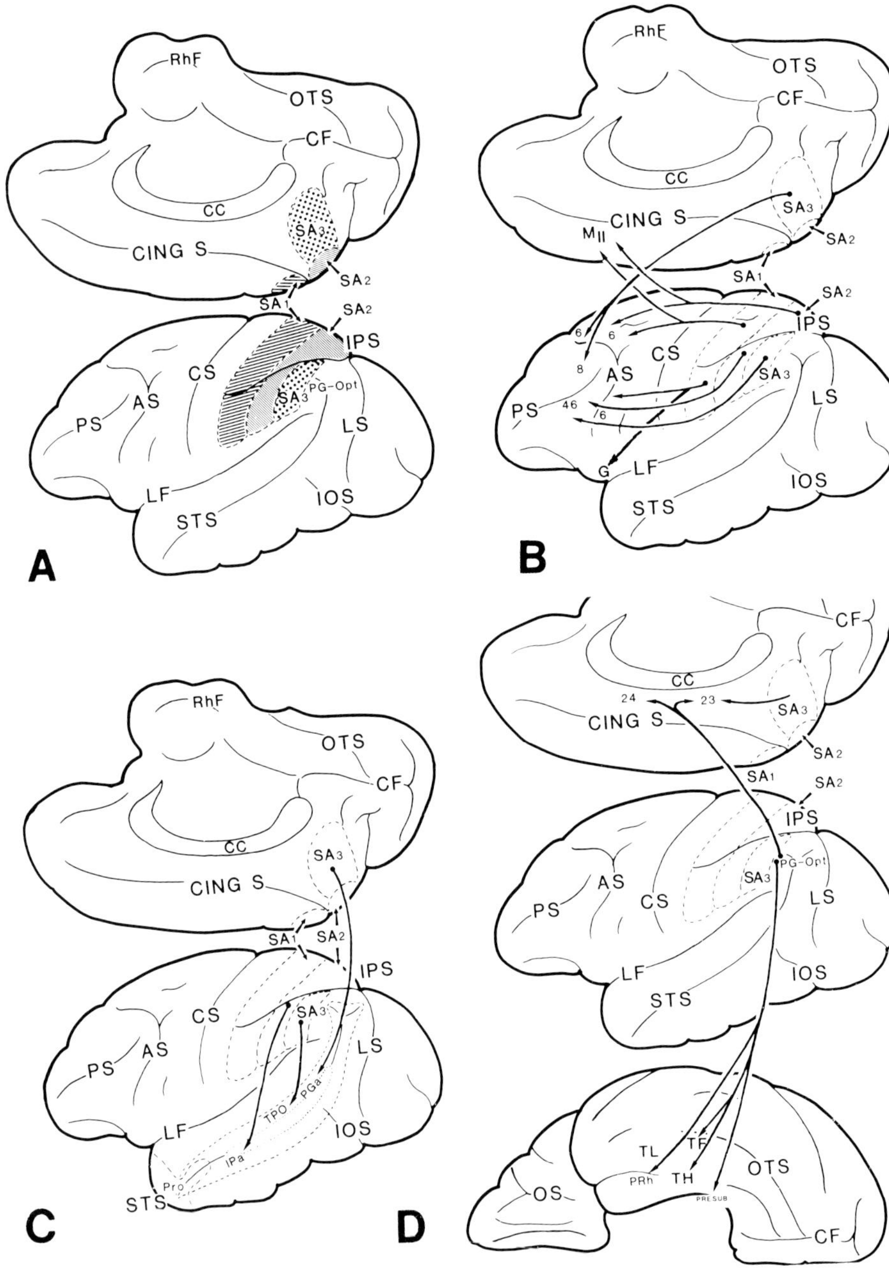

Figure 17. Diagrams showing three major subdivisions of the somatosensory association areas of the posterior parietal cortex (A) and their frontal (B), superior temporal sulcus (C), and paralimbic (D) connections.

of an area in the lower lip of the IPS. This area projects predominantly to ventral area 46, below the principal sulcus (Fig. 17B). The next parietal association sectors are designated SA3. The dorsal SA3 area is comprised of area PGm on the medial surface of the parietal lobe and projects to rostral area 6 in and above the upper limb of the arcuate sulcus and to adjacent area 8. Unlike the areas SA1 and SA2 of the SPL, area SA3 lacks a projection to MII. Area SA3 of the IPL consists of area PFG and rostral area PG and projects mainly to rostral area 46 in the ventral bank of the principal sulcus (Fig. 17B). The caudalmost portion of IPL consists of caudal area PG and area Opt. This area is situated at the junction of the parietal and occipital association regions and can be considered as equivalent to the angular gyrus, or area 39, of the human brain. Although this region is within the parietal lobe, functionally as well as connectionally it is quite complex. The frontal lobe connections of this region are directed to dorsal area 8 and area 46 (Fig. 26B).

3.3.2. Parietotemporal Connections

Like the auditory association areas, the parietal association regions have connections to specific areas of the superior temporal sulcus (Jones and Powell, 1970a; Seltzer and Pandya, 1978, 1984). The STS projections from the SPL are only from area SA3, and are directed to the caudal portion of areas TPO and PGa (Fig. 17C). Likewise, the projections from the IPL to the STS are given to the caudal TPO–PGa area from SA3. In addition, however, area SA2 of the IPL has connections to the rostral portion of the STS, namely area IPa. All of these STS projections are restricted and topographically distinct. Neither areas SA1 and SA2 of the SPL nor area SA1 of the IPL project to the STS. Finally, the caudalmost IPL (caudal area PG and area Opt) has quite extensive projections to area TPO as well as area PGa of the STS (Fig. 26B). Area SA1 of the IPL has distinct projections to a specific region in the lower bank of the IPS, area POa (Seltzer and Pandya, 1980). As described below (Section 6.1), area POa also receives projections from the first-order visual association areas (Fig. 26A)

3.3.3. Paralimbic Connections

The paralimbic projections from the parietal association areas originate mainly from the caudal regions, namely area SA3 of both SPL and IPL (Pandya and Kuypers, 1969; Jones and Powell, 1970a; Petras, 1971; Seltzer and Pandya, 1978, 1984; Van Hoesen, 1982). These paralimbic projections, however, are differentially distributed. Thus, area SA3 of the SPL sends projections to area 23 of the cingulate gyrus, whereas area SA3 of the IPL has projections to both areas 23 and 24 of the cingulate gyrus (Fig. 17D). In addition, area SA3 of the IPL has connections with the ventral paralimbic regions, namely the perirhinal region (area 35), lateral parahippocampal gyrus (areas TF and TL), and presubiculum (area 27). The caudalmost portion of the IPL, i.e., caudal area PG and area Opt, has projections to the paralimbic areas similar to those of area SA3 of the IPL (Fig. 26B). The only difference is that these projections are more extensive. As in the auditory system, most of these long association connections are reciprocal (Figs. 25, 29A, and 30).

3.4. Overview

Somatosensory association areas can be differentiated on the basis of their connections. These connections become more widespread as one proceeds from the rostral to the caudal association cortex in the parietal lobe. Thus, SA1 areas are more directly related to somatosensory and motor systems by virtue of their connections with SI, SII, SSA, MII, and the caudal premotor region (area 6). The SA2 areas, on the other hand, do not have direct connections with SI, but are related to SII and SSA as well as MII and the rostral premotor regions. In comparison with the SA1 and SA2 areas, the SA3 areas have more extensive connectivity to other cortical association regions and appear to be the least directly connected with primary and secondary somatosensory areas (SI and SII). Thus, the SA3 areas project to the prefrontal as well as the premotor cortex, to the STS (areas TPO and PGa), and to medial and ventral paralimbic regions (cingulate and parahippocampal gyri). Finally, caudal area PG and area Opt, although within the parietal lobe, stand apart as a multimodal region on connectional grounds (see Section 6.1).

On the basis of their cortical connectivity, the association areas of the SPL and IPL appear to be involved in somewhat different aspects of the somatosensory sphere. Areas SA1, SA2, and SA3 of the SPL comprise a sequence of projection areas originating from the trunk and extremity representations of SI, ultimately leading to the cingulate gyrus and dorsal premotor, supplementary motor, as well as prefrontal regions. In contrast, areas SA1, SA2, and SA3 of the IPL form a series of projection regions which begin in the face, head, and neck representations of SI and ultimately relate to the cingulate and parahippocampal gyri, the ventral premotor and prefrontal regions, as well as the gustatory area in the frontal operculum (Pandya *et al.,* 1980). This differential arrangement coincides with the proposed dual nature of origin of somatosensory related regions (Sanides, 1972), and corresponds to the dual nature of frontal lobe phylogenetic development, where the principal sulcus represents a dividing line between upper and lower architectonic trends (see Section 5.1).

The above notion receives further support when one considers the reciprocal connections of these two trends. Thus, each of these association areas is connected with its rostrally adjacent region. These connections are specifically given to layer I. Although the current information in terms of these reciprocal connections is far from complete, one might suggest that SPL association areas are linked mainly with cingulate proisocortex via SSA, SA3, SA2, and SA1, whereas the association areas of the IPL are connected primarily with the parinsular proisocortex via SII, SA3, SA2, and SA1. It should be pointed out that these two trends of the parietal lobe are interrelated at certain levels. Thus, areas SII and SSA are known to be interconnected (Jones and Powell, 1969; Pandya and Kuypers, 1969; Pandya and Seltzer, 1982a).

4. Visual Association Areas

The cortical visual system has long been known to be located in the occipital lobe and the inferotemporal region. The primary visual area lies within the occipital lobe and is related to a specific thalamic nucleus, the lateral geniculate body (Hubel and Wiesel, 1972; Winfield *et al.,* 1975; Benevento and Yoshida, 1981; Yukie and Iwai, 1981; Weller and Kaas, 1983; Doty, 1983; Bullier and Kennedy, 1983). This region, the so-called striate cortex, has a characteristic architecture (Fig. 6), and has been designated area 17 by Brodmann (1909) and area OC by von Bonin and Bailey (1947). Physiological studies have shown that it is involved in the analysis of visual input as it first reaches the cortical level (Hubel and Wiesel, 1968). Daniel and Whitteridge (1961) have shown that the midline of central vision is located at the very edge of striate cortex (juxtastriate), whereas the peripheral visual field is represented in the remainder of the striate cortex. In recent years, on the basis of physiological as well as anatomical observations, several investigators have described multiple visual representations in the surrounding peristriate belts (Zeki, 1978; Allman *et al.,* 1981; Van Essen and Maunsell, 1983; see Van Essen, Volume 3 of this treatise). The visual association areas include areas 18 and 19, the circumstriate belt which surrounds the striate cortex, and extends into the inferior temporal regions, areas TE1, TE2, and TE3 or areas 20 and 21 (Figs. 6 and 18). Unlike the primary visual area (area 17), the thalamic connections of these regions are predominantly with associative nuclei, especially the inferior pulvinar nucleus (Clark, 1936; Benevento and Rezak, 1976). These association areas have been shown to be involved in sequential processing of visual information (Mishkin, 1972; Gross *et al.,* 1981; Ungerleider and Mishkin, 1982; Mishkin *et al.,* 1983).

4.1. Architectonics

Just as the somatosensory system, the visual system perhaps has evolved from both primordial moieties, i.e., archicortical and paleocortical. The recent analysis of architecture in primates carried out by Rosene and Pandya (1983; Fig. 21A) along with the observations of Galaburda (1983) indicate that two distinct architectonic trends can be traced within the cortical visual system. As shown in Fig. 18, both trends originate from the proisocortex of the temporal lobe. One trend progresses ventrolaterally with sequential modification of granularization through areas TE1, TE2, and TE3 in the inferotemporal region and areas OAa (designated area VII—second visual area—and corresponding to area MT of New World primates), OA, OB, and OC in the occipital lobe. Similarly, the other trend progresses ventromedially in sucessive steps through the parahippocampal gyrus (areas TH, TL, TF and the prostriate region) and the occipital lobe regions (areas OA, OB, and OC). Although these observations need to be elaborated upon in further studies, it can be suggested that the differential function in terms of central and peripheral vision at the cortical level may be reflected in these two trends. Thus, in the ventromedial trend, the prostriate area can be considered as the prokoniocortical stage related primarily to areas of central vision. In the ventrolateral trend, area OAa in the STS (area MT) may represent the prokoniocortical step in the phylogenetic development of areas of peripheral vision.

4.2. Intrinsic Connections

A number of studies have demonstrated a sequential flow of connections within the visually related areas of the occipital and inferotemporal cortices (Kuypers *et al.,* 1965; Tigges *et al.,* 1973; Seltzer and Pandya, 1976; Zeki, 1978; Rockland and Pandya, 1979, 1981; Tigges *et al.,* 1981; Ungerleider and Mishkin, 1982; Van Essen and Maunsell, 1983; Rosene and Pandya, 1983). Thus, the primary visual area projects to the surrounding prestriate area 18 in a topographic manner (Fig. 19). This striate-recipient area 18 in turn is connected with area 19, again in a topographic manner. Area 19 in turn is connected with area 20 of the inferotemporal cortex, which itself projects to rostral inferotemporal

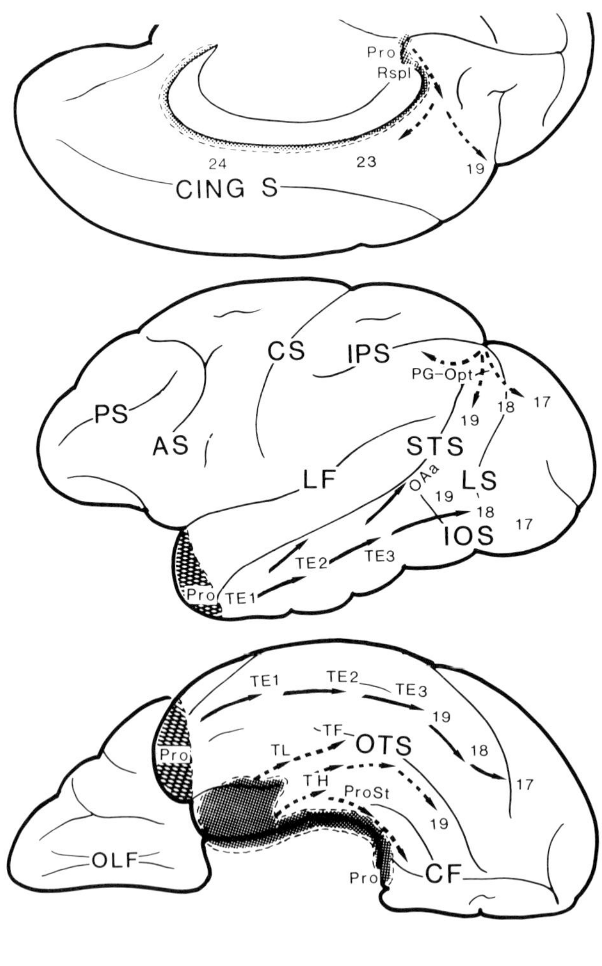

Figure 18. Diagrammatic representation of dual architectonic trends of visually related areas of the inferotemporal and occipital cortices. Note that solid lines with arrows indicate a lateral or temporal polar trend, which leads toward areas representing the central visual field. Arrows with dashed lines indicate a medial or hippocampal trend which develops into areas representing the peripheral visual field.

area 21. Area 21 sends projections to the proisocortex of the temporal pole. In addition, just as for area 17, areas 18 and 19 project to an architectonically distinct region in the caudal portion of the STS, i.e., area OAa (or MT). Just as in the auditory and somatosensory systems, in visually related areas there is a systematic stepwise pattern of reciprocal connectivity. Moreover, as in other sensory systems, there is a specific organization of laminar origin and termination of these intrinsic connections. Rostrally directed connections originate mainly from neurons in layer III and terminate in and around layer IV of adjacent areas. In contrast, connections in the reverse direction (caudally directed, or backgoing, connections) originate from neurons in layers V and VI and to a lesser extent IIIa and terminate mainly in layer I (Fig. 19).

4.3. Long Association Connections

As described above, the visual association areas surround the primary visual cortex, area 17, and receive topographically organized connections from striate cortex. On the basis of long cortical connections, just as the auditory and somatosensory association areas, the visual association regions can be divided into three distinct sectors (Chavis and Pandya, 1976). The first visual association area, VA1, consists of areas 18 and 19, while the second, VA2, is comprised of area 20 of Brodmann (area TE3, as described by Seltzer and Pandya, 1978). The third visual association area, VA3, consists of area 21, or TE2 and TE1 regions

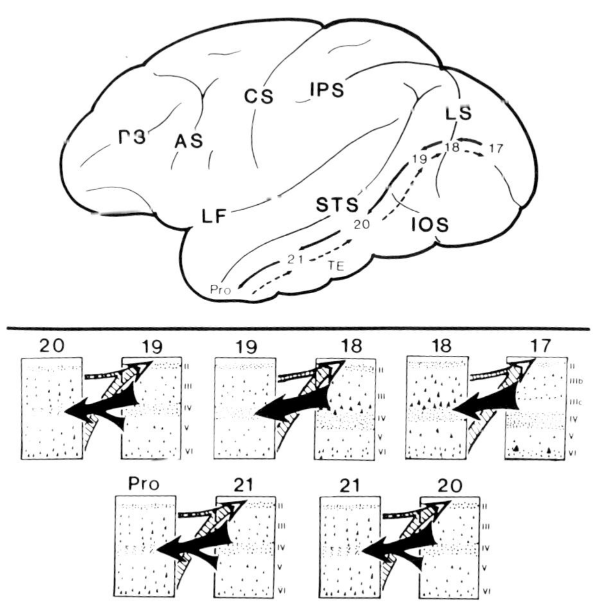

Figure 19. (Upper) The intrinsic reciprocal connections of visually related areas of the occipital lobe and the inferotemporal region. (Lower) The cells of origin and laminar terminations of forward and backgoing connections of these areas.

(Fig. 20A). Each of these regions sends distinct connections to the frontal lobe, to parietotemporal regions, and to paralimbic areas, similar to the projections described from the auditory and somatosensory association areas.

4.3.1. Frontal Lobe Connections

Both medial and lateral prestriate cortex, VA1, are shown to project to the premotor region (periarcuate cortex—area 8). Area VA2 of the caudal inferotemporal region, on the other hand, sends projections to the premotor region (rostral area 8) and the prearcuate area (area 46) below the principal sulcus. The rostral inferotemporal area VA3 (area 21) projects to the same frontal area as VA2, but in addition has a distinct connection to area 11 on the orbital surface (Fig. 20B). Thus, visual association areas VA1, VA2, and VA3 show a sequential distribution of projections to the frontal lobe, beginning in the periarcuate region

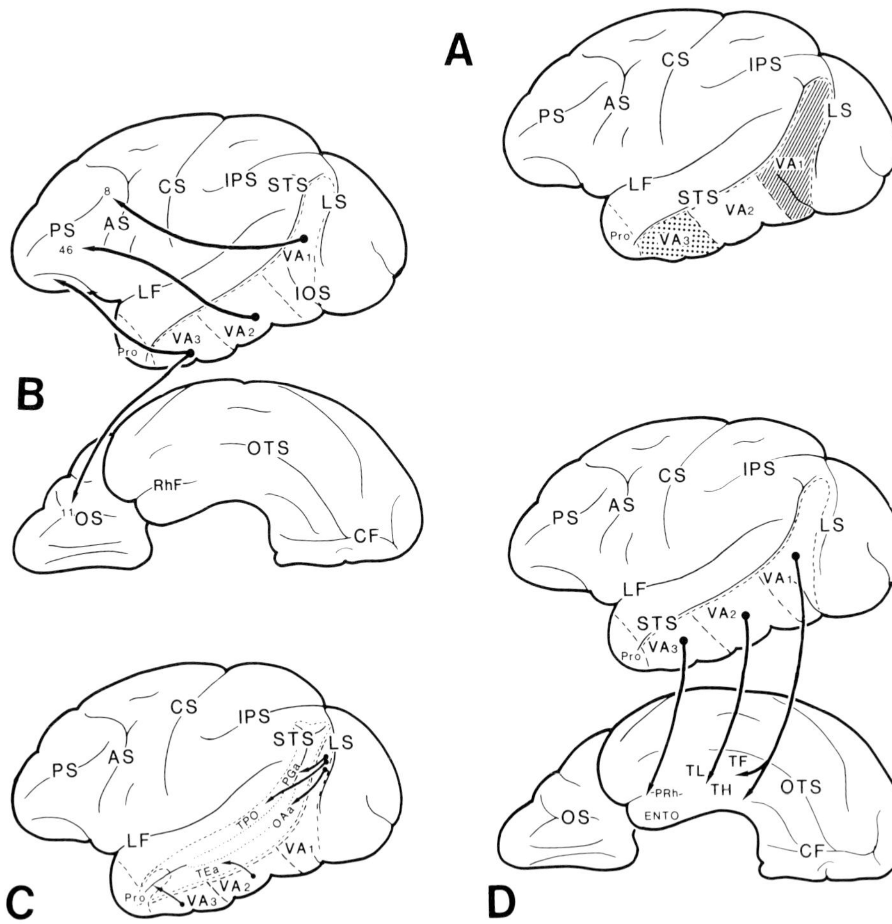

Figure 20. Diagrams showing three major divisions of visual association areas, VA1, VA2, and VA3 (A), and their connections to the frontal lobe (B), superior temporal sulcus (C), and paralimbic regions (D).

and progressing rostrally through the prearcuate and orbitofrontal regions (Kuypers *et al.*, 1965; Chavis and Pandya, 1976; Barbas and Mesulam, 1981; Van Hoesen, 1982; Rosene and Pandya, 1983).

4.3.2. Parietotemporal Connections

Each of the visual association areas sends projections to distinct architectonic regions in the STS (Kuypers *et al.*, 1965; Zeki, 1978; Seltzer and Pandya, 1978; Ungerleider and Mishkin, 1979; Gross *et al.*, 1981; Van Essen and Maunsell, 1983). Thus, area VA1, as described above, projects to area OAa (area VII, or MT of New World monkeys) in the caudal portion of the STS. Area VA2, on the other hand, has projections to area TEa in the midventral bank of the STS. Finally, area VA3 projects to the rostral portion of the STS, in the ventral bank, area Pro (Fig. 20C). Therefore, within the STS, the visual association areas maintain a caudorostral topography of connections. Area VA1 in addition has connections to a distinct architectonic area (area POa as described by Seltzer and Pandya, 1980) in the intraparietal sulcus (Fig. 26A).

4.3.3. Paralimbic Connections

Each of the visual association areas has distinct paralimbic connections (Kuypers *et al.*, 1965; Jones and Powell, 1970a; Van Hoesen *et al.*, 1972; Seltzer and Pandya, 1976; Rockland and Pandya, 1981; Van Hoesen, 1982). Area VA1 sends connections to the posterior portion of the parahippocampal region (areas TF, TL, and TH), whereas VA2 is connected with rostral areas TF, TL, and TH. VA3 projects to perirhinal as well as prorhinal regions (Fig. 20D). Thus, the paralimbic connections from each of the visual association areas show a distinct sequential distribution.

4.4. Overview

Information from the visual association areas is distributed to the frontal lobe, parietotemporal region, and paralimbic areas, in a stepwise manner. Thus, at the first stage, efferents from VA1 are to area 8 in the premotor cortex, to area OAa in the STS, and to caudal parahippocampal regions. In the next stage, area VA2 is related to the prefrontal cortex (area 46), to area TEa in the STS, and to the rostral parahippocampal regions. Finally, in the third stage, area VA3 conveys its information to the orbitofrontal cortex (area 11), to the rostral STS (area Pro), and to the pro- and perirhinal cortices.

As in the auditory system, each stage of visual association areas is related to frontal, parietotemporal, and paralimbic regions which appear to occupy similar levels of architectual organization. Thus, for example, area VA1, which has well-developed supragranular layers, is connected with areas 8, OAa, and the caudal parahippocampal region, each of which has an emphasis on the supragranular layers. In contrast, area VA3, which shows dysgranularity and an emphasis on infragranular layers, is related to area 11, area Pro in the STS, and the pro- and perirhinal cortices, each of which shows basically similar architectonic features. Area VA2, which has architectonic characteristics intermediate to those of areas VA1 and VA3, is related to those cortical regions with similar characteristics.

In addition to the connections of visual association areas described above, area VA1 has been shown to have a distinct projection to the intraparietal sulcus, area POa (Seltzer and Pandya, 1980). Area POa receives connections particularly from the VA1 area which is the striate-recipient zone of peripheral visual input (Rockland and Pandya, 1981; Ungerleider and Mishkin, 1982). This is in contrast to area OAa in the STS which is more related to that part of area VA1 which is predominantly a striate-recipient zone of central visual input (Gattass and Gross, 1981; Rockland and Pandya, 1981; Ungerleider and Mishkin, 1982). As mentioned above, area POa also receives input from the somatosensory association area related to the head, neck, and face (area SA1 of the IPL).

At the beginning of the description of visual association areas, we suggested that this system might have evolved from two moieties, the temporal proisocortex, and the prostriate region of the medial parahippocampal gyrus. The connections described above pertain mainly to the trend which is related to the temporal proisocortex (Fig. 21C). Recent studies of the prostriate region (Rosene and Pandya, 1983), although preliminary in nature, have shown that unlike the connections of the polar proisocortical trend, the prostriate area is connectionally related to the perirhinal, entorhinal, and presubicular regions on the one hand, and to lateral and medial area OA, to the cingulate gyrus, and to area Opt on the other (Fig. 21B). All of these latter regions except for the cingulate gyrus are related to peripheral vision. Thus, the dichotomous patterns of projections within the visually related areas support the concept of the dual nature of cortical visual systems with one trend relating to the peripheral visual field and the other to central vision.

5. Frontal Association Areas

Traditionally, the frontal lobe cortex has been divided into three broad regions (Fig. 22A). The precentral gyrus (area 4, or MI) is the motor cortex, while areas 6 and 8 in front of the motor cortex are considered as premotor regions. The supplementary motor area on the medial surface (MII) is also thought to be part of the premotor region. Finally, rostral to the premotor region is a wide expanse of cortex on the lateral, medial, and ventral surfaces of the hemisphere known as prefrontal cortex. Whereas the MI and MII areas are implicated in somatic motor activity, the premotor and prefrontal regions are involved in complex integrative functions and are considered as frontal association areas (Woolsey, 1958; Rosvold and Mishkin, 1961; Mishkin, 1964; Nauta, 1971; Butters *et al.*, 1972; Petrides and Iversen, 1976; Damasio, 1979; Fuster, 1980; Van Hoesen *et al.*, 1980; Wiesendanger, 1981; Brinkman and Porter, 1983). The thalamic connections also differentiate these regions (Walker, 1940; Akert, 1964; Johnson *et al.*, 1968; Goldman and Rosvold, 1970; Tobias, 1975; Kievit and Kuypers, 1977; Künzle, 1978; Siwek and Pandya, 1984). Thus, the precentral areas are connected mainly with the ventrolateral (VL) thalamic nu-

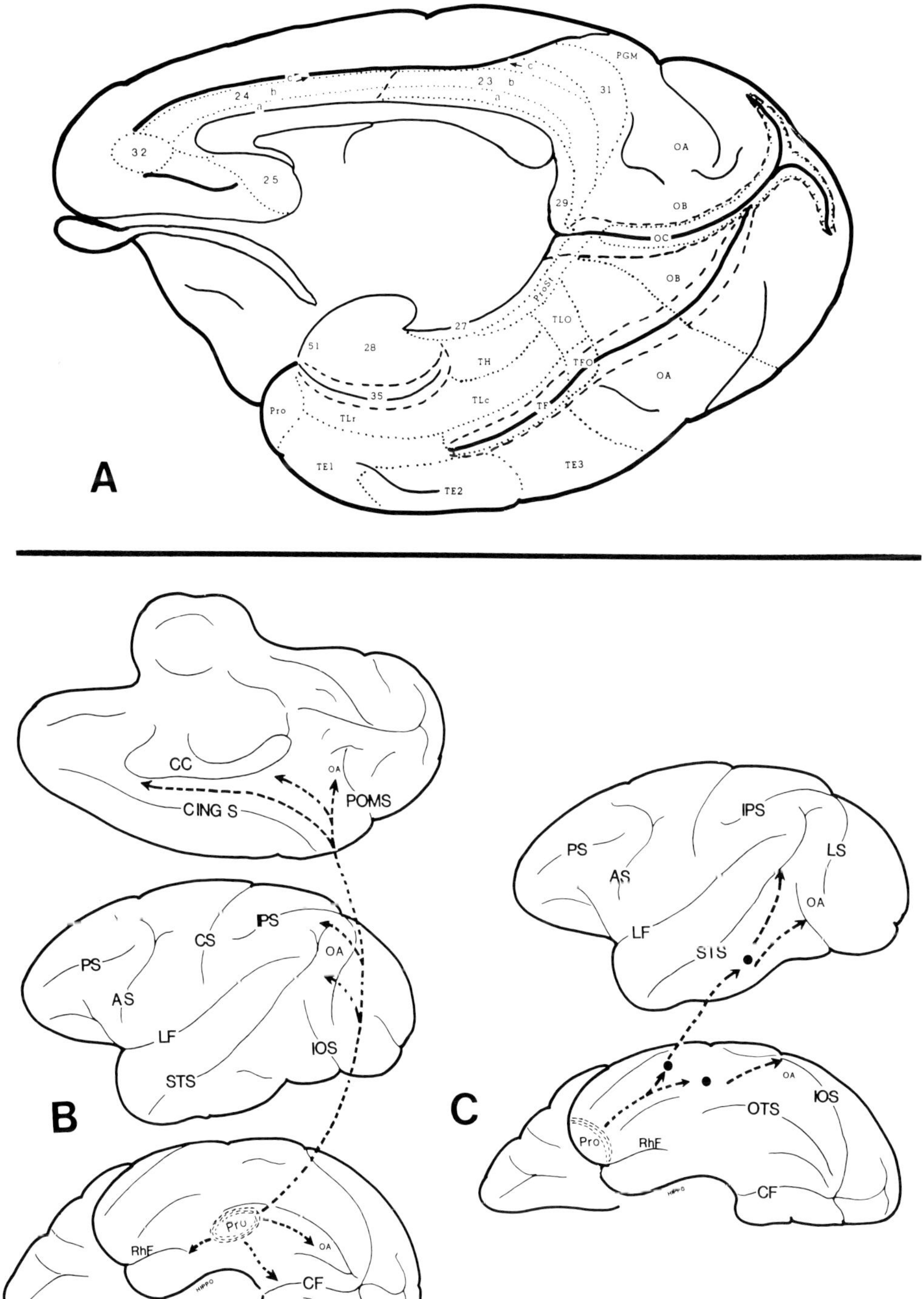

Figure 21. (A) Diagram of the medial surface of the cerebral cortex showing the architectonic parcellation of the parahippocampal gyrus as described by Rosene and Pandya (1983). (B and C) Summary diagrams showing two projection trends, one from the temporal polar proisocortex and the other from the ventromedial proisocortex.

cleus, whereas the premotor region is also connected with the ventroanterior (VA) thalamic nucleus. The main thalamic connections of the prefrontal region are with the dorsomedial (MD) thalamic nucleus.

5.1. Architectonics

Architectonic studies have parcellated the frontal association cortex (premotor and prefrontal) into several different regions. Thus, according to Brodmann (1909) and Walker (1940), the premotor region consists of areas 6 and 8, whereas the prefrontal region contains areas 9, 46, 10, and 12 on the lateral surface, areas 13, 14, and 11 on the orbital surface, and areas 9, 25, 32, and 14 on the medial surface (Figs. 22A and 24A). According to Sanides (1969), the premotor and precentral areas have developed from two distinct architectonic moieties, ventrally from paleocortex and dorsally from archicortex. Just as in the somatosensory system, further steps can be followed from these two primordial areas in two directions (Fig. 22B). Thus, one of these lines seems to

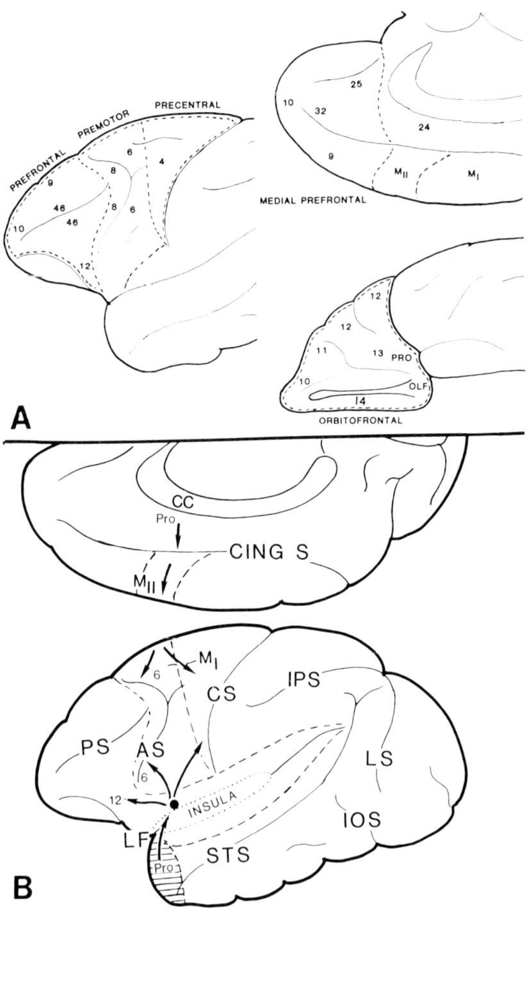

Figure 22. (A) Diagrams showing architectonic subdivisions of the frontal areas (precentral, premotor, and prefrontal) on the lateral, medial, and orbital surfaces. (B) Diagrammatic representation of medial (cingulate) and ventral (insular) architectonic trends leading to the motor and premotor areas (shown by arrows).

progress from ventral or insular proisocortex and develop into parinsular areas and finally into ventral areas 6 and 4. The second trend can be traced on the medial surface from the cingulate gyrus and passes through the paralimbic proisocortex and the supplementary motor region. From this stage further progression leads to the development of dorsal areas 6 and 4. According to recent architectonic analysis, similar dual trends are observed in the prefrontal cortex (Barbas and Pandya, 1982). One part of this region begins in the periallocortex surrounding the olfactory tubercle on the orbital surface, then differentiates into proisocortex with bilaminated appearance and further develops into progressive isocortical areas (Fig. 24B). These isocortical regions, according to their degree of acquisition of granule cells in layer IV and pyramidal neurons in layer III, can be traced in order as Pro → 13 → 12 → ventral 10 → ventral 46 → ventral 8 (Fig. 24B). Similiarly, the medial periallocortical region around the rostrum

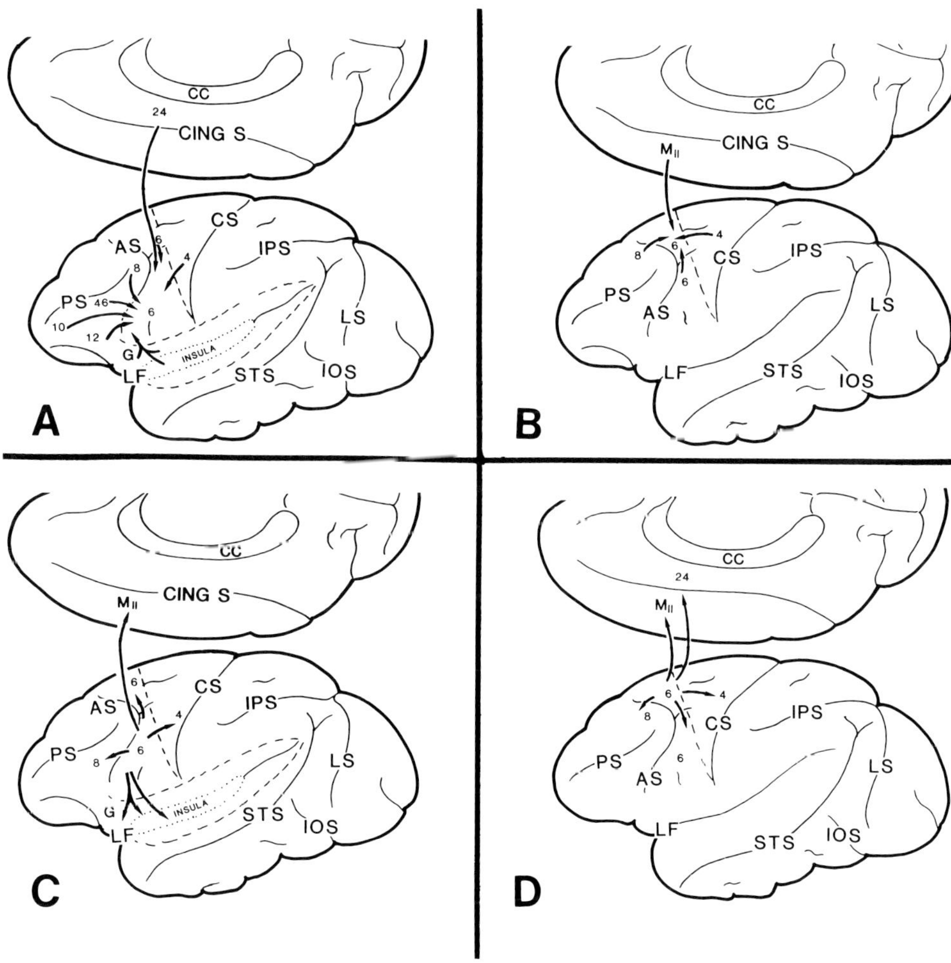

Figure 23. Diagrams showing intrinsic connections of the premotor areas. (A, C) Afferent and efferent connections, respectively, of ventral area 6; (B, D) similar connections of dorsal area 6.

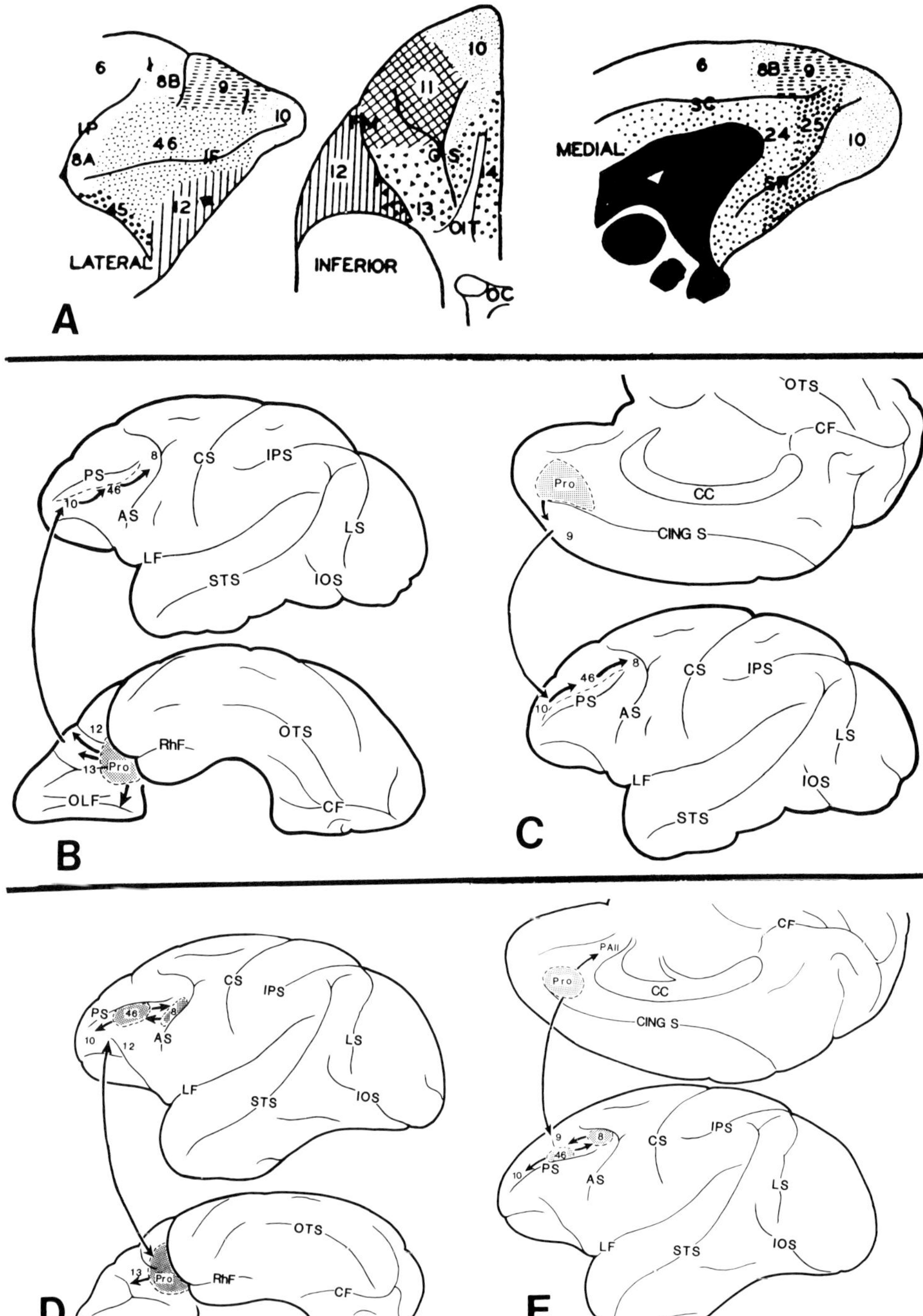

Figure 24. (A) Architectonic parcellation of the prefrontal cortex according to Walker (1940). (B) Progressive architectonic steps from orbital proisocortex leading to ventral area 8. (C) Similar steps from medial proisocortex leading to dorsal area 8. (D, E) Intrinsic connections of subregions within the ventral and dorsal trends.

of the corpus callosum further develops into proisocortex with bilaminated appearance (areas 25 and 32). From this proisocortical region, as in the ventral prefrontal regions described above, a series of dorsal areas can be delineated with progressive granularization of layer IV and acquisition of pyramidal neurons in layer III, namely Pro → 9 → dorsal 10 → dorsal 46 → dorsal 8 (Fig. 24C). These architectonic trends for the premotor and prefrontal regions provide a framework for understanding the intrinsic and long connections of the frontal association areas.

5.2. Intrinsic Connections

The intrinsic connections of the premotor regions have been described recently by Barbas and Pandya (1981) and Muakkassa and Strick (1979). Ventral area 6 receives projections from the frontal opercular region as well as the ventral precentral region (MI). In addition, it receives connections from the ventral prefrontal cortex (areas 8, 12, 46, and 10) and the depth of the cingulate sulcus (Fig. 23A). This area in turn projects to the frontal opercular region (gustatory area) and rostral insula as well as ventral precentral regions. It also projects to area 8 in the prefrontal region and to the supplementary motor cortex (Fig. 23C). Dorsal area 6, in contrast, has limited intrinsic connections. Thus, it receives and sends projections to dorsal areas 4 and 8 on one hand and the supplementary motor area on the other (Fig. 23B,D). These two parts of area 6 are mutually interrelated. The intrinsic connections of area 6 support the dual nature of architectural development as proposed above, i.e., dorsal area 6 is related to the dorsal or cingulate trend whereas ventral area 6 is associated with the ventral or parinsular trend. The immediate relationship of area 6 with the motor cortex lends support to its being a motor association area.

The intrinsic connections of the prefrontal cortex also are organized according to the notion of dual architectonic trends (Barbas and Pandya, 1982). Thus, the ventral proisocortical area projects mainly to areas 13 and 12 on one hand and area 14 on the other. Area 13 in turn projects to the proisocortical area as well to areas 12 and 10. Area 12 projects to areas 10 and 13 on one hand and ventral area 46 on the other. Ventral area 46 sends its connections to area 12 and 10 rostroventrally and to ventral area 8 caudally. Area 8 projects to ventral area 46 rostrally and to ventral area 6 caudally (Fig. 24D). Similarly, the medial proisocortical region projects to areas 25 and 14 ventrally and areas 9 and 10 dorsally. Area 9 in turn has connections with proisocortex medially and dorsal area 10 and area 46 laterally. Area 46 projects to areas 9 and 10 on one hand and to dorsal area 8 on the other. Finally, dorsal area 8 projects to dorsal area 46 rostrally and to dorsal area 6 caudally (Fig. 24E). It is also noteworthy that these two trends are interconnected at certain levels. Thus, the ventral and medial proisocortical areas are interconnected, as are areas 9 and 12 (Pandya *et al.*, 1971). Finally, the dorsal and ventral portions of area 8 are also interconnected. These observations indicate that within each trend, a given cortical region projects to an architectonically less differentiated area as well as to a region with a more developed cortical laminar organization.

5.3. Long Association Connections

The long association connections of premotor regions including area 6 and 8 indicate that these areas are connected with the first-order sensory association regions. Thus, area 6 is topographically connected with the rostral parietal association region, SA1 (Fig. 25A). Likewise, the premotor region in front of the arcuate sulcus, area 8, is connected with the first-order auditory association area, AA1. The connections from area 8 to the visual association area are limited (Jones and Powell, 1970a; Pandya and Vignolo, 1971; Künzle and Akert, 1977; Deacon *et al.,* 1982).

The long connections of the prefrontal cortex (Fig. 25B) are given to the posterior parietal cortex, to the temporal lobe (superior and inferior temporal regions including the STS), and to the paralimbic areas (Nauta, 1964; Pandya and Kuypers, 1969; Jones and Powell, 1970a; Pandya *et al.,* 1971). The posterior parietal projections originate from caudal areas 46 and 9, and are directed to areas SA2 and SA3 of the IPL and SPL including the IPS in a topographic manner. Dorsal areas 46 and 8 are also connected with area Opt. The temporal lobe connections from prefrontal regions are organized as follows: Dorsal areas 46 and 10 project to the AA2 region as well as to the adjacent STS, whereas ventral areas 46 and 10 are connected with the VA2 area and with the adjacent STS. The orbitofrontal regions, i.e., areas 12 and 13 as well as medial prefrontal area 25, are connected with AA3 and VA3. Finally, the paralimbic connections

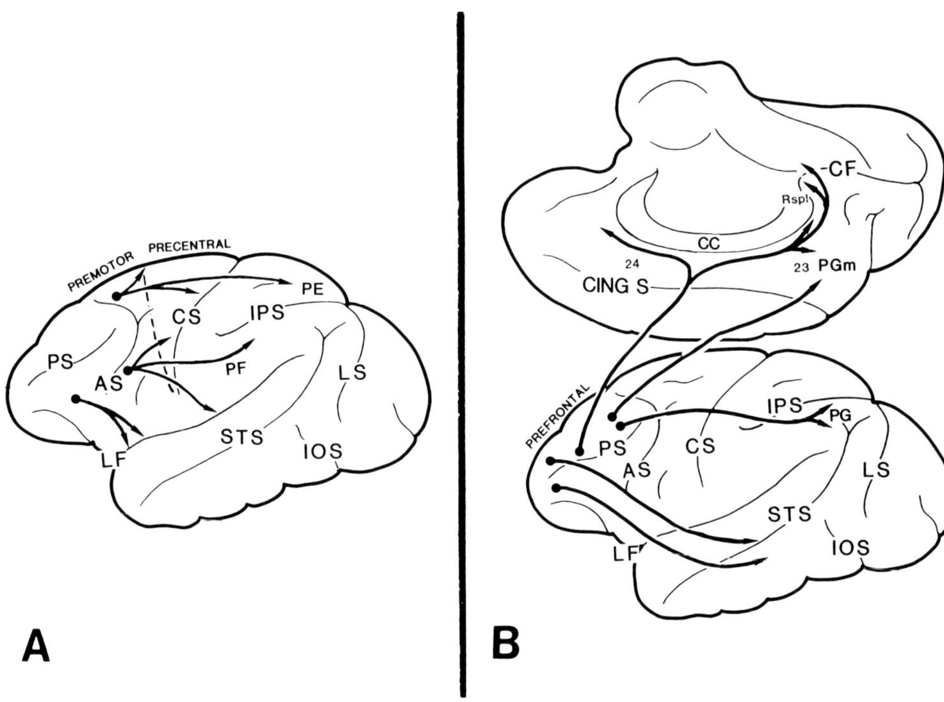

Figure 25. Diagrams showing long association connections of premotor regions (A) and prefrontal areas (B).

originate from the dorsal and ventral prefrontal regions. Thus, the orbital proisocortical region and area 13 are connected with the temporal polar proisocortical area, as well as the perirhinal region and the parahippocampal gyrus, areas TH and TL (Fig. 29C). The connections from dorsal areas 46, 9, and medial proisocortex (area 32) are directed mainly to the cingulate gyrus and the retrosplenial cortex (Fig. 25B).

The long association connections of the prefrontal regions appear to be directed to those sensory association areas from which the frontal region receives afferent connections. Consequently, most of the connections between pre- and post-Rolandic association areas are reciprocal in nature. The frontal projections to these sensory association areas are organized such that those from the premotor region are related to first-order association areas. Those from the lateral prefrontal cortex are connected with second-order sensory association areas, and those from the orbital and medial prefrontal regions are related to third-order sensory association areas. The third-order projections to somatosensory association areas originate from the rostral prefrontal region. The paralimbic connections of the frontal region indicate a dichotomy, such that the ventral trend regions, especially ventral proisocortex as well as surrounding areas, are related to the ventral temporal paralimbic regions. In contrast, the dorsal proisocortex and its immediately surrounding regions project to the cingulate gyrus and retrosplenial cortex. These connectional and architectural dichotomies of prefrontal regions were originally suggested by Nauta (1964) and Sanides (1972), and further emphasize the dual nature of cortical development.

6. Multimodal Association Areas

So far we have described the post-Rolandic parasensory association areas and the frontal lobe association areas. The post-Rolandic regions described are strictly modality-specific, although they show architectural and connectional heterogeneity. At the junction of these unimodal association areas are found the specialized cortical regions which receive input from more than one sensory modality (Fig. 4). Similarly, the paralimbic region (described below) as well as the frontal lobe also contain regions which receive connections from more than one modality (Fig. 4). We will outline these areas according to their location and the nature of their cortical input.

6.1. Post-Rolandic Multimodal Areas

There are several locations where distinct multimodal areas are found in the parietotemporal region; in the intraparietal sulcus (Fig. 26A), the caudal inferior parietal lobule (Fig. 26B), the caudal superior temporal gyrus (Fig. 26C), the medial parietal region (Fig. 26D), and the STS (Fig. 27). The mid-lower bank of the IPS contains an area which has been demarcated on the bases of architecture and connections by Seltzer and Pandya (1980). This region (Fig. 26A), termed area POa, receives input from the first-order somatosensory association area (SA1) containing the head, face, and neck representations. It also receives strong connections from the first-order visual association area (VA1) of the preoccipital gyrus, representing mainly the peripheral visual field (Kuypers *et al.*, 1965; Rockland and Pandya, 1981; Ungerleider and Mishkin, 1982; Van Essen and Maunsell, 1983). Additionally, area POa receives projections from the cortex at the rostral tip of the IPS, which is known to contain a cortical vestibular representation (Fredrickson *et al.*, 1966; Büttner and Lang, 1979). Thus, by virtue of its relationship with somatosensory, visual, and vestibular areas, area POa is considered a multimodal region and may be involved in integrating sensory information about head position with peripheral visual input. Such integrated information might allow for orientation of the head and neck with reference to the visual periphery (see Fuster, this volume). While the precise function of area POa remains to be investigated, it is interesting to note that this

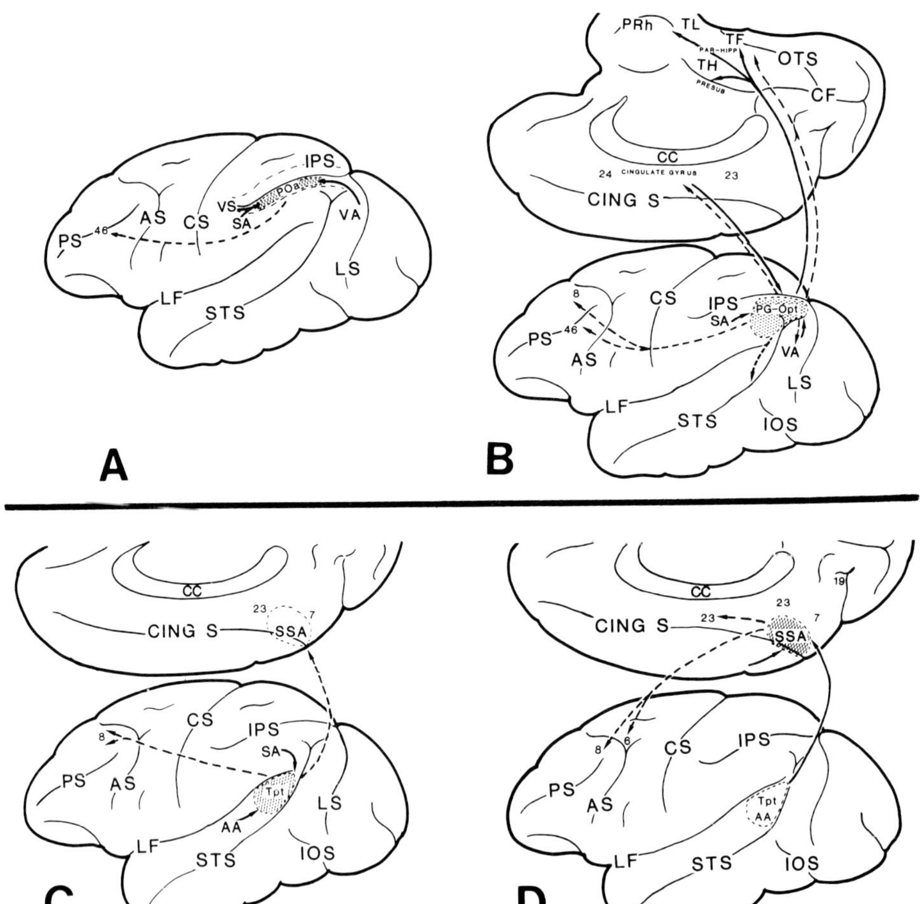

Figure 26. Diagrams showing locations of four post-Rolandic multimodal areas and their connections. (A) Area POa in the intraparietal sulcus; (B) area PG-Opt in the inferior parietal lobule; (C) area Tpt in the caudal portion of the superior temporal gyrus; (D) area SSA in the medial parietal cortex.

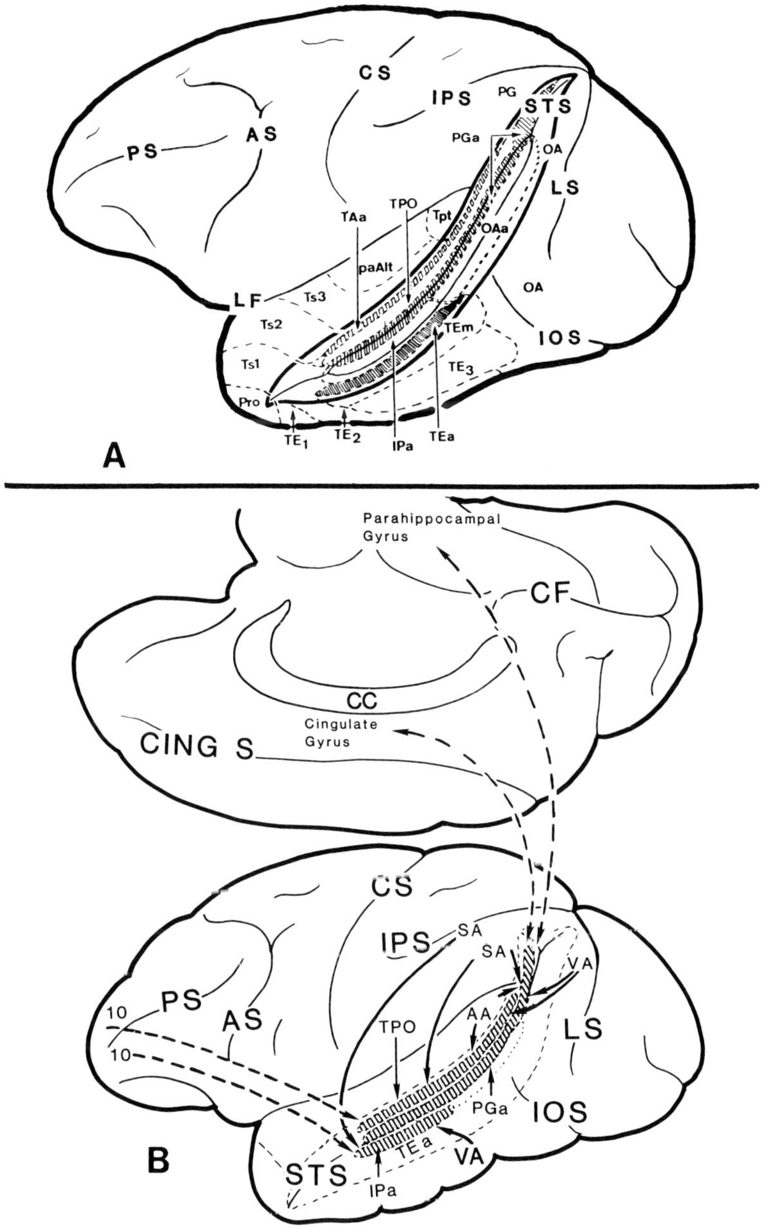

Figure 27. (A) Diagram showing the architectonic parcellation of the superior temporal sulcus and surrounding cortex according to Seltzer and Pandya (1978). (B) Diagrams showing two multimodal areas (TPO and PGa) and their connections.

area is reciprocally connected with the frontal eye field region (Pandya *et al.*, 1971; Mesulam *et al.*, 1977; Barbas and Mesulam, 1981).

The cortex of the caudal IPL is known to contain two distinct architectonic areas (caudal area PG and area Opt). This region (Fig. 26B) receives connections from both somatosensory association areas, area SA3 of both SPL and IPL, and from the visual association area, VA1 (Kuypers *et al.*, 1965; Pandya and Seltzer, 1982a; Ungerleider and Mishkin, 1982). Additionally, this area receives paralimbic input from both the cingulate gyrus (area 23) and the parahippocampal region (Divac *et al.*, 1977; Mesulam *et al.*, 1977; Pandya *et al.*, 1981; Van Hoesen, 1982; Rosene and Pandya, 1982). Thus, this area is considered another multimodal region and may allow for the integration of higher-order somatosensory, visual, and limbic information (e.g., Mesulam, 1983). In terms of efferent connections, this region is connected with the other multimodal areas of the STS, the paralimbic regions (cingulate gyrus and parahippocampal gyrus), as well as the prefrontal cortex (Jones and Powell, 1970a; Chavis and Pandya, 1976; Seltzer and Pandya, 1976; Petrides and Pandya, 1983). By virtue of this connectivity, the caudal IPL region may be presumed to be involved, in part, in visuospatial attention and orientation especially as pertaining to stimuli which are of significance to the organism. Indeed, several experimental as well as clinical studies have pointed to such a role for the caudal inferior parietal region in monkeys, and for a homologous region, the angular gyrus, in humans (Critchley, 1953; Denny-Brown and Chambers, 1958; Heilman *et al.*, 1970; Mountcastle *et al.*, 1975; Robinson and Goldberg, 1978; Heilman, 1979; Lynch, 1980; Sakata *et al.*, 1981; Hyvärinen, 1982; Mesulam, 1983).

Another region at the parietotemporal junction, termed area Tpt (Fig. 26C), receives input from both the somatosensory association areas as well as the auditory association region (Pandya and Kuypers, 1969; Jones and Powell, 1970a). This area in turn projects strongly to the dorsal premotor region of the frontal lobe (Jones and Powell, 1970a; Chavis and Pandya, 1976), as well as to the posterior cingulate gyrus (Pandya *et al.*, 1969). Given this connectivity, area Tpt may be involved in the integration of auditory input and somatosensory information regarding face, head, and neck. Just as visual somatosensory integration in caudal IPL allows for visuospatial attention and orientation, the auditory–somatosensory integration in area Tpt might subserve a similar audiospatial function. As suggested by Hyvärinen (1982), area Tpt may therefore be involved in head turning toward significant auditory sounds, by virtue of its efferent connections with the premotor and cingulate regions. It seems, then, that whereas the primary auditory region is concerned mainly with the analysis of spectral properties and tonotopic relationships (Merzenich and Brugge, 1973), area Tpt may be crucial for the mechanisms relating to sound source localization (Hyvärinen, 1982).

An additional auditory–somatosensory multimodal region is located in the medial parietal cortex below the cingulate sulcus (Fig. 26D). This area is known to receive input from auditory association area AA1 (Pandya *et al.*, 1969) and somatosensory input from SPL, areas SA1 and SA2 (Pandya and Seltzer, 1982a). The functional role of this region remains to be determined. However, it may have a function similar to that of area POa. That is, by virtue of its efferents leading to the dorsal premotor area, this region may subserve orientation of the trunk toward sound sources.

The multimodal nature of the STS was first described by Jones and Powell (1970a). According to these investigators and others, the cortex of the STS contains subregions with bimodal and trimodal convergence of sensory input (Seltzer and Pandya, 1978). This is a linear sector in the upper bank of the STS, and is identified as area TPO on the basis of its architecture (Fig. 27A). It receives connections from the auditory association areas, AA (AA1, AA2, and AA3), somatosensory association area, SA (SA3), and visual association areas, VA (VA2 and VA3) (Fig. 27B). Similarly, another architectonically distinct area in the depth of STS is called area PGa. This area, like area TPO, seems to receive input from the sensory association areas; however, it is related predominantly to the somatosensory modality. Each sensory association area has a unimodal projection zone in the STS surrounding the multimodal areas TPO and PGa. Thus, auditory association areas project to a region called TAa, the visual association areas (VA2 and VA3) project to area TEa, and the somatosensory association region projects to areas IPa and PGa (Fig. 27B).

The exact nature of sensory convergence in the STS regions as well as their efferent cortical projections remain to be determined. The multimodal quality of this region has been documented in both behavioral and physiological investigations (Heilman *et al.*, 1970; Petrides and Iversen, 1978; Bruce *et al.*, 1981; Leonard *et al.*, 1983). The fact that most of the input to the STS is derived from the third-order sensory association areas and other polymodal areas such as the caudal IPL (Pandya and Kuypers, 1969; Jones and Powell, 1970a) as well as the prefrontal cortex (Nauta, 1964; Jones and Powell, 1970a; Pandya *et al.*, 1971), and the paralimbic (cingulate gyrus and parahippocampal gyrus) regions (Pandya *et al.*, 1981; Van Hoesen, 1982; Amaral *et al.*, 1983) suggests that these multimodal regions of the STS may be involved in the highest level of cortical integration of sensory and limbic information.

6.2. Paralimbic Multimodal Areas

The caudal part of the parahippocampal gyrus receives input from the various sensory association areas (Fig. 28A) and hence has been considered as a multimodal region (Jones and Powell, 1970a; Seltzer and Pandya, 1976). This paralimbic region contains distinct architectonic areas designated TF, TL, and TH (von Bonin and Bailey, 1947; Rosene and Pandya, 1983). It receives input from visual association areas (VA1 and VA2) as well as auditory association areas (AA2). The somatosensory input to this region, according to recent studies, seems to be indirect, i.e., via the multimodal area of the caudal IPL (Seltzer and Pandya, 1984). Within this region, both bimodal and trimodal sensory convergence have been identified. Moreover, this area, as will be discussed below, receives input from other paralimbic regions such as the cingulate gyrus and the caudal orbitofrontal cortex, and is further connected with limbic regions such as the hippocampus via entorhinal cortex (Van Hoesen *et al.*, 1972; Baleydier and Mauguiere, 1980; Pandya *et al.*, 1981). The complex multimodal sensory and limbic connections of this region suggest its involvement in functions such as learning and memory (Van Hoesen, 1982).

6.3. Frontal Lobe Multimodal Areas

The frontal lobe contains several regions where the convergence of sensory inputs has been shown to occur (Pandya and Kuypers, 1969; Jones and Powell, 1970a; Chavis and Pandya, 1976). The premotor region around the arcuate sulcus is a site of sensory convergence from the first-order sensory association areas (Fig. 28B). Thus, area AA1 and area SA1 of the SPL send connections to the dorsal periarcuate region in a partly overlapping manner. Similarly, area VA1 and area SA1 of the IPL have partially overlapping projections in the ventral periarcuate region. Finally, trimodal convergence of all first-order sensory association areas takes place in the central portion of the arcuate sulcus. In addition to multimodal areas of the premotor region, the ventral prefrontal cortex also has a zone of bimodal and trimodal convergence. Unlike premotor multimodal areas, the prefrontal region receives input from second-order sensory association areas. Thus, ventral area 46 receives input from the AA2, VA2, and SA2 regions

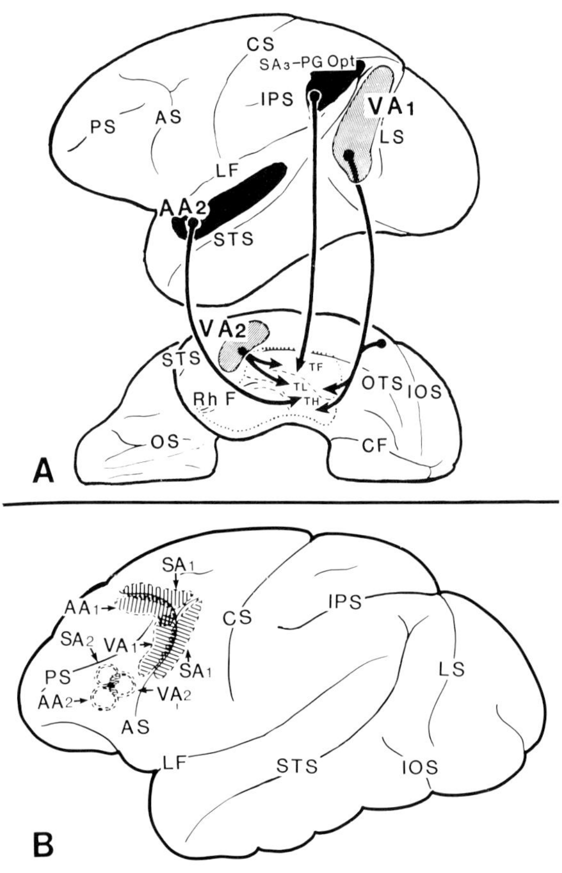

Figure 28. (A) Diagram showing the location and connections of the multimodal area of the parahippocampal gyrus. (B) Diagram showing the location and connections of premotor and prefrontal multimodal areas.

in an overlapping fashion (Fig. 28B). Although the full functional significance of these frontal multimodal regions remains to be worked out, several physiological and behavioral studies point to a role for these areas in complex processes relating to more than one sensory modality. With regard to the premotor region, single-unit recordings have revealed polysensory characteristics among the neurons in the periarcuate region (Bignall and Imbert, 1969; Nelson and Bignall, 1973; Vaadia *et al.*, 1983). Additionally, lesion–behavioral studies have revealed the importance of the periarcuate region in visual, auditory, and somatosensory tasks as well as behavior contingent upon the integration of input from more than one modality (Welch and Stuteville, 1958; Petrides and Iversen, 1976; Van Hoesen *et al.*, 1980). Likewise, the prearcuate trimodal convergence zone also is implicated in behavior contingent upon the synthesis of information from more than one modality (Passingham, 1972; Passingham and Ettlinger, 1972).

7. Paralimbic Association Areas

These regions are situated between the sensory association areas and the limbic regions and have direct connections with limbic structures (Figs. 5, 29, and 30). The paralimbic association areas include the cingulate region (areas 23, 24, 25, and 32), the caudal orbitofrontal cortex (area 13), the temporal pole, and the parahippocampal area (areas TF, TL, and TH). Unlike the other association areas described above, which have well-developed six-layered cortex, the paralimbic regions in general have architectonic features intermediate to those of isocortex and allocortex (Yakovlev *et al.*, 1966; Sanides, 1972; Braak, 1980). In terms of cortical connections, these regions are connected with the cortical association areas on one hand and the allocortical structures of the limbic system on the other. For example, the cingulate gyrus (Fig. 29A) has reciprocal connections with the prefrontal association cortex, the posterior parietal region, and the insula (Nauta, 1964; Jones and Powell, 1970a; Pandya *et al.*, 1971, 1981; Petras, 1971; Mesulam *et al.*, 1977; Balcydier and Mauguiere, 1980; Mesulam and Mufson, 1982; Mufson and Mesulam, 1982). Its limbic efferent connections are directed mainly to the amygdala, the presubiculum, and the perirhinal and retrosplenial cortices (Baleydier and Mauguiere, 1980; Pandya *et al.*, 1981). Similarly, the temporal pole (Fig. 29B) receives input from second-order visual and auditory association areas and projects to the insula, medial frontal and orbitofrontal cortex, amygdala, and the perirhinal region (Jones and Powell, 1970a; Van Hoesen *et al.*, 1972; Herzog and Van Hoesen, 1976; Seltzer and Pandya, 1976). The orbitofrontal cortex (Fig. 29C) is connected with other prefrontal areas on one hand and perirhinal, entorhinal, and parahippocampal regions on the other (Pandya and Kuypers, 1969; Jones and Powell, 1970a; Van Hoesen *et al.*, 1972; Barbas and Pandya, 1982). Finally, the parahippocampal area receives (Fig. 30) afferent connections from second-order parasensory association areas of the visual, auditory, and somatosensory modalities as well as orbitofrontal cortex (Jones and Powell, 1970a; Seltzer and Pandya, 1976; Amaral *et al.*, 1983), and has efferent connections to these areas as well as to the hippocampus by way of the entorhinal cortex (Van Hoesen, 1982).

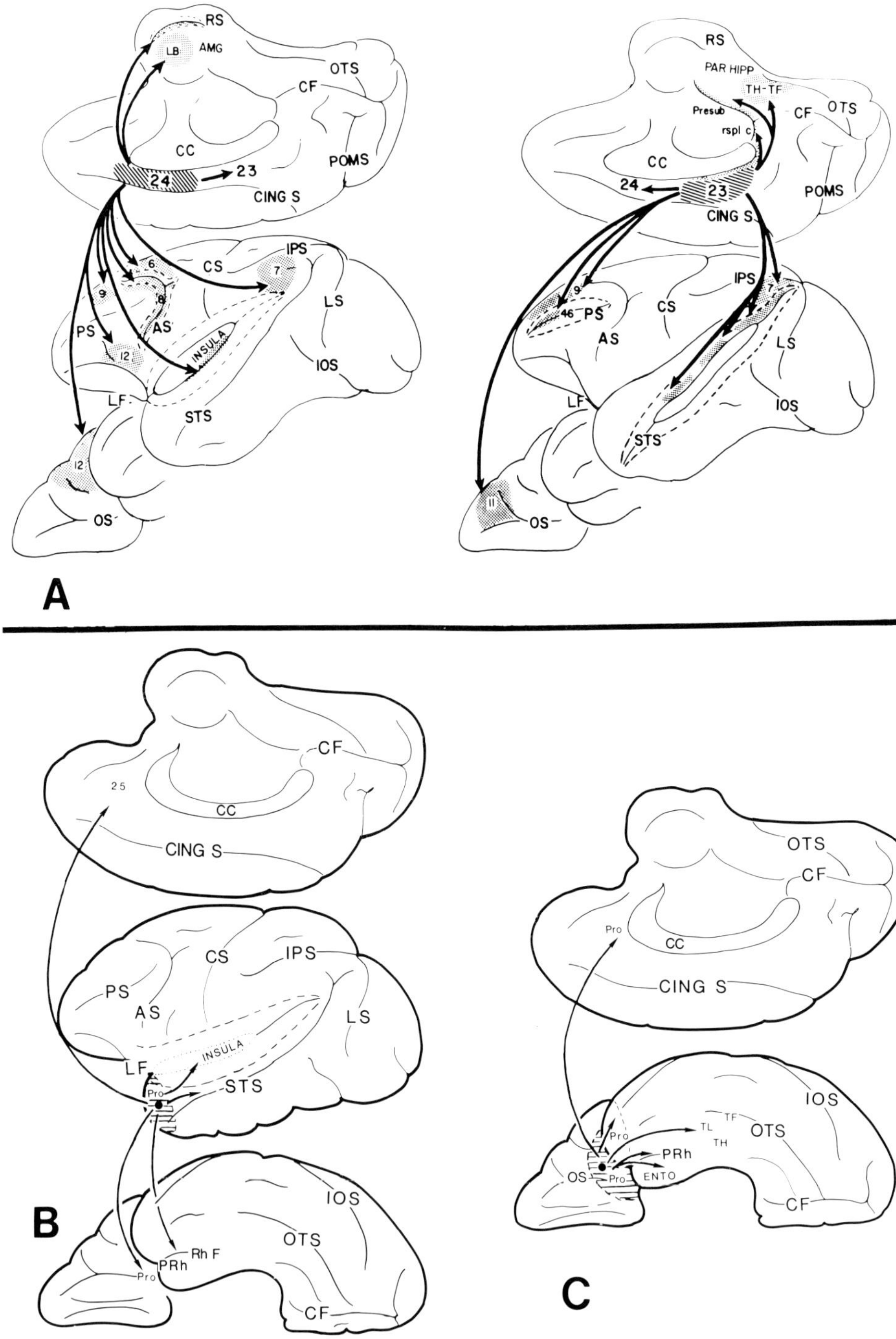

Figure 29. (A) Summary diagram of efferent projections of the cingulate gyrus (areas 23 and 24). (B, C) Efferent connections of the temporal polar and orbitofrontal paralimbic regions.

There are also interconnections among these paralimbic regions. For example, the temporal pole and the caudal orbitofrontal cortex are reciprocally connected as are the cingulate gyrus and the parahippocampal region (Bailey *et al.*, 1943; Pandya and Kuypers, 1969; Jones and Powell, 1970a; Pandya *et al.*, 1971, 1981; Baleydier and Mauguiere, 1980). While the detailed efferent connections of the paralimbic regions remain to be delineated, a clear dichotomy appears to exist. Thus, the cingulate gyrus is connected predominantly with the lateral prefrontal, premotor, and parietal regions, whereas the parahippocampal region is connected mainly with the caudal orbitofrontal, ventral prefrontal, and temporal polar cortices (Figs. 29A and 30). Therefore, the cingulate region ties in with the dorsal or hippocampal trend as described above, whereas the parahippocampal region seems to be related to the ventral or paleocortical trend (Sanides, 1972; Galaburda, 1984).

A distinguishing connectional feature of the paralimbic areas is that all of their cortical input is derived from the second- and third-order sensory association areas, and the prefrontal and orbitofrontal cortices with which they are reciprocally interconnected. Thus, the paralimbic areas have no direct relationship with either parasensory association areas or primary sensorimotor regions. This would suggest that these areas deal with highly processed information. Moreover, the strong connectivity between paralimbic and limbic structures implies a role in relating such information to motivational and emotional states (e.g., Mesulam, 1983).

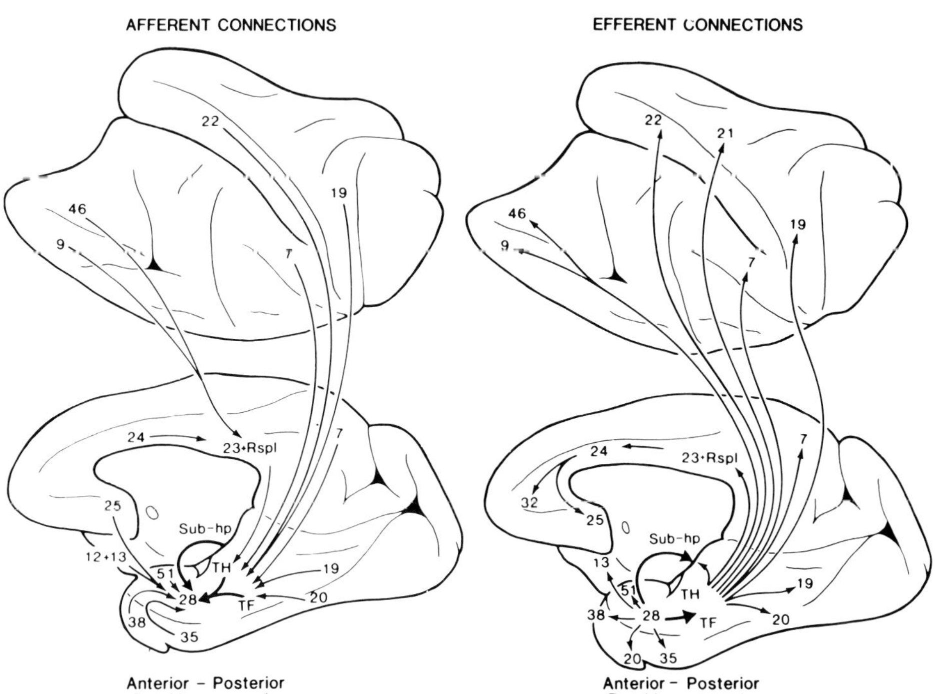

Figure 30. Summary diagram of afferent and efferent connections of the parahippocampal paralimbic region according to Van Hoesen (1982).

8. Discussion

The cortical association areas comprise the most extensive part of the brain in primates. Evolutionarily, these areas represent the focus of greatest differentiation between the brains of primate and nonprimate species, and have long been viewed as crucial to higher cognitive and behavioral functions. The exact morphological and functional properties of these areas have been the subject of intensive inquiry in recent years. For the systematic evaluation of the functions of these regions, using either an experimental approach or clinical data, it is extremely important to have an understanding of the organizational principles underlying the morphology of these areas. In particular, knowledge of the architecture and connections of these regions may provide a framework for further investigations and interpretations of functional significance.

Recent architectonic studies have put forward an important concept which addresses the issue of the evolutionary development of cortical areas and the possible meaning behind the arrangement of the neurons within each of these regions. For example, the work of Sanides (1972) has provided a major impetus in the concept of the dual nature of cortical evolution. As described earlier, the neocortex is viewed as having developed from two primordial moieties, archicortical and paleocortical, which proceeded through respective proisocortical stages to culminate in the isocortices. Accordingly, each post-Rolandic sensory system is assumed to have evolved as an outgrowth of either paleocortical alone or paleocortical and archicortical trends combined. Likewise, the frontal cortices—motor, premotor, and prefrontal—are assumed to have developed from the two moieties. Furthermore, it is proposed that the sequential modification of existing cortical cell layers may involve common patterns of differentiation among the various systems. For example, the temporal polar region which contains a bilaminar-appearing proisocortex, as a next step from the paleocortex, gives rise to a stepwise progression of regions with increasing development of the supragranular layers (layers II, III, and IV) and a diminution and separation of infragranular layers (Galaburda and Pandya, 1983). With this principle as a basis, within the STR a rostrocaudal sequence of four cortical complexes is recognized, each consisting of a medial root field, a central core field, and a lateral belt field. The most rostral of these complexes appears to be structurally most primitive (proisocortical), and is therefore thought to represent a first phase in the evolution of the auditory cortex. This evolution reaches its highest plateau in the third stage, where the core field is formed by the primary auditory cortex. Four successive rostrocaudal belt areas of the STG represent the auditory association areas for each stage. Although comparable architectonic analyses have not been carried out for the other sensory systems, preliminary observations suggest that their development may follow similar trends (Pandya and Seltzer, 1982; Rosene and Pandya, 1983).

Thus, the association areas of each post-Rolandic sensory system may be grouped in a sequential manner on the basis of their architectonic features. The progression of stages within each of these systems may follow a basically similar pattern of laminar development. This is suggested by a comparison of the general features of first-order, second-order, and third-order parasensory association areas of the somatosensory, visual, and auditory systems. The third-order association areas in each modality are characterized by relatively less laminar differentiation, with prominent infragranular neurons. In contrast, the first-order areas are characterized by greater laminar differentiation, with marked emphasis on supragranular layers. The second-order association areas exhibit intermediate features in their laminar organization.

Recent architectonic analyses have indicated that the frontal lobe association areas have evolved in analogous stages, where the proisocortical areas on the orbital and medial surfaces represent the first stage of differentiation, and the periarcuate region represents the final level (Barbas and Pandya, 1981, 1982). This architectonic concept suggests an orderly arrangement for all cortical association areas. Additionally, there are a number of cortical regions, some of which are situated at the borders of neighboring parasensory association regions and have distinct architectonic features, which are considered as multimodal regions since they receive input from more than one sensory modality.

We fully realize that an organizational schema for the association areas based on an evolutionary architectonic conceptualization represents only one particular way of systematizing these heterogeneous regions. However, the connectional data suggest a close correlation between specific architectonic subdivisions and both their local and distant connections. Viewed in this manner, a number of general principles of organization become evident.

One important principle regarding the intrinsic connections of the cortical association areas has emerged in recent years. Available data, though fragmentary for certain systems, indicate that adjacent association areas in each sensory modality are interconnected in a sequential fashion, with a specific and uniform laminar origin and termination: outgoing (forward) connections from a given parasensory area originate from supragranular layer III and terminate in the adjacent association area in and around layer IV in a columnar manner. In contrast the backgoing (reciprocal) connections from a given region originate from infragranular layers V and VI and terminate in the precursor region in the first layer. These patterns have been traced fully in auditory association areas as well as visual association regions (Tigges *et al.*, 1973; Rockland and Pandya, 1979; Tigges *et al.*, 1981; Galaburda and Pandya, 1983). Information for somatosensory association areas is incomplete; however, there are indications that this system might follow the same basic principle (Jones *et al.*, 1978; Vogt and Pandya, 1978; Pandya and Seltzer, 1982a; see Burton, this volume).

The functional significance of this local circuitry with its laminar specificity and reciprocal connections is not fully understood. The peripheral input to a given region is derived from thalamic afferents and is relayed to layer IV. This information is advanced to the next region by layer III neurons, and relayed in and around layer IV. This type of organization will forward the successively processed information about the periphery all the way up to proisocortex and beyond into the limbic system. A parallel system of reciprocal connections starts from limbic structures and proceeds through the proisocortex and isocortical regions, originating in each sequential stage mainly from the infragranular layers and terminating in the first layer of the next region. This arrangement might provide a means by which incoming information from the periphery could be integrated with information about the internal state of the organism emanating from the limbic system via the proisocortex. Such integration might provide a way whereby incoming information may be associated with motivational and emotional states to subserve processes such as selective attention and memory formation and retrieval. While this functional concept based on intrinsic circuitry clearly does not take into account all of the connectional features of a given sensory association area, it may help to generate future experimental paradigms regarding sensory–limbic integration.

Another principle common to all three sensory modalities is observed in regard to the long association connections. As pointed out above, on the basis of architecture as well as their connections to the frontal cortices, the parasensory association areas in each modality can be divided into three major sectors. The first-order association areas in each modality send projections predominantly to the periarcuate (premotor) regions, whereas second-order areas are most strongly connected with prearcuate (prefrontal) regions. Finally, the third-order visual and auditory association areas project preferentially to the orbitofrontal region, while the third-order somatosensory association areas project to the rostral prefrontal region. It is of great interest to note that each sector of the sensory association areas is connected with a frontal lobe region which has basically similar architectonic features. In other words, each sector is connected with a portion of the frontal lobe which appears to occupy a similar stage of architectonic differentiation. This would imply that each sensory association sector, from first-order through second-order to third-order, may have developed in parallel with a specific frontal lobe region, and with that region may constitute a functional subsystem within the cerebral cortex. Added support for this concept is provided by the existence of reciprocal connections between each frontal region and its respective post-Rolandic sensory association areas.

The other long association connections of the post-Rolandic areas are to the parietotemporal cortex of the STS and to the paralimbic areas (cingulate and parahippocampal gyri). These connections are derived mainly from the third-order sensory association areas, although all three auditory association areas are connected with the STS and likewise all visual association areas project to the parahippocampal gyrus.

The functional significance of these long association connections is not fully understood. However, by considering the organizational pattern of these connections along with behavioral data, one can speculate as to their functional role. For example, the first-order association areas project to the premotor region of the frontal lobe which in turn has reciprocal connections with the motor cortex. This circuitry then may be involved with the use of sensory information to influence motor behavior. Stimulation of premotor cortex induces an orienting response, whereas lesions of this area produce a deficit in responsing to contralateral sensory stimulation (Lilly, 1958; Welch and Stuteville, 1958; Fuster, 1980; Wiesendanger, 1981). Furthermore, the overlapping nature of cortical sensory input to premotor cortex implies a role for this area in behavior which is contingent upon the ability to integrate information from different sensory modalities. Behavioral studies demonstrate that lesions in this area produce deficits in learning a motor task only when it involves making an intermodal association and not when it depends simply upon a unimodal sensory discrimination (Petrides and Iversen, 1976; Van Hoesen *et al.*, 1980; Petrides, 1982). The projections to the prefrontal and orbitofrontal cortices are from the second- and third-order parasensory areas as well as paralimbic regions, and suggest a somewhat different functional role from that of the premotor cortex. The prefrontal regions are less directly connected with the motor cortex and receive highly integrated information along with paralimbic input. This pattern of connectivity suggests that this region allows the organism to take into consideration both complex sensory data and motivational and emotional factors prior to executing a motor act. Humans with frontal lobe damage show alterations in affect, one of which is an apparent state of indifference (Hecaen, 1964; Nauta, 1971; Blumer and Benson, 1975; Damasio, 1979). At the same time, some of these patients have a tendency to act impulsively and appear to be unable to carry out organized behavi or which calls for shifting strategies. These deficits may arise, in part, from an inability to relate behavior to internal states. This would result in a tendency to perseverate in an original strategy despite its being inappropriate in a different context. Following lesions of the prefrontal cortex in monkeys, a similiar perseverative phenomenon may be observed, e.g., a failure to follow a reinforced stimulus in a delayed-alternation paradigm (Mishkin, 1964).

Thus, the differential post-Rolandic connectivity of the premotor and prefrontal regions suggests that these areas make different contributions to the generation of motor behavior in the context of stimulus input. The close relationship of the premotor cortex with motor cortices, considered along with the fact that the premotor cortex receives connections from first-order sensory association areas, implies a functional role for this region in orienting to a specific stimulus prior to executing a motor act. In contrast, the prefrontal cortex, which receives highly processed sensory and limbic input and is much less directly related to the motor cortex, may serve in decision-making and sequencing processes in which external (parasensory) and internal (limbic) factors are integrated to produce adaptive responding (Nauta, 1971).

Unlike the parasensory association areas which are located adjacent to the primary sensory cortices, the post-Rolandic multimodal regions are situated at the junctions of the sensory association areas. Other multimodal regions are found in the premotor and prefrontal cortices and in the parahippocampal gyrus. These regions can be divided into three different types, according to their architecture and the relative complexity of their cortical input. The parahippocampal sensory convergence region (areas TH, TF, and TL) can be viewed as a more primitive multimodal zone because of its relatively rudimentary architecture and the fact that it receives main cortical afferents from the third- and second-order sensory association areas. Convergence areas of a second type are located in the STS and in the prefrontal cortex, and can be considered to have intermediate architectonic features. They receive predominant input from second- and third-order sensory association areas which themselves have intermediate architectonic characteristics. The third type of sensory convergence zone occurs in the parietal lobe (area POa in the IPS, areas PG and Opt in the IPL, and area Tpt in the STG) and in the premotor regions. These areas have highly developed cortical lamination (with emphasis on supragranular layers) and receive input mainly from the first-order sensory association areas. It should be pointed out that within each of these architecturally distinct sensory convergence zones, there is a central region of maximum convergence surrounded by regions of bimodal and unimodal input. Another organizational characteristic is that several of these multimodal areas are interconnected and have strong ties to the limbic system. While the functional significance of these regions remains unclear, in order to adapt behavior successfully to a sensory environment which seldom consists of stimuli in only a single modality, it would seem essential to integrate information between and among sensory systems. The limbic input to such a region would provide an added dimension relating to the organism's internal state.

As discussed previously, apart from the sensory convergence areas having ties with the paralimbic regions, the second- and third-order sensory association areas as well as frontal regions other than sensory convergence zones have connections with the paralimbic regions. These paralimbic connections from the association areas can be grouped into two broad systems, one relating mainly to the cingulate gyrus and the other to the parahippocampal and temporal polar proisocortical regions. The cingulate region is connected with those association areas which are related primarily to the truncal and extremity representations of somatosensory and motor regions, to the dorsal prefrontal region, and to the association areas related to the peripheral visual field. The parahippocampal and temporal polar regions, on the other hand, are connected mainly with auditory association areas, with the visual association areas relating to central vision, with the somatosensory association areas for the head, face, and neck representations, and with the ventral prefrontal and orbitofrontal regions. These dichotomous connectional systems of paralimbic regions may have their origin in the two distinct primordial moieties, namely the archicortical and paleocortical. This is suggested by the fact that the cingulate region itself relates to the presubiculum and in turn to the hippocampus, whereas the parahippocampal and temporal polar regions relate to the orbital and insular regions and finally to temporofrontal paleocortex.

On the basis of these differential connections, a general functional dichotomy may be suggested. The cingulate region may be involved preferentially in processes of bodily awareness and spatial orientation and in basic motivation of goal-directed spatial behavior. The parahippocampal–temporal polar regions, with their specific sensory and limbic connections, may have a preferential involvement in stimulus identification and precise discrimination of stimuli according to their emotional and motivational contexts. In other words, the cingulate region and its connections might have predominantly a localization function, whereas the parahippocampal–temporal polar region would have an identification function. It should be realized that these hypothetical functional and connectional dichotomies are not intended to imply mutually exclusive roles for these two paralimbic entities. Rather, given the reciprocal connections between the cingulate and parahippocampal regions, these two trends may function in unison to bring about fully integrated behavior.

In summary, the cortical connections of the association areas of the cerebral cortex are organized in a systematic manner, and there is a close correlation between the connections and architecture of these regions. Based on cortical connections, the following generalization can be made for the overall functional role of the association areas. As sensory input arrives at the cerebral cortex, it first passes through successive stages of intramodality elaboration, allowing progressively more complex discrimination of stimulus features. Subsequently, by a series of further interconnections, this already elaborated complex information is conveyed to multimodal areas for intersensory integration. In turn, the information is relayed to paralimbic and limbic structures for investment with emotional tone and for memory consolidation. Finally, it is the frontal lobe regions where both sensory and limbic influences are integrated in preparation for the organism to execute behavior appropriate to environmental and internal factors.

ACKNOWLEDGMENTS. We are highly grateful to Dr. W. J. H. Nauta for his most useful comments and suggestions. Also we would like to express our sincere thanks to Mr. Brian Butler and Ms. Lydia Kenton for excellent technical assistance and Marcia Williams for art work. This study is supported by the Veterans Administration, Edith Nourse Rogers Memorial Hospital, Bedford, Massachusetts, NIH Grant 16841, and Colby College Social Science Grants A22116 and A22132.

9. References

Ades, H. W., and Felder, R., 1942, The acoustic area of the monkey (*Macaca mulatta*), *J. Neurophysiol.* **5:**49–54.

Akert, K., 1964, Comparative anatomy of frontal cortex and thalamo-frontal connections, in: *The Frontal Granular Cortex and Behavior* (J. M. Warren and K. Akert, eds.), McGraw–Hill, New York, pp. 372–396.

Allman, J. M., Baker, J. F., Newsome, W. T., and Petersen, S. E., 1981, Visual topography and function: Cortical visual areas in the owl monkey, in: *Cortical Sensory Organization,* Volume 2 (C. N. Woolsey, ed.), Humana Press, Clifton, N.J., pp. 171–185.

Amaral, D. G., Insausti, R., and Cowan, W. M., 1983, Evidence for a direct projection from the superior temporal gyrus to the entorhinal cortex in the monkey, *Brain Res.* **275:**263–277.

Bailey, P., von Bonin, G., Garol, H. W., and McCulloch, W. S., 1943, Long association fibers in cerebral hemispheres of monkey and chimpanzee, *J. Neurophysiol.* **6:**129–134.

Baleydier, C., and Mauguiere, F., 1980, The duality of the cingulate gyrus in monkey, *Brain* **103:**525–554.

Barbas, H., and Mesulam, M.-M., 1981, Organization of afferent input to subdivisions of area 8 in the rhesus monkey, *J. Comp. Neurol.* **200:**407–432.

Barbas, H., and Pandya, D. N., 1981, Frontal lobe afferent input to area 6 in the rhesus monkey, *Soc. Neurosci. Abstr.* **7:**414.

Barbas, H., and Pandya, D. N., 1982, Cytoarchitecture and intrinsic connections of the prefrontal cortex of the rhesus monkey, *Soc. Neurosci. Abstr.* **8:**933.

Benevento, L. A., and Rezak, M., 1976, The cortical projections of the inferior pulvinar and adjacent lateral pulvinar in the rhesus monkey (*Macaca mulatta*): An autoradiographic study, *Brain Res.* **108:**1–24.

Benevento, L. A., and Yoshida, K., 1981, The afferent and efferent organization of the lateral geniculo-prestriate pathways in the macaque monkey, *J. Comp. Neurol.* **203:**455–474.

Bignall, D. E., and Imbert, M., 1969, Polysensory and cortico-cortical projections to frontal lobe of squirrel and rhesus monkeys, *Electroencephalogr. Clin. Neurophysiol.* **26:**206–215.

Blumer, D., and Benson, D. F., 1975, Personality changes with frontal and temporal lobe lesions, in: *Psychiatric Aspects of Neurologic Disease* (D. F. Benson and D. Blumer, eds.), Grune & Stratton, New York, pp. 151–170.

Bowker, R. M., and Coulter, J. D., 1981, Intracortical connectivities of somatic sensory and motor areas: Multiple cortical pathways in monkeys, in: *Cortical Sensory Organization,* Volume 1 (C. N. Woolsey, ed.), Humana Press, Clifton, N.J., pp. 205–242.

Braak, H., 1980, *Architectonics of the Human Telencephalic Cortex,* Springer-Verlag, Berlin.

Brinkman, C., and Porter, R., 1983, Supplementary motor area and premotor area of monkey cerebral cortex: Functional organization and activities of single neurons during performance of learned movement, in: *Motor Control Mechanisms in Health and Disease* (J. E. Desmedt, ed.), Raven Press, New York, pp. 393–420.

Brodmann, K., 1908, Beiträge zur histologischen Lokalisation der Grosshirnrinde. VI. Mitteilung: Die Cortexgliederung des Menschen, *J. Psychol. Neurol.* **10:**231–246.

Brodmann, K., 1909, *Vergleichende Lokalisationlehre der Grosshirnrinde in ihren Prinzipien dargestellt auf Grund des Zellenbaues,* Barth, Leipzig.

Bruce, C., Desimone, R., and Gross, C. G., 1981, Visual properties of neurons in a polysensory area in superior temporal sulcus of the macaque, *J. Neurophysiol.* **46:**369–384.

Bullier, J., and Kennedy, H., 1983, Projection of the lateral geniculate nucleus onto cortical area V2 in the macaque monkey, *Exp. Brain Res.* **53:**168–172.

Butters, N., Pandya, D., Stein, D., and Rosen, J., 1972, A search for the spatial engram within the frontal lobes of monkeys, *Acta Neurobiol. Exp.* **32:**305–329.

Büttner, U., and Lang, W., 1979, The vestibulocortical pathway: Neurophysiological and anatomical studies in the monkey, in: *Reflex Control of Posture and Movement* (R. Granit and O. Pompeiano, eds.), Elsevier, Amsterdam, pp. 581–588.

Campbell, A. W., 1905, *Histological Studies on the Localisation of Cerebral Function,* Cambridge University Press, London.

Chavis, D. A., and Pandya, D. N., 1976, Further observations on corticofrontal connections in the rhesus monkey, *Brain Res.* **117:**369–386.

Clark, W. E. L., 1936, The thalamic connections of the temporal lobe of the brain in the monkey, *J. Anat.* **70:**447–464.

Critchley, M., 1953, *The Parietal Lobes,* Arnold, London.

Damasio, A., 1979, The frontal lobes, in: *Clinical Neuropsychology* (K. M. Heilman and E. Valenstein, eds.), Oxford University Press, London, pp. 360–412.

Daniel, P. M., and Whitteridge, D., 1961, The representation of the visual field on the cerebral cortex in monkeys, *J. Physiol. (London)* **159:**203–221.

Deacon, T. W., Rosenberg, P., Eckert, M. K., and Shank, C. E., 1982, Afferent connections of the primate inferior arcuate cortex, *Anat. Rec.* **8:**933.

Denny-Brown, D., and Chambers, R. A., 1958, The parietal lobe and behavior, *Res. Publ. Assoc. Res. Nerv. Ment. Dis.* **36:**35–117.

Divac, I., LaVail, L. H., Rakic, P., and Winston, K. R., 1977, Heterogeneous afferents to the inferior parietal lobule of the rhesus monkey revealed by retrograde transport method, *Brain Res.* **123:**197–207.

Doty, R. W., 1983, Nongeniculate afferents to striate cortex in macaques, *J. Comp. Neurol.* **218:**159–173.

Duffy, F. H., and Burchfiel, J. L., 1971, Somatosensory system: Organizational hierarchy from single units in monkey area 5, *Science* **172:**273–275.

Ferrier, D., 1876, *The Functions of the Brain,* Smith, Elder, London.

Fitzpatrick, K. A., and Imig, T. J., 1980, Auditory cortico-cortical connections in the owl monkey, *J. Comp. Neurol.* **192:**589–610.

Fredrickson, J. M., Figge, U., Scheid, P., and Kornhuber, H. H., 1966, Vestibular nerve projection to the cerebral cortex of the rhesus monkey, *Exp. Brain Res.* **2:**318–327.

Fuster, J. M., 1980, *The Prefrontal Cortex,* Raven Press, New York.

Galaburda, A. M., 1984, Anatomy of language; lessons from comparative anatomy, in: *Neurolinguistics* (D. Kaplan, A. R. Lecours, and J. Marshall, eds.), MIT Press, Cambridge, Mass., pp. 398–415.

Galaburda, A. M., and Pandya, D. N., 1983, The intrinsic architectonic and connectional organization of the superior temporal region of the rhesus monkey, *J. Comp. Neurol.* **221:**169–184.

Gattass, R., and Gross, C. G., 1981, Visual topography of striate projection zone (MT) in posterior superior temporal sulcus of the macaque, *J. Neurophysiol.* **46:**621–638.

Geschwind, N., 1965a, Disconnexion syndromes in animals and man. Part I, *Brain* **88:**237–294.

Geschwind, N., 1965b, Disconnexion syndromes in animals and man. Part II, *Brain* **88:**585–644.

Goldman, P. S., and Rosvold, H. E., 1970, Localization of function within the dorsolateral prefrontal cortex of the rhesus monkey, *Exp. Neurol.* **27:**291–304.

Gross, C. G., Bruce, C. J., Desimone, R., Fleming, J., and Gattass, R., 1981, Cortical visual areas of the temporal lobe: Three areas in the macaque, in: *Cortical Sensory Organization,* Volume 2 (C. N. Woolsey, ed.), Humana Press, Clifton, N.J., pp. 187–216.

Hecaen, H., 1964, Mental symptoms associated with tumors of the frontal lobe, in: *The Frontal Granular Cortex and Behavior* (J. M. Warren and K. Akert, eds.), McGraw–Hill, New York, pp. 335–352.

Heilman, K. M., 1979, Neglect and related disorders, in: *Clinical Neuropsychology* (K. M. Heilman and E. Valenstein, eds.), Oxford University Press, London, pp. 268–307.

Heilman, K. M., Pandya, D. N., and Geschwind, N., 1970, Trimodal inattention following parietal lobe ablations, *Trans. Am. Neurol. Assoc.* **95:**259–261.

Herzog, A. G., and Van Hoesen, G. W., 1976, Temporal neocortical afferent connections to the amygdala in the rhesus monkey, *Brain Res.* **115:**57–69.

Hubel, D. H., and Wiesel, T. N., 1968, Receptive fields and functional architecture of monkey striate cortex, *J. Physiol (London)* **195:**215–243.

Hubel, D. H., and Wiesel, T. N., 1972, Laminar and columnar distribution of geniculocortical fibers in the macaque monkey, *J. Comp. Neurol.* **146:**421–450.

Hurst, E. M., 1959, Some cortical association systems related to auditory functions, *J. Comp. Neurol.* **112:**103–119.

Hyvärinen, J., 1982, *The Parietal Cortex of Monkey and Man,* Springer-Verlag, Berlin.

Jacobson, S., and Trojanowski, J. Q., 1977, Prefrontal granular cortex of the rhesus monkey. I. Intrahemispheric cortical afferents, *Brain Res.* **132:**209–233.

Johnson, T. N., Rosvold, H. E., and Mishkin, M., 1968, Projections from behaviorally-defined sectors of the prefrontal cortex to the basal ganglia, septum, and diencephalon of the monkey, *Exp. Neurol.* **21:**20–34.

Jones, E. G., and Burton, H., 1976, Areal differences in the laminar distribution of thalamic afferents in cortical fields of the insular, parietal and temporal regions of primates, *J. Comp. Neurol.* **168:**197–248.

Jones, E. G., and Powell, T. P. S., 1969, Connexions of the somatic sensory cortex in the rhesus monkey. I. Ipsilateral cortical connexions, *Brain* **92:**477–502.

Jones, E. G., and Powell, T. P. S., 1970a, An anatomical study of converging sensory pathways within the cerebral cortex of the monkey, *Brain* **93:**793–820.

Jones, E. G., and Powell, T. P. S., 1970b, Connexions of the somatic sensory cortex of the rhesus monkey. III. Thalamic connexions, *Brain* **93:**37–56.

Jones, E. G., and Wise, S. P., 1977, Size, laminar and columnar distribution of efferent cells in the somatic-sensory cortex of monkeys, *J. Comp. Neurol.* **175:**391–438.

Jones, E. G., Coulter, J. D., and Hendry, S. H. C., 1978, Intracortical connectivity of architectonic fields in the somatic sensory, motor, and parietal cortex of monkeys, *J. Comp. Neurol.* **181:**291–348.

Jones, E. G., Wise, S. P., and Coulter, J. D., 1979, Differential thalamic relationships of sensory-motor and parietal fields in monkeys, *J. Comp. Neurol,* **183:**833–882.

Kaas, J. H., 1983, What, if anything, is SI? Organization of first somatosensory area of cortex, *Physiol. Rev.* **63:**206–231.

Kaas, J. H., Sur, M., Nelson, R. J., and Merzenich, M. M., 1981, The postcentral somatosensory cortex: Multiple representations of the body in primates, in: *Cortical Sensory Organization,* Volume 1 (C. N. Woolsey, ed.), Humana Press, Clifton, N.J., pp. 29–45.

Kennedy, T. T. K., 1955, An Electrophysiological Study of the Auditory Projection Areas of the Cortex in Monkey (*Macaca mulatta*), Thesis, University of Chicago.

Kievit, J., and Kuypers, H. G. J. M., 1977, Organization of the thalamocortical connexions to the frontal lobe in the rhesus monkey, *Exp. Brain Res.* **29:**299–322.

Künzle, H., 1978, An autoradiographic analysis of the efferent connections from the premotor and adjacent prefrontal regions (areas 6 and 9) in *Macaca fascicularis, Brain Behav. Evol.* **15:**185–234.

Künzle, H., and Akert, K., 1977, Efferent connections of cortical area 8 (frontal eye field) in *Macaca fascicularis:* A reinvestigation using the autoradiographic technique, *J. Comp. Neurol.* **173:**147–164.

Kuypers, H. G. J. M., Szwarcbart, M. K., Mishkin, M., and Rosvold, H. E., 1965, Occipitotemporal corticocortical connections in the rhesus monkey, *Exp. Neurol.* **11:**245–262.

Lassen, N. A., Ingvar, D. H., and Skinhøj, E., 1978, Brain function and blood flow, *Sci. Am.* **239:**62–71.

Leinonen, L., Hyvärinen, J., and Sovijarvi, A. R. A., 1980, Functional properties of neurons in the temporo-parietal association cortex of awake monkey, *Exp. Brain Res.* **39:**203–215.

Leonard, C. M., Rolls, E. T., Baylis, G. C., Wilson, F. A. W., Williams, G. V., Griffiths, C., and Murzi, E., 1983, Response properties and distribution of neurons which respond to faces in the monkey, *Soc. Neurosci. Abstr.* **9:**958.

Lilly, J. C., 1958, Correlations between neurophysiological activity in the cortex and short-term behavior in the monkey, in: *Biological and Biochemical Bases of Behavior* (H. F. Harlow and C. N. Woolsey, eds.), University of Wisconsin Press, Madison, pp. 83–100.

Locke, S., Angevine, J. B., Jr., and Yakovlev, P. I., 1964, Limbic nuclei of the thalamus and connections of the limbic cortex. VI. Thalamocortical projection of lateral dorsal nucleus in cat and monkey, *Arch. Neurol.* **11:**1–12.

Lynch, J. C., 1980, The functional organization of the posterior parietal association cortex, *Behav. Brain Sci.* **3:**485–499.

Merzenich, M. M., and Brugge, J. F., 1973, Representation of the cochlear partition on the superior temporal plane of the macaque monkey, *Brain Res.* **50:**275–296.

Merzenich, M. M., Sur, M., Nelson, R. J., and Kaas, J. H., 1981, Multiple cutaneous representations in areas 3b and 1 of the owl monkey, in: *Cortical Sensory Organization,* Volume 1 (C. N. Woolsey, ed.), Humana Press, Clifton, N.J., pp. 67–119.

Mesulam, M.-M., 1983, The functional anatomy and hemispheric specialization for directed attention, *Trends Neurosci.* **6:**384–387.

Mesulam, M.-M., and Mufson, E. J., 1982, Insula of the Old World monkey. III. Efferent cortical output and comments on function, *J. Comp. Neurol.* **212:**38–52.

Mesulam, M.-M., and Pandya, D. N., 1973, The projections of the medial geniculate complex within the Sylvian fissure of the monkey, *Brain Res.* **60:**315–333.

Mesulam, M.-M., Van Hoesen, G. W., Pandya, D. N., and Geschwind, N., 1977, Limbic and sensory connections of the inferior parietal lobule (area PG) in the rhesus monkey: A study with a new method for horseradish peroxidase histochemistry, *Brain Res.* **136:**393–414.

Mishkin, M., 1964, Perseveration of central sets after frontal lesions in monkeys, in: *The Frontal Granular Cortex and Behavior* (J. M. Warren and K. Akert, eds.), McGraw–Hill, New York, pp. 219–241.

Mishkin, M., 1972, Cortical visual areas and their interactions, in: *Brain and Human Behavior* (A. G. Karczmar and J. C. Eccles, eds.), Springer-Verlag, Berlin, pp. 187–208.

Mishkin, M., Ungerleider, L. G., and Macko, K. A., 1983, Object vision and spatial vision: Two cortical pathways, *Trends Neurosci.* **6:**414–417.

Mountcastle, V. B., Lynch, J. C., Georgopoulos, A., Sakata, H., and Acuna, C., 1975, Posterior parietal association cortex of the monkey: Command functions for operations within extrapersonal space, *J. Neurophysiol.* **38:**871–909.

Muakkassa, K. F., and Strick, P. L., 1979, Frontal lobe inputs to primate motor cortex: Evidence for four somatotopically organized premotor areas, *Brain Res.* **177:**176–182.

Mufson, E. J., and Mesulam, M.-M., 1982, Insula of the Old World monkey. II. Afferent cortical input and comments on the claustrum, *J. Comp. Neurol.* **212:**23–37.

Murray, E. A., and Coulter, J. D., 1981, Supplementary sensory area: The medial parietal cortex in the monkey, in: *Cortical Sensory Organization,* Volume 1 (C. N. Woolsey, ed.), Humana Press, Clifton, N.J., pp. 167–195.

Myers, R. E., 1967, Cerebral connectionism and brain function, in: *Brain Mechanisms Underlying Speech and Language* (C. H. Millikan and F. L. Darley, eds.), Grune & Stratton, New York, pp. 61–72.

Nauta, W. J. H., 1964, Some efferent connections of the prefrontal cortex in the monkey, in: *The Frontal Granular Cortex and Behavior* (J. M. Warren and K. Akert, eds.), McGraw–Hill, New York, pp. 397–409.

Nauta, W. J. H., 1971, The problem of the frontal lobe: A reinterpretation, *J. Psychiatr. Res.* **8:**167–187.

Nelson, C. N., and Bignall, K. E., 1973, Interactions of sensory and non-specific thalamic inputs to cortical polysensory units in the squirrel monkey, *Exp. Neurol.* **40:**189–206.

Pandya, D. N., and Kuypers, H. G. J. M., 1969, Cortico-cortical connections in the rhesus monkey, *Brain Res.* **13:**13–36.

Pandya, D. N., and Sanides, F., 1973, Architectonic parcellation of the temporal operculum in rhesus monkey, and its projection pattern, *Z. Anat. Entwicklungsgesch.* **139:**127–161.

Pandya, D. N., and Seltzer, B., 1982a, Intrinsic connections and architectonics of posterior parietal cortex in the rhesus monkey, *J. Comp. Neurol.* **204:**196–210.

Pandya, D. N., and Seltzer, B., 1982b, Association areas of the cerebral cortex, *Trends Neurosci.* **5:**386–390.

Pandya, D. N., and Vignolo, L. A., 1971, Intra- and interhemispheric projections of the precentral, premotor and arcuate areas in the rhesus monkey, *Brain Res.* **26:**217–233.

Pandya, D. N., Hallett, M., and Mukherjee, S. K., 1969, Intra- and interhemispheric connections of the neocortical auditory system in the rhesus monkey, *Brain Res.* **14:**49–65.

Pandya, D. N., Dye, P., and Butters, N., 1971, Efferent cortico-cortical projections of the prefrontal cortex of the rhesus monkey, *Brain Res.* **31:**35–46.

Pandya, D. N., Mufson, E. J., and McLaughlin, T. J., 1980, Some projections of the frontal operculum (gustatory area) in the rhesus monkey, *Anat. Rec.* **196:**143A.

Pandya, D. N., Van Hoesen, G. W., and Mesulam, M.-M., 1981, Efferent connections of the cingulate gyrus in the rhesus monkey, *Exp. Brain Res.* **42:**319–330.

Passingham, R. E., 1972, Visual discrimination learning after selective prefrontal ablations in monkeys (*Macaca mulatta*), *Neuropsychologia* **10:**27–39.

Passingham, R. E., and Ettlinger, G., 1972, Tactile discrimination learning after selective prefrontal ablations in monkeys (*Macaca mulatta*), *Neuropsychologia* **10:**17–26.

Paul, R. L., Merzenich, M. M., and Goodman, H., 1975, Mechanoreceptor representation and topography of Brodmann's areas 3 and 1 of *Macaca mulatta*, in: *The Somatosensory System* (H. H. Kornhuber, ed.), Thieme, Stuttgart, pp. 262–269.

Petras, J. M., 1971, Connections of the parietal lobe, *J. Psychiatr. Res.* **8:**189–201.

Petrides, M., 1982, Motor conditional associative-learning after selective prefrontal lesions in the monkey, *Behav. Brain Res.* **5:**407–413.

Petrides, M., and Iversen, S. D., 1976, Cross-modal matching and the primate frontal cortex, *Science* **192:**1023–1024.

Petrides, M., and Iversen, S. D., 1978, The effect of selective anterior and posterior association cortex lesions in the monkey on performance of a visual–auditory compound discrimination task, *Neuropsychologia* **16:**527–538.

Petrides, M., and Pandya, D. N., 1983, Projections to the frontal cortex from the posterior parietal region in the rhesus monkey, *Soc. Neurosci. Abstr.* **9:**40.

Powell, T. P. S., and Mountcastle, V. B., 1959, Some aspects of the functional organization of the postcentral gyrus of the monkey: A correlation of findings obtained in a single unit analysis with architecture, *Bull. Johns Hopkins Hosp.* **105:**133–162.

Randolph, M., and Semmes, J., 1974, Behavioral consequences of selective subtotal ablations in the postcentral gyrus of *Macaca mulatta*, *Brain Res.* **70:**55–70.

Robinson, D. L., and Goldberg, M. E., 1978, Sensory and behavioral properties of neurons in posterior parietal cortex of the awake, trained monkey, *Fed. Proc.* **37:**2258–2261.

Rockland, K. S., and Pandya, D. N., 1979, Laminar origins and terminations of cortical connections of the occipital lobe in the rhesus monkey, *Brain. Res.* **179:**3–20.

Rockland, K. S., and Pandya, D. N., 1981, Cortical connections of the occipital lobe in the rhesus monkey: Interconnections between areas 17, 18, 19 and the superior temporal gyrus, *Brain Res.* **212:**249–270.

Rosene, D. L., and Pandya, D. N., 1983, Architectonics and connections of the posterior parahippocampal gyrus in the rhesus monkey, *Soc. Neurosci. Abstr.* **9:**222.

Rosvold, H. E., and Mishkin, M., 1961, Non-sensory effects of frontal lesions on discrimination learning and performance, in: *Brain Mechanisms and Learning* (J. F. Delafresnaye, ed.), Blackwell, Oxford, pp. 555–576.

Sakata, H., 1975, Somatic sensory responses of neurons in the parietal association area (area 5) in monkeys, in: *The Somatosensory System* (H. H. Kornhuber, ed.), Thieme, Stuttgart, pp. 250–261.

Sakata, H., Shibutani, H., and Kawano, K., 1981, Neural correlates of space perception in the parietal association cortex of the monkey, in: *Advances in the Physiological Sciences*, Volume 17 (G. Adam, I. Meszaras, and E. I. Banyai, eds.), Akademiai Kiado, Budapest, pp. 291–298.

Sanides, F., 1969, Comparative architectonics of the neocortex of mammals and their evolutionary interpretation, *Ann. N.Y. Acad. Sci.* **167:**404–423.

Sanides, F., 1972, Representation in the cerebral cortex and its areal lamination patterns, in: *Structure and Function of Nervous Tissue*, Volume 5 (G. F. Bourne, ed.), Academic Press, New York, pp. 329–453.

Seltzer, B., and Pandya, D. N., 1976, Some cortical projections to the parahippocampal area in the rhesus monkey, *Exp. Neurol.* **50:**146–160.

Seltzer, B., and Pandya, D. N., 1978, Afferent cortical connections and architectonics of the superior temporal sulcus and surrounding cortex in the rhesus monkey, *Brain Res.* **149:**1–24.

Seltzer, B., and Pandya, D. N., 1980, Converging visual and somatic sensory cortical input to the intraparietal sulcus of the rhesus monkey, *Brain Res.* **192:**339–351.

Seltzer, B., and Pandya, D. N., 1983, The intraparietal sulcus of rhesus monkey: Intrinsic connections and architectonics, *Soc. Neurosci. Abstr.* **9:**352.

Seltzer, B., and Pandya, D. N., 1984, Further observations on parieto-temporal connections in rhesus monkey, *Exp. Brain Res.* **55:**301–312.

Siwek, D., and Pandya, D. N., 1984, Corticothalamic connections of prefrontal cortex in the rhesus monkey, *Anat. Rec.* **208:**188–189.

Tigges, J., Spatz, W. B., and Tigges, M., 1973, Reciprocal point-to-point connections between parastriate and striate cortex in the squirrel monkey (*Saimiri*), *J. Comp. Neurol.* **148:**481–490.

Tigges, J., Tigges, M., and Perachio, A. A., 1977, Complementary laminar terminations of afferents to area 17 originating in area 18 and in the lateral geniculate nucleus in the squirrel monkey, *J. Comp. Neurol.* **176:**87–100.

Tigges, J., Tigges, M., Anschel, S., Cross, N. A., Letbetter, W. D., and McBride, R. L., 1981, Areal and laminar distribution of neurons interconnecting the central visual cortical areas 17, 18, 19 and MT in squirrel monkey (*Saimiri*), *J. Comp. Neurol.* **202:**539–560.

Tobias, T. J., 1975, Afferents to prefrontal cortex from the thalamic mediodorsal nucleus in the rhesus monkey, *Brain Res.* **83:**191–212.

Trojanowski, J. Q., and Jacobson, S., 1975, A combined horseradish peroxidase–autoradiographic investigation of reciprocal connections between superior temporal gyrus and pulvinar in squirrel monkey, *Brain Res.* **85:**347–353.

Ungerleider, L. G., and Mishkin, M., 1979, The striate projection zone in the superior temporal sulcus of *Macaca mulatta:* Location and topographic organization, *J. Comp. Neurol.* **188:**347–366.

Ungerleider, L. G., and Mishkin, M., 1982, Two cortical visual systems, in: *Advances in the Analysis of Visual Behavior* (D. J. Ingle, M. A. Goodale, and R. J. W. Mansfield, eds.), MIT Press, Cambridge, Mass., pp. 549–586.

Vaadia, E., Goldstein, M. H., Jr., Hienz, R. D., and Benson, D. A., 1983, Single unit activity in frontal cortex of monkeys localizing sound sources, *Soc. Neurosci. Abstr.* **9:**1084.

Van Essen, D. C., and Maunsell, J. H. R., 1983, Hierarchical organization and functional streams in the visual cortex, *Trends Neurosci.* **6:**370–375.

Van Hoesen, G. W., 1982, The parahippocampal gyrus, *Trends Neurosci.* **5:**345–350.

Van Hoesen, G. W., Pandya, D. N., and Butters, N., 1972, Cortical afferents to the entorhinal cortex of the rhesus monkey, *Science* **175:**1471–1473.

Van Hoesen, G. W., Vogt, B. A., Pandya, D. N., and McKenna, T. M., 1980, Compound stimulus differentiation behavior in the rhesus monkey following periarcuate ablation, *Brain Res.* **186:**365–378.

Vogt, B. A., and Pandya, D. N., 1978, Corticocortical connections of somatic sensory cortex (areas 3, 1 and 2) in the rhesus monkey, *J. Comp. Neurol.* **177:**179–192.

Vogt, C., and Vogt, O., 1919, Allgemeinere Ergebnisse unserer Hirnforschung, *J. Psychol. Neurol.* **25:**279–461.

von Bonin, G., and Bailey, P., 1947, *The Neocortex of Macaca mulatta,* University of Illinois Press, Urbana.

von Economo, C., and Horn, L., 1930, Uber Windungsrelief, Masse und Rindenarchitektonik der Supratemporalflache, ihre individuellen und ihr Seitenunterschiede, *Z. Gesamte Neurol. Psychiatr.* **130:**678–857.

von Economo, C., and Koskinas, G. N., 1925, *Die Cytoarchitektonik der Hirnrinde des erwachsenen Menschen,* Springer, Berlin.

Walker, A. E., 1938, *The Primate Thalamus,* University of Chicago Press, Chicago.

Walker, A. E., 1940, A cytoarchitectural study of the prefrontal area of the macaque monkey, *J. Comp. Neurol.* **73:**59–86.

Walzl, E. M., and Woolsey, C. N., 1943, Cortical auditory areas of the monkey as determined by electrical excitation of nerve fibers in the osseous spiral lamina and by click stimulation, *Fed. Proc.* **2:**52.

Watson, R. T., Heilman, K. M., Cauthen, J. C., and King, F. A., 1973, Neglect after cingulectomy, *Neurology* **23:**1003–1007.

Weber, J. T., and Yin, T. C. T., 1984, Subcortical projections of the inferior parietal cortex (area 7) in the stumptail monkey, *J. Comp. Neurol.* **224:**206–230.

Wegener, J. G., 1969, Auditory discrimination behavior of brain damaged monkey, *J. Aud. Res.* **4:**227–239.

Wegener, J. G., 1976, Auditory and visual discrimination following lesions of the anterior supratemporal plane in monkeys, *Neuropsychologia* **14:**161–174.

Weiskrantz, L., and Mishkin, M., 1958, Effects of temporal and frontal cortical lesions on auditory discrimination in monkeys, *Brain* **81:**406–414.

Welch, K., and Stuteville, P., 1958, Experimental production of neglect in monkeys, *Brain* **81:**341–347.

Weller, R. E., and Kaas, J. H., 1983, Retinotopic patterns of connections of area 17 with visual areas V-II and MT in macaque monkeys, *J. Comp. Neurol.* **220:**253–279.

Whitlock, D. G., and Nauta, W. J. H., 1956, Subcortical projections from the temporal neocortex in *Macaca mulatta, J. Comp. Neurol.* **106:**183–212.

Wiesendanger, M., 1981, Organization of secondary motor areas of cerebral cortex, in: *Nervous System,* Volume 2 (Part 2) (V. B. Brooks, ed.), American Physiological Society, Bethesda, pp. 1121–1147.

Winfield, D. A., Gatter, K. C., and Powell, T. P. S., 1975, Certain connections of the visual cortex of the monkey shown by the use of horseradish peroxidase, *Brain Res.* **92:**456–461.

Woolsey, C. N., 1958, Organization of somatic sensory and motor areas of the cerebral cortex, in: *Biological and Biochemical Bases of Behavior* (H. F. Harlow and C. N. Woolsey, eds.), University of Wisconsin Press, Madison, pp. 63–81.

Woolsey, C. N., and Fairman, D., 1946, Contralateral, ipsilateral, and bilateral representation of cutaneous receptors in somatic areas I and II of the cerebral cortex of pig, sheep and other mammals, *Surgery* **19:**684–702.

Yakovlev, P. I., Locke, S., and Angevine, J. B., 1966, The limbus of the cerebral hemisphere, limbic nuclei of the thalamus, and the cingulum bundle, in: *The Thalamus* (D. P. Purpura and M. D. Yahr, eds.), Columbia University Press, New York, pp. 77–91.

Yeterian, E. H., and Pandya, D. N., 1982, Cortico-thalamic connections of the posterior parietal cortex in rhesus monkeys, *Anat. Rec.* **202:**211A.

Yukie, M., and Iwai, E., 1981, Direct projection from the dorsal lateral geniculate nucleus to the prestriate cortex in macaque monkeys, *J. Comp. Neurol.* **201:**81–98.

Zeki, S. M., 1978, Functional specialization in the visual cortex of the rhesus monkey, *Nature* **274:**423–428.

2

The Extrastriate Visual Cortex

A Historical Approach to the Relation between the "Visuo-Sensory" and "Visuo-Psychic" Areas

IRVING T. DIAMOND, DAVID FITZPATRICK, and JAMES M. SPRAGUE

1. Introduction

At one time the visual cortex meant exclusively the "calcarine" area of Campbell which was later called area 17 by Brodmann (1905) or the striate area by many others. The idea of an extrastriate visual cortex is a more recent concept developed to account for increasing evidence of visual sensory functions for the cortical areas lying rostral to the striate area. In the traditional view the extrastriate visual areas were called association cortex, suggesting a higher level of function and the absence of modality specificity. There is one slight qualification: the region lying immediately adjacent to striate was indeed considered a part of association cortex and served the higher function of association, but according to Campbell it performed this role more or less exclusively with the information received from striate cortex. To suggest the modality-specific but higher function

IRVING T. DIAMOND and DAVID FITZPATRICK • Departments of Psychology and Anatomy, Duke University, Durham, North Carolina 27706. JAMES M. SPRAGUE • Department of Anatomy, University of Pennsylvania, Philadelphia, Pennsylvania 19104.

of the regions investing the striate area, Campbell used the term "visuo-psychic" area, while the striate area itself was called "visuo-sensory."

In the 19th century before Campbell's histological definitions of the visuo-sensory area, a pathway had been traced by Ramón y Cajal (1906) and others from the eye to the lateral geniculate to the cortex. And even before Ramón y Cajal, experimental ablation methods were used to define a visual sensory area. For example, Schafer reported in 1888 that removal of the occipital lobe in the monkey led to a blindness in the contralateral half of the visual field. There was also evidence of transneuronal degeneration of the visuo-sensory area after long-standing atrophy of the optic nerve. Campbell felt that the method of ablation was a crude way to define the borders and that his method of marking changes in the laminar pattern of cortex provided a precise way to localize the visuo-sensory area.

Campbell's view of the visuo-psychic area is especially significant for our purpose since it became the first principle in the study of functional localization in cortex. Campbell argued that "the mere existence of an area contiguous with the striate cortex suggests that it is concerned with sorting out and elaborating the visual impressions produced by the striate cortex" (Campbell, 1905). There were several reasons for the confidence with which Campbell assigned functions to the visuo-sensory and visuo-psychic areas: First, Flechsig's study of the development of myelin in different cortical areas showed that the visuo-psychic area in man was immature months after the striate area had developed (Flechsig, 1901). Campbell as well as Flechsig took this to mean that the function of elaborating on simple sensations is just not required in a newborn infant. Second, and even more significant, the visuo-psychic area did not appear to receive fibers from the visual pathway, or indeed from any nucleus in the thalamus, but only from the striate area or other cortical areas. Third, the symptoms of neurological patients pointed to a relation between the loss of recognition of the meaning of objects (visual agnosia), and lesions of the extrastriate cortex. The fourth reason is not based on either anatomical or clinical evidence, but instead on the unquestioned prevailing view of what is a "higher" function and what is a "lower" function. No one would doubt for a moment that conscious sensation is the simplest function of cortex and simple sensation is transformed to object perception by learning or association. This more complex role was assigned to the *association* cortex intercalated between the sensory areas.

In the decade following Campbell's publication of his findings on localization in the cortex (1905), the work of comparative anatomists, notably Elliot-Smith, supported the idea that sensory cortex served the function of simple sensation while association cortex had the more complex role of learning (1910). In a word, the sensory areas remained stable in the evolution of mammals while the association cortex expanded dramatically in the evolution of higher primates.

The application of the methods of Gasser and Erlanger (1927) to the study of cortex further supported Campbell's view of sensory and association cortex. When Adrian, Woolsey, and their collaborators first recorded evoked potentials from the cortex with the cathode-ray oscilloscope (e.g., Adrian, 1941; Woolsey, 1947), the responsive areas coincided exactly with the "sensory" areas of Campbell or the koniocortex of the German school of architectonics [e.g., Brodmann and the Vogts (see Polyak, 1957)]. The association areas were silent. In the beginning, the results of evoked potential studies of cortex depended on deeply anesthetized animals. Later, when the anesthesia was lightened, or Nembutal was replaced by chloralose, evoked potentials were also found in the regions adjacent to sensory cortex. The areas first discovered consisted of a topographic representation of the retina, cochlea or body; the areas lying adjacent to and beyond the first areas were also topographically organized. But it was not clear if the second areas were secondary in the sense of being dependent on cortical fibers from the first areas or if they were the targets of sensory relay nuclei in the thalamus. So as not to prejudge the question of their dependence on the cortical areas discovered first, these new areas were given neutral names, AII, VII, and SII. Thus, visual area I is equivalent to striate cortex, and visual area II is equivalent to some portion of the visuo-psychic cortex.

Here it is necessary and important to say that between the time of Campbell and the time of the first studies of evoked potentials, the method of thalamic retrograde degeneration (see, e.g., Walker, 1938, and Clark, 1932) showed that the association areas received projections from the thalamus, but from nuclei such as the pulvinar nucleus or the medial dorsal nucleus which do not receive fibers from the optic tract. Rose and Woolsey (1949) used the term "intrinsic" to classify those thalamic nuclei that do not receive fibers from sources outside the thalamus, but rely instead on the sensory relay or extrinsic thalamic nuclei for sensory impulses. In this way, Rose and Woolsey preserved the idea that the function of those cortical areas intercalated between the primary areas serve a higher-level function which depends on the sensory pathways to primary areas.

This brief historical review takes the inquiry into the functional relation between sensory and association areas to 1949. It serves as a good introduction since our principal aim is to try to describe how this issue has fared in the light of more recent studies and newer methods. The chief question posed is: To what extent does the function of the association areas *depend* on the sensory areas? We will postpone the related issue of "higher" and "lower" functions until our final summary section.

2. The Ablation Method

2.1. Introduction

This is perhaps the oldest method for studying the visual pathways going back to Flourens in 1824, but it remains a viable one, as we shall show. As late as 1950 there was no reason to doubt that removal of any primary sensory area would result in a complete or drastic sensory deficit. Indeed, the ablation studies of Klüver (Klüver and Bucy, 1937, 1938; Klüver, 1942) supported the conclusions of Munk and Schafer more than 50 years earlier (Munk, 1881; Schafer, 1888). After the removal of the striate cortex, a monkey is unable to discriminate depth, color, pattern, or brightness. To be sure, the lesions in Klüver's experiments were not confined to area 17 alone and the animals were not totally blind because they could distinguish between different energies of light. But even a photoelectric cell is capable of detecting differences in luminous flux and Klüver's results were no cause for doubt about the traditional view.

2.2. Auditory Cortex in Cat

The first studies raising doubts about the relation between sensory areas and their surrounding association areas were directed to the auditory cortex. Bilateral removal of the auditory sensory area in Campbell's sense of the term, or AI as defined by evoked potentials, had little or no effect on the cat's ability to discriminate differences in the intensity or pitch of a tone or even the ability to localize a sound source or distinguish between temporal patterns of tones. However, when the lesion was extended to include all of the cortex between auditory area I (AI) and the rhinal fissure, then a number of deficits emerged (Neff and Diamond, 1958). The point is that if auditory area II (AII), the posterior ectosylvian gyrus, the temporal area, and the insular cortex depended entirely on AI, the lesion of AI alone should suffice to produce any deficit caused by the larger lesion. It seemed that either there was some difference between cats and primates or some difference between auditory cortex and visual cortex (or both could be true). The apparent explanation was soon at hand: the areas surrounding AI received projections from subdivisions of the medial geniculate–posterior complex including the magnocellular division of the medial geniculate nucleus and the anterior and posterior divisions of the posterior nuclear group. These, in turn, were targets of auditory pathways that originated largely outside of the central nucleus of the inferior colliculus. Eventually, a picture emerged of several distinct pathways projecting to a large field of cortex (see Fig. 1). Most of the pathways projected to more than one cortical subdivision and one pathway projected to all divisions. There is no cortical area without more than one pathway and where pathways overlap they terminate in different layers of cortex. This picture brings to mind earlier evidence disquieting for the traditional view: first, the evidence of widespread, and perhaps diffuse projections from the reticular formation and intralaminar thalamic nuclei to neocortex (Moruzzi and Magoun, 1949); second, the evidence of thalamic fibers terminating in layer I and traveling horizontally across several cortical subdivisions (Lorente de Nó, 1938).

2.3. Visual Cortex in *Tupaia*

The visual system remained the bastion of the traditional view and this seemed reason enough to reinvestigate the effects of removing the striate and extrastriate cortex. We have reviewed the reasons for selecting the tree shrew as well as the principal results in several places. In a word, we found that the tree shrew discriminates depth, color, pattern, and movement in the absence of the entire striate area. However, after removal of a large part of the extrastriate cortex with the result of severe degeneration in a substantial part of the pulvinar nucleus, the animals lost the habit of discriminating between simple patterns, and showed a deficit in relearning (see Fig. 2 from Killackey *et al.*, 1971). Figure 2 also shows that when the triangles were embedded in annuli, the deficit was exacerbated. With removal of *both* striate and extrastriate areas, the tree shrew suffered a severe deficit in the simplest habit of discriminating between vertical and horizontal stripes (Snyder and Diamond, 1968). These results demonstrate that the extrastriate cortex in the tree shrew must receive visual impulses that are not relayed through the striate cortex (see Lund *et al.*, Volume 3 of this treatise). Further, since the lesion of striate cortex above produced a complete retrograde degeneration of the lateral geniculate nucleus, the second path to the extrastriate cortex could hardly relay through the geniculate. We first postulated, and later demonstrated by the anterograde degeneration method, the existence of a tectopulvinar path in the tree shrew (Harting *et al.*, 1972). The significance of these results would be diminished if the tree shrew were unique among mammals. However, the tectopulvinar pathway has since been demonstated in a wide variety of mammals, including cat and primate (Benevento and Fallon, 1975; Graham, 1977; Lin and Kaas, 1979).

2.4. Visual Cortex in Cat

Behavioral studies of the cat also show that area 17 is not necessary to perceive form and pattern. After removing areas 17 and 18, cats show perfect retention and normal learning of both pattern and form discrimination tasks

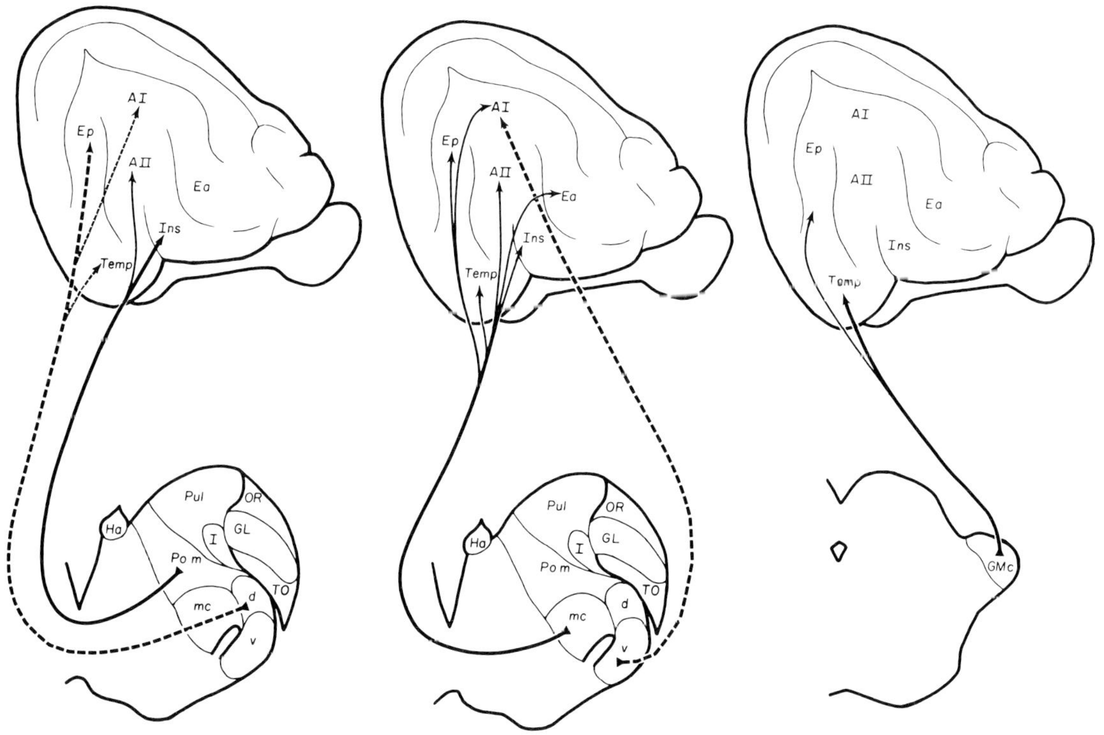

Figure 1. Diagrams showing parallel auditory pathways in the cat. The auditory sensory area in the traditional sense is confined to a subdivision of the whole field, AI; the chief point is that the "association" cortex below the auditory sensory area is in receipt of sensory fibers from the medial geniculate nucleus and is not solely dependent on AI. GMc, caudal portion of medial geniculate nucleus; Pom, posterior complex, medial division; portions of medial geniculate nucleus: magnocellular (mc), dorsal (d), ventral (v).

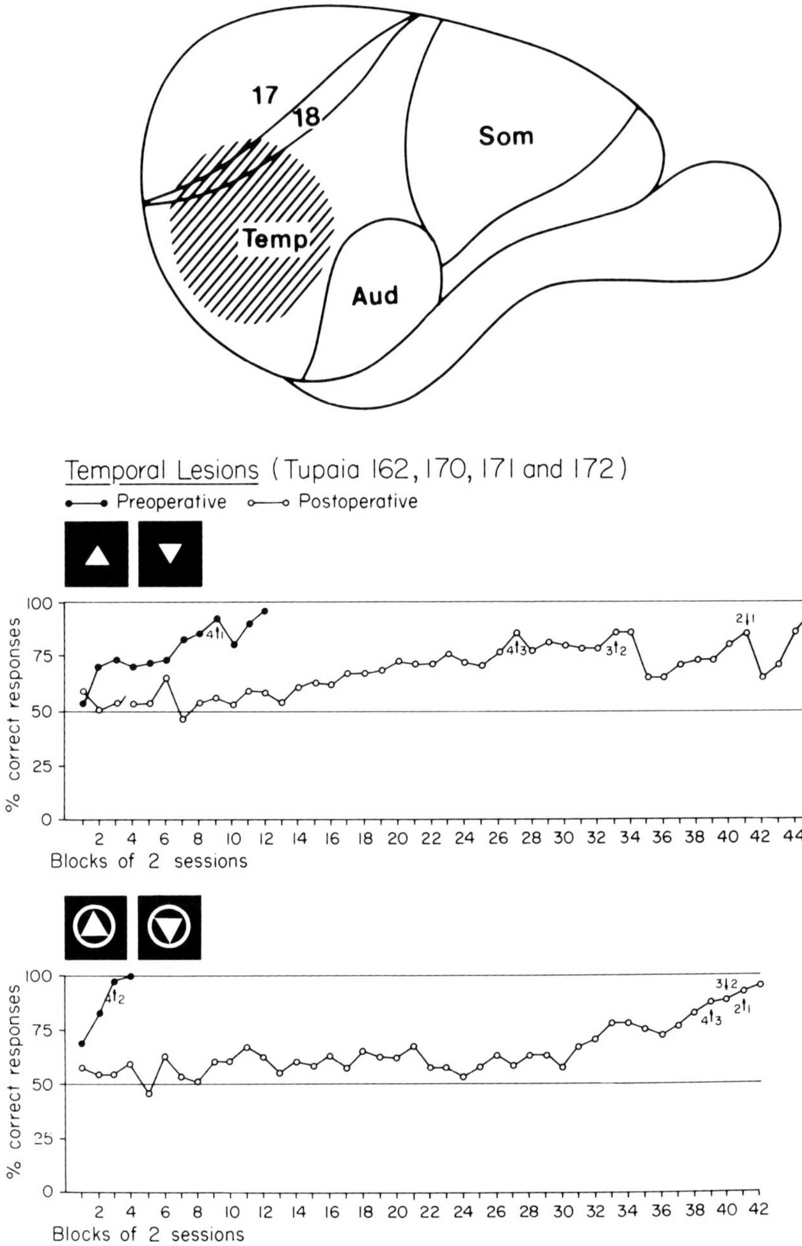

Figure 2. Diagram showing the behavioral effect of removing the temporal extrastriate area in tree shrew. The lesion is depicted by shading on the diagram of the cerebral hemisphere. Of course, the lesions were made bilaterally, and in four animals. Before surgery, all four animals learned upright versus inverted triangles by 25 sessions. After surgery, the average performance was greatly retarded. This deficit could be interpreted as either a learning deficit or a sensory deficit; we cannot distinguish between them.

After learning the first triangle discrimination, the normal animals showed transfer of training to the discrimination of triangles embedded in annuli. But the average performance after surgery hovered about the chance level for over 60 sessions. From Killackey *et al.* (1971).

(Berkley and Sprague, 1979). But deficits in retention and learning can be produced by lesions of area 19 and the cortex of the middle suprasylvian gyrus including the Clare–Bishop area. These results indicate that a double dissociation of deficits is produced by removal of either the striate or extrastriate visual cortex: Lesions limited to areas 17 and 18 produce deficits in acuity, but not in form perception, whereas the removal of area 19 and the suprasylvian cortex produces deficits in form, but not in acuity (Berlucchi and Sprague, 1981; Sprague *et al.*, 1981).

The extrastriate visual areas are also implicated in transfer of form learning from one hemisphere to the opposite hemisphere in a split-chiasm preparation (Sprague *et al.*, 1981). Bilateral removal of the middle suprasylvian gyrus abolishes transfer from one side to the other. If the lesion is made only in one hemisphere, transfer is absent from the eye on the side of the intact cortex to the opposite side, but transfer is intact if it is tested in the opposite direction. This means that the engram must be reconstructed in the suprasylvian cortex of the recipient side. The commissural connections of area 17 and 18 do not appear to be involved in interhemispheric transfer of form discrimination.

In summary and conclusion, ablation studies of the cat show that the extrastriate visual cortex is involved in visual discrimination and visual learning and the pathways for these functions do *not* require the integrity of the striate area. Campbell's view of the relation between visuo-sensory and visuo-psychic is not supported, and further, it does not seem that the surprisingly mild deficits after striate removal can be attributed to the relative importance of subcortical "reflex" centers. On the contrary, in both cat and tree shrew it is the "sensory" functions of the extrastriate visual areas that account for the preservation of visual capacity in the absence of striate cortex.

2.5. Striate and Extrastriate Visual Cortex in Monkey

As mentioned previously, the landmark papers by Schafer in 1888 and Klüver in 1942 showed that a monkey deprived of striate cortex is severely impaired in visual discrimination. While not totally blind, the impaired monkeys misjudged distance, did not follow moving objects, and showed marked deficits in pattern and object vision.

The severe sensory loss after removal of the striate area in the monkey fulfilled Campbell's expectations: the visuo-psychic area, being entirely dependent on the striate, could hardly preserve any visual capacity in the absence of the striate. Campbell, however, could not have anticipated the full extent of the extrastriate visual area and, in particular, the presence of visual areas in the temporal lobe remote from area 17. Visual functions in the temporal lobe were discovered unexpectedly by Klüver and Bucy (1938). Complete removal of the temporal lobe led to a number of bizarre symptoms, but for our purpose, one is particularly significant: the monkeys lost habits based on visual discrimination that they had learned before surgery, and if the operation preceded any training, they had great difficulty in learning for the first time simple discriminations such as selecting a circle or square. They even showed symptoms that could be interpreted as a lack of recognition of objects. For example, when the deprived animal was presented with a snake, it might put it in its mouth. Given the evidence that sensory and memory functions can be affected separately in man, researchers beginning with Munk were on the lookout for a purely visual memory loss—psychic blindness. Klüver, however, was skeptical about the diagnosis of agnosia (Klüver and Bucy, 1938, p. 61):

> A review of the clinical literature indicates that the more carefully a case is studied the more difficult it may be to decide whether there is really an "agnostic" or merely a "visual" defect, that is, whether the variety of "agnostic" symptoms can be reduced to disturbances of "elementary" or "higher" visual functions.

Further studies of the temporal cortex in monkey showed that the deficit in memory and learning was surprisingly specific to vision and that habits based on tactual, auditory, and olfactory discriminations were unaffected. These studies also tried to localize more precisely than Klüver did the region of temporal cortex related to vision, with the result that the cortex below the temporal sulcus was identified as the visual area. The next problem was to try to identify the pathway from the geniculostriate system—which was taken as a synonym for *the* visual path—to this remote area in the temporal lobe. At first the path did not seem to travel over corticocortical fibers originating in the striate cortex: Lashley (1948) found that large lesions of areas 18 and 19 that deeply undercut the white matter had little or no effect on visual habits.

In 1966 Mishkin found evidence for the role of a path from striate cortex in an ingenious way. In a first stage he removed the inferotemporal area on one side only. Naturally, there was little or no behavioral effect. Then in a second stage, he removed the striate cortex on the opposite hemisphere. Still there was no deficit. In a final stage, Mishkin cut the corpus callosum and produced the deficit. This suggests that the pathway by which impulses reach the inferotemporal area began in the intact striate cortex, passed next to the adjacent belt of prestriate cortex, area 18, crossed via the corpus callosum to the homologous area in the opposite hemisphere, where it was relayed finally to an intact inferotemporal area on that side (see Mishkin, 1966, p. 112).

In sum, the tree shrew and cat differ from the monkey inasmuch as at least one subdivision of the extrastriate visual cortex in monkey seems to be entirely dependent on the striate area. What is the significance of these species differences?

In some biological inquiries a deeper understanding can be achieved by ignoring species differences and instead concentrating on one species. For example, very few species reproduce by self-fertilization—yet this kind of reproduction was essential for Mendel's "pure-line" method. On the other side, Harvey exploited the difference between fish and mammals to show that the pulmonary circulation was only a detour in the transference of blood from veins to arteries.

Our approach to the puzzling differences between tree shrew, cat, and monkey is to seek some common plan in order to be able to view the differences as variations on a theme. One obvious way to explain differences once homology (the common plan) has been established is to relate them to the mode of life. This works especially well when the habitat of two closely related species differ in *one* conspicuous way—such as a nocturnal or diurnal habitat, or feeding on the surface or the bottom of a lake. Another way to view "departures" from the prototypic organization is to identify a phylogenetic trend, such as the increase in the size and development of the dorsal thalamus in vertebrate evolution. To test the idea that the extrastriate visual cortex becomes more dependent on the striate cortex in primate evolution, we studied the prosimian, *Galago*. While *Galago* is not, of course, a representative of the ancestors of monkeys, it may be closer to the prototypical primate. In any case, the neocortex in *Galago* is poorly differentiated and the striate area, in particular, is primitive in comparison to that found in monkeys.

We removed all of the striate cortex in two animals with only minimal damage to extrastriate (Atencio *et al.*, 1975). In both cases the lesions extended into area 18 (see Fig. 3). We studied every section of the lateral geniculate and no part was spared from severe retrograde degeneration. Under high-power magnification, small pale neurons could still be identified, but our best view is that they are interneurons which do not project to cortex. One animal failed to discriminate between a simple difference in form—an O versus X—but was able to discriminate $\frac{1}{2}$-inch horizontal versus vertical stripes (see Fig. 4). This same animal had a profound deficit in locating stationary objects or following moving objects (Atencio *et al.*, 1975).

The second animal, *Galago* 171, was able to follow moving stimuli, to locate small objects, and to learn simple form discriminations. On the task involving horizontal and vertical stripes, this case succeeded at $\frac{1}{4}$ inch, but failed in 500 trials at $\frac{1}{8}$ inch. A normal animal has no difficulty with $\frac{1}{16}$ inch (see Fig. 4). On the task involving O versus X, the animal succeeded with large figures, but failed when the size was reduced (see Fig. 5). If we could be absolutely certain that all the striate cortex was removed, the behavior of the second case would be much more significant than the first. After all, if you find a single animal that can see without the visual centers thought to be necessary for vision, then the original belief is overturned. To the best of our knowledge the lesions in the two cases were complete and identical. A subsequent study by Caldwell and Ward (1982) supports the idea that the *Galago* deprived of striate cortex can localize small objects, follow moving objects, and discriminate horizontal from vertical stripes so long as they are wide and the animals can approach to a distance of 15 cm from the stimuli.

Suppose for the argument that our histological analysis of the lesion was correct: it follows that the extrastriate areas in *Galago* are not as dependent on the striate cortex as they are in the monkey. If that is so, we should be able to identify areas in the extrastriate cortex with visual functions that depend on thalamic projections. As it turned out, we did identify an area of the temporal lobe of *Galago* the removal of which produced a visual deficit.

With the aim of finding a precursor or homolog to the inferotemporal area in the monkey, our first lesions were located below the middle temporal area (Atencio *et al.*, 1975) (see Fig. 6). Here we should insert that the subdivisions of extrastriate visual cortex in *Galago* were not completely known at the time. Area 18 was clearly one exception since it can be defined on grounds of both electrophysiology and architectonics. In the temporal lobe the middle temporal area also stands out since it too can be defined either by mapping the retina or as an architectonic area [e.g., it contains an unusually dense population of myelinated

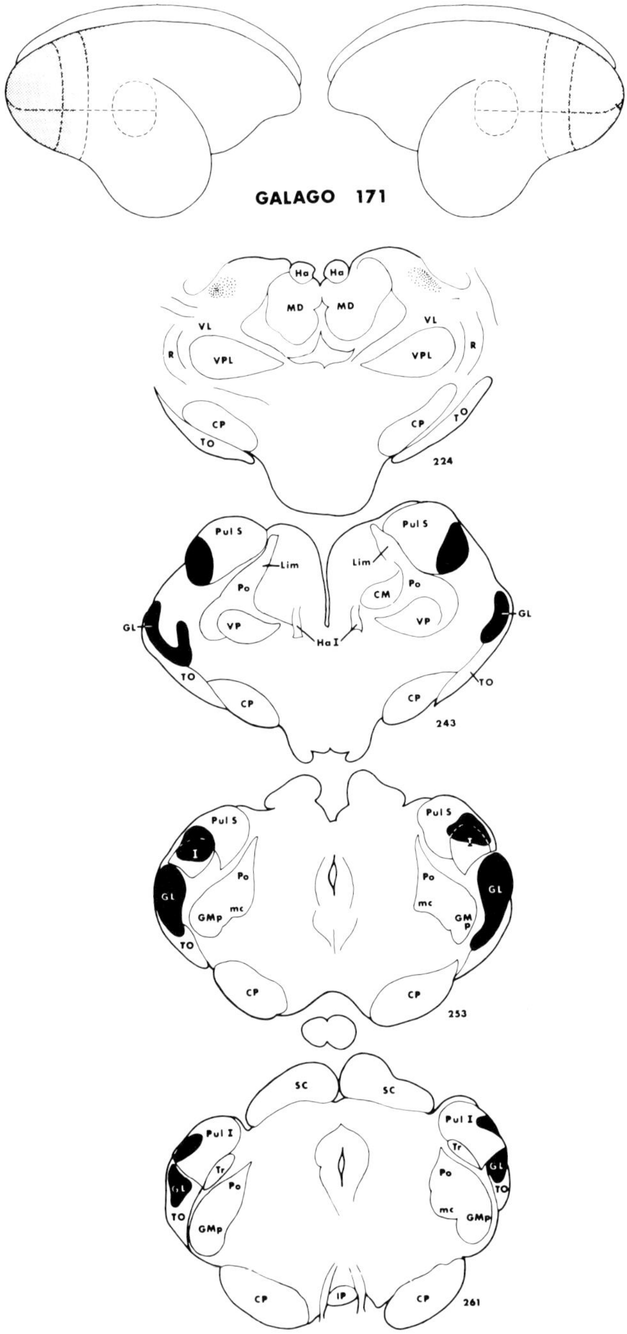

Figure 3. Diagrams showing the lesion of striate cortex in *Galago* 171. The cortical lesion is depicted by shading. Severe thalamic retrograde degeneration is depicted by black. Note that all of the lateral geniculate nucleus underwent complete degeneration. Note further that there is a central region of the pulvinar nucleus—the dorsal part of pulvinar I and the ventral part of pulvinar S—that shows severe degeneration. This is the striate-recipient zone of the pulvinar nucleus and it projects to areas 18 as well as 17. From Atencio *et al.* (1975). (See Fig. 5 for diagrams showing the animal could learn simple form discrimination.)

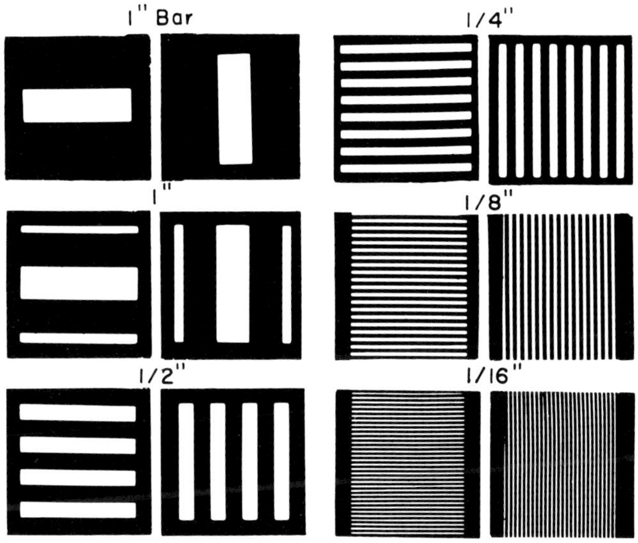

Figure 4. Tests for acuity. *Galago* 171 succeeded at $\frac{1}{4}$ inch, but failed at $\frac{1}{8}$ inch. A normal animal has no difficulty with $\frac{1}{16}$ inch. From Atencio *et al.* (1975).

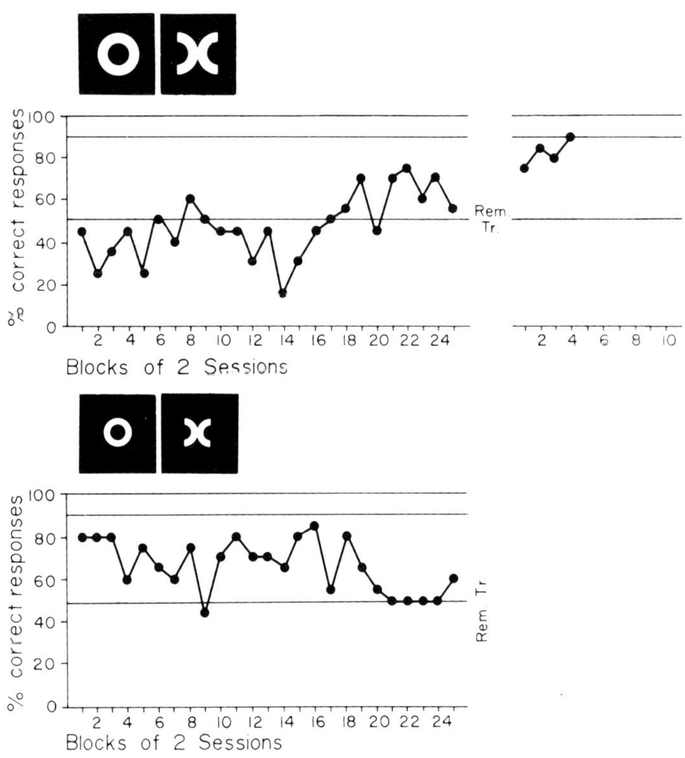

Figure 5. Learning curves for *Galago* 171. This animal was able to relearn the discrimination O versus X but only after remedial training. When the size was further decreased, the animal showed some transfer of training, but its performance deteriorated and training was discontinued. From Atencio *et al.* (1975). (See Fig. 3 for the location of the lesion in striate cortex.)

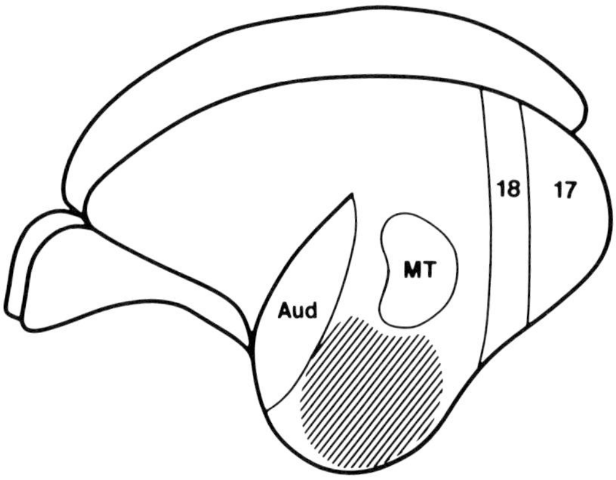

Figure 6. Diagram showing the locus of the lesion in the temporal lobe of *Galago*. The shaded area is the region removed bilaterally in four animals.

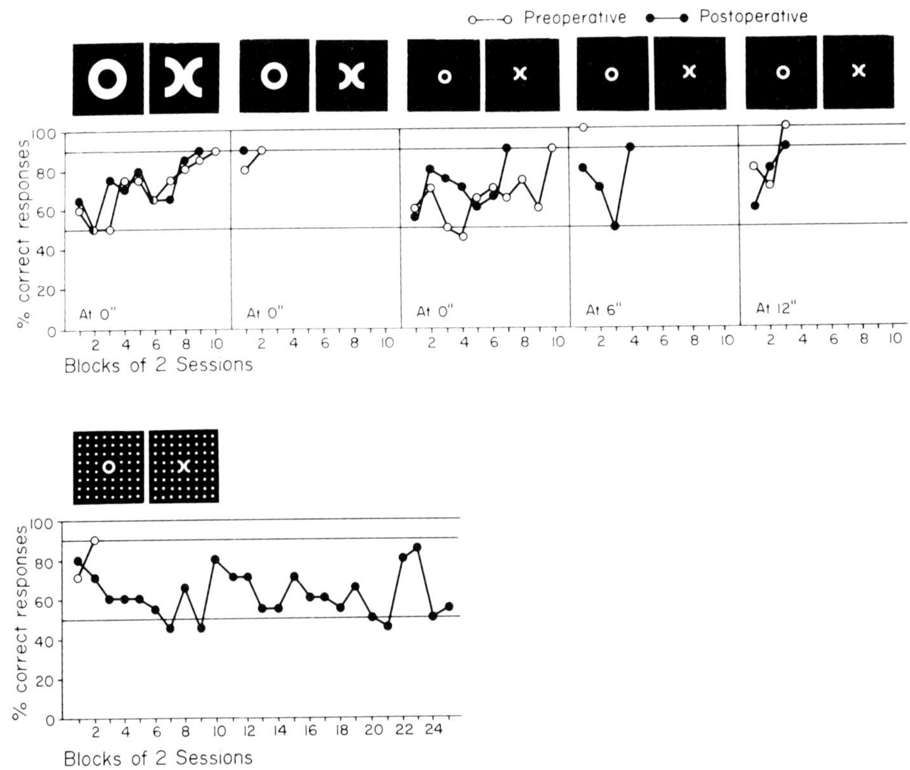

Figure 7. Behavioral results of *Galago* 241 before and after the temporal lesion shown in Fig. 6. After surgery, the animal could discriminate a small O versus X from a distance of 12 inches. However, the addition of white dots distributed uniformly on the black card disrupted the performance. The normal animal paid little or no attention to the dots. From Atencio *et al.* (1975).

fibers in the superficial layers (Glendenning *et al.*, 1975)]. In the next section we will consider the projections of the pulvinar nucleus as a way of defining subdivisions of the temporal lobe. For the present, suffice it to say that the region lying ventral to the middle temporal region appears to be located in a position approximately similar to the inferotemporal area of monkeys and consequently seemed like a good place to begin the behavioral study. Among the various symptoms found after the lesion shown in Fig. 6 were: (1) a failure to discriminate between simple forms (O vs. X) when these patterns were placed on a background of potentially distracting dots (see training record for case 241, Fig. 7); (2) deficits in learning to discriminate between simple forms [where normal animals would learn to discriminate a small O versus an X (the behavioral method was to train animals on successively smaller patterns), in 430 trials on the average, the original learning of animals with lesions required three to seven times that long (see training record for case 185, Fig. 8)].

In summary, the idea that *Galago* may reflect a step on the way to the evolution of monkeys remains viable. First, the deficit after striate cortex lesions in *Galago* is not as severe as in monkey; and second, the visual functions of the temporal lobe may depend on a parallel tectopulvinar pathway.

The question of whether the pathway to the temporal areas significant for visual learning and discrimination in *Galago* requires a relay in the striate cortex

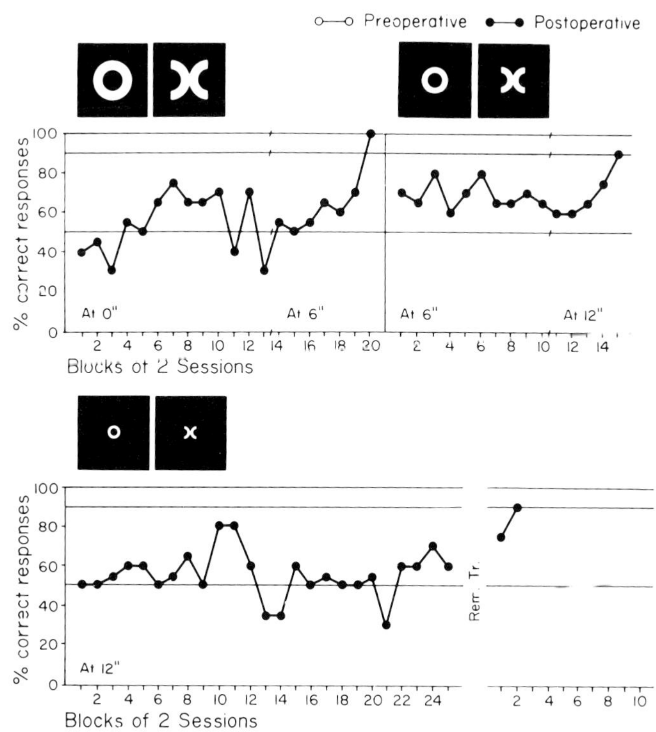

Figure 8. Behavioral record of another temporal lobe case, *Galago* 185. In this case the animal had great difficulty in discriminating a small O versus X from a distance of 12 inches. From Atencio *et al.* (1975).

or depends on the tectopulvinar pathway parallel to the geniculostriate system will be discussed in the next section.

3. The Comparative Neurology of the Connections between Striate and Extrastriate Cortex

We began with behavioral studies, not because function is prior to structure in inquiry; on the contrary, it is necessary to have some idea of an organ such as striate cortex in order to test its role with ablation. Rather, we tried to set the stage for the chief question of this chapter: Is the extrastriate visual cortex, or the association areas investing the striate area, entirely dependent on the striate? At a first approximation, the answer seems to be "yes" in the monkey, and "no" in the cat and tree shrew. *Galago* seems to be intermediate between tree shrew and monkey. Now the problem is to try to understand these species differences.

The connections of the striate and extrastriate provide the starting point for the solution. The views of Campbell and the experimental studies of Mishkin require a projection from striate to extrastriate cortex. Such corticocortical fibers were known to Campbell, and more recently, their targets have been identified precisely in the monkey by current transport methods (Tigges *et al.,* 1973; Zeki, 1978; Rockland and Pandya, 1979). Thus, the striate cortex projects to area 18 and the middle temporal area in monkey (see Pandya and Yeterian, this volume). By itself this does not account for the differences in species: in the cat, tree shrew, and *Galago,* the striate cortex also projects to extrastriate areas. It does not help to say this path is more "important" in the primate, unless we can offer concrete evidence for a difference between monkey and other primates in the organization and nature of the projections of striate to other cortical areas.

The results of ablation studies in *Tupaia* require a pathway from retina to extrastriate visual cortex that bypasses both the lateral geniculate nucleus and the striate cortex. Again the mere presence of this path cannot in some obvious way account for the differences between species, since it has been identified in cat and monkey and prosimian. We have looked for some differences in this pathway, again with a view to identifying some fundamental change in the organization of the path in primate evolution. The superior colliculus in *Tupaia* is unusually well developed and its projection to the pulvinar nucleus is robust (Harting *et al.,* 1972). In primates the superior colliculus has not kept pace with the development of the geniculostriate system while the pulvinar nucleus has continued to expand and differentiate. It seemed reasonable to believe that only one subdivision or only a part of one subdivision of the pulvinar nucleus in primates receives tectal fibers.

The Tectopulvinar Path in *Galago* and Monkey

We have studied the tectopulvinar pathway in *Galago* by several methods: anterograde degeneration, retrograde transport of HRP, and anterograde transport of WGA-HRP (Glendenning *et al.,* 1975; Raczkowski and Diamond, 1981).

Figure 9 is designed to convey the main results by showing the target of the projections from three loci in the superior colliculus in a sagittal section of the pulvinar nucleus. In the same section of the pulvinar nucleus we also depict the pattern of cells labeled by a series of HRP injections of the striate cortex beginning on the lateral surface and continuing around the caudal pole to the calcarine fissure (the medial view is not drawn). Since it is almost a certainty that descending projections from the striate area to the pulvinar nucleus precisely reciprocate ascending projections from the pulvinar nucleus to the striate area, we can identify the bands of cells as the "striate-recipient zone" (Symonds and Kaas, 1978; Carey *et al.*, 1979). We have actually traced both pulvinar-striate and tectopulvinar projections in the same animal and the figure is intended to summarize exactly these experiments. The chief point is that the tecto-recipient zone and the striate-recipient zone are separate and distinct. The tecto-recipient zone lies at the caudal and ventral extremity of the inferior division, while the striate-recipient zone occupies a central zone on both sides of the fiber band that divides inferior from superior divisions.

The next question for the present argument is: What is the difference in the cortical targets of the tecto- and striate-recipient zones? The striate-recipient

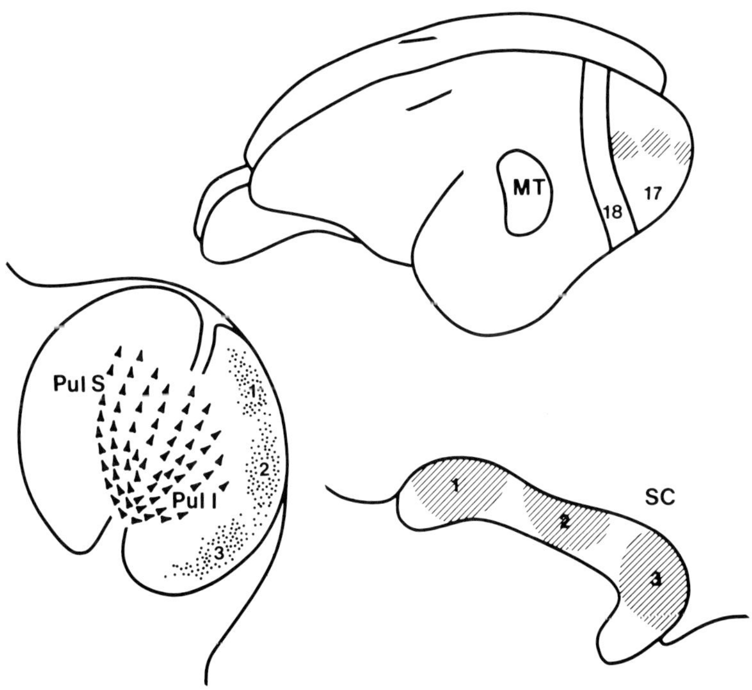

Figure 9. Diagram to show the relation between tecto-recipient and striate-recipient zones of the pulvinar nucleus in *Galago*. The shaded areas (labeled 1, 2, and 3) depict three WGA-HRP injections of the superior colliculus and the dots in the pulvinar nucleus indicate terminals resulting from anterograde transport. In the same sagittal section labeled cells (depicted by triangles) reflect the retrograde transport from striate cortex injections of HRP. The text explains that the precise reciprocity permits using the location of labeled cells as a "marker" for terminals from striate cortex to pulvinar.

zone projects to layer I of striate cortex, to area 18, and to the middle temporal area (MT) (Symonds and Kaas, 1978; Carey *et al.,* 1979, 1985; Raczkowski and Diamond, 1981). It has not escaped our notice that areas 18 and MT are just the areas of extrastriate cortex that are direct targets of the striate cortex. Thus, these two areas must be entirely dependent on striate cortex. On the other hand, the tecto-recipient zone of the pulvinar projects to the region ventral to MT (see Fig. 10; Raczkowski and Diamond, 1980), just the location of the lesions described in the previous section.

If the inferotemporal area of monkey is not a target of the tectopulvinar path, we have a simple explanation for the difference between *Galago* and ma-

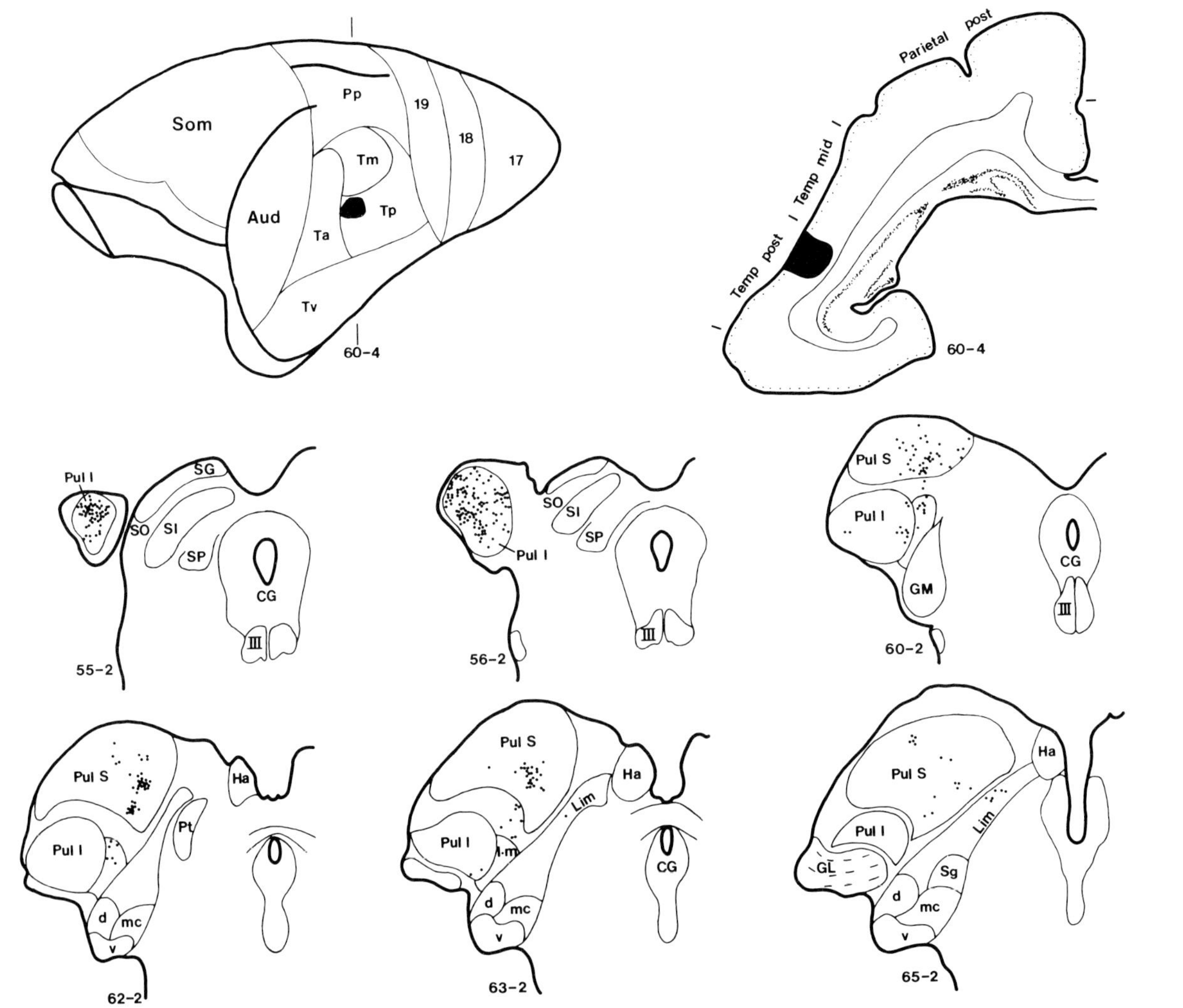

Figure 10. Retrograde transport of HRP from the inferotemporal region of *Galago* 1611L to the pulvinar nucleus. The injection site is depicted by black in the diagrams of the cortex. The labeled cells in thalamus are shown by dots. The chief point is that the densest population lies in the caudal part of the inferior division of the pulvinar (Pul). From Raczkowksi and Diamond (1980).

caque. We have tried to reach some view of this question by reviewing studies of the tectopulvinar path in monkeys. In the early behavioral studies, very large lesions were made in the cortex below the superior temporal sulcus resulting in retrograde degeneration in the caudal pulvinar, in the so-called "lateral division" (Chow, 1950; Mishkin, 1972). On the other hand, the tecto-recipient zone of the pulvinar nucleus—at least that part that receives the most robust projection—lies in the "inferior division" far rostral to the site of retrograde degeneration. Our best but tentative conclusion is that the inferotemporal area is not a target of the tectopulvinar path. We say tentative because of the following reservation: (1) to our knowledge there are no HRP transport studies of the inferotemporal area in macaque, which means that the projections from collaterals have yet to be identified; (2) we have for the present purpose selected the target of the clearest and most intense projection from the superior colliculus to the pulvinar. Subject to the above qualifications, we suggest that the prosimian may lack the homolog of the inferotemporal area in macaque and that the dependence of this area on the striate cortex in macaque is a consequence of the evolution of Old World monkeys.

In summary, the inferotemporal area, although special in its role in shape recognition, is just one of several areas of the extrastriate visual cortex. The relation between the striate area to the whole of the extrastriate depends, at least in part, on how many of these subdivisions receive their major projections from the striate-recipient zones of the pulvinar nucleus. Clearly, the higher this number, the greater is the amount of temporal cortex deprived of visual sensory impulses by striate cortex ablation. Our guess based on our work in *Galago* and the work of others in macaque is that the amount of extrastriate cortex dependent on the striate area is greater in macaque than in *Galago*. In the next section we will look at some differences between prosimian and simian in the nature of the projections of striate cortex to extastriate cortex.

4. Comparative Study of Striate Cortex and Its Projections to Temporal Lobe

In this section we will present evidence for a significant difference between striate cortex of simian and prosimian that may contribute to an understanding of the relation between striate and extrastriate in these two species. Figure 11 illustrates the difference in cytoarchitecture of striate cortex in *Galago* and two monkeys, *Aotus* (owl monkey) and *Saimiri* (squirrel monkey) (Diamond *et al.*, 1985). The chief point of the comparison is that there are more layers in the monkey and these are more distinct. Note, for example, the layer labeled IIIC in each species of monkey; no comparable layer can be identified in the prosimian. We wondered, "What is the functional significance of the increased laminar differentiation in the evolution of monkeys?" This question is especially significant since laminar differentiation does not keep pace with the expansion of neocortex in every mammalian line of descent. Thus, in carnivores the neocortex has expanded beyond the stage represented by a generalized insectivore; yet cortical layers III and IV are not sharply segregated or differentiated, and this is so in AI or VI. Even better evidence that expansion and differentiation are separate processes in evolution comes from the study of the whales and dolphins (Peter Morgane, personal communication). In these groups the neocortex is huge and highly convoluted, but the laminar differentiation is primitive—hardly different from insectivores.

A starting point for answering our question is the pattern of projections from the lateral geniculate nucleus. The distribution of the terminations from the entire geniculate is shown for the three primate species in Fig. 11. In a word, the geniculate projections in *Galago* are more compressed, and the cortex of *Galago* lacks layer IIIC. We were struck by the fact that IIIC is spared from geniculate terminals in both monkeys. So the next question was: "What are the connections of IIIC?" The answer is contained in Fig. 12. Every pyramidal cell, or almost every cell in layer IIIC of *Aotus,* is filled by the retrograde transport of HRP from area MT of the temporal lobe (Diamond *et al.,* 1985). In *Galago* layer III also projects to the temporal lobe, but the cells are not aligned in a single row. If the geniculate projections terminate in distinct layers, and if the intrinsic projections within striate are also organized by layer, the alignment of the IIIC pyramids is a way of organizing influences on the apical dendrites. To be specific, we know the magnocellular layers of the geniculate project to IVα and IVα projects to IIIC, so the magnocellular layers have the greater opportunity of influencing the cells projecting to the middle temporal area than the parvocellular layers. Electrophysiological studies of single cells in the middle temporal area (MT) support the idea that the MT is more clearly related to the "Y" or magnocellular path than the "X" or parvocellular path. Thus, cells in the MT are especially sensitive and responsive to movement and have a crucial role in eye movements following moving stimuli (R. Wurtz, personal communication).

We can offer further examples of the functional significance of the greater differentiation of the layers of striate cortex in monkeys and they all come down to the segregation of separate pathways. This conclusion is congenial to the many studies of Zeki showing that different subdivisions of extrastriate visual cortex

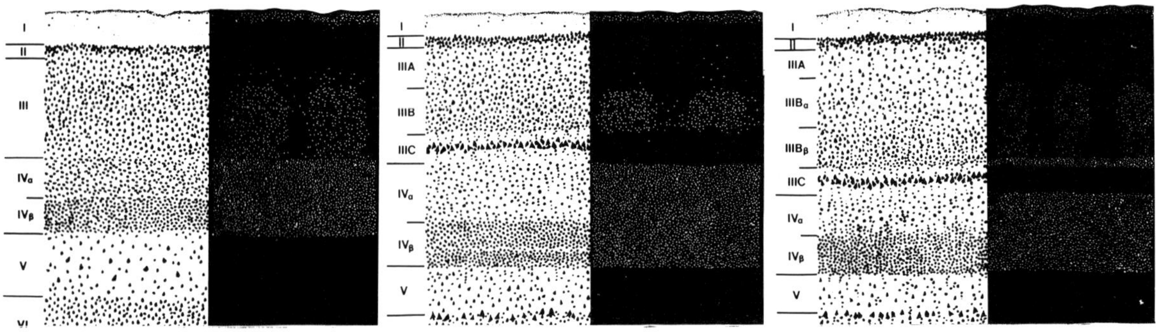

Figure 11. Diagrams depicting the laminar organization of the striate cortex in three primates in relation to the distribution of terminals from the lateral geniculate nucleus.

which are direct targets of striate cortex (MT and several subdivisions of area 18) are distinguished by the domination of one or another path: Thus, Zeki finds the cells in V5 (or MT) are predominantly motion sensitive (Zeki, 1983), while one of the subdivisions of 18 is predominantly color sensitive (Zeki, 1974; Livingstone and Hubel, 1984; see also Van Essen, Volume 3 of this treatise).

5. Projections of the Lateral Geniculate Nucleus to Striate and Extrastriate Cortex

While the tectopulvinar path may contribute to the understanding of the species differences described earlier, this contribution will not be easy to describe in a few words. The projections of the lateral geniculate nucleus may be the best opportunity for explaining at least some of the puzzling differences. In the traditional view, this would probably have been the last on the list of possible candidates for explaining cortical differences since the sensory relay nuclei were considered stable in the evolution of mammals, in contrast to the expansion of the intrinsic thalamic nuclei and association cortex (see Rose and Woolsey, 1949). Thus, at the time of Campbell, the striate cortex was believed to be the sole

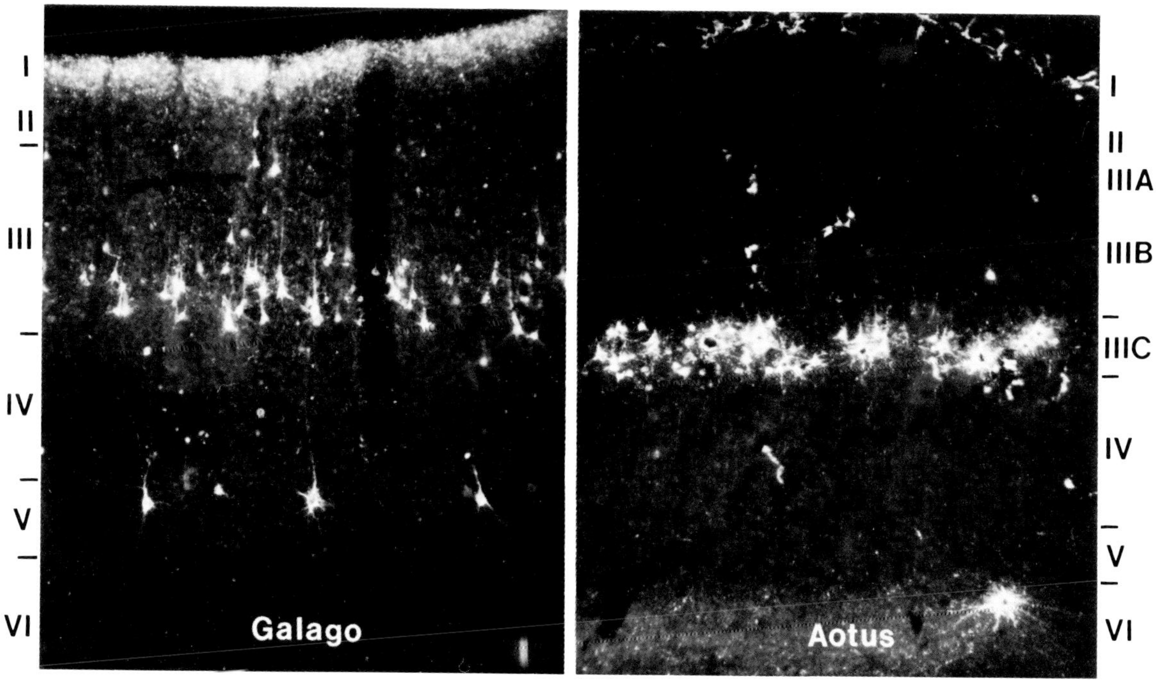

Figure 12. Photomicrographs revealing the distribution of labeled cells in the striate cortex of *Galago* and *Aotus* (owl monkey) after HRP injections into the middle temporal area. The chief point is that while the labeled terminals are found in layer III of both species, they are more compactly organized in the monkey. We interpret this to mean a more uniform relation between the influence of the parvocellular and magnocellular pathways on these cells. The result is that the middle temporal areas will be influenced to a greater extent by the magnocellular pathway.

target of the lateral geniculate nucleus in all mammals. In 1963 we published a study of the projections of the lateral geniculate nucleus in the opossum and offered evidence suggesting that the geniculate in that marsupial projected to the extrastriate visual cortex (Diamond and Utley, 1963). Indeed, we also found evidence that the pulvinar nucleus projects to the striate area, so our final view was that the projections of these two nuclei overlap extensively in striate as well as extrastriate areas (see Fig. 13). The best interpretation of the conclusion at the time seemed to depend on the fact that the opossum is a marsupial and may reflect an earlier stage of evolution before the origin of eutherian mammals. It turns out that the cat also has a significant projection from its geniculate nucleus to the extrastriate visual cortex (Raczkowski and Rosenquist, 1983). This projection consists of two of the three retinogeniculate pathways—the so-called Y and W pathways. Only the X path is restricted to the striate area, which goes a long way toward explaining the behavioral results inasmuch as the X path is defined by cells with small receptive fields and sustained responses—just the sort of properties one might associate with acuity.

Naturally, we looked for evidence for a geniculate projection to extrastriate visual cortex in *Tupaia* and *Galago.* None was found in *Tupaia;* in *Galago* we observed a few labeled small cells in the interlaminar regions of the geniculate after HRP injections in the temporal lobe. Cells of the interlaminar regions of

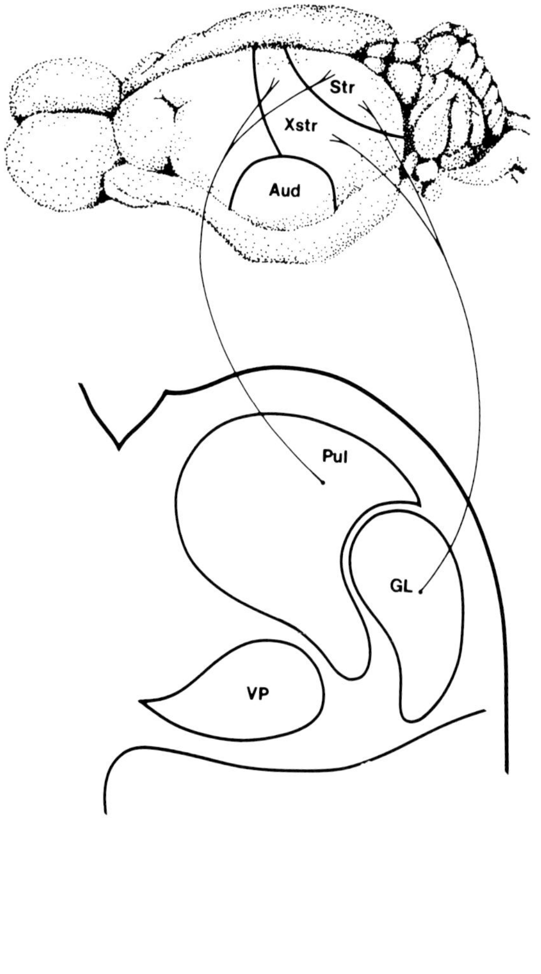

Figure 13. Diagrams showing the overlap of the lateral geniculate nucleus and pulvinar nucleus projections in both striate and extrastriate cortex of the opossum. A similar picture would apply to the hedgehog, so it is not reckless to suggest that this could resemble the prototypical mammal.

the geniculate in monkey also project outside striate cortex (Benevento and Yoshida, 1981; Yukie and Iwai, 1981). We have speculated that the small cells in the interlaminar regions and in the intercalated layers of the lateral geniculate nucleus of primates may reflect an early stage of geniculate evolution (Fitzpatrick *et al.*, 1983).

6. Recapitulation and Conclusion

We began with Campbell's views because he defined the basic issue as clearly as anyone in the study of localization of function in cortex: What is the relation between sensory areas and association areas? According to Campbell, the visual association area depends entirely on the visual sensory area; thus, the loss of the visual sensory area amounts to a loss of association areas because the latter are deprived of all visual impulses.

However, the results from behavioral ablation studies in tree shrew and cat raise doubts about Campbell's views. The extrastriate visual cortex in neither cat nor tree shrew uniquely depends on striate. The behavioral results are significant not simply because they take issue with a long-standing concept. It is easy to find flaws in a traditional idea and difficult to replace it with a better idea. Toward a better idea we showed that the behavioral results in *Tupaia* led to the discovery of visual pathways that parallel the geniculostriate path and find targets in the extrastriate areas. We emphasized the tectopulvinar path, but recognize this as one of a larger class. The similar behavioral results in the cat set the stage for the discovery that in addition to the tectopulvinar path, the geniculate itself projects to association cortex. While the monkey has similar pathways, these are not sufficient to preserve more than a rudimentary visual capacity in the absence of striate cortex. Pathways from striate cortex to subdivisions in the extrastriate cortex in the monkey are essential for the behavioral functions of the middle temporal and inferotemporal areas. Thus, the dependence postulated by Campbell is realized in the monkey. We presented evidence to show that the separation of sensory pathways in the monkey striate cortex is maintained in the areas beyond the striate.

We cannot rest content simply to say that the relation of striate and extrastriate visual cortex differs in different species. Species differences for our purpose offer the opportunity of achieving a deeper understanding if they can be viewed as departures from a basic common plan. We believe that the common plan is reflected in some species still extant—generalized insectivores and marsupials. For example, in opossum and hedgehog the projections of the lateral geniculate and the pulvinar nuclei overlap through all of the visual cortex—striate and extrastriate. In the carnivore line the extrastriate projections of the geniculate developed even further and became specialized to deal with motion, form perception, changes in attention, and learning. The striate path differentiated to concentrate on fine detail and acuity. In the end, the Y cells of the lateral geniculate nucleus have a widespread target equivalent to all of the visual cortex, while the X cells project just to striate.

A different sort of specialization is found in *Tupaia:* The overlap between the lateral geniculate and pulvinar projections in cortex is lost entirely, with the result that the temporal lobe is a tectopulvinar target and the striate area is a geniculate target. Each has a considerable degree of independence, but, of course, they remain closely linked by reciprocal corticocortical projections.

The primate line represents still another specialization, and indeed, may represent several different sorts of modifications of the basic scheme. The pulvinar nucleus still retains a powerful projection to the striate area and these terminate in layer I. How this path is related functionally to the several pathways that relay in the lateral geniculate nucleus remains an unsolved problem. We showed evidence for substantial differences between prosimian and simian in the role of the tectopulvinar path.

It only remains to review Campbell's concept of higher and lower function. For Campbell the meaning of "higher" was clear and precise; it was based on the epistomology of Hume and Locke: a "perception" is higher than a "sensation." None of the experimental evidence since Campbell in any one of the species discussed supports the idea that the extrastriate visual cortex is "higher" in Campbell's sense or in any other clear sense of the term.

First, there is no reason to believe that the brain works in the way required by Hume's philosophy. Learning is not necessarily a sign of a higher function—learning can occur in a single interneuron in a cockroach. Second, there is no reason to believe that "sensation" occurs before object recognition. Vertebrates without any cortex certainly adapt to food and mates and it is unlikely that they must learn in order to transform sensation to the meaning of such objects. Arguments can be made that even in man the simple element of consciousness (i.e., sensation) is not so simple after all and is a sophisticated experience depending on a sophisticated training method (the elimination of the object error). Size and color constancies, rather than being products of learning in association cortex, may be relatively simple achievements, at least in the sense that the lowest vertebrate usually does not confuse "close" and "big."

For all of the reasons above, we took Campbell's view of dependence as the key issue. We recognize that if the association cortex were indeed athalamic, and if corticocortical projections went only in the direction from sensory to association area, and if association areas expand while sensory areas are stable in primate evolution, the temptation to add "higher" to "dependent" might be hard to resist. But none of these views of anatomy and evolution can be supported. Cortical subdivisions are reciprocally interconnected and most, perhaps all, of the neocortex receives thalamic fibers; and the laminar differentiation of striate cortex reaches a more advanced level than that of any other cortical area. Even the distinction in the neurological clinic between sensory loss and agnosia can be interpreted quite differently from the way Munk and Campbell viewed it. Scotoma in man after lesions of striate could reflect the dependence of all other visual centers on striate including the tectum, without insisting that the behavioral function of striate cortex is simpler, and can be identified with sensation.

Is there any other way of defining higher and lower functions that avoids the philosophic issues between Hume on the one side and Lorenz and Kohler on the other? Of course, phylogeny comes first to mind. It is fair to insist that cortical functions are higher than tectal functions, with the qualification that the tectum of primates is hardly equivalent functionally to the tectum of a frog.

To our knowledge there is one part of the visual system where an unambiguous judgment about higher and lower function can be related to dependence. In the retina it is established that retinal ganglion cells depend solely for their function on the impulses they receive from amacrine and bipolar cells—which in turn depend on rods and cones and horizontal cells. The landmark experiments of Hartline (1938), Kuffler (1952), and Werblin and Dowling (1969) led to the concept that the receptive field of ganglion cells, and, in particular, the antagonism between one and another part of the receptive field, can be explained as an integrative product of the activity of units lower in the chain. In the clear sense this antagonism, as well as the off-response and the transient response, is a by-product of a sustained on-response and inhibitory interneurons; therefore, it is fair to conclude that the ganglion cell has a higher function. The receptive field concept, however, has not contributed in a comparable way to the relation between striate and extrastriate cortex. The receptive fields of cells in extrastriate visual cortex cannot be built from striate cells in the way that the receptive field of the retinal ganglion cells can be explained by the activity of amacrine and bipolar cells. We concede that in the primate different subdivisions of the extrastriate visual cortex can be related to different "features" found in the cells of the striate area. Thus, area MT is clearly involved in detecting motion. To our knowledge, there is just one example of a difference in receptive fields of striate and extrastriate cortex that would satisfy Campbell's distinction between visuo-sensory and visuo-psychic. In extrastriate area V4 (of Zeki) cells sensitive to wavelength show color constancy, while the cells of the striate area respond simply to the energy of the wavelength (S.M. Zeki, personal communication). Since the receptive fields of the cells of the lateral geniculate nucleus do not differ significantly from the cells of the retina, the receptive field method does not appear to be a good way to study the integrative functions of the lateral geniculate nucleus. From the geniculate to the striate cortex, there is a dramatic change in receptive field that very likely offers the hope for a new way to define "complexity." It remains for further study to explain how the center–surround organization found in the lateral geniculate nucleus is transformed to slits or lines of various orientation.

ACKNOWLEDGMENTS. This work was supported by NIMH Research Grant 04849, NIH Research Grant 19206, NSF Research Grant BNS-8209081 (I.T.D.), and NSF Research Grant BNS-8411964 (D.F.).

7. References

Adrian, E. D., 1941, Afferent discharges to the cerebral cortex from peripheral sense organs, *J. Physiol. (London)* **100:**159–191.

Atencio, F. W., Diamond, I. T., and Ward, J. P., 1975, Behavioral study of the visual cortex of *Galago senegalensis*, *J. Comp. Physiol. Psychol.* **89:**1109–1135.

Benevento, L. A., and Fallon, J. H., 1975, The ascending projections of the superior colliculus in the rhesus monkey (*Macaca mulatta*), *J. Comp. Neurol.* **160:**339–362.

Benevento, L. A., and Yoshida, K., 1981, The afferent and efferent organization of the lateral geniculo-prestriate pathways in the macaque monkey, *J. Comp. Neurol.* **203:**455–474.

Berkley, M. A., and Sprague, J. M., 1979, Striate cortex and visual acuity functions in the cat, *J. Comp. Neurol.* **187:**679–702.

Berlucchi, G., and Sprague, J. M., 1981, The cerebral cortex in visual learning and memory, and in interhemispheric transfer in the cat, in: *The Organization of the Cerebral Cortex* (F. O. Schmitt, F. G. Worden, G. Adelman, and J. G. Dennis, eds.), MIT Press, Cambridge, Mass.

Brodmann, K., 1905, Beiträge zur histologischen Lokalisation der Grosshirnrinde. Dritte Mitteilung: Die Rinderfelder der niederen Affen, *J. Psychol. Neurol.* **4:**177–226.

Caldwell, R. B., and J. P. Ward, 1982, Central visual field representation in striate–peristriate cortex as the functional unit of pattern and form discrimination in bushbaby (*Galago senegalensis*), *Brain Behav. Evol.* **21:**161–174.

Campbell, A. W., 1905, *Histological Studies on the Localization of Cerebral Function,* Cambridge University Press, London.

Carey, R. G., Fitzpatrick, D., and Diamond, I. T., 1979, Layer I of striate cortex of *Tupaia glis* and *Galago senegalensis:* Projections from thalamus and claustrum revealed by retrograde transport of horseradish peroxidase, *J. Comp. Neurol.* **186:**393–438.

Carey, R. G., Fitzpatrick, D., and Diamond, I. T., 1985, Tectal projections to pulvinar in *Galago crassicaudatus*, *J. Comp. Neurol.* (in press).

Chow, K. L., 1950, A retrograde cell degeneration study of the cortical projection field of the pulvinar in the monkey, *J. Comp. Neurol.* **93:**313–340.

Clark, W. E. L., 1932, The structure and connections of the thalamus, *Brain* **55:**406–470.

Diamond, I. T., and Utley, J. D., 1963, Thalamic retrograde degeneration study of sensory cortex in opossum, *J. Comp. Neurol.* **120:**129–160.

Diamond, I. T., Conley, M., Itoh, K., and Fitzpatrick, D., 1985, Laminar organization of geniculocortical projections in *Galago senegalensis* and *Aotus trivirgatus*, *J. Comp. Neurol.* (in press).

Elliot-Smith, G., 1910, Some problems relating to the evolution of the brain, *Lancet* **1:**1–6, 147–153, 221–227.

Fitzpatrick, D., Itoh, K., and Diamond, I. T., 1983, The laminar organization of the lateral geniculate body and the striate cortex in squirrel monkey (*Saimiri sciureus*), *J. Neurosci.* **3:**673–702.

Flechsig, P., 1901, Developmental (myelogenetic) localization of the cerebral cortex in the human subject, *Lancet* **2:**1027–1029.

Flourens, P., 1824, *Investigations of the Properties and the Functions of the Various Parts Which Compose the Cerebral Mass* (G. von Bonin, trans.), Thomas, Springfield, Ill., 1960.

Gasser, H. S., and Erlanger, J., 1927, The role played by the sizes of the constituent fibers of a nerve trunk in determining the form of its action potential wave, *Am. J. Physiol.* **80:**522–547.

Glendenning, K. K., Hall, J. A., Diamond, I. T., and Hall, W. C., 1975, The pulvinar nucleus of *Galago senegalensis*, *J. Comp. Neurol.* **161:**419–458.

Graham, J., 1977, An autoradiographic study of the efferent connections of the superior colliculus of the cat, *J. Comp. Neurol.* **173:**629–654.

Harting, J. K., Hall, W. C., and Diamond, I. T., 1972, Evolution of the pulvinar, *Brain Behav. Evol.* **6:**424–452.

Hartline, H. K., 1938, The response of single optic nerve fibers of the vertebrate eye to illumination of the retina, *Am. J. Physiol.* **121:**400–415.

Killackey, H., Snyder, M., and Diamond, I. T., 1971, Function of striate and temporal cortex in the tree shrew, *J. Comp. Physiol. Psychol.* **74:**1–29.

Killackey, H., Wilson, M., and Diamond, I. T., 1972, Further studies of the striate and extrastriate visual cortex in the tree shrew, *J. Comp. Physiol. Psychol.* **81:**45–63.

Klüver, H., 1942, Functional significance of the geniculo-striate system, *Biol. Symp.* **7:**253–299.

Klüver, H., and Bucy, P. C., 1937, "Psychic blindness" and other symptoms following bilateral temporal lobectomy in rhesus monkeys, *Am. J. Physiol.* **119:**352–353.

Klüver, H., and Bucy, P. C., 1938, An analysis of certain effects of bilateral temporal lobectomy in the rhesus monkey, with special references to "psychic blindness," *J. Psychol. (Paris)* **5:**33–54.

Kuffler, S. W., 1952, Neurons in the retina: Organization, inhibition and excitation problems, *Cold Spring Harbor Symp. Quant. Biol.* **17:**281–292.

Lashley, K. S., 1948, The mechanism of vision. XVIII. Effects of destroying the visual "association areas" of the monkey, *Genet. Psychol. Monogr.* **37:**107–166.

Lin, C. S., and Kaas, J. H., 1979, The inferior pulvinar complex in owl monkeys: Architectonic subdivisions and patterns of input from the superior colliculus and subdivisions of visual cortex, *J. Comp. Neurol.* **187:**655–678.

Livingstone, M. S., and Hubel, D. H., 1984, Anatomy and physiology of a color system in the primate visual cortex, *J. Neurosci.* **4:**309–356.

Lorente de Nó, R., 1938, Cerebral cortex: Architecture, intracortical connections, motor projections, in: *Physiology of the Nervous System* (J. F. Fulton, ed.), Oxford University Press, London, pp. 291–325.

Mishkin, M., 1966, Visual mechanisms beyond the striate cortex, *Frontiers in Physiological Psychology*, Academic Press, New York.

Mishkin, M., 1972, Cortical visual areas and their interactions, in: *Brain and Human Behavior* (A. G. Karczmar and J. C. Eccles, eds.), Springer-Verlag, Berlin.

Moruzzi, G., and Magoun, H. W., 1949, Brain stem reticular formation and activation of the EEG, *Electroencephalogr. Clin. Neurophysiol.* **1:**455–473.

Munk, H., 1881, *On the Functions of the Cortex* (G. von Bonin, trans.), Thomas, Springfield, Ill., 1960.

Neff, W. D., and I. T. Diamond, 1958, The neural basis of auditory discrimination, in: *The Biological and Biochemical Basis of Behavior* (C. Woolsey and H. Harlow, eds.), University of Wisconsin Press, Madison.

Polyak, S. L., 1957, *The Vertebrate Visual System* (H. Klüver, ed.), University of Chicago Press, Chicago.

Raczkowski, D., and Diamond, I. T., 1980, Cortical connections of the pulvinar nucleus in *Galago*, *J. Comp. Neurol.* **193:**1–40.

Raczkowski, D., and Diamond, I. T., 1981, Projections from the superior colliculus and cortex to the pulvinar nucleus in *Galago*, *J. Comp. Neurol.* **200:**231–254.

Raczkowski, D., and Rosenquist, A. C., 1983, Connections of the multiple visual cortical areas with the lateral posterior–pulvinar complex and adjacent thalamic nuclei in the cat, *J. Neurosci.* **3:**1912–1942.

Ramón y Cajal, S., 1906, The structure and connections of neurons, in: *Physiology or Medicine: Nobel Lectures Including Presentation Speeches and Laureates' Biographies 1901–1921*, Nobel Foundation, Volume I, Elsevier, Amsterdam, 1967, pp. 220–253.

Rockland, K. S., and Pandya, D. N., 1979, Laminar origins and terminations of cortical connection of the occipital lobe in the rhesus mokey, *Brain Res.* **179:**3–20.

Rose, J. E., and Woolsey, C. N., 1949, Organization of the mammalian thalamus and its relationships to the cerebral cortex, *Electroencephalogr. Clin. Neurophysiol.* **1:**391–404.

Schafer, E. A., 1888, Experiments on special localization in the cortex cerebri of the monkey, *Brain* **10:**362–380.

Snyder, M., and Diamond, I. T., 1968, The organization and function of the visual cortex in the tree shrew, *Brain Behav. Evol.* **1:**244–288.

Sprague, J. M., Hughes, H. C., and Berlucchi, G., 1981, Cortical mechanisms in pattern and form perception, in: *Brain Mechanisms and Perceptual Awareness* (O. Pompeiano and C. A. Marsan, eds.), Raven Press, New York.

Symonds, L. L., and Kaas, J. H., 1978, Connections of striate cortex in the prosimian *Galago crassicaudatus*, *J. Comp. Neurol*, **181:**477–512.

Tigges, J., Spatz, W. B., and Tigges, M., 1973, Reciprocal point-to-point connections between parastriate and striate cortex in the squirrel monkey (*Saimiri*), *J. Comp. Neurol.* **148:**481–490.

Walker, A. E., 1938, *The Primate Thalamus*, University of Chicago Press, Chicago.

Werblin, F. S., and Dowling, J. E., 1969, Organization of the retina of the mudpuppy, *Necturus maculosus*. II. Intracellular recording, *J. Neurophysiol.* **32:**339–355.

Woolsey, C. N., 1947, Patterns of sensory representation in the cerebral cortex, *Fed. Proc.* **6:**437–441.

Yukie, M., and Iwai, E., 1981, Direct projection from the dorsal lateral geniculate nucleus to the prestriate cortex in macaque monkeys, *J. Comp. Neurol.* **201:**81–97.

Zeki, S. M., 1974, Functional organization of a visual area in the posterior bank of the superior temporal sulcus of the rhesus monkey, *J. Physiol. (London)* **242:**827–841.

Zeki, S. M., 1978, The cortical projections of foveal striate cortex in the rhesus monkey, *J. Physiol. (London)* **277:**227–244.

Zeki, S. M., 1983, The distribution of wavelength and orientation selective cells in different areas of monkey visual cortex, *Proc. R. Soc. London Ser. B* **217:**449–470.

3

Cingulate Cortex

BRENT ALAN VOGT

1. Introduction

Each cortical region has unique functional properties which depend upon its cytological composition and connections. Cingulate cortex stands out as being of particular interest for a growing number of reasons. First, it has been implicated in emotion (Papez, 1937) and affective responses to pain (White and Sweet, 1969). It has also been linked directly to the pain/opiate system with receptor binding, enkephalin immunohistochemical, and connectional procedures. Second, laminar and cytological variations are present in cingulate cortex which cast light on general questions regarding the nature of cortical transition and structural equivalencies of cortical areas among different species. Third, thalamocortical connections have a unique pattern of termination and histochemistry. Fourth, cingulate cortex is a major component of the cholinergic system as early defined with an assay for acetylcholinesterase activity (Lewis and Shute, 1967) and more recently with receptor binding and choline acetyltransferase immunohistochemistry. Fifth, the medial location of cingulate cortex above the corpus callosum provides a unique opportunity for removing slices of cortex together with a single afferent for electrophysiological analyses of the properties of neurons and the synaptic activity of isolated excitatory inputs to cortical cells.

This chapter is primarily a synthesis of the cytology, connections, pharmacology, and electrophysiological properties of neurons in posterior cingulate cortex and the anterior thalamic nuclei. This part of cingulate cortex has been chosen because it has an interesting structural organization in comparison to

BRENT ALAN VOGT • Departments of Anatomy and Physiology, Boston University School of Medicine, Boston, Massachusetts 02118.

what is generally accepted as characteristic of neocortex. The posterior cingulate region is also being evaluated because no systematic reviews are available which integrate the wide range of neuroscientific investigations into this part of the cerebral cortex.

2. Definition of Cingulate Cortex

2.1. General Considerations

A number of morphological criteria have been used to define cingulate cortex including topographic, cytoarchitectural, and connectional ones. Broca (1878) first recognized that the cingulate gyrus is part of a group of medial structures which he referred to as the "grand limbic lobe." This and other topographical and comparative studies (e.g., Kappers *et al.*, 1967) note that the cingulate gyrus surrounds the corpus callosum and in higher mammals is limited ventrally by the rostral and terminal sulci, dorsally by the cingulate and splenial sulci, and posteriorly by the calcarine sulcus. In lissencephalic animals the callosal sulcus limits the ventral border of cingulate cortex, while dorsal borders with lateral neocortex are established on cytoarchitectural grounds.

Three investigators published cytoarchitectural schemes early in this century which specified how the cingulate gyrus might be subdivided: Brodmann (1909), Rose (1912, 1927, 1931), and von Economo (1929). Although each account contains a differing number of subdivisons, there is uniformity in the sense that an anterior agranular and two major posterior granular subdivisions are recognized. Cytoarchitecturally, cortex of the cingulate gyrus has thicker infra- than supragranular layers and layers II–IV tend to be poorly differentiated in comparison to isocortical structures.

A number of influential articles subsequently appeared which provided both connectional and theoretical bases for distinguishing cingulate from other cortices. First, Clark and Boggon (1933) employed retrograde cell degeneration procedures to evaluate thalamocortical connections and recognized that the anterior thalamic nuclei selectively project to cingulate cortex. Later, Papez (1937) theorized that cingulate cortex and the anterior thalamic nuclei are actually part of a larger connectional system that "may elaborate the functions of central emotion, as well as participate in emotional expression." This conceptualization has had a profound influence on subsequent structural and functional studies of cingulate cortex.

A third influential article was that of Rose and Woolsey (1948) in which both thalamocortical connections and cytoarchitecture were analyzed to characterize each subdivision of the cingulate gyrus. On the basis of this and later articles it is now generally agreed that anterior cingulate cortex receives input mainly from the anteromedial and mediodorsal thalamic nuclei. In contrast, posterior and ventral cingulate cortex receives input from the anteroventral and anterodorsal thalamic nuclei, while posterior and dorsal cingulate cortex receives input mainly from the laterodorsal and lateroposterior nuclei. In addition, a number of midline and intralaminar thalamic nuclei project throughout cingulate cortex.

In recent years the cytoarchitectural parcellation of cingulate cortex by Rose and Woolsey (1948) has been most frequently employed in studies of rabbit and cat, while that of Brodmann (1909) is usually applied to studies of rat and monkey. These cytoarchitectural approaches are not mutually exclusive and do not reflect fundamental differences among the cingulate cortices of these species. Rose and Woolsey state explicitly how Brodmann's areas in the rabbit brain are consolidated to form the larger areas that they enumerate. The present analysis employs the Brodmann parcellation scheme, because it embodies one of the first comprehensive and comparative analyses of the cerebral cortex. The individual areas are usually large enough to encompass physiologically characterized fields, although little is presently known about the function of individual cingulate areas, and the map is easily modified to account for new observations.

2.2. Cytoarchitecture and Cortical Transition in the Monkey

A number of investigators have observed that different cortical areas represent a progressive elaboration of one or two basic layering schemes (Brodmann, 1909; von Economo, 1929; Lorente de Nó, 1938; Sanides, 1970; Sanides and Sanides, 1972; Vogt, 1976; Braak, 1979; Galiburda and Pandya, 1983). Progressive cytoarchitectural, cytological, and connectional changes between cortical areas are referred to as cortical transitions. It is particularly important to understand transitional features in cingulate cortex because major structural alterations occur in this region.

The basic principles of cortical transition are as follows. A base layer of large pyramidal neurons is present in all areas, and major cytoarchitectural changes involve the addition of layers to this basic layer, while minor changes include intralaminar alterations in cell packing density. Each of these changes is the result of the addition or modification of individual cell types and so Golgi preparations are required for evaluation of the cytological basis of cortical transition. It is also important to know the afferent connections and efferent projection patterns of neurons in each layer, because these projections often confirm equivalencies between the layers of different cortices.

Transitional features are most clear in cortex surrounding the splenium of the corpus callosum of the monkey because here there is continuity in the layer of large pyramidal cells between adjacent cortical areas (Fig. 1, asterisks). This layer is densely packed in the indusium griseum, more dispersed in subicular cortex, and is continuous with layer V in areas 29, 30, and 23. Two major changes occur in area 29. First, a dense and undifferentiated external pyramidal layer is added to the base layer V. Second, layer VI separates from the base layer. Further transitional features in areas 29 and 30 include separation of layers II, III, and IV and subdivision of layer V into a layer Va of large pyramidal neurons and a less cell-dense layer Vb. Finally, in area 23a, which is a ventral subdivision of Brodmann's area 23, layer IV becomes more prominent and layer III less cell dense, as compared with the adjacent area 30.

Three basic patterns of cortical differentiation in suprasplenial cortex have been enumerated by Sanides (1970) and Sanides and Sanides (1972). These classifications are useful in characterizing similar moieties of cortex both within and among species. Cortices with a single ganglionic layer of predominantly pyramidal neurons are classified as allocortex and these include the indusium griseum and subiculum. Cortices with a granular and relatively undifferentiated external pyramidal layer are periallocortices and in the suprasplenial region they include much of area 29. Finally, cortices with a clearly differentiable but thin layer IV are proisocortices and these include areas 29d, 30, and 23a.

Two major cytological alterations are associated with cortical transition in this region. These changes involve elaborations of the basal dendritic trees of pyramidal cells and the addition and differentiation of new cell types. Neurons of the indusium griseum either lack basal dendrites or have a limited number of obliquely projecting dendrites (Fig. 2). The greater development of subpial dendritic tufts as compared with the development of basal dendrites is a characteristic of poorly differentiated or "extraverted" pyramidal neurons (Sanides and Sanides, 1972). In the adjacent subiculum and area 29, on the other hand, pyramidal neurons in the base layer have larger numbers of basal dendrites

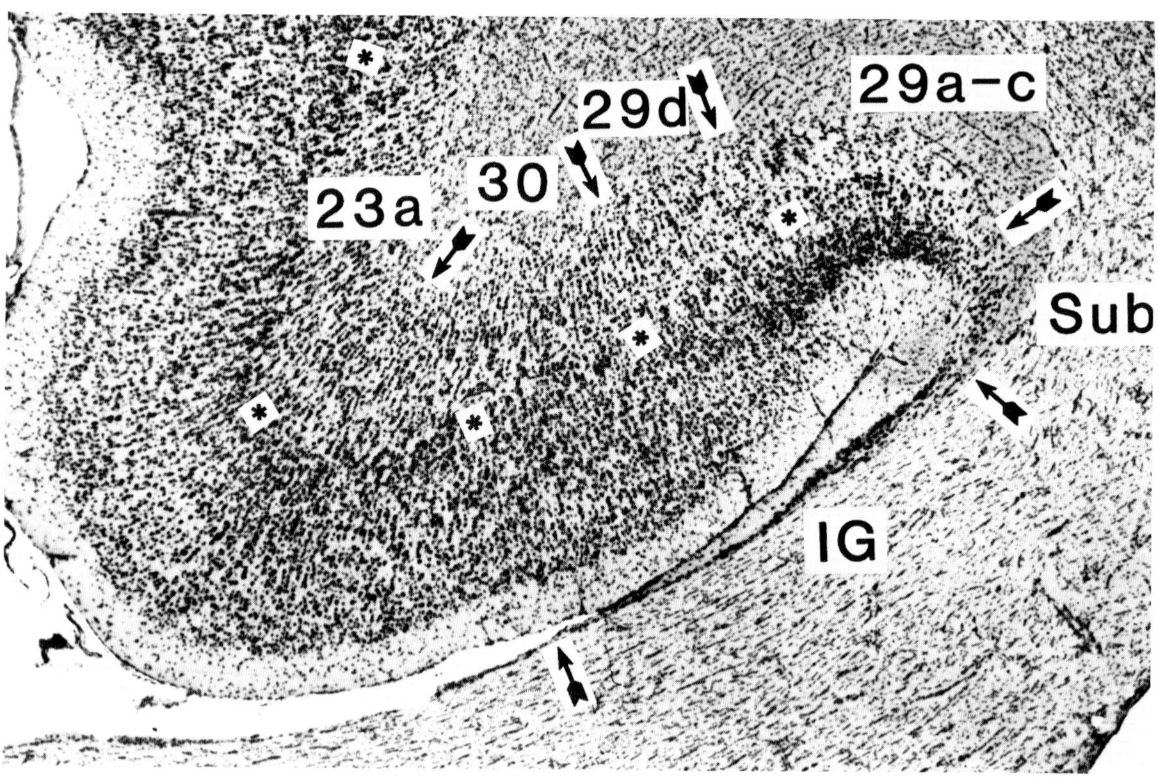

Figure 1. Cytoarchitecture of posterior cingulate cortex in monkey. Area 29a–c is a retrosplenial granular subdivision and areas 29d and 30 are dysgranular retrosplenial cortices. IG, indusium griseum; Sub, subiculum; asterisks, layer V; cresyl violet, ×45.

and these project from the somata with a horizontal orientation. This progressive elaboration of the basal dendrites is associated with reduced cell packing densities.

A major cytological change occurs in area 29, where an external pyramidal cell layer first appears. This granular layer is homogeneous in terms of cell density, directly abuts layer V, and contains a large number of star pyramidal neurons. It has been referred to as layer III(IV) (von Economo, 1929) or layer III (Braak, 1979) to emphasize its relation with pyramidal neurons in layer III of neocortex. However, since it receives substantial thalamic input (Section 3.2.2), which is generally considered to be characteristic of layer IV in neocortex, and since it is undifferentiated, the external pyramidal cell layer might also be referred to as layer II–IV. This layer serves as a primary focal point for determining structural equivalencies among species.

2.3. Cytoarchitecture of Posterior Cingulate Cortex in Rodents

The number and distribution of subdivisions in posterior cingulate cortex in rodents and rabbit has been a matter of dispute throughout most of this century. Brodmann (1909), Rose (1912, 1927), Krieg (1946), Rose and Woolsey (1948), Sanides and Sanides (1972), Caviness (1975), Zilles *et al.* (1980), and Vogt and Peters (1981) each arrived at a somewhat different set of conclusions. The issues to be considered regarding the parcellation of this cortex include clarification of which cytoarchitectural moieties are present, the lamination scheme,

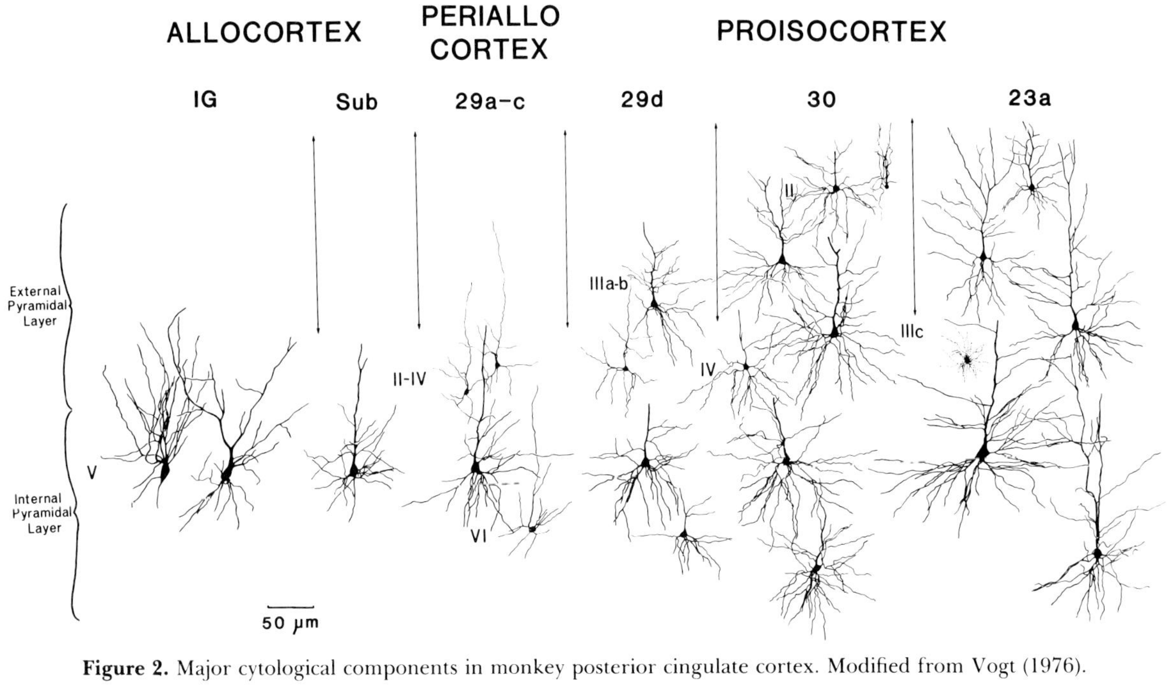

Figure 2. Major cytological components in monkey posterior cingulate cortex. Modified from Vogt (1976).

whether or not an area 23 can be discerned, identification of its dorsolateral border, and the cytology and connections of each subarea.

As in the monkey, posterior cingulate cortex in the rat is composed of both periallo- and proisocortices. The external layers are more differentiated in the rat and so three subdivisions of periallocortex are present, including areas 29a, b, and c, while proisocortex is limited to one subdivision, area 29d. Each of these areas has been thoroughly described (Vogt and Peters, 1981) and their distribution on the medial surface of the rat brain is presented in Fig. 3.

In brief, the most ventral area, 29a, abuts the subiculum and postsubiculum which is also referred to as the dorsal presubiculum or area 48. Area 29a has an undifferentiated external pyramidal layer II–IV, while the adjacent area 29b has a granular layer which is composed of dense (layer II–III) and sparse (layer IV) subdivisions. Although layer V is homogeneous in area 29b, in area 29c it divides into a superficial Va with medium-sized pyramids and a deep Vb with large pyramids. The granular layer is composed of dense and sparse subdivisions, as in area 29b. Finally, layer VI is poorly defined in area 29a, but progressively expands in width in each subsequent area moving in the dorsolateral direction.

In area 29d layer II–III is much thicker than in area 29c and a thin layer IV can be discerned (Fig. 4). This cortex has traditionally been termed "agranular," but, since it has a limited layer IV, this designation is misleading and should be replaced with the term "dysgranular" (Sanides, 1970; Vogt, 1976; Braak, 1979). Apparently, area 29d is the only proisocortex in the posterior cingulate region of rodents. Rose (1912) was unable to discern an area 23 in the mouse and neither has this area been observed by a number of subsequent investigators (Caviness, 1975; Zilles *et al.,* 1980; Vogt and Peters, 1981). Only Krieg (1946) claims that area 23 exists in the rat, but what he actually describes

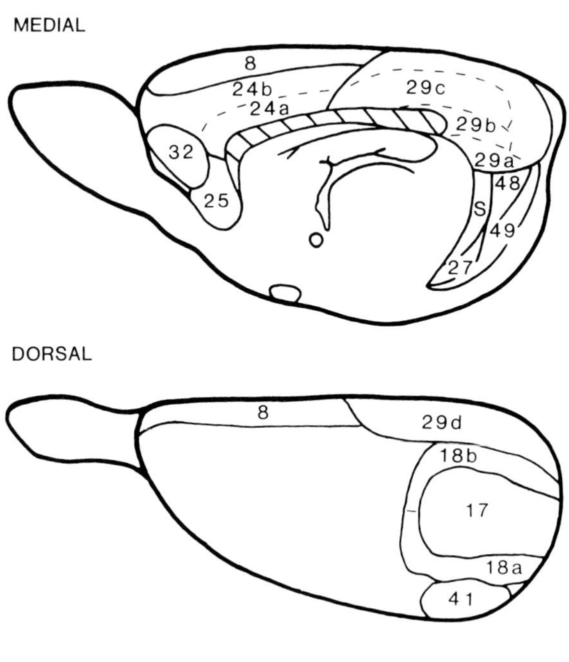

Figure 3. Topographic distribution of cytoarchitectural areas in rat cingulate cortex including ventral parahippocampal and lateral neocortical areas. S, subiculum. Modified from Vogt and Miller (1983).

is a transitional area between areas 24 and 29 which does not have a well-defined layer IV (i.e., more clear than that in area 29d).

Defining the dorsolateral border of cingulate cortex in rodents is difficult without the dorsally limiting cingulate sulcus, but definition of this border is particularly important because interpretation of certain corticocortical connections (i.e., visual/cingulate) depends upon where this border is located. Caviness (1975) places area 29d and Zilles *et al.* (1980) an area 29D lateral to the area 29d/18b border which is presented in Fig. 3. However, visual field maps of Montero *et al.* (1973) and Espinoza and Thomas (1983) localize two representations of the temporal visual field medial to area 17 at levels which reach the

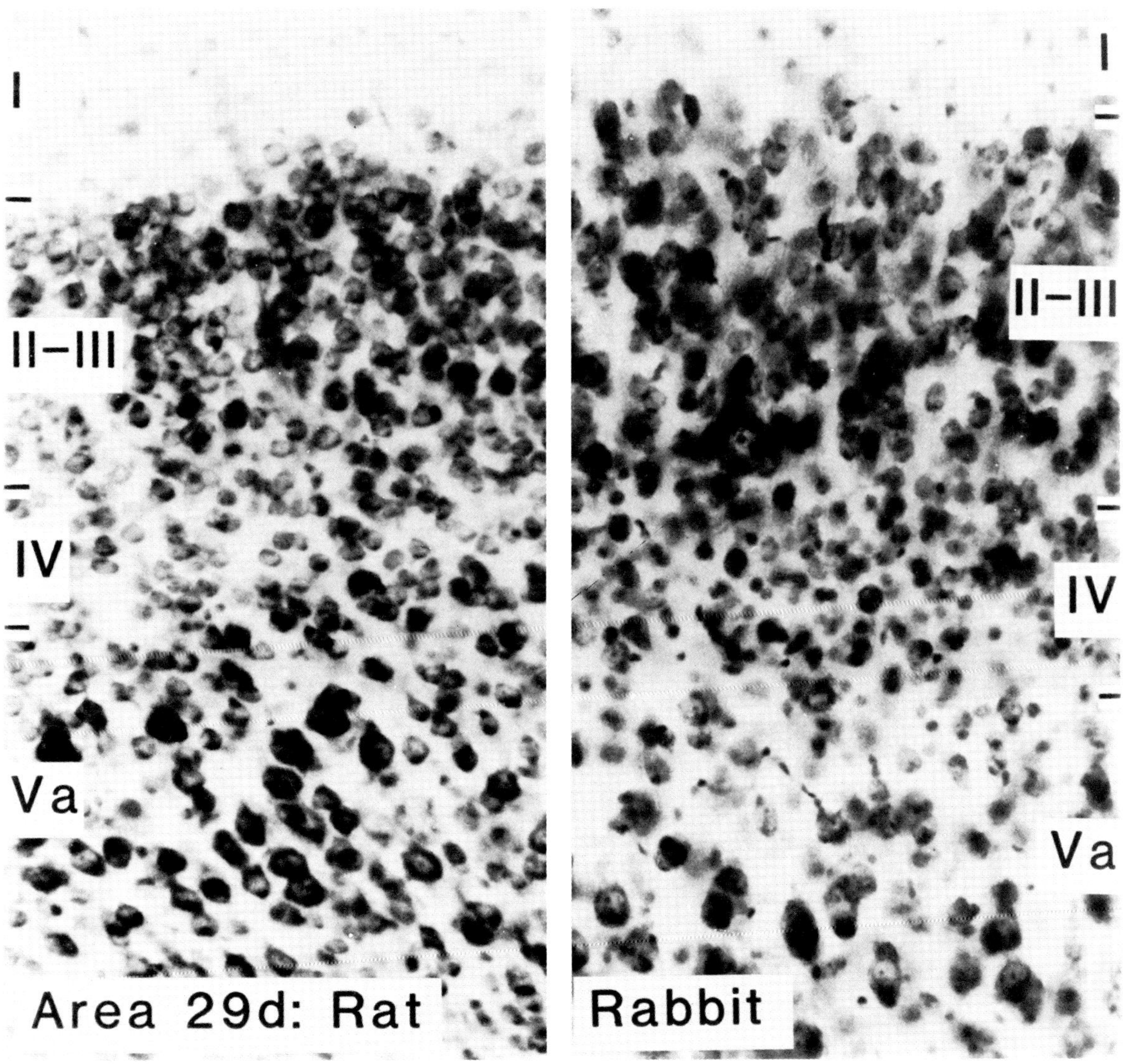

Figure 4. Comparison of superficial layers of area 29d in rat and rabbit. Photographs taken at levels shown by stars in Fig. 5; cresyl violet, ×350.

area 29d/18b border. It appears, therefore, that secondary visual cortex directly abuts area 29d of cingulate cortex in the rat.

2.4. Cytoarchitectural Equivalencies among Species

The cytoarchitectural subdivisions of cingulate cortex vary between species with regard to laminar differentiation, cytological composition, and connectivity. However, examination of the criteria for characterizing each area suggests that there are relative morphological similarities between certain areas. These similarities can form the basis for specifying equivalencies, thus circumventing the need for a separate parcellation scheme and nomenclature for each species.

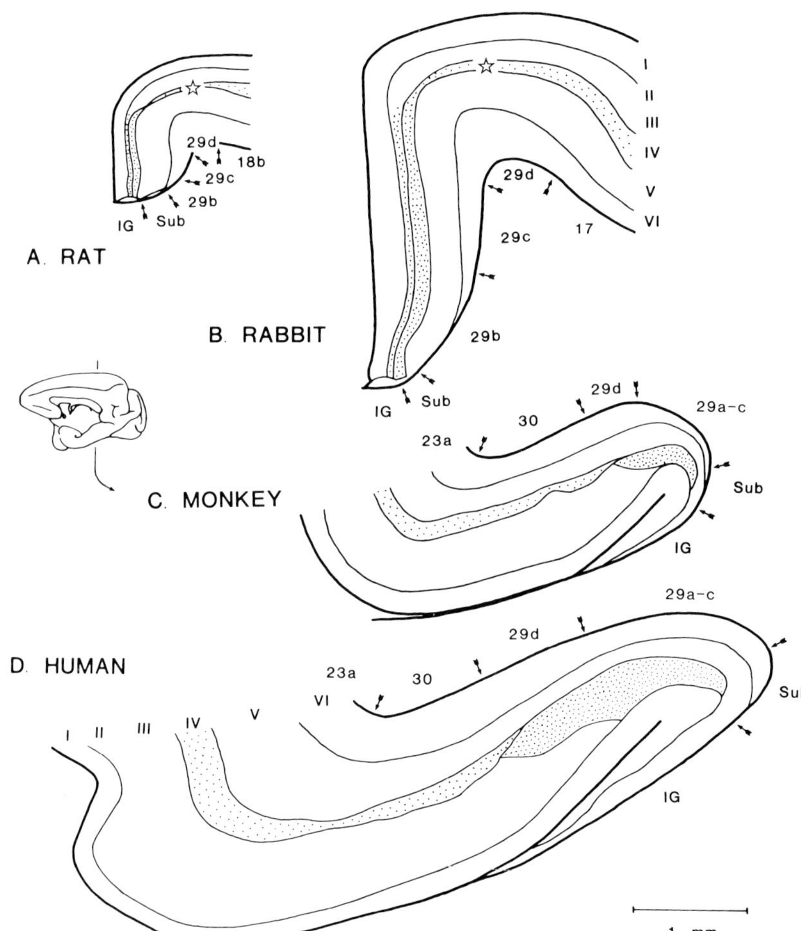

Figure 5. Cytoarchitectural composition of suprasplenial cortex in four species. An example of the level chosen is presented for the monkey (C). The undifferentiated or poorly differentiated granular layers II–IV are indicated by dense stippling. The stars in area 29d of (A) and (B) represent levels from which Fig. 4 photographs were taken.

One basic similarity in all mammals is a common trend in cytoarchitectural differentiation in suprasplenial cortex. As noted above, for the monkey and rat there is a progression from allocortex (indusium griseum and subiculum) to periallocortex (retrosplenial granular areas 29a–c) to proisocortex (areas 29d or 30 and/or 23a) and finally to isocortex (areas 23b and c and 31 of cingulate cortex as well as visual and parietal neocortices). Figure 5 presents drawings of transverse sections through suprasplenial cortex in four species (see also Stephan, 1975). Although the least differentiated part of area 29 ventral to the corpus callosum in the rat and rabbit is not represented, the cytoarchitectural progression in each species is apparent. These drawings raise a number of questions regarding the exact nature of the granular layer and certain generally accepted associations among the areas of different species.

There are two reasons for equating the granular layer in each species (dense stippling in Fig. 5). First, the cells are small and densely packed and the granularity is due mainly to small star and fusiform pyramids in both rat and monkey (Vogt, 1976; Vogt and Peters, 1981). Second, thalamocortical connections from the anteroventral and anterodorsal nuclei of the thalamus terminate directly in this layer (Section 3.2). Thus, periallocortical areas in different species which have this layer might be homologous. In the rat areas 29a–c have a granular layer, while previous studies of primates have identified two retrosplenial areas which possess this layer, including LE_1 and LE_2 (von Economo, 1929). RSg1 and RSg2 (Vogt, 1976), and rsl and rsi (Braak, 1979). In terms of the nomenclature used in rodents, areas 29a and 29b have the least differentiated granular layers II–IV and layer V, while both external and internal pyramidal layers are clearly subdivided in area 29c. Therefore, although Brodmann did not subdivide area 29 in primates, it appears that the lateral granular area in the monkey may be equivalent to areas 29a–c in the rat and the more differentiated medial division similar to area 29d. These designations have been applied in Figs. 1, 2, and 5.

Medial to area 29 in primates, Brodmann identifies area 30 or area retrolimbica agranularis. This area is very similar in position and structure to what is also known as LD (von Economo, 1929), RSag (Vogt, 1976), and rsm (Braak, 1979). This is the area in which a "neocortical-like" layer IV can first be identified, although it is thin and often intermingles with the large cells of layer III.

As noted at the outset, no area in cingulate cortex has exactly the same features in all species. Some of the major differences include the proportions of certain cell types in the granular layer, the extent of granularity of layer IV in proisocortices, and the distribution of anterior thalamic afferents in layers I–IV. Thus, although small and star pyramidal neurons characterize the granular layer in the rat and monkey, rats have a greater density of fusiform pyramids than do monkeys. Layer II–III in areas 29a–c in the rat is composed mainly of fusiform pyramids, while this cell type is less frequent in the monkey and layer III of area 29d in the monkey is actually composed mainly of small and medium-sized pyramidal neurons. Furthermore, a comparison of neurons in the granular layer of each species suggests that small pyramids become larger and are primarily medium-sized pyramidal neurons in humans. Another difference among species is that there is an overall increase in the thickness and granularity of layer IV in proisocortices. This is clear, for example, in area 29d of the rat and rabbit (Fig. 4), since layer IV in the rabbit has more small neurons than does the same layer in the rat. In primates, layer IV of area 23a also increases in width, although layer IV is more pronounced in humans than in the monkey. The fact that area 23 is not even present in rodents is a major contrast with primate species.

Another difference among species is the laminar distribution of anterior thalamic afferents. As will be detailed in Section 3.2, thalamic input in the rat terminates mainly in layer Ia with smaller inputs to layers Ib–c and IV. In contrast, almost all of these afferents terminate directly in the granular layer of the monkey with only a very small contribution to layer I.

In conclusion, similarities in the progression of cortical transition, cytology, and connections appear to warrant drawing equivalencies among specific areas of posterior cingulate cortex in a number of species. This approach produces a less confusing view of cingulate cortex because it is not necessary to use a different nomenclature for each species.

3. Cytology of Rat Cingulate Cortex

The cytology of cingulate cortex in rodents has been thoroughly analyzed with Golgi preparations (Ramón y Cajal, 1911, 1922; Schulz and Schönheit, 1974; Herrmann and Schulz, 1978; Iwahori and Mizuno, 1981a,b; Kuscika and Schulz, 1981; Vogt and Peters, 1981; Schönheit and Schulz, 1982). In light of the coverage in Volume 1 of this treatise, the morphology of each neuronal type will not be fully restated; rather, each type will be considered briefly, but with emphasis on those characteristics which are unique to cingulate cortex.

3.1. Pyramidal Neurons

A neuron which displays two or more of the following features is generally categorized as a pyramidal cell: orienting apical dendrite with an apical tuft, moderate to dense concentrations of dendritic spines, and an axon that projects into the white matter. As in other cortical regions, layers II–Va contain typical pyramids and layers Vb–VI have both typical and inverted pyramidal neurons, although Ramón y Cajal (1922) recognized that areas 29a–c also have a number of modifications in pyramidal cell morphology which are unique to cingulate cortex (Fig. 6).

3.1.1. Extraverted Pyramids

Ramón y Cajal (1922) first observed "large stellate" neurons in layer II which have dendrites that radiate from around the cell body (Fig. 6, B). These cells

Figure 6. Reproduction of Ramón y Cajal's original drawing of neurons in area 29c. A, neuron with a horizontal axon; B, large stellate cells; C and D, small pyramids; E, vertical fusiform cell; J and K, deep vertical fusiform cells. From Ramón y Cajal (1922) with permission.

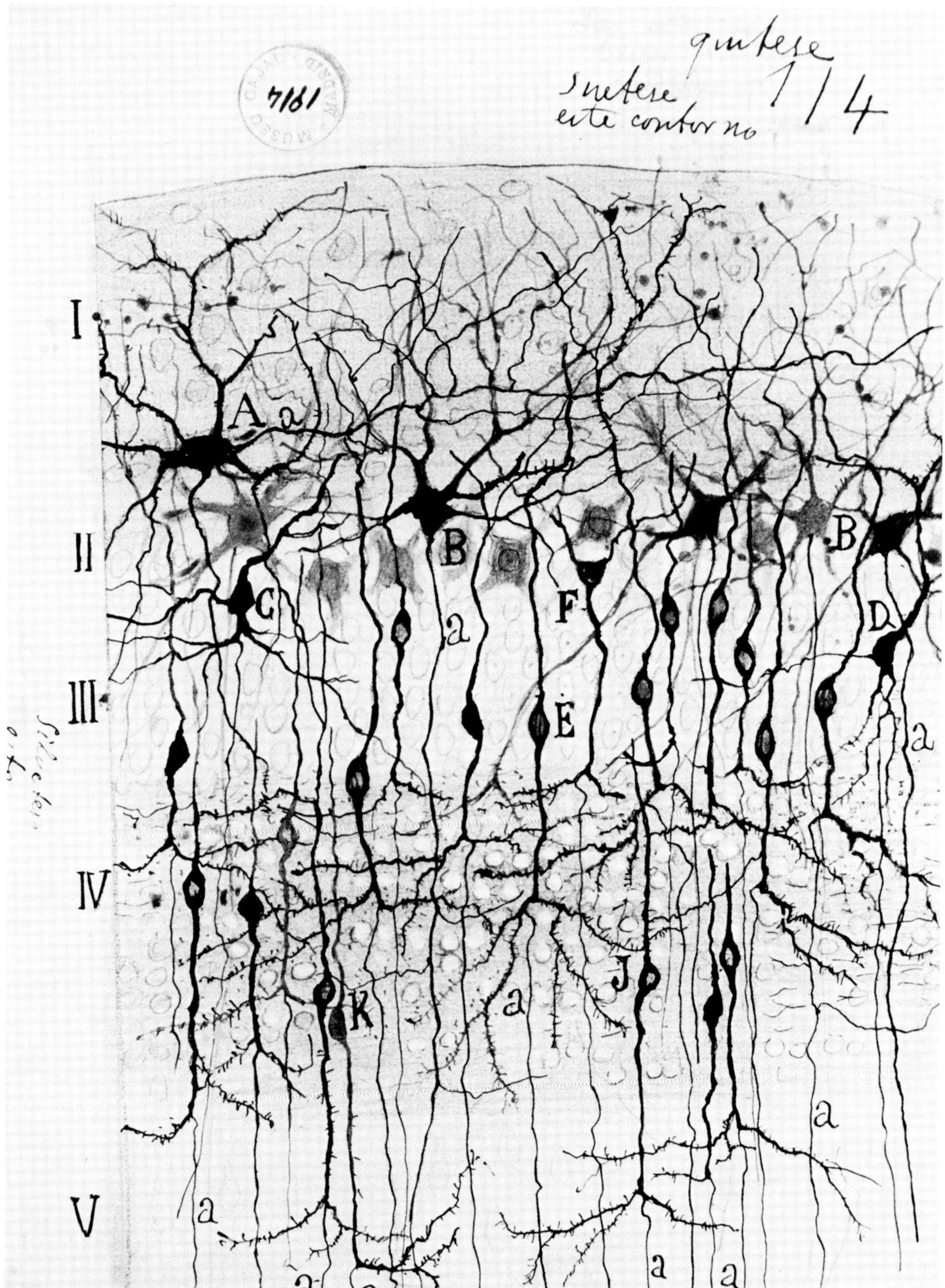
I
II
III
IV
V
A
B
C
D
E
F
J
K
a

do not have a single orienting apical dendrite, but their dendrites are very spiny and their axons enter the white matter. These cells are similar to star cells in layer II of entorhinal cortex (Ramón y Cajal, 1911; Lorente de Nó, 1933). However, they do not make up a separate layer II in the rat, but are interspersed with other cells of the granular layer II–III. Also, these cells should be given a different name, since the term "stellate" is generally used to refer to multipolar interneurons, while the "large stellate" cells conform to most of the criteria set for pyramids. Figure 7 illustrates an example of a large multipolar cell with smooth dendrites (cell a) and a "large stellate" neuron (cell b). In fact, the dendrites rarely radiate symmetrically from around the soma. Rather, they have a preferential spread into layer I, usually with only one to three branches projecting into layer IV. Large stellate neurons are an example of what Sanides and

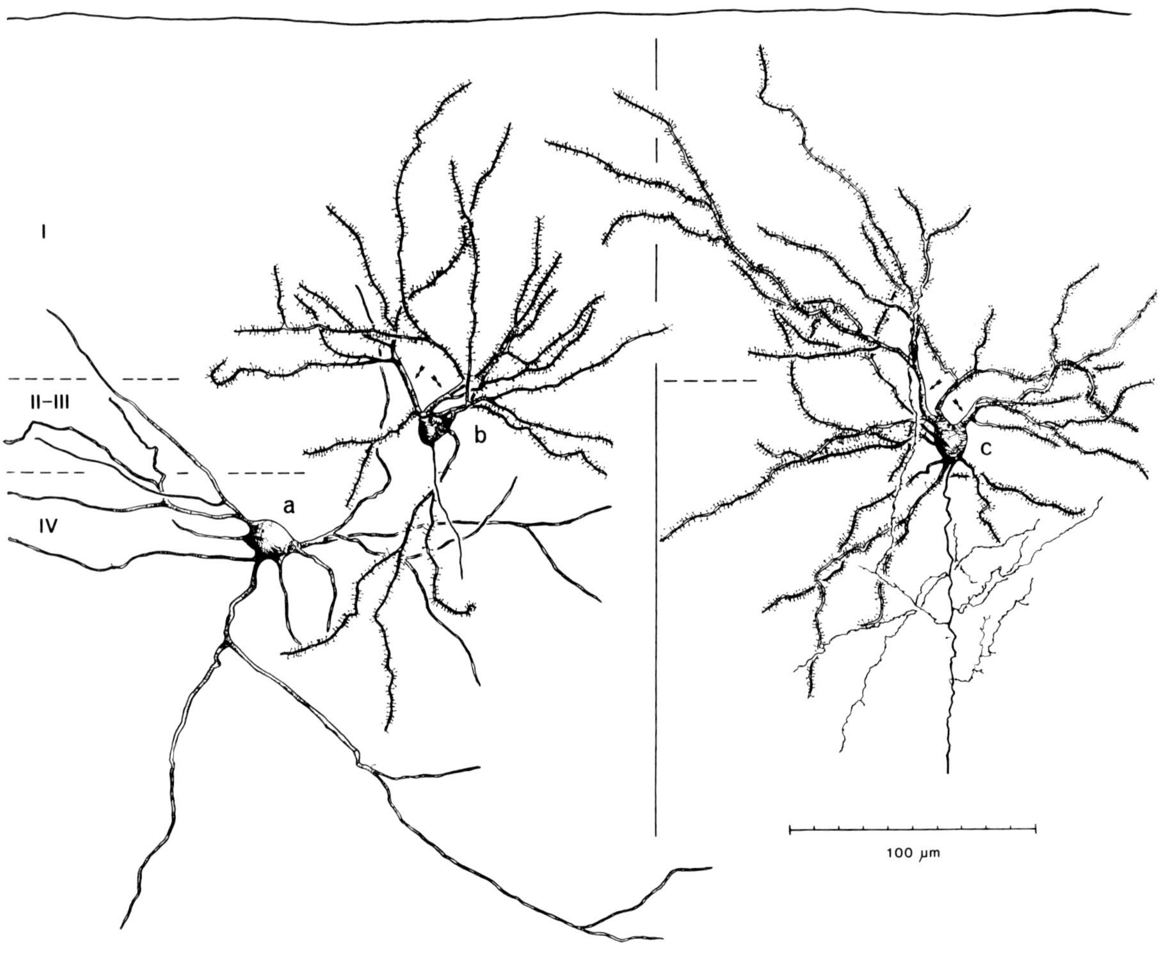

Figure 7. Extraverted pyramidal neurons in layer II–III of area 29c (b) and layer II of area 29d (c). A large multipolar neuron in layer IV (a) is presented for comparison. Arrows point to primary apical processes.

Sanides (1972) referred to as extraverted pyramidal neurons. These pyramids are found in layer II of periallo- and proisocortices such as cingulate cortex and have a subpial dendritic arborization which predominates over a limited basal dendritic tree. An example of an extraverted pyramid from area 29d is presented in Fig. 7 (cell c) for comparison with the type of cell found in area 29c. Note that both cells have spherical somata and that although neither cell has an orienting apical dendrite, each has a pair of medium-sized dendrites of similar size (arrows) projecting into layer I, as well as thinner dendrites projecting into both layers I and IV.

In addition to extraverted pyramidal cells there are a number of moderately sized, typical pyramids in layers II–IV and a few pyramids which have apical dendrites running horizontally along the layer I/II–III border instead of in a vertical direction.

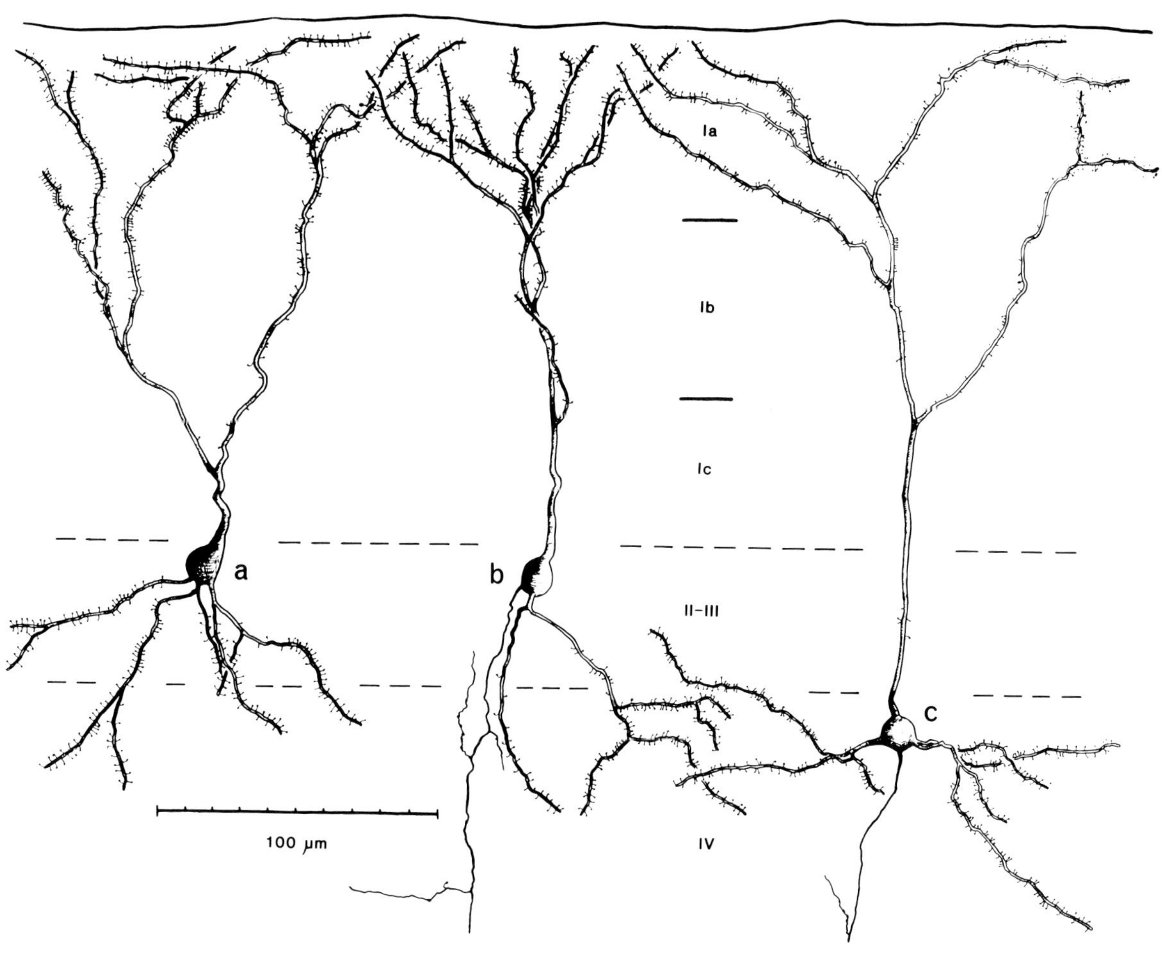

Figure 8. Fusiform (a, b) and small (c) pyramidal neurons in area 29c.

3.1.2. Fusiform Pyramids

Cingulate cortex is the only cortex in which fusiform pyramids have been reported and these cells result from the cytoarchitectural and connectional characteristics of this region. "Vertical fusiform" cells of Ramón y Cajal (1922) or fusiform pyramids (Vogt and Peters, 1981) have small, round or fusiform somata and are mainly present in layer II–III in the rat. One to three basal dendrites project vertically from the base of the cell body and their branches distribute horizontally throughout layer IV. Each cell has a conical dendrite which may bifurcate (Fig. 8, cells a and b) and has very few spines. In all instances a subpial, apical tuft is formed in layer Ia. These spiny, apical tufts (Fig. 9) are distributed in the layer which is the primary recipient of thalamocortical afferents. It is possible that there are orientation preferences in the spread of these apical dendrites, such that some may be oriented in the rostrocaudal plane (Fig. 8, cell b) and others in the dorsoventral plane (cell a). Finally, some neurons with similar somal shapes and basal dendritic trees have also been observed in the granular layer of monkey retrosplenial cortex (Vogt, 1976). It appears that the morphology of these neurons is constrained by the dense packing of neurons in layer II–III and preferential termination of anterior thalamic afferents in layer Ia.

3.1.3. Small Pyramids

Small pyramidal neurons probably account for the largest proportion of neurons in layer IV of the rat, but they are also occasionally present in layer II–III. They are similar to small pyramids in other cortices in having round somata, a limited number of horizontally oriented basal dendrites, and an apical dendrite. The unique characteristic of these cells in areas 29b and 29c is that the apical dendrites form a spiny tuft in layer Ia (Fig. 8, cell c), as do the fusiform pyramids. This tuft distinguishes these cells from the star pyramids described by Lorente de Nó (1938), for the latter have poorly branched apical dendrites in layer I.

3.1.4. Layer V Pyramids

In area 29c layer V has two divisions, an outer layer Va composed of medium-sized pyramids and an inner layer Vb containing the largest pyramids in this cortex. These neurons are similar to other neocortical pyramids in somal shape, dendritic tree distribution, and spine density, except for distal parts of the apical dendrite. The distal apical dendrites of these neurons differ from their neocortical counterparts in two respects: branching of the apical tuft and density of spines. Apical tuft branching occurs at, or just below, the border of layers I and II–III. These branches are distributed evenly throughout layer I without selective termination in layer Ia as is the case for fusiform and small

Figure 9. The apical tufts of fusiform and small pyramidal neurons are concentrated in dense aggregations in layer Ia; Rapid Golgi, ×850.

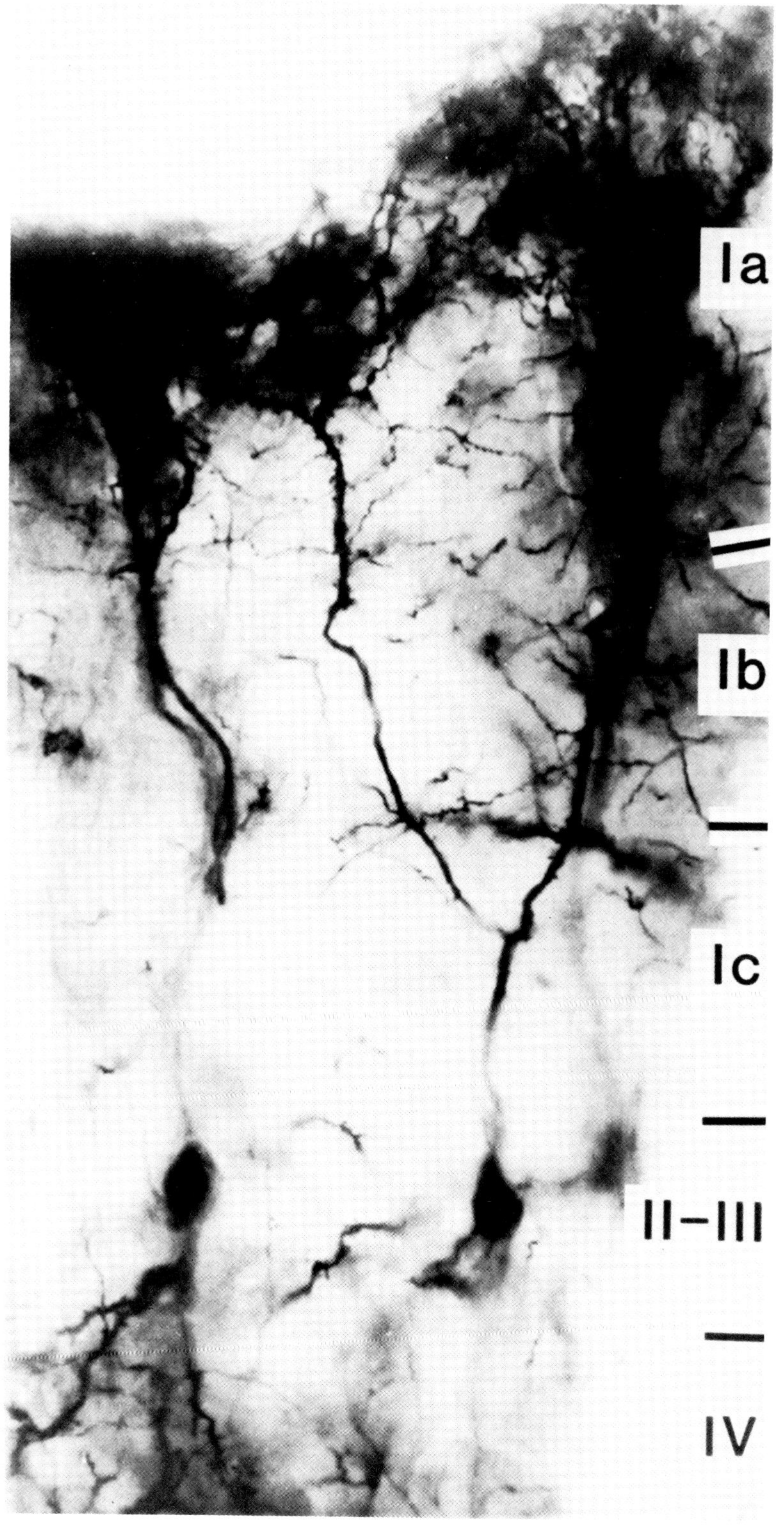
Ia
Ib
Ic
II–III
IV

pyramids. In area 29d, anterior area 24, and other neocortical areas, the branching is initiated either within layer II or at the top of layer III.

The density of spines on the apical dendrites of neocortical pyramids varies with the distance from the cell body (Marin-Padilla, 1967; Globus and Scheibel, 1967; Schierhorn *et al.,* 1972; Kemper *et al.,* 1973). The more proximal portion of the apical dendrite is free of spines, but beyond the first oblique branch there is an increase in the frequency of spines, which usually reaches a peak in layer IV. The density of spines then decreases to low numbers on apical tuft branches. It has been postulated that variation in the density of spines is determined by the distribution of afferents to pyramidal neurons, and in particular those from the thalamus (Valverde, 1967; Marin-Padilla and Stibitz, 1968; Parnavelas *et al.,* 1973). This hypothesis may be confirmed in posterior cingulate cortex.

Most layer V neurons in areas 29b and 29c have a bimodal distribution of spines on their apical dendrites. This is demonstrated by counting visible spines protruding from each side of 10-μm lengths of the apical dendrites and branches of Golgi-impregnated pyramids. In general, the spine densities peak in layers IV and V, and drop precipitously in layer II–III, just before the branching of the apical tuft, and increase again in layer I (Fig. 10). However, there are differences in the distribution of spines, depending on where the soma is located in layer V. Thus, the spine densities of superficial pyramids reach peak values in layer IV and return to densities in layer I that are similar to those in layer IV. Pyramids deeper in layer Va and superficial Vb have the greatest numbers of spines as they pass through layers V and VI, and these densities are 50%

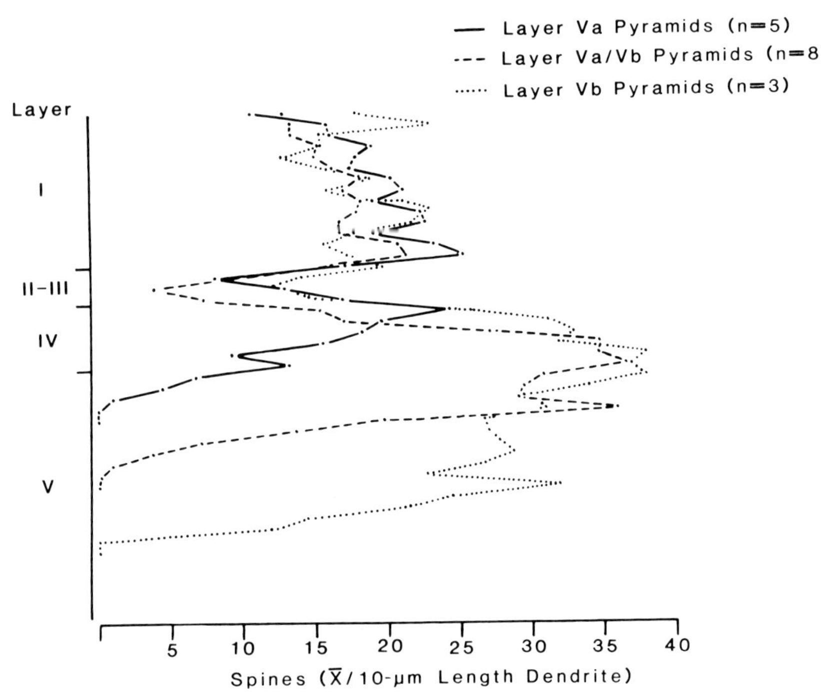

Figure 10. Distribution of spines on the apical dendrites of pyramidal neurons with somata at three different depths of layer V.

higher than the peak densities reached by superficial cells (Fig. 10; 36 spines/10-μm length vs. 24 for deep vs. superficial neurons, respectively). Further, they fail to return to peak layer V densities in layer I.

3.2. Nonpyramidal Neurons

In general, nonpyramidal cells lack apical dendrites, have few to moderate numbers of dendritic spines, and axons which do not enter the white matter but rather terminate in the vicinity of the cell body and dendritic tree. It appears that no nonpyramidal neuron is so frequent that it determines the cytoarchitectural characteristics of any layer in cingulate cortex. For example, layers II–IV in areas 29a–c are dominated by small and fusiform pyramids, while the cytological features of layer V are due mainly to medium and large pyramids. However, nonpyramids are scattered throughout all layers and have been described by a number of investigators (Ramón y Cajal, 1911, 1922; Herrmann and Schulz, 1978; Kusicka and Schulz, 1981; Iwahori and Mizuno, 1981b; Vogt and Peters, 1981).

There are many different varieties of nonpyramidal cells. In addition to the stellate cell, there are a number of spindle-shaped interneurons. These cells are most conveniently categorized into three basic types on the basis of dendritic tree morphology: multipolar, bipolar, bitufted. The limits of these classes are not strict and there are cells which overlap two categories, such as a multipolar cell with a slight vertical orientation which might give it the appearance of a bitufted cell. There may also be functional overlap, since peptide immunohistochemistry often labels more than one class of nonpyramidal neuron. However, the usefulness of dendritic criteria is enhanced by the association of specific spine densities and axonal arbors for most of these classes. For example, both dendritic and axonal trees of bipolar cells are pencil-shaped. Finally, there are some axonal specializations which require further subcategories and are associated with more than one type of dendritic tree. The chandelier axon, for example, can be formed by both bitufted and multipolar neurons.

3.2.1. Multipolar Neurons

These cells have dendritic trees which radiate from a spherical or polygonal-shaped soma. Since spine density and axonal tree distribution are quite variable, it is helpful to further subdivide this class into small, medium, and large multipolar cells. Small multipolars have small somata (8–12 μm in diameter) and dendrites that are limited in spread to one or two layers. Of all nonpyramidal cells, including bipolar and bitufted varieties, the small multipolar cell has the highest density of dendritic spines, and the axon is restricted to a single layer in which it forms dense arcades usually within the sphere of the dendritic tree. In areas 29b–c these cells are located mainly in layer I, while in area 29d they are found throughout layers I–III.

Medium multipolar neurons have intermediate-sized somata (10–15 μm in diameter) and generally have fewer primary dendrites than small or large multipolars (3–5 vs. 4–8, respectively). These dendrites have a very wide distribution, traverse three or four layers, and are lightly spinous. The axons of these cells usually ascend, emitting branches in the layer of the soma as well as in layers I and IV.

Large multipolar neurons have somata that can be as large (12–16 μm in diameter) as medium to large pyramidal cell somata (Fig. 7, a). The dendrites are virtually spine-free and extend through one-half to three-fourths of the cortical thickness. The axons of large multipolar cells in layers IV–VI contribute to a major interlaminar connection. In addition to projecting collaterals within the layer of axonal origin, many medium and large multipolar and bitufted cells have axons that terminate in layers I and IV (Ramón y Cajal, 1911, 1922; Iwahori and Mizuno, 1981b; Vogt and Peters, 1981). Although some superficial multipolar and bitufted cells in layers I–IV have a few axon collaterals which project into deeper layers, it appears that in areas 29b–c the interlaminar projections of these interneurons are not reciprocal. There are, however, many collaterals of superficial pyramidal neurons which are distributed throughout deep layers. Thus, local circuit neurons project extensively from deep to superficial layers, while superficial pyramidal cells are responsible for projections from superficial to deep layers.

3.2.2. Bipolar Neurons

Bipolar neurons have two primary dendrites projecting from both ends of the fusiform somata (8–12 μm wide). The dendrites of these cells are restricted in the horizontal plane, so that the tree forms a cylindrical or pencil-shaped domain. Their primary axis is almost always oriented perpendicular to the pial surface and the dendrites usually span three to five cortical layers. The axon arises from the soma or one of the primary dendrites and either ascends or descends, emitting many collaterals along its course. The axonal arbor is restricted in its horizontal spread, as are the dendrites, such that both conform to a cylindrical and overlapping domain.

3.2.3. Bitufted Neurons

Bitufted neurons have fusiform somata (9–13 μm wide) with two or more primary dendrites originating from the ends of the soma. The critical feature of these cells is a wide horizontal dispersion of the light to moderately spinous dendrites so that they are contained within an hourglass-shaped domain. As already noted, the axons of many deep-layer bitufted cells contribute to a major interlaminar connection which terminates in layers I and IV.

Some bitufted and, occasionally, medium multipolar cells in layers II–III of areas 29d and 24 have axons of the chandelier variety. These axons form a horizontally oriented plexus in the outer layers and have many short, vertical branches. They contain glutamic acid decarboxylase and so are probably GABAergic and selectively form symmetric synapses with the initial segments of other axons, in particular those of pyramidal neurons (Somogyi, 1977; Peters *et al.*, 1982). Chandelier-type axons have not yet been encountered in granular retrosplenial areas, although symmetric synapses are formed with the initial segments of small and fusiform pyramids in layers II–IV (unpublished observations). There are, however, multipolar neurons in layer IV which have axonal arbors intrinsic to this layer and which have short, vertically oriented branchlets (Iwahori and Mizuno, 1981b). These cells may be similar to multipolar neurons with chandelier axons.

3.2.4. Histochemistry of Nonpyramidal Neurons

There is evidence for the presence of GABAergic interneurons in cingulate cortex. Thus, a high density of GABA receptors is present in area 29 (Palacios *et al.,* 1981) and some neurons accumulate [^{3}H]-GABA (unpublished observations). In addition, there are large numbers of cells which stain for newly synthesized GABA-transaminase, the enzyme which metabolizes GABA and is thought to be concentrated presynaptically (Nagai *et al.,* 1983). Since smooth multipolar neurons are known to accumulate [^{3}H]-GABA in visual cortex (Somogyi *et al.,* 1981, 1984) and form symmetric synapses with pyramidal neurons (Peters and Proskauer, 1980), it is quite likely that these cells form inhibitory synapses in cingulate cortex.

A number of peptides have been localized to neurons in cingulate cortex including vasoactive intestinal polypeptide (Lorén *et al.,* 1979), cholecystokinin (Innis *et al.,* 1979), somatostatin (Bennett-Clarke *et al.,* 1980; Roberts *et al.,* 1984), enkephalin (Sar *et al.,* 1978; Khachaturian *et al.,* 1983), and neurotensin (Hara *et al.,* 1982; Roberts *et al.,* 1984). In general, these peptides are found primarily in bipolar cells in layers II–IV of area 29 with fewer in layer V. This localization is not pure since a few small and medium multipolar cells also contain peptides. In addition, neurotensin is the only peptide localized mainly to pyramidal neurons of cingulate cortex (Hara *et al.,* 1982). Neurotensin-positive cells appear to be located in layer VI and, since there are neurotensin-positive axons in the anteroventral thalamic nucleus, it is possible that layer VI corticothalamic projection neurons contain neurotensin.

Of all the peptidergic systems, that of cholecystokinin (CCK) has been most completely characterized in cingulate cortex. High levels of CCK have been detected in cingulate cortex (Beinfeld *et al.,* 1981). Immunohistochemical procedures have been used to identify CCK-positive neurons which are mainly in the outer layers II–IV and have been characterized as bipolar and bitufted in form (Innis *et al.,* 1979; Peters *et al.,* 1983). It may be presumed that these superficial cells have axons which are associated with CCK receptors, of which there are also high levels in area 29 (Zarbin *et al.,* 1983). Thus, outer layer I and layers II–IV have a moderate density of CCK receptors while layer V has very few. The highest binding levels, however, occur in layer VI. Although it is possible that a group of bipolar/bitufted cells has axons which selectively descend to layer VI, these cells have not been observed in Golgi preparations. Alternatively, it is possible that these receptors are associated with axons of extrinsic origin. Peters *et al.* (1983) have noted white matter and layer VI CCK-positive axons in visual cortex. Since the lateral hypothalamus contains CCK-positive neurons (Innis *et al.,* 1979), it is possible that these cells have axons which project to cingulate cortex and activate CCK receptors in the deep layers.

4. Thalamic Connections

4.1. Subdivisions of the Anterior Thalamus

The anterior thalamus is subdivided cytoarchitecturally into dorsal (AD), ventral (AV), and medial (AM) divisions. Since each of these subdivisions has differential projections within cingulate cortex, as well as intrinsic morphological and pharmacological variations, each of them can be expected to contribute differentially to the function of cingulate cortex. Therefore, before detailing the cortical projections of these nuclei, some of the intrinsic differences between each will be enumerated.

There are a greater number of small, local circuit interneurons in AV/AM than in AD. The small neurons in each of these nuclei have primarily intrinsic axonal arborizations, while the larger cells are thalamocortical projection neurons (Somogyi *et al.,* 1973; Somogyi *et al.,* 1979; Keefer *et al.,* 1980). Somogyi *et al.* (1973) have plotted the long diameter of ATN neurons against the logarithm of nuclear volume and found that AD has a 4 : 1 ratio of large to small cells. In contrast, this ratio is 2 : 1 for AV/AM. Small, local circuit neurons in the thalamus are frequently GABAergic. Penny *et al.* (1983) have shown that small neurons in cat ventral posterior nucleus are immunoreactive for the enzyme glutamic acid decarboxylase (GAD) which catalyzes the synthesis of GABA. In addition, larger neurons, which have more extensive dendritic trees, project to cerebral cortex, but are not GAD-positive. Differences in GAD concentrations and GABA receptor binding in the ATN suggest that there are differences in GABAergic control over these nuclei. Higher concentrations of GAD occur in AV (234 ± 25 pmoles/μg protein per hr ± S.E., Tappaz *et al.,* 1976) than in AD (170 ± 18 μmoles/g protein per hr, Massari *et al.,* 1976). There are also more high-affinity GABA receptors in AV/AM than in AD (Palacios *et al.,* 1981). Apparently, AD is under less GABAergic control than are AV and AM.

Aspects of the cholinergic innervation of the ATN also vary among these subdivisions. In diisopropylfluorphosphate (DFP)-treated animals, the highest concentrations of newly synthesized acetylcholinesterase (AChE) in the thalamus are in somata of AD neurons, while AV has the least activity and AM is intermediate between these extremes with sparse, lightly labeled neurons (Parent and Butcher, 1976). As will be discussed in Section 8, a major AChE-positive input can be demonstrated to posterior cingulate cortex from the ATN and this input probably originates mainly from AD. In addition, the highest overall binding of the muscarinic acetylcholine receptor antagonist propylbenzilylcholine mustard is in AV, while AD and AM have lower levels of binding (Rotter *et al.,* 1979). The AV nucleus also has higher concentrations of acetylcholine and choline acetyltransferase (ChAT) activity than either AD or AM (Hoover and Baisden, 1978). Finally, a major source of AChE-positive input to the anterior thalamus originates in the laterodorsal tegmental nucleus (LDT) (Hoover and Baisden, 1980). Ablations of the LDT greatly reduce AChE activity in AV but not in AD or AM. Also, this afferent is probably cholinergic, because neurons in the LDT are ChAT-positive (Satoh *et al.,* 1983). Thus, neurons in AD synthesize greater amounts of AChE than do AV or AM, but AV has the greatest amount of muscarinic acetylcholine receptor binding and the greatest proportion of extrinsic input from the LDT.

There are a number of other anterior thalamic afferents in addition to those from the LDT which have differential projections among the various anterior subdivisions. Of the mammillary nuclei, the medial division projects to AV/AM and the lateral division to AD (Cowan and Powell, 1954; Cruce, 1975). Noradrenergic input from the locus coeruleus terminates mainly in AV, less in AD, and least in AM (Swanson and Hartman, 1975).

Finally, Keefer *et al.* (1980) backfilled ATN neurons with HRP and observed that thalamocortical projection neurons in AD have longer somata (16 × 26.5 μm in diameter) and relatively smooth dendrites as previously observed in Golgi material (Somogyi *et al.*, 1973). Neurons in AV/AM have similarly sized somata (14.5 × 22 and 16 × 20 μm, respectively), while the dendrites of neurons in AV are heavily spinous, and those in AM are intermediate in dendritic spine density.

4.2. Thalamocortical Connections: Light Microscopy

Early retrograde cell degeneration studies (Clark and Boggon, 1933; Waller, 1934; Rose and Woolsey, 1948) demonstrated that the ATN project differentially to cingulate cortex. Thus, AM projects to anterior cingulate cortex, AV projects to dorsal posterior cortex (areas 29c and 29d), and AD projects to ventral posterior cortex (areas 29a and 29b). Recent studies have extended these observations in three main ways. First, projections to the posterior region originate from a larger number of thalamic nuclei, including the intralaminar, AM, mediodorsal, reuniens, and lateral nuclei (Domesick, 1972; Jones and Leavitt, 1974; Herkenham, 1978; Vogt *et al.*, 1979). Second, the extent of overlap of these afferents is substantial (Robertson and Kaitz, 1981). Third, the laminar termination sites of anterior thalamic afferents has been determined (Domesick, 1972; Berger *et al.*, 1980; Vogt *et al.*, 1981).

4.2.1. Rat, Rabbit, Cat

Large injections of HRP into areas 29c–d (Jones and Leavitt, 1974; Fig. 11) result in extensive retrograde transport to neurons in the following nuclei: AD, AV, AM, laterodorsal, and lateroposterior. A substantial number of HRP-positive neurons also occur in the centrolateral and superior centrolateral nuclei and the ventroanterior nucleus, with only a few in the mediodorsal nucleus. These injections, however, include two areas with very different cytoarchitectural characteristics. Degeneration studies in the rat (Domesick, 1972) and transport studies in the cat (Robertson and Kaitz, 1981) indicate that although dorsal posterior cingulate cortex receives input from each subdivision of the anterior nuclei with most from AV, it is the laterodorsal and lateroposterior nuclei which are the primary source of input to area 29d. Furthermore, although it is difficult to limit HRP injections to area 29d in the rat, the rabbit area 29d is quite large and cortical injections of HRP can be more easily restricted to this area. In this

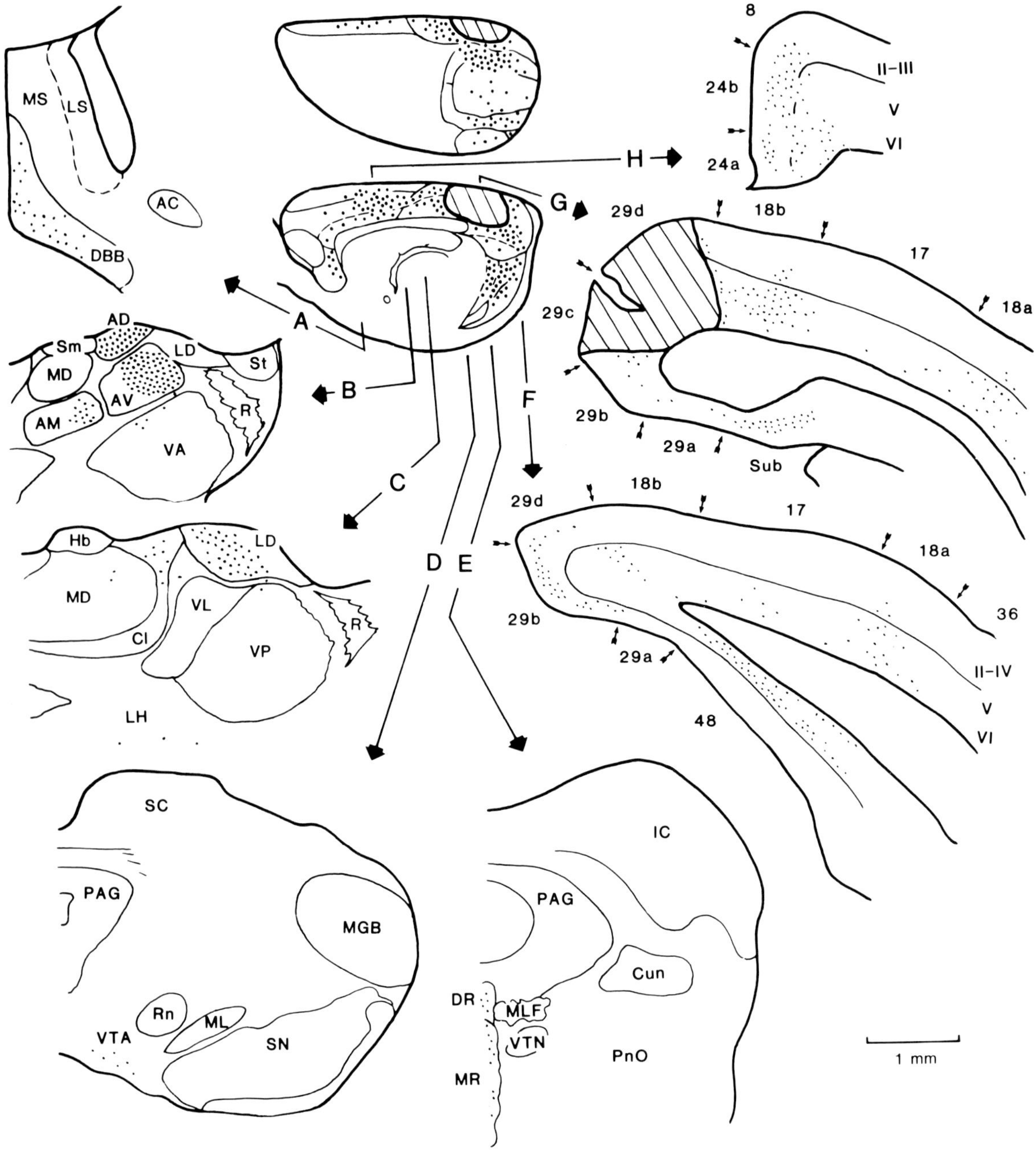

Figure 11. Distribution of HRP-positive somata following an injection into areas 29c–d of the rat. AC, anterior commissure; Cun, cuneiform nucleus; DBB, diagonal band of Broca; DR, dorsal raphé; Hb, habenula; IC, inferior colliculus; LH, lateral hypothalamus; LS, lateral septum; ML, medial lemniscus; MLF, medial longitudinal fasciculus; MR, median raphé; MS, medial septum; PAG, periaqueductal gray; PnO, oral part of the pontine reticular formation; Rn, red nucleus; SC, superior colliculus; Sm, stria medullaris; SN, substantia nigra; St, stria terminalis; Sub, subiculum; VTA, ventral tegmental area; VTN, ventral tegmental nucleus. Thalamic nuclei: AD, anterodorsal; AM, anteromedial; AV, anteroventral; Cl, centrolateral; LD, laterodorsal; MD, mediodorsal; MGB, medial geniculate body; R, reticular; VA, ventroanterior; VL, ventrolateral; VP, ventroposterior.

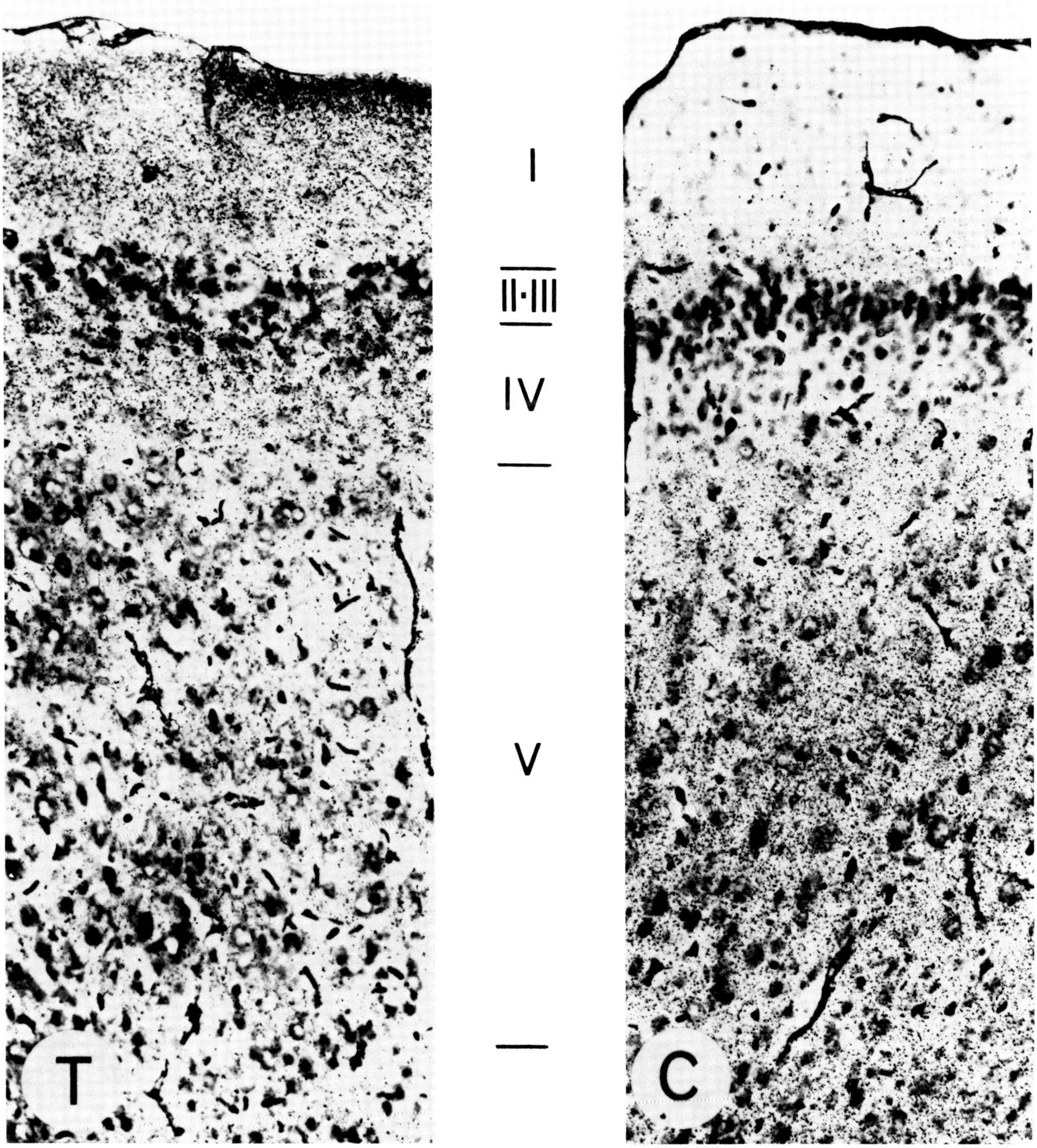

Figure 12. Terminal degeneration in area 29c of the rat following anterior thalamic (T) and callosal (C) ablations; Fink–Heimer stain, ×170.

instance there is also retrograde transport to neurons in both AD and AV, but most HRP-positive cells are in the laterodorsal and lateroposterior nuclei (unpublished observations). The differential distribution of anterior and lateral thalamic connections within posterior cortex is also supported by observations of corticothalamic connections (Section 3.4).

Anterior thalamic input to the granular retrosplenial areas 29a–c terminates mainly in layer Ia and less in layers Ib–c and IV (Domesick, 1972; Vogt *et al.*, 1981; Figs. 12 and 13). The preferential termination in the outer one-third of layer I is unique to ATN afferents to this granular cortex in rodents. Although post- and presubicular cortices receive AV input, it spreads evenly throughout layer I (unpublished observations). Also, granular areas in the rabbit (Berger *et al.*, 1980) and cat (Robertson and Kaitz, 1981) do not display this particular specialization.

In conclusion, AD and AV have very similar cortical targets in areas 29a–c. However, since there are substantial differences in the GABAergic, cholinergic, and other extrinsic afferent innervation of these two nuclei, it is likely that they both contribute differentially to cellular activity in the granular retrosplenial areas. Also, although rodent ATN afferents terminate mainly in layer Ia, which contains the dendritic tufts of small and fusiform pyramids, this sublaminar specialization does not occur in higher species. Finally, although there are minor projections of AD and AV to area 29d, the primary origins of thalamic input to area 29d are the laterodorsal and lateroposterior nuclei.

4.2.2. Monkey

Studies of thalamocortical connections in primates (Vogt *et al.*, 1979; Baleydier and Mauguiere, 1980) have shown that posterior cingulate cortex receives input from each subdivision of the ATN, some of the intralaminar nuclei, the laterodorsal, lateroposterior, and pulvinar nuclei. Except for the pulvinar projection, posterior cingulate cortex in the monkey appears to be very similar to other species in terms of thalamic afferents. This conclusion requires further consideration in light of the extensive cytoarchitectural elaboration of proiso- and isocortical structures in the monkey cingulate gyrus.

A retrogarde cell degeneration study by Locke *et al.* (1964) suggests that ablations of parts of posterior cingulate cortex in monkeys can produce degeneration in the lateral nucleus, without a similar loss in the anterior nuclei. Injections of HRP that are restricted to area 23 (Fig. 14) result in a pattern of labeling in the thalamus which is different from that reported for larger injections also involving areas 29 and 30. In the former case, HRP-positive neurons are mainly in the medial pulvinar nucleus with less in the mediodorsal nucleus and still fewer in the AM nucleus, less in the lateroposterior nucleus, and few in the centrodensocellular nucleus of Olszewski (1952). It is interesting that in contrast to an area 29 injection, area 23 injections do not produce retrograde labeling in the AD and AV nuclei, nuclei which are generally considered the primary source of thalamic input to posterior cingulate cortex.

Another distinction between the monkey and lower species is the termination site of the ATN input to area 29. An injection of ^{3}H-labeled amino acids into the AV and ventroanterior nuclei (Fig. 13, C and D) produces anterograde

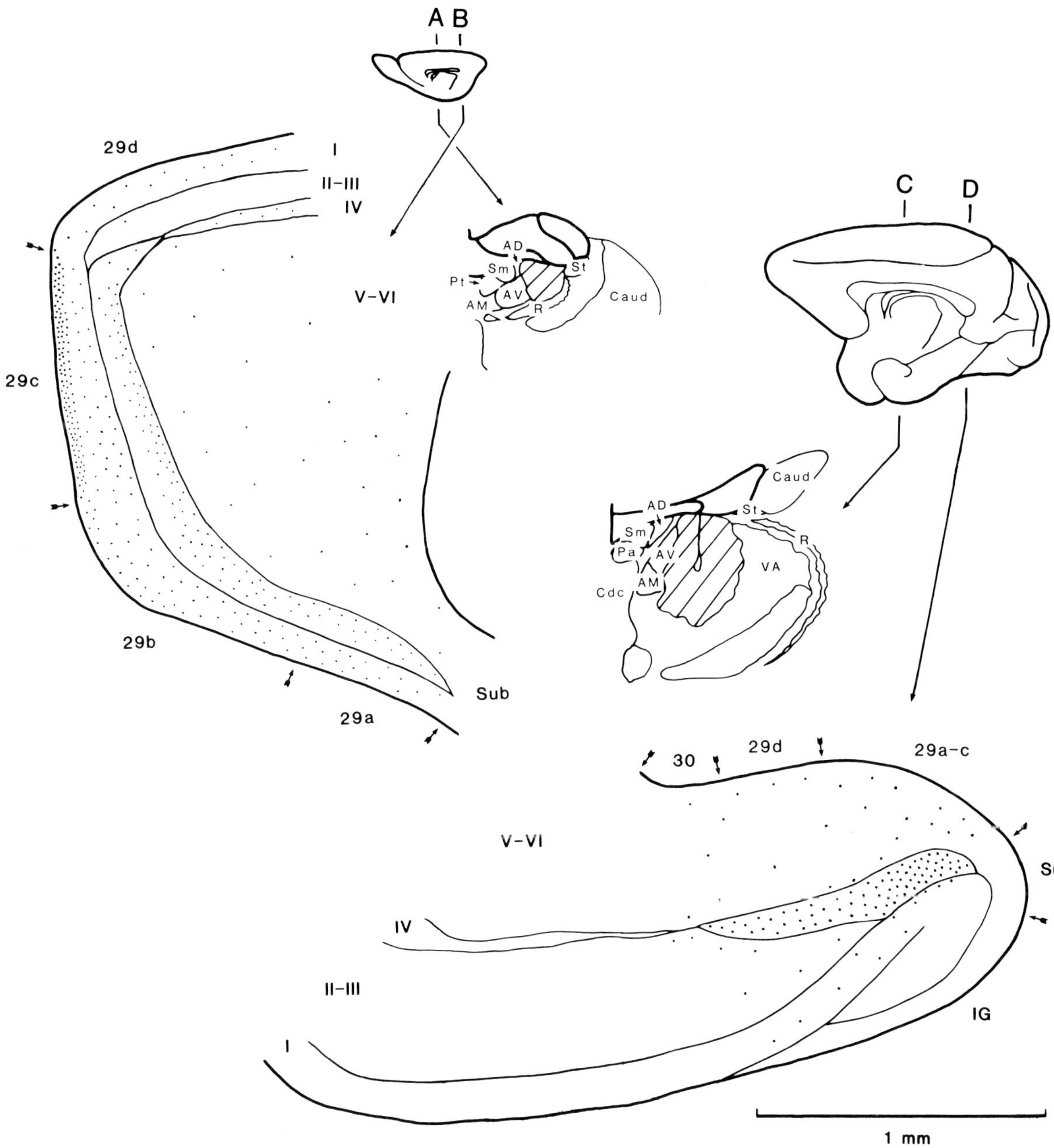

Figure 13. Distribution of thalamic afferents in suprasplenial cortex. (A) ^{3}H-labeled amino acid injection site (hatched) in rat. (B) Location of grains over area 29. (C) ^{3}H-labeled amino acid injection site (hatched) in rhesus monkey thalamus. (D) Distribution of grains over area 29. In addition to abbreviations in Fig. 11: Caud, caudate; IG, indusium griseum. Thalamic nuclei: Cdc, centrodensocellular; Pa, paraventricular; Pt, paratenial.

labeling mainly over the granular layer in area 29, for most grains over deeper layers probably result from ^{3}H-labeled amino acids in axons of passage. Very little labeling is present in layer I and none over area 30 or 23.

The following conclusions can be drawn regarding thalamocortical connections in the rat, rabbit/cat, and monkey. First, progressive cytoarchitectural elaboration of cingulate cortex is directly associated with changes in thalamic input. In lower species, in which area 29 dominates posterior cingulate cortex, thalamic projections mainly arise from the anterior nuclei. In higher species there is a

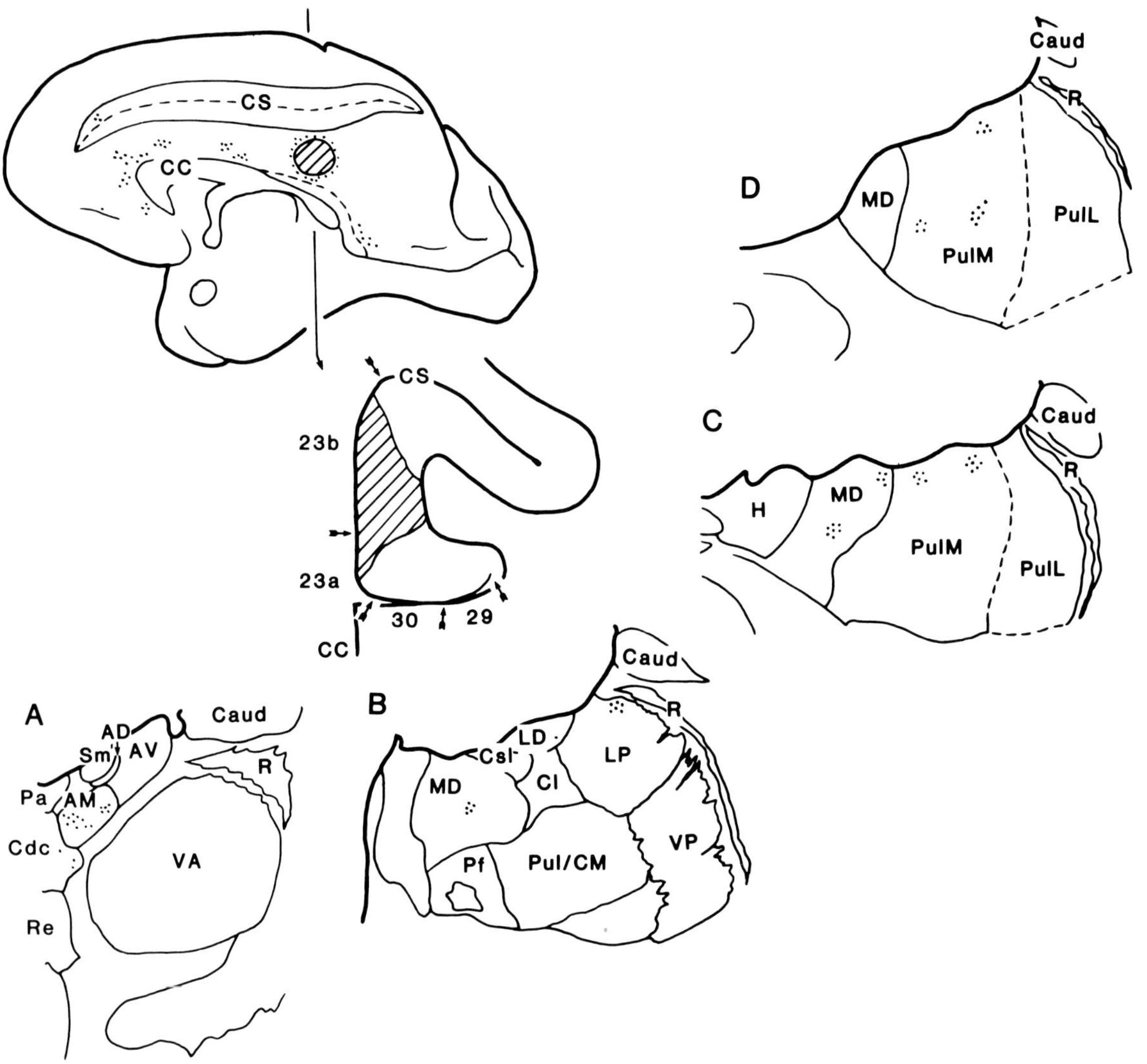

Figure 14. Location of HRP-positive somata following an injection into area 23 (hatched) of the rhesus monkey. In addition to abbreviations in Figs. 11 and 13: CC, corpus callosum; CS, cingulate sulcus; H, habenular. Thalamic nuclei: Csl, superior centrolateral; Pf, parafascicular; Re, reuniens; Pul/CM, pulvinar/centromedian; PulL, lateral pulvinar; PulM, medial pulvinar.

greater contribution from the lateral thalamic nuclei as area 29d is more expanded. In the monkey the pulvinar nucleus projects to much of the posterior cingulate cortex and this is due to the presence of area 23, which is unique to primates. Second, there is a progressive shift in the laminar preference of ATN input in area 29. In the rat most of these terminals are located in layer Ia with fewer in layer IV. In the rabbit and cat there is an even distribution between layers I and IV, while in the monkey most of these terminals end in layer II–IV, with only a minimal contribution to layer I.

4.3. Thalamocortical Connections: Electron Microscopy

Ultrastructural analysis of degenerating thalamic afferents to rat area 29c confirms light microscopic observations that most of them terminate in layer Ia and fewer in deeper layers I and IV. Two particularly interesting features of limbic corticothalamic connections emerge from this analysis (Vogt *et al.*, 1981). First, virtually no afferents form synapses in layer VI and this is in contrast to neocortex, in which a major thalamic termination field occurs in this layer (Ribak and Peters, 1975). Second, the axon terminals that form asymmetric synapses with spines and smooth dendritic processes in layer Ia are often three times larger than those synapsing in layers Ib–c and IV. A comparison of these two types of degenerating axon terminals and a large normal terminal, possibly of thalamic origin, are presented in Fig. 15. It is possible that, although these terminals form synapses on the distal portions of the dendritic trees of deeper-lying neurons, the efficiency of this input in effecting neuronal discharge is particularly enhanced by the larger size and greater number of these terminals in layer Ia. Although they have not yet been analyzed ultrastructurally in primates, it is unlikely that this type of compensation is necessary in higher species, since these terminals form synapses closer to the cell bodies.

4.4. Corticothalamic Connections

It is well established that many layer VI and a few layer V pyramids of cingulate cortex and deep-lying pyramidal neurons of the subiculum project to the ATN (Sikes *et al.*, 1977; Meibach and Siegel, 1977; Berger *et al.*, 1980; Kaitz and Robertson, 1981; Royce, 1983; Donovan and Wyss, 1983). A number of reports have documented which cytoarchitectural areas project to the various ATN subdivisions. The subicular complex projects to AM and AV (Meibach and Siegel, 1977). Areas 29a and 29b project to AV in the rat (Meibach and Siegel, 1977) and cat (Kaitz and Robertson, 1981), but in the cat there are additional projections to AM and the laterodorsal nucleus. Area 29c projects mainly to dorsal AV and lightly to AD (Domesick, 1969), while area 29d projects mainly to LP (lateroposterior) and to a lesser extent AM and AV.

Analyses of the projections of cingulate cortex to AD have suggested that there is a paucity of termination in AD when compared to that in AV (Domesick, 1969; Dekker and Kuypers, 1976b; Kaitz and Robertson, 1981). It is unclear whether this is because AD actually receives less cortical input or whether the cortical lesions/injections have been too small to detect the full extent of this AD afferent. A large injection of ^{3}H-labeled amino acids into posterior cingulate cortex (Fig. 16) that involves much of area 29d and most of areas 29b and 29c confirms that there is a relatively small cortical input to AD. In this case there is extensive labeling over AV, but very little over AD. Thus, AD stands out as a unique thalamic nucleus for many reasons. In contrast to AV and other thalamic nuclei, it has few interneurons and the largest number of AChE-containing cells. Furthermore, although large numbers of AD neurons project to cingulate cortex, this is a "minimally reciprocal" connection, for AD receives very little cortical input.

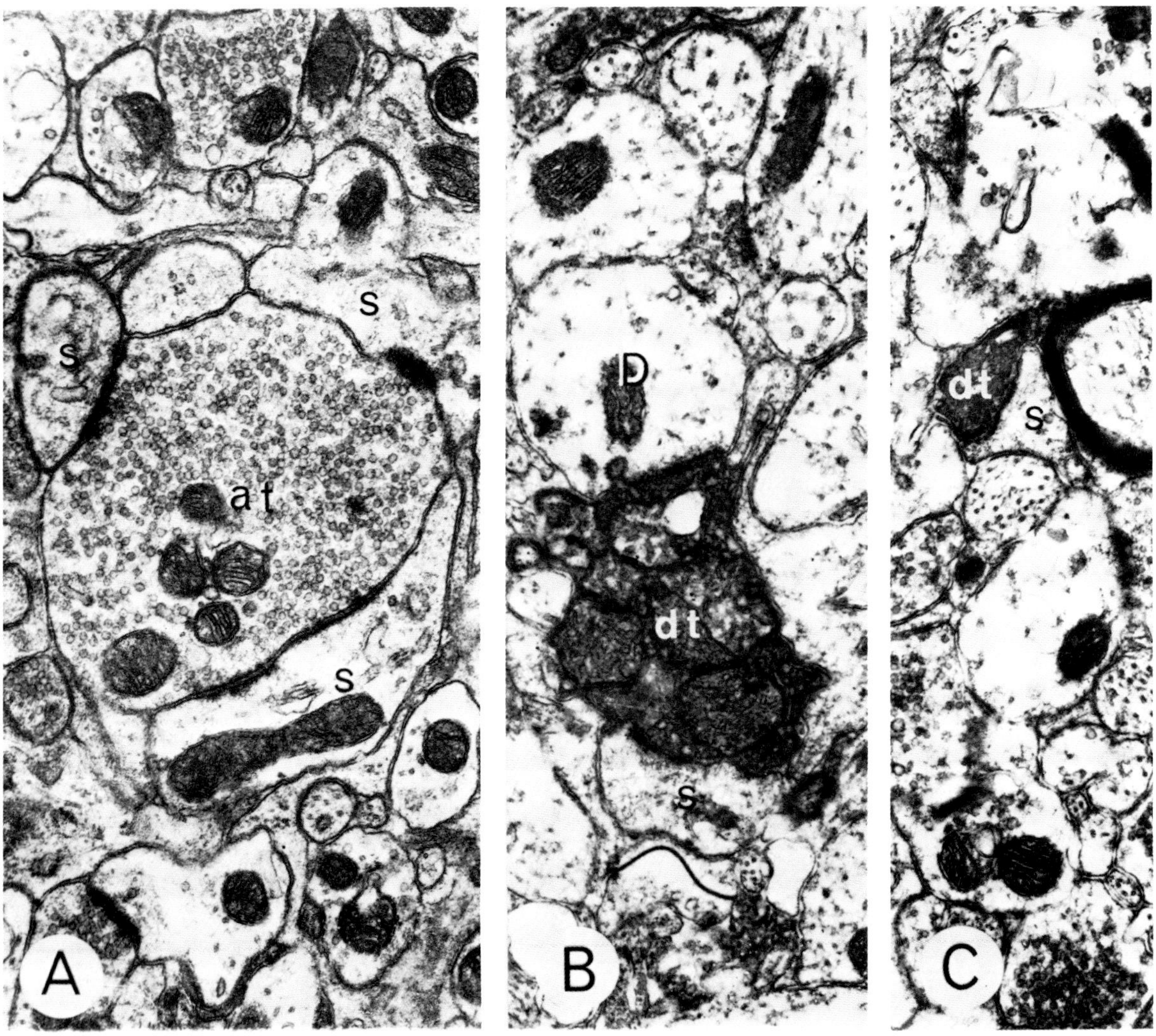

Figure 15. Normal (at) and degenerating (dt) axon terminals in rat area 29c. (A) Large normal axon terminal in layer Ia forming synapses with three spines (s). (B) Large degenerating terminal in layer Ia following an ATN lesion which forms asymmetric synapses with a spine and smooth dendrite (D). (C) Small degenerating terminal in layer IV following an ATN lesion; ×28,000.

The anteromedial nucleus receives topographically dispersed projections from anterior and posterior cingulate cortex and the subicular complex (Domesick, 1969; Meibach and Siegel, 1977; Kaitz and Robertson, 1981). The diffuse nature of its afferents is in accord with its similarly diffuse projections to cingulate cortex in rat, rabbit, cat, and monkey (Rose and Woolsey, 1948; Beckstead, 1976; Benjamin *et al.*, 1978; Baleydier and Mauguiere, 1980; Robertson and

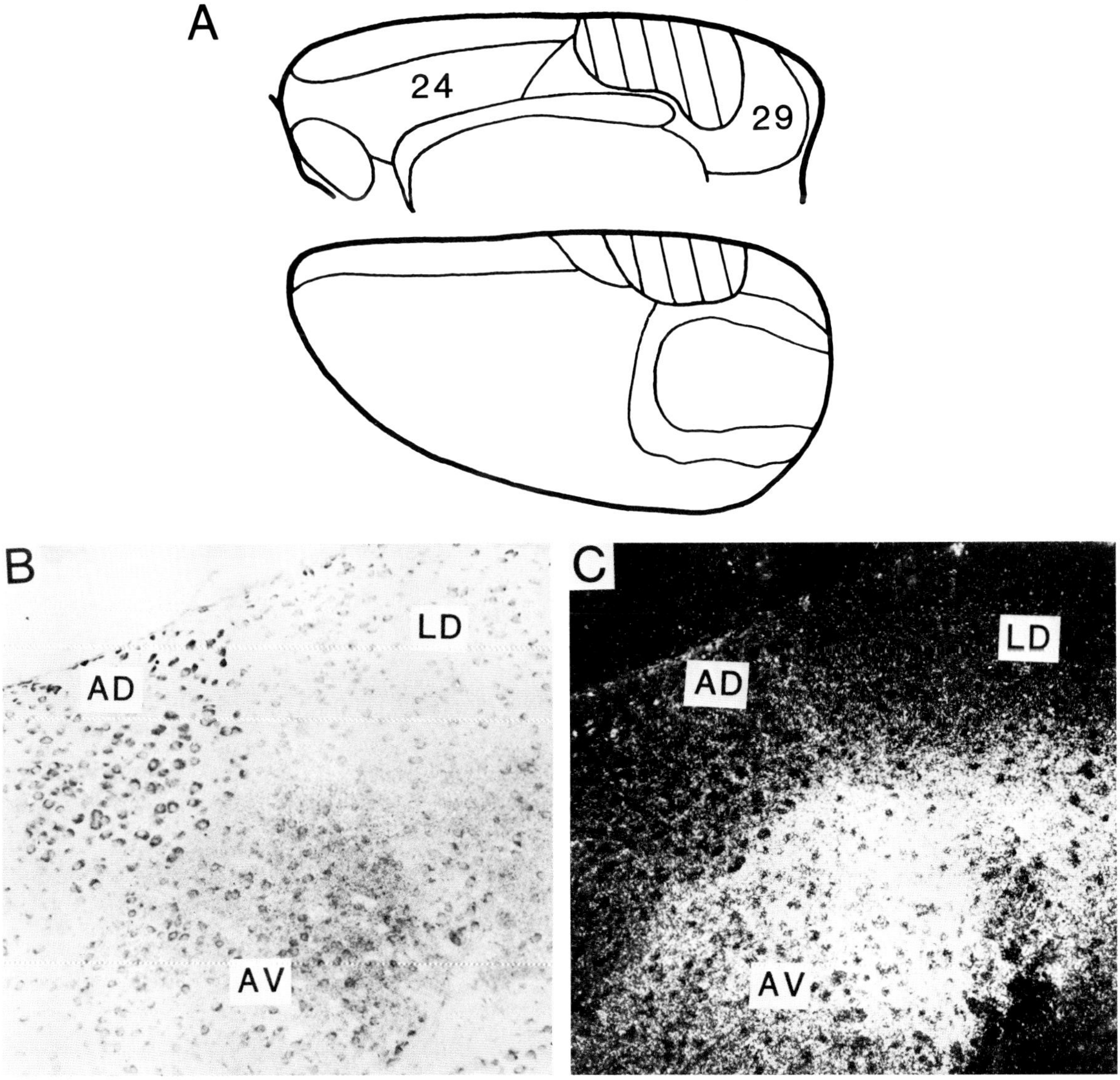

Figure 16. Corticothalamic connections in the rat. (A) Large ^{3}H-labeled amino acid injection site (hatched) in area 29. (B) Cytoarchitecture of anterior thalamic nuclei. Abbreviations as in Fig. 11. (C) Distribution of grains over these same nuclei; darkfield photomicrograph, ×80.

Kaitz, 1981). It is the dispersion of reciprocal cortical and thalamic connections which distinguishes AM from AD and AV.

Neurons of the ATN form different types of synapses with each afferent and these synapses are spatially segregated on the dendritic tree. Symmetric and probably inhibitory synapses are formed by intrinsic neurons with the somata and initial axon segments of AV neurons. Large, asymmetric synapses are formed with the proximal dendrites of AD and AV neurons by axon terminals of mammillary body origin (Somogyi *et al.*, 1977; Dekker and Kuypers, 1976a,b). Small, asymmetric synapses are formed with more distal secondary and tertiary dendritic branches by axons of layer VI pyramidal cells in cingulate cortex. Therefore, it is in the anterior nuclei that a major interaction occurs between the hippocampus via the mammillary bodies and cingulate cortex. As noted previously, the lateral division of the mammillary bodies projects to AD where there are few interneurons and from which there are extensive projections to areas 29a–c. It can be predicted that activity in the lateral mammillary nuclei is transmitted directly to cingulate cortex virtually unmodified by either intrinsic or cortical feedback activity. In contrast, activity of the projections from the medial division of the mammillary bodies must undergo extensive transformation as it passes via the AV/AM nuclei to cingulate cortex, since there are many more inhibitory connections in AV/AM and cingulate cortical feedback projections are extensive.

5. Afferent Connections

5.1. Corticocortical

One major source of cortical afferents to posterior cingulate cortex is area 24. In the rat and rabbit, neurons in layers III and V of area 24 project mainly to layer I of areas 29a, b, and d (Beckstead, 1979; Bassett and Berger, 1982; Vogt and Miller, 1983; Fig. 11). In contrast, area 24 in the monkey projects mainly to area 23, and there appears to be no projection to area 29 (Pandya *et al.*, 1981). There has been a question about whether the most rostral areas 32 and 25 are cingulate or frontal cortices. One criterion for considering these rostral cortical areas as a part of cingulate cortex is their corticocortical connections with posterior cingulate cortex. Since area 32 has no such connections in either the monkey (Vogt *et al.*, 1979; Pandya *et al.*, 1981) or the rat (Vogt and Miller, 1983), it may be more appropriate to treat it as a subdivision of frontal rather than cingulate cortex. Area 25, on the other hand, does have limited connections with posterior cingulate cortex and probably should be considered a part of cingulate cortex.

A second major source of cortical afferents is the subicular and postsubicular cortices of the hippocampal system. In all species analyzed including the rat (Meibach and Siegel, 1977; Vogt and Miller, 1983; Finch *et al.*, 1984b), guinea pig (Sørensen, 1980), and monkey (Rosene and Van Hoesen, 1977), the subiculum projects to retrosplenial granular cortex. The subicular projection in the rat and guinea pig is particularly interesting for a number of reasons. First, axons of subicular pyramids course directly through the cortex including layer I in addition to passing through the white matter. Second, they terminate in areas 29a–c, but not area 29d. Third, these axons terminate directly in the granular layer II–III as well as inner layers I and IV. No other extrinsic afferent is known to terminate in layer II–III; in fact, other inputs to areas 29b and 29c tend to selectively straddle layer II–III, e.g., callosal, anterior thalamic, and raphé afferents. Fourth, it is one of the few cortical connections which is not reciprocal. Although cingulate cortex projects to the postsubiculum, it does not appear to terminate in the subiculum itself.

Parahippocampal projections in the rat are reciprocal and widespread. Injections of HRP into cingulate or visual cortices retrogradely label neurons in the internal pyramidal cell layer of the postsubiculum (Vogt and Miller, 1983; Fig. 11). In the monkey, connections arise from neurons in the proisocortical areas TH and TF (Vogt *et al.*, 1979; Baleydier and Mauguiere, 1980) and terminate mainly in area 23.

A third major source of input to the posterior cingulate region is from lateral neocortex. In the rat and rabbit this input is limited mainly to visual cortex. Figure 11 indicates that neurons of area 18b, ones mainly in layer V, project to area 29 with less coming from area 18a and least from area 17. An analysis of both autoradiographic and HRP connection cases has led to the conclusion that areas 29d and 18b are the cortical subdivisions which are primarily responsible for reciprocal visual/cingulate connections (Vogt and Miller, 1983). In contrast, although there is also a projection from medial visual cortex (area 19) to area 23 in the monkey (Vogt and Pandya, 1986), primates have extensive neocortical inputs from association cortices which may provide numerous alternative routes by which visual and other stimuli might activate cingulate neurons. These afferents originate in both dorsolateral and orbital frontal cortex, the inferior parietal lobe, and cortex in the depths of the superior temporal sulcus (Pandya *et al.*, 1971; Petras, 1971; Künzle, 1978; Vogt *et al.*, 1979; Baleydier and Mauguiere, 1980; Goldman-Rakic *et al.*, 1984).

5.2. Callosal

The callosal system in rodent retrosplenial granular cortex is organized quite differently than that in any other neocortical region (Jacobson, 1970; Yorke and Caviness, 1975; Vogt *et al.*, 1981). In addition, the close proximity of cingulate cortex to the corpus callosum and the homotopic character of much of this input makes it an ideal system for electrophysiological analysis using 0.5-mm-thick coronal slices (*vide infra*).

The two largest populations of neurons that have callosal projecting axons are the small pyramids of layers II–IV and medium to large pyramids of layer V. The main layers of termination in areas 29b and 29c, identified with Fink–Heimer-stained preparations, are layers V and VI although there is a small component to layer Ic (Fig. 12). Area 29a receives an additional input to the granular layer. Finally, the callosal afferent input to area 29c is separated from most other major inputs, as summarized in Fig. 17.

The callosal connections of area 29d differ from those of the granular areas.

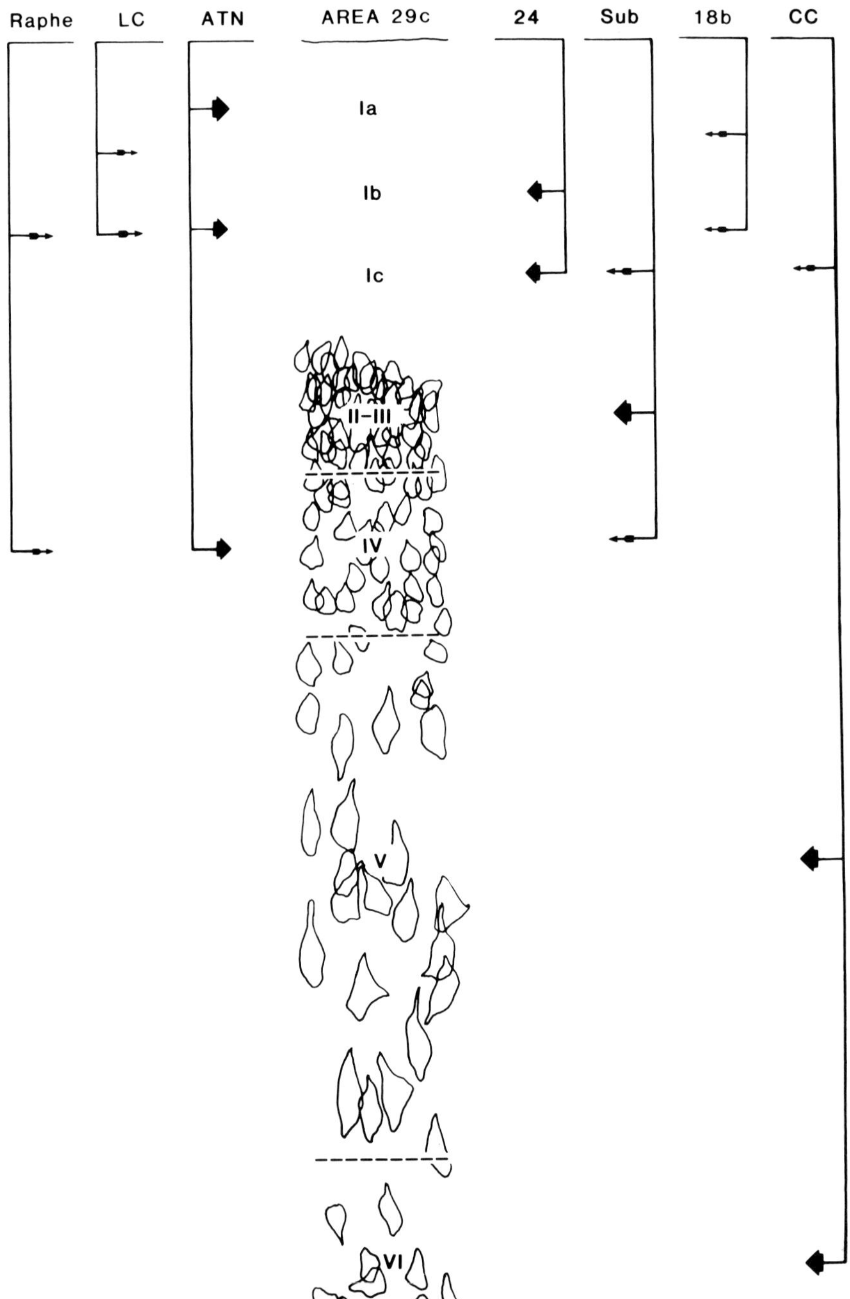

Figure 17. Summary of some of the major afferent connections to area 29c in the rat. Arrow size is an approximate reflection of current estimates of the density of each input. ATN, anterior thalamic nuclei; CC, corpus callosum; LC, locus coeruleus; Sub, subiculum.

Thus, callosal projecting axons originate mainly from neurons in layers III and V and terminate in layers III–VI. In addition, area 29d has a small heterotopic projection which originates from layer V neurons that project to contralateral areas 17 and 18b (Miller and Vogt, 1984b).

Ultrastructural analyses of the callosal afferents (Vogt *et al.*, 1981; Vogt and Gorman, 1982) have confirmed light microscopic observations of the laminar distribution of this input. These axon terminals form asymmetric synapses primarily with dendritic spines, but also occasionally with smooth dendrites and rarely with neuronal somata. Finally, these synapses involve the proximal apical and basal dendrites of pyramidal neurons in area 29c. The functional implications of this organization will be considered in Section 10.

5.3. Claustral and Diagonal Band of Broca

The claustrum contains neurons which respond to both auditory and visual field stimulation (Spector *et al.*, 1974) and may be a major source of sensory input to cingulate cortex. Injections of HRP into area 29 retrogradely label neurons in the claustrum (Vogt and Miller, 1983).

The diagonal band of Broca (DBB) is a major source of cholinergic input to cingulate cortex (Section 8). Bigl *et al.* (1982) have shown that neurons in both the vertical and horizontal limbs of the DBB are retrogradely labeled by fluorescent dye injections into cingulate cortex (see also Fig. 11).

5.4. Locus Coeruleus

The locus coeruleus (LC) appears to be the only major afferent to posterior cingulate cortex which is also shared by the ATN. It is generally thought that, since the LC has diffuse, "nonspecific" connections, its functional role is generalized and that the diffuse projections produce the same effects in all regions. Consideration of the area 29/ATN system suggests that, although LC projections are diffuse, there are primary target sites heavily innervated by it.

The LC projects to anterior and posterior cingulate cortices more densely than it does to lateral neocortical areas (Jones and Moore, 1977). The notion that the LC preferentially innervates medial cortex is supported by the facts that the greatest number of α-adrenergic receptors are in cingulate cortex (U'Prichard *et al.*, 1977) and that there is a denser innervation of layer I of cingulate cortex by noradrenergic fibers than of layer I in lateral neocortex (Swanson and Hartman, 1975; Morrison *et al.*, 1978). Within the thalamus one of the densest termination sites is the anteroventral and anteromedial nuclei in contrast to other parts of the thalamus.

The LC is responsible for maintaining wakefulness (e.g., Cespuglio *et al.*, 1982) and its neuronal discharges are modified by visual, somesthetic, and noxious stimuli (Nakamura, 1977; Watabe, *et al.*, 1982; Aghajanian and Vandermaelen, 1982). All evidence presently available suggests that the noradrenergic afferents to cingulate cortex reduce neuronal activity. Thus, LC stimulation reduces deoxyglucose uptake in cingulate cortex (Abraham *et al.*, 1979) and the discharge rate of spontaneously active cells in cingulate cortex is blocked by α-receptor antagonists (Bunney and Aghajanian, 1976; Olpe *et al.*, 1980). The mechanism of this depression is not yet fully understood but may involve activation of a $Na^{+}K^{+}$ pump and/or intrinsic GABAergic neurons (Taylor and Stone, 1980; Segal, 1981; Moroni *et al.*, 1982). It might be concluded that activation of the LC by a sensory stimulus results in depressing cingulate cortical activity such that afferents relaying specific receptive field or other information about a particular stimulus might then control cingulate cortical neuron discharge. In other words, the LC enhances the signal-to-noise ratio for sensory stimuli (Woodward *et al.*, 1979).

5.5. Raphé, Ventral Tegmental, Hypothalamic, Summary

Both the dorsal and median raphé nuclei project to area 29 (Bobillier *et al.*, 1979; Fig. 11). Lidov *et al.* (1980) have demonstrated that there are large-diameter, serotonergic axons in layer Ic as well as axons dispersed throughout the remainder of layers I and IV. Although it is not understood what role these afferents play in cingulate function, it is known that raphé neurons are driven by visual and auditory stimuli (Trulson and Trulson, 1982) and that the ongoing activity of 50–70% of the neurons in area 29 can be inhibited with electrical stimulation of the dorsal raphé nucleus (Olpe, 1981). In addition, the activity of 92% of the cells in area 29 can be directly inhibited by serotonin iontophoresis (Olpe, 1981).

A number of neurons are labeled throughout the lateral hypothalamus and ventral tegmental area (VTA) following HRP injections into posterior cingulate cortex (Fig. 11). The VTA has a major projection to anterior cingulate cortex (Beckstead *et al.*, 1979; Porrino and Goldman-Rakic, 1982) and is probably dopaminergic (Lindvall *et al.*, 1974; Lewis *et al.*, 1979). It has not been generally appreciated, however, that a small VTA input also terminates in area 29.

Figure 17 summarizes the laminar distribution of some of the afferents to area 29c. The major afferents originate from the ATN, area 24, the subiculum, and contralateral cortex and the primary termination site of each is different. However, layer Ic stands out as being unique, since it receives at least some input from all afferents.

6. Sources of Visual Input to Area 29

There have been a number of reports documenting visual-evoked responses in cingulate cortex of both rabbit and monkey (Cuénod *et al.*, 1965; Vinogradova, 1975). Although the exact pathway by which these flash-evoked responses reach cingulate cortex is not known, it is possible that cingulate cortex is pivotal in mediating visual-evoked responses throughout much of the limbic system. For example, Vastola (1982) has abolished visual-evoked responses in rat hippocampus with lesions in posterior cingulate cortex.

There are now four known pathways by which visual input could reach cingulate cortex. The main route would appear to be directly from visual cortex. HRP injections into areas 17 and 18 of rat visual cortex result in retrograde transport to layer V pyramidal neurons, mainly in area 29d but also in area 29b (Vogt and Miller, 1983). Tritiated amino acid injections confirm that areas 18a and 18b project into cingulate cortex (Vogt and Miller, 1983). The reciprocal pathway is demonstrated with injections of HRP into cingulate cortex (areas 29c–d; Fig. 11), since this shows layer V neurons in areas 18a and 18b to be HRP-positive with a few other HRP-positive neurons in layers V and VI of area 17. There is a similar organizational pattern in the rabbit, with most cingulate and visual cortical connections occurring between area 29d and the second visual area (Montero, 1981; Sikes and Vogt, 1985; Vogt *et al.*, 1986).

Although the exact nature of visual cortical input to cingulate cortex is not known, one might speculate that it is the receptive field properties of layer V neurons in visual cortex that could most dominate cingulate neuronal activity. The receptive field characteristics of layer V neurons in rodents and rabbit that may be relevant include binocularity (Diao *et al.*, 1983); simple, complex, and hypercomplex properties (Chow *et al.*, 1971; Murphy and Berman, 1979; Parnavelas *et al.*, 1983); directional selectivity (Chow *et al.*, 1971); and large field, nonorientation (Mangini and Pearlman, 1980; Parnavelas *et al.*, 1983).

Another significant source of visual afferents may originate in the claustrum. The claustrum has a major projection to posterior cingulate cortex and receives area 18b afferents (Vogt and Miller, 1983; Miller and Vogt, 1984a). It is known from studies of the cat that claustral neurons respond to visual stimuli (Spector *et al.*, 1974), that electrical stimulation of the claustrum produces excitatory and inhibitory responses in visual cortical neurons (Tsumoto and Suda, 1982), and that the claustrum contributes to length selectivity of area 17 neurons (Sherk and LeVay, 1983). Whether or not the claustrum is an excitatory afferent to cingulate cortex is not known, but it may operate separately or in concert with other sources of visual inputs to produce flash and other visual-evoked responses.

The LP nucleus provides a major thalamic afferent to area 29d in the rat and rabbit (Towns *et al.*, 1982). Since this nucleus receives both pretectal (Weber and Harting, 1980; Robertson *et al.*, 1980, 1983) and visual cortical (Giolli *et al.*, 1978; Lent, 1982) afferents, it is likely that the projection of the LP nucleus to area 29d is another source of visual afferents.

Finally, retinal afferents (Conrad and Stumpf, 1975; Itaya *et al.*, 1981) are an indirect source of visual input to posterior cingulate cortex. This retinal efferent terminates in the anterior thalamus, but is minor as judged on morphological grounds (e.g., number of neurons from which the projection originates).

7. Projections to Motor System

Extensive studies have documented the fact that electrical stimulation of anterior cingulate cortex produces autonomic and somatic motor responses, including changes in blood pressure, cardiac and respiratory rates, pupillary responses, piloerection, and vocalization (Kremer, 1947; Ward, 1948; Pool and Ransohoff, 1949; Kaada, 1951; Anand and Dua, 1954; Talairach *et al.*, 1973;

Buchanan and Powell, 1982). In contrast, very few of these responses are produced by stimulation of posterior cingulate cortex. Since area 29 has a number of cortical and subcortical inputs to the motor system in addition to reciprocal connections with anterior cingulate cortex, it might be presumed that the role of posterior cortex is related to modifying motor responses that are acquired during learning processes, rather than direct "prewired" responses.

It has been suggested that area 29d in the rat is a sensorimotor association cortex (Vogt and Miller, 1982, 1983). One direct way by which area 29d might effect motor activity is via its connections with area 8. Area 8 receives visual cortical input (Miller and Vogt, 1984c) and is part of the motor system, since electrical stimulation of this medial agranular field can evoke eye movements (Hall and Lindholm, 1974; Donoghue and Wise, 1982). Another cortical-level motor influence of area 29d could be effected via projections to area 24b, although the contribution of each anterior cingulate subdivision to the various electrically stimulated responses has not been determined.

Three subcortical regions receive cingulate input, and each of them is probably associated with motor activity. First, posterior cingulate cortex projects to the caudate nucleus in the rat (Domesick, 1969), cat (Royce, 1982), and monkey (Yeterian and Van Hoesen, 1978; Powell, 1978). Layer II–IV and V pyramidal neurons are primarily responsible for this connection (Schwab *et al.*, 1977; Royce, 1982). Second, corticopontine input in the rat also originates mainly from layer V neurons, but primarily in cortex which is transitional between areas 24 and 29, although there is also a direct input from area 29 (Wiesendanger and Wiesendanger, 1982a,b). These cells have a substantial contralateral projection which terminates mainly in the periphery of the pontine gray matter in both rat and monkey (Vilensky and Van Hoesen, 1981; Wyss and Sripanidkulchai, 1984). Finally, posterior cingulate cortex has direct projections into the deepest layer of the superior colliculus and the periaqueductal gray (Domesick, 1969; Powell, 1978). Many of the autonomic responses noted following electrical stimulation of area 24 can also be evoked by stimulation of the periaqueductal gray including piloerection, pupillary dilation, and changes in cardiovascular parameters (Skultety, 1963; Flynn *et al.*, 1970). Since these same responses are difficult to produce or are negligible following electrical stimulation of area 29, it might be concluded that these mesencephalic projections are relatively weak as judged by responses obtained in acute preparations.

8. Cytoarchitectural Schemata in Light of Connectivity

Brodmann's (1909) parcellation of posterior cingulate cortex in the rabbit defined five area 29 subdivisions and an area 23. Rose and Woolsey (1948) later consolidated these areas such that their cingular area included areas 29d and 29c of Brodmann and their retrosplenial field contained Brodmann's areas 29a and 29b. This section will evaluate Rose and Woolsey's consolidation plan, as well as the modified Brodmann scheme of Vogt and Peters (1981), in light of recent cytoarchitectural and connectional studies.

Areas 29c and 29d are fundamentally different types of cortex and should not be treated as a single field for the following reasons. First, these areas represent different stages of cortical differentiation with area 29c being a periallocortex and area 29d a proisocortex. Layer II–III in area 29d is more sparse and the cells larger than in area 29c. The cytological basis for these changes in area 29d includes a large reduction in the number of fusiform and small pyramids and an increase in the number of medium-sized pyramidal neurons.

Second, although Rose and Woolsey state that there are architectural differences between these areas, they do not recognize differences in thalamocortical connections to support such a distinction. As noted in Section 3, although there is much overlap in connections between cortex and thalamus, the AD/AV divisions of the anterior nuclei project primarily to area 29c, while LP of the lateral nuclei projects mainly to area 29d.

Third, the laminar termination of certain afferents is quite different in the two subdivisions. For example, in the rat ATN input ends mainly in layer Ia of area 29c, while anterior and lateral thalamic termination in area 29d is about evenly distributed in layers I and IV. In addition, acetylcholine receptors are associated with ATN axons and produce a heterogeneous pattern of binding in area 29c which is not present in area 29d (Section 8). The callosal afferents in the rat also terminate in different layers with most afferents terminating in layers V and VI of area 29c, while most of the input to area 29d ends in the superficial layers.

Fourth, these two areas receive different cortical afferents. One of the major cortical inputs to area 29c originates in the subiculum, yet area 29d does not receive subicular afferents. Also, although there are direct connections between visual cortex and the granular parts of area 29, it is area 29d which has the most extensive connections with lateral neocortex and not area 29c.

Since the position of Rose and Woolsey and Vogt and Peters is similar regarding area 23 and its disposition, the only other major issue is whether or not areas 29a and 29b should be consolidated. Area 29a is located posterior and ventral to the corpus callosum and has a homogeneous granular layer, i.e., it is not divisible into a dense layer II–III and a sparse layer IV. Since this area is so small, it is difficult to analyze with most current connectional technologies. Although it does have an additional callosal input to layer II–IV when compared to area 29b, other connections to area 29a including those from the ATN and subiculum are shared by both. In addition, distinct areas 29a and 29b have not been discerned in the monkey. In conclusion, it is important that areas 29c and 29d be identified as separate architectural entities. It appears to be less crucial, however, whether areas 29a and 29b are treated individually or as a single field.

9. Cholinergic Architecture

There is now substantial evidence that cingulate cortex is a major component of the medial cholinergic system as first described by Lewis and Shute (1967) on the basis of AChE activity. Each molecular component of the cholinergic system has been localized in cingulate cortex, including acetylcholine (Jacobowitz and Goldberg, 1977; Hoover *et al.*, 1978), choline acetyltransferase (ChAT) (Johnston *et al.*, 1981), and muscarinic and nicotinic acetylcholine receptors

(AChR). The focus of this section is the cellular localization of each of these molecules.

9.1. Cholinergic Connections

Two sources of acetylcholine are known for cingulate cortex: extrinsic projections from the basal forebrain and intrinsic connections of bipolar neurons. Neurons of the DBB are cholinergic and contain ChAT in their somata (Kimura *et al.*, 1981) and axon terminals, since lesions of the DBB reduce ChAT activity in cingulate cortex (Kamiya *et al.*, 1981). Another line of evidence which indicates that the input from the DBB is cholinergic is the joint labeling of DBB neurons for AChE activity and retrogradely transported fluorescent dyes following dye injection into posterior cingulate cortex (Bigl *et al.*, 1982). Direct evidence is not available as to which laminae in cingulate cortex receive DBB input. However,

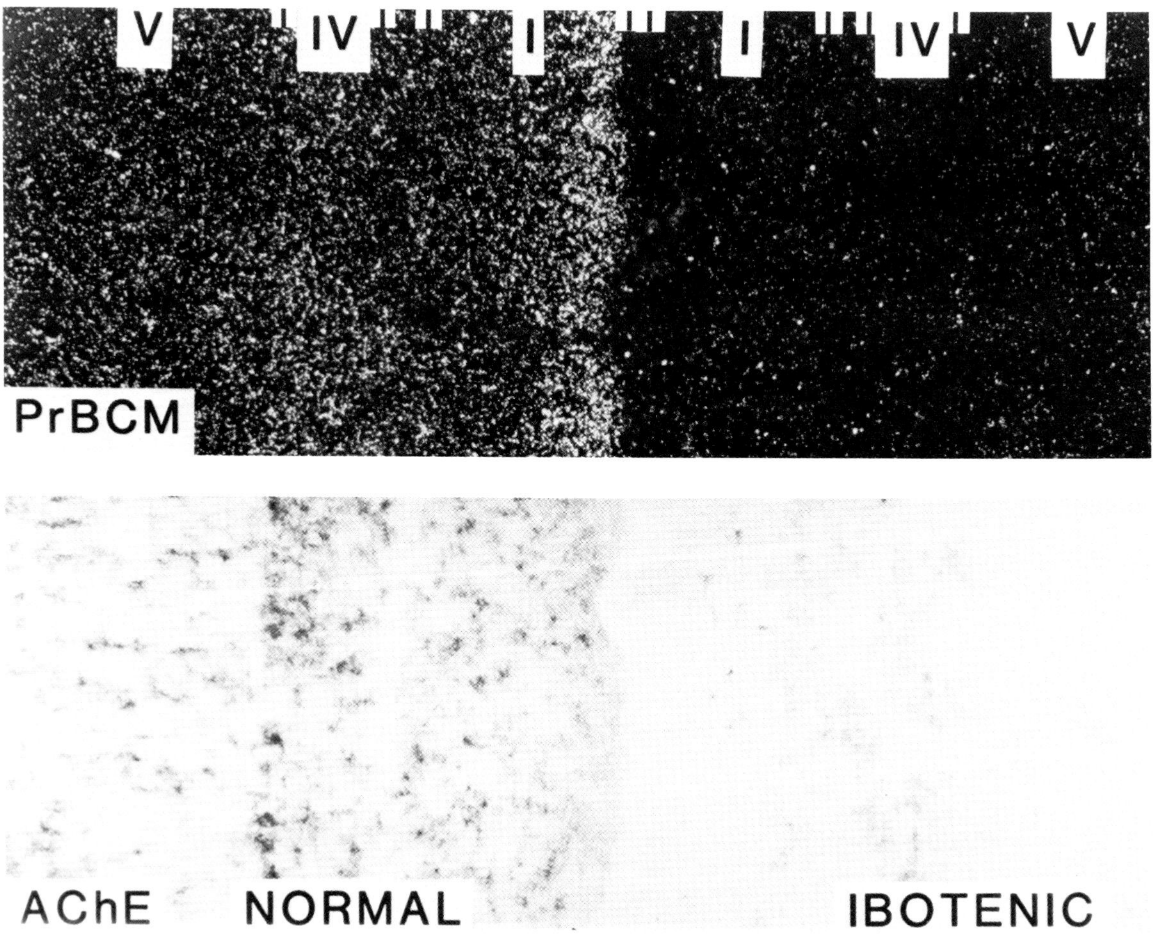

Figure 18. Distribution of acetylcholine receptors (PrBCM) and AChE activity in normal cortex and contralateral cortex which received ibotenic acid injections in both area 29c and the ATN; ×170.

one hint as to where it terminates may be inferred from examining the AChE activity which remains following ablation of the only other known AChE-positive connections of cingulate cortex, namely those from the ATN and intrinsic neurons. Following ablation of both of these connections with the neurotoxin ibotenic acid, AChE activity remains in inner layer I and layer IV (Fig. 18).

The intrinsic source of acetylcholine is from bipolar neurons, which are both ChAT-positive (Eckenstein and Thoenen, 1983; Houser *et al.*, 1983; Armstrong *et al.*, 1983) and AChE-positive (Johnston *et al.*, 1981). These cells are located mainly in layers II–IV, although there are also a few in layer V.

9.2. α-Bungarotoxin Binding

Putative nicotinic AChR have been localized in cingulate cortex. Intermediate to high levels of nicotine and curare displaceable binding of $[^{125}I]\alpha$-bungarotoxin to homogenates of cingulate cortex have been reported by Segal *et al.* (1978) and Morley and Kemp (1981). Laminar localization of specific binding following intraventricular injection of α-bungarotoxin or applied *in vitro* to cryostat cut sections indicates that receptors in layers I and VI are responsible for most of this binding (Hunt and Schmidt, 1978; Arimatsu *et al.*, 1981. In addition, following intraventricular injections this ligand is internalized by large neurons in layer II–III. The size, shape, and location of these cells suggest that they may be extraverted pyramidal neurons. It is not clear what role these receptors play in neurotransmission, since α-bungarotoxin does not block acetylcholine-activated conductance changes in the central nervous system (reviewed by Morley and Kemp, 1981; Fibiger, 1982). No known afferents to posterior cingulate cortex terminate selectively in layers I and VI, and so it is also unclear with which intrinsic and/or extrinsic connections this receptor binding is associated. In contrast to the α-bungarotoxin findings, a study of nicotine binding (Clarke *et al.*, 1984) indicates that, although binding of nicotine occurs throughout cingulate cortex, there are peaks in binding in layers I and IV (their layer III). Without more specific cellular localization and functional information for both α-bungarotoxin and nicotine binding it is difficult to assess the lack of correspondence between these binding profiles.

9.3. Muscarinic Acetylcholine Receptor Binding

Administration of the muscarinic agonist oxotremorine *in vivo* elevates 2-deoxy-D-glucose uptake by rat cingulate cortex (Dam *et al.*, 1982). This response may be mediated by specific binding of oxotremorine to muscarinic AChR which have been demonstrated in the rat (Kuhar and Yamamura, 1976; Kobayashi *et al.*, 1978; Kamiya *et al.*, 1981; McKinney and Coyle, 1982). Muscarinic AChR have also been identified in the cingulate cortices of the dog (Hiley and Burgen, 1974), monkey (Yamamura *et al.*, 1974), and human (Lang and Henke, 1983). These binding sites are located on two major intracortical elements: dendritic and axonal. The laminar distribution of muscarinic antagonist binding to area 29c *in vitro* is particularly provocative, because it suggests a number of cellular localizations which are not apparent in other neocortical structures.

Binding of the muscarinic AChR antagonist [^{3}H]propylbenzilylcholine mustard (PrBCM) is heterogeneous in area 29c. Specific binding of PrBCM (total binding minus nonspecific binding remaining after treatment with 1 μM atropine) is highest in layer Ia (116 ± 7.2 grains/2500 μm^2), lower in layer IV (75 ± 3.1), and relatively constant in layers Ic, II–III, Va, Vb, and VI (55 ± 1.7) (Vogt, 1984). Peak or excess binding of PrBCM in layers Ia and IV suggests that some muscarinic receptors might be associated with ATN afferents, since these are the same layers in which most ATN input terminates. A number of experiments demonstrate that axonal AChR are associated with ATN input and that the ATN are the main source of axonal AChR.

Half of the excess binding in layers Ia and IV is abolished with ATN lesions, providing confirmation that a major population of receptors may be associated with the axons of ATN neurons (Vogt, 1984). Since deafferentation lesions can alter receptor binding via transneuronal degeneration (Hattori and Fibiger, 1982; McKinney and Coyle, 1982), it is not clear from thalamic ablation-induced changes alone that there are axonal (presynaptic?) receptors on this afferent. An alternative strategy is to remove all neurons with the neurotoxin ibotenic acid, so that axons of extrinsic origin remain intact (Köhler *et al.*, 1979; Schwarcz *et al.*, 1979; Köhler and Schwarcz, 1983). Without laminar architecture following neuron removal it can be difficult to determine where to make measurements, especially since cortical shrinking is associated with neuronal losses. In one case a small island of neurons was spared in area 29c (Fig. 19A). The lamination in this cortex can be used as a guide in taking measurements from the ablated tissue. Two peaks of binding remain in layers Ia and IV of this case, as they do in three other cases where estimates of laminar positions were made with reference to binding in layer Ia, contralateral cytoarchitecture, and distances between the pia and white matter. Specific PrBCM binding in the ablated hemispheres was as follows: Ia, 130 ± 9.3; Ic, 70 ± 8.0; II–III, 69 ± 13.4; IV, 103 ± 10.9; Va, 73 ± 15.6; Vb, 73 ± 16.8; VI, 81 ± 4.1. This persistance in binding following ibotenic acid ablations confirms the hypothesis that part of the excess binding in layers Ia and IV is associated with axons of the ATN. There is also evidence for an axonal localization of muscarinic AChR in other regions of the nervous system (Aguilar *et al.*, 1982; Zarbin *et al.*, 1982; Mash *et al.*, 1985).

If the ATN were the only extrinsic source of termination selective for layers Ia and IV, it could be concluded from the above observations that muscarinic receptors are located only on these axons. However, subicular, raphé, and DBB afferents also terminate in layer IV (Section 4). Ablations which totally isolate superficial layers I–Va of area 29c produce a reduction in binding both in layers Ia and IV which is similar in magnitude to that which occurs following ATN ablations (Vogt, 1984). Thus, the primary and possibly sole afferent with muscarinic receptors originates in the anterior thalamus.

A second major location of muscarinic receptors is on neurons of cingulate cortex. This is suggested by the stability of PrBCM binding in layers which do not receive thalamic input following undercut lesions and the large reductions following ibotenic acid treatment.

Two classes of muscarinic receptors have now been identified using the nonclassical antagonist pirenzipine (PZ) (Hammer *et al.*, 1980; Luthin and Wolfe,

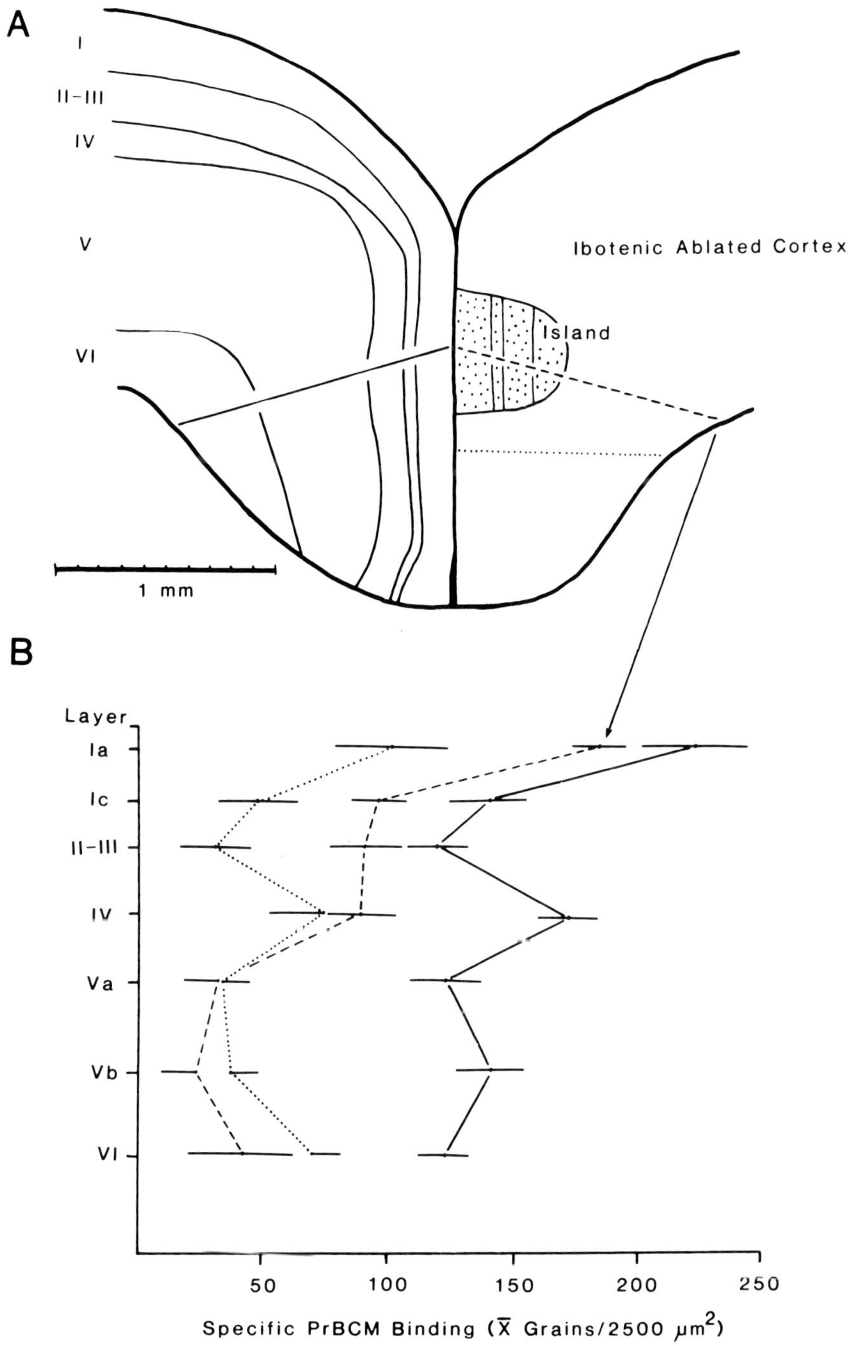

Figure 19. Specific binding of the muscarinic AChR antagonist PrBCM in normal area 29c (solid line), partially ablated with ibotenic acid (dashed line through island of partially preserved cortex), and almost completely ablated tissue (dotted line). The plane of measurement is made to conform approximately to an orientation parallel to the apical dendrites of layer V pyramidal neurons.

1984; Wamsley *et al.*, 1984) which are present in approximately equal proportions in neocortex: high-affinity or M_1 and low-affinity or M_2 receptors. Since M_2 muscarinic receptors may be presynaptic (Raiteri *et al.*, 1984; Mash *et al.*, 1985), it is possible that ATN afferents have M_2 receptors. The distribution of specific [^{3}H]-PZ binding has been analyzed in normal hemispheres and in those which have received ATN ablations (Vogt *et al.*, 1985). In normal hemispheres specific PZ binding is essentially homogeneous with least binding in layer Ia (274 ± 15 grains/2500 μm^2) and most in layer Ic (354 ± 19 grains/2500 μm^2). Since no significant changes occur in PZ binding following ATN lesions, it would appear that previously noted reductions in PrBCM binding following ATN lesions reflect changes in M_2 receptor density. A more direct test of this observation is to block [^{3}H]-PrBCM binding with 1 μM of cold PZ to directly analyze changes in M_2 binding. Under these conditions ATN lesions produce a 38% change in PrBCM binding in layer Ia in comparison to normal hemispheres, providing confirmation that ATN afferents are associated with M_2 muscarinic receptors.

Most muscarinic receptors associated with cortical neurons are of the M_1 subtype. Thus, ibotenic acid ablations of cingulate cortex reduce [^{3}H]-PZ binding by 80%, while in the same cases there is a 70% reduction in [^{3}H]-PrBCM binding. The distribution of muscarinic receptors on specific neurons can be directly assessed in dissociated cingulate cortical cells. Using this technique most receptors are on the secondary and tertiary dendrites of pyramidal neurons with less specific binding to the apical dendritic and somatic membrane of these neurons.

10. The Medial Pain System

The pain system is generally thought to be dissociated into two major components (e.g., Mehler *et al.*, 1960; Kerr, 1975; Price and Dubner, 1977; Willis *et al.*, 1979; Boivie, 1979). The lateral pain system is involved in stimulus localization. It has projections from the spinal cord via the neospinothalamic tract, synapses primarily in the ventroposterior thalamic nucleus, and from there projects to somatosensory cortex. In contrast, the medial pain system is involved in affective responses to painful stimuli. It has spinal cord projections which pass via the paleospinothalamic tract to the midline and intralaminar thalamic nuclei, which project to cingulate cortex.

10.1. Medial and Intralaminar Thalamic Nuclei

The medial and intralaminar thalamic nuclei include the parafascicular, limitans, centrolateral, and superior centrolateral nuclei. There are a number of reasons for implicating these nuclei in "pain-related" functions. First, they contain neurons which respond to noxious stimuli. These neurons have large, often bilateral receptive fields, a nonsomatotopic organization, and may be polymodal (Kruger and Albe-Fessard, 1960; Perl and Whitlock, 1961; Casey, 1966; Dong *et al.*, 1978). Second, electrical stimulation of the parafascicular nucleus produces analgesia in rats (Rosenfeld and Holzman, 1978) and alleviates pain in humans (Richardson and Akil, 1977). Third, ablation of the medial thalamus also alleviates pain by an unknown mechanism (Mark *et al.*, 1963; Uematsu *et al.*, 1974) and interferes with escape responses of cats from tooth pulp stimulation (Kaelber *et al.*, 1975). Fourth, the medial thalamus may contain a higher density of opiate receptors than the lateral thalamus (Kuhar *et al.*, 1973). Fifth, application of morphine to the medial thalamus and parafascicular nucleus produces analgesia (Pert and Yaksh, 1974).

These thalamic nuclei and the centrodensocellular nucleus of Olszewski (1952), reuniens and pars densocellularis of the mediodorsal nucleus project to anterior and to a lesser extent posterior cingulate cortex in rat, cat, and monkey (Jones and Leavitt, 1974; Macchi *et al.*, 1977; Herkenham, 1978; Vogt *et al.*, 1979; Robertson and Kaitz, 1981). Although the medial and intralaminar thalamic nuclei project to other cortical areas (e.g., Jones and Leavitt, 1974; Trojanowski and Jacobson, 1976; Macchi *et al.*, 1977; Herkenham, 1978), the following section describes the reasons for considering cingulate cortex as one of the primary cortical termination sites of the medial pain system.

10.2. Cingulate Cortex

In addition to evidence that the thalamic nuclei which project to cingulate cortex are involved in "pain-related" functions, there is direct evidence for believing that cingulate cortex is also involved in such functions. For example, it has been known for some time that cingulumotomy can alleviate a patient's affective responses to painful stimuli, without interrupting his or her ability to localize a noxious stimulus (Foltz and White, 1962; White and Sweet, 1969; Turnbull, 1972; Amromin *et al.*, 1975). There is also particularly high opiate receptor binding and enkephalin immunoreactivity in cingulate cortex (Hiller *et al.*, 1973; Kuhar *et al.*, 1973; Sar *et al.*, 1978; Lewis *et al.*, 1983). Immunoreactive enkephalin terminals are found in posterior cingulate cortex and surround individual cell bodies (Sar *et al.*, 1978). In colchicine-treated animals, enkephalin-positive somata are found in only one brain-stem nucleus which is known to project to area 29, and that is the lateral hypothalamus. Since some neurons in cingulate cortex are also immunoreactive for enkephalin (Khachaturian *et al.*, 1983), potentially there are two sources of enkephalin in cingulate cortex.

Another reason for implicating cingulate cortex in responses to pain is that there are direct projections from anterior cingulate cortex to the periaqueductal gray (PAG). The PAG is a key link in the descending inhibitory pain system (e.g., Mayer and Liebeskind, 1974; Giesler and Liebeskind, 1976; Messing and Lytle, 1977; Yeung *et al.*, 1977; Yaksh, 1979) and layer V neurons of area 24 have extensive projections into the PAG (Domesick, 1969; Hardy and Leichnetz, 1981; Morrell *et al.*, 1981; Mantyh, 1982). One aspect of conscious (cortical?) responses to painful stimuli might involve this pathway.

In conclusion, reasonable evidence is available implicating cingulate cortex in the medial, opiate pain system. However, frontal cortex shares many of the pharmacological and connectional characteristics noted for cingulate cortex, so it is unlikely that the cingulate region is the only cortex involved in "pain-related" functions.

10.3. Role of Cingulate Cortex and Noxious Stimuli in Learning

Behavioral deficits which result from cingulate cortex ablation may be interpreted in light of the role of cingulate cortex in producing affective responses to noxious stimuli. Numerous studies have shown that ablations of cingulate cortex produce deficiencies in active, shock avoidance learning in rats and cats (Peretz, 1960; Thomas and Slotnick, 1963; Lubar, 1964; Lubar and Perachio, 1965; Thomas *et al.*, 1968). Since these animals do not display sensory (visual or auditory) or motor deficits, it is possible that the behavioral deficiency arises from the animal's inability to produce an association between conditioned (light or tone) and unconditioned (noxious footshock) stimuli. The role of cingulate cortical neurons in extracting the significance code from conditioned stimuli has been reviewed by Gabriel *et al.* (1980b).

Behavioral studies have emphasized the importance of anterior cingulate cortex in establishing active avoidance learning, and the above-noted observations on receptor binding and connections of the midline thalamic nuclei with area 24 lead to the conclusion that initial associative changes involved in pairing conditioned and unconditioned stimuli may occur in anterior cingulate cortex. Multiple unit activity recorded in cingulate cortex during the acquisiton of this behavior confirms that neurons in area 24 acquire a significance code, i.e., develop differential responsiveness to positive and negative conditioned stimuli, before neurons in posterior cingulate cortex (Gabriel *et al.*, 1982). Since neurons in layers V and VI of area 29 acquire a significance code which temporally follows those in area 24, it is possible that corticocortical connections between areas 24 and 29 mediate the development of the significance code in area 29.

There is another indication that cingulate cortex mediates behavioral responses to noxious stimuli. Electrical stimulation of anterior cingulate cortex can produce shrill vocalization (Smith, 1945; Kaada, 1951; Showers, 1959; Robinson, 1967), which might be associated with escape responses, and ablation of this cortex is accompanied by a loss of phonatory performance (Sutton *et al.*, 1974). Phonation responses may also involve projections of anterior cingulate cortex to the PAG (Müller-Preuss and Jürgens, 1976), since electrical stimulation of the PAG elicits vocalization (Jürgens and Pratt, 1979).

In conclusion, a tight linking of anterior cingulate cortex with the medial pain system is one explanation for why ablations of cingulate cortex lead to deficiencies in active avoidance learning. In addition, projections into the descending inhibitory pain system and vocalization centers via the PAG implicate cingulate cortex in affective responses to pain. Finally, associative neuronal changes which initially take place in anterior cingulate cortex may be relayed to posterior area 29 through corticocortical connections.

11. Electrophysiological Properties of Cingulate Neurons

Renewed interest in the membrane properties and synaptic-evoked activity of neocortical neurons has been generated by analysis of *in vitro* preparations. Since cingulate cortex is directly adjacent to the corpus callosum (CC) and has strong homotopic projections through the CC, 0.5-mm-thick, coronal slices can be made of cingulate cortex and the CC in which this one extrinsic afferent system is retained essentially intact (Vogt and Gorman, 1982). Figure 20 is a photograph of such a preparation viewed by transillumination. The CC, underlying white matter (cingulum bundle), and area 29 are each clearly defined, as are most layers in area 29c. Thus, it is possible to make direct, visually guided placements of stimulating and recording electrodes in the CC and various cortical laminae. The virtues of the *in vitro* preparation are this direct visual placement, mechanical stability, and potential for electrical stimulation of a single extrinsic afferent. However, one drawback is that it may not completely reflect *in vitro* activity, since there is reduced tonic synaptic input which may result in limited spontaneous activity and more negative resting membrane potentials than normal.

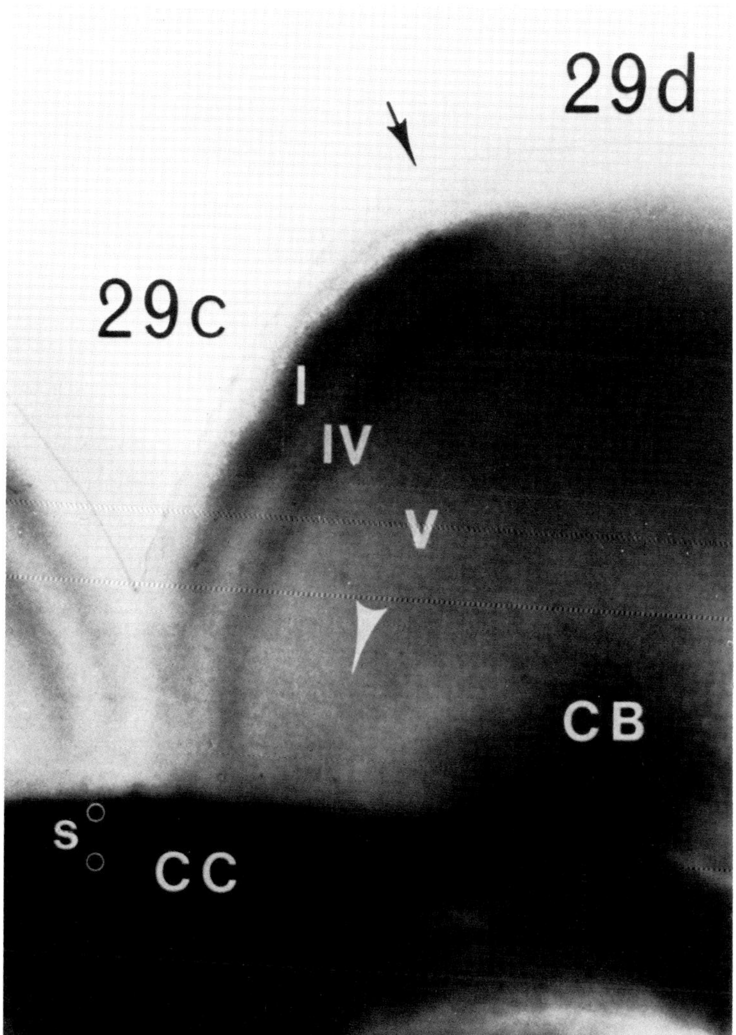

Figure 20. Transilluminated slice preparation. S, bipolar stimulating electrodes; CC, corpus callosum; CB, cingulum bundle; ×55.

11.1. Membrane Properties and Synaptic-Evoked Responses

The electrophysiological properties and responses of cortical neurons are influenced by circuitry alterations produced during slice preparation. Variation in slice thickness (e.g., Schwartzkroin and Knowles, 1984) and orientation may partially account for some of the differences which have been observed between *in vitro* and *in vivo* preparations. In *in vitro* preparations, neocortical neurons rest at more negative membrane potentials (V_m) than do those *in vivo* (Scholfield, 1978; Connors *et al.*, 1982), which is also the case for cingulate neurons which rest at a V_m of 60 mV in the slice (Vogt and Gorman, 1982) and 48 mV *in vivo* (Finch *et al.*, 1984a). More negative resting V_m of neurons in slices result in lower rates of spontaneous activity and may be responsible for higher-amplitude excitatory postsynaptic potentials (EPSPs). In addition, spontaneous bursting occurs more frequently in neurons *in vivo* than in slices.

The passive membrane properties of cingulate neurons are similar to those of neocortical cells. The membrane time constant of 4.4 msec is somewhat shorter than has been reported for other neocortical neurons (e.g., Lux and Pollen, 1966; Ogawa *et al.*, 1981; Connors *et al.*, 1982), while the input resistance of 47 MΩ is close to the 42.1 MΩ value reported for visual cortical cells (Ogawa *et al.*, 1981).

EPSPs evoked by electrical stimulation of the CC *in vitro* (Vogt and Gorman, 1982) and the subicular complex and thalamus *in vivo* (Finch *et al.*, 1984a) have been characterized for layer V neurons of area 29. The types of layer V neurons analyzed in both studies might be quite similar, since the largest and most frequent types of neuron in layer V are efferent-projecting pyramidal cells from which antidromic spikes can be evoked with electrical stimulation and which have been injected intracellularly with HRP. Other large neurons which occur much less frequently in layer V are the large multipolar and bitufted cells. These cells may make a small contribution to nonantidromically activated and otherwise unidentified cell groups, but it is unlikely that the very small and infrequent layer V bipolar neurons have been intracellularly penetrated for prolonged analysis.

Electrical stimulation of the subicular complex evokes EPSPs with an average duration of 17.4 msec and amplitude of 5.6 mV. Although callosal-evoked EPSPs have a longer mean duration and amplitude, 26 msec and 11.5 mV, respectively, it is likely that these EPSP differences in part reflect the more hyperpolarized state of slice neurons. Thus, the amplitude of callosal-evoked EPSPs can be increased with hyperpolarizing current, and variation of EPSP amplitude is an approximately linear function of V_m with a calculated reversal potential of 0 to −20 mV.

Spike initiation thresholds can be estimated by varying the amplitude of stimulation to the CC. Increasing current pulses to the CC evokes higher-amplitude EPSPs and longer-latency spikes (Fig. 21B). The mean EPSP amplitude required to evoke a single action potential in 15 neurons recorded from layer V is 8.4 ± 0.8 mV (mean ± S.E.). Since the mean resting V_m for these neurons is 58.9 ± 2.9 mV, the first spike discharge occurs at a V_m of 50.5 mV. This value is similar to that for cells in sensorimotor cortex, since Connors *et al.* (1982) report that neurons with a mean resting V_m of 76.5 mV discharge an initial spike to injected square current pulses 25.0 mV above rest or at a V_m of 51.5 mV. Stimulation of the CC at high intensities often evokes two or more action potentials and each of these additional spikes usually has a higher threshold. Neurons in layer V of area 24 have spike thresholds which increase linearly with each successive spike (Fig. 21A), while those for posterior area 29 have lower and more stable thresholds. In addition to having higher-amplitude spike thresholds, spike durations broaden and the latencies for each successive spike shorten.

The high-amplitude EPSPs produced by high-intensity CC stimulation are generally associated with reduced-amplitude action potentials. Although there are instances in which increasing EPSP amplitude has little effect on spike amplitude (Fig. 22A), or only reduces longer-latency spike amplitudes (Fig. 22B), in most cases all spikes are reduced in amplitude when they are associated with high-amplitude EPSPs (Fig. 22C). Variation in spike amplitude also occurs in bursting cells in cingulate cortex *in vivo* (Finch *et al.*, 1984a), in the hippocampus (Kandel and Spencer, 1961; Schwartzkroin, 1975), and in sensorimotor cortex following intracellular depolarizing pulses (Connors *et al.*, 1982).

There are at least three types of callosal-evoked EPSPs in the slice preparation. First, there are low-amplitude (6–8 mV), monophasic EPSPs. Since these EPSPs often follow antidromic spikes, it might be suggested that these are depolarizing afterpotentials (DAPs; Wong and Prince, 1981). However, higher-intensity stimulation enhances the amplitude of these potentials, often leading to generation of further spike discharge. In addition, spike discharges produced by intracellular depolarizing pulses at a similar V_m have only low-amplitude (2–4 mV) DAPs associated with them. Also, monophasic EPSPs without associated spike discharge or longer-latency hyperpolarizations are evoked from layer V pyramids *in vivo* following subicular stimulation. A second type of EPSP has a biphasic rise to peak and is usually of higher amplitude (15–20 mV) than are the other two types. Third, low-amplitude (5–10 mV) and long-duration (50–90

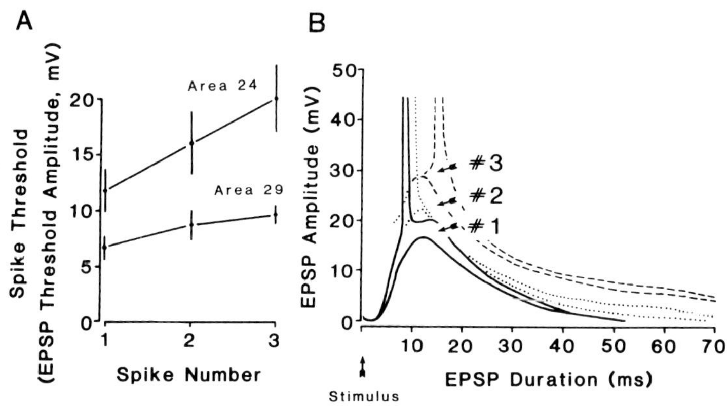

Figure 21. (A) Mean (± S.E.) spike thresholds for the first three spikes evoked by CC stimulation in five area 24 and 10 area 29 neurons. (B) Example of how thresholds are determined using both sub- and suprathreshold CC stimulations for each subsequent spike in a neuron from area 24. Arrows indicate points at which threshold measurements are made.

msec) EPSPs can be evoked and may reflect the activity of multisynaptic, excitatory interneurons in the slice preparation.

Both temporal and spatial summation of EPSPs occur in posterior cingulate cortex. Temporal summation is demonstrated in the slice by reducing the duration between paired callosal stimuli. As this duration is shortened, the EPSP amplitude evoked by the second stimulus increases progressively. In one instance when the delay was 30 msec or greater, the EPSP evoked by the second stimulus was 9 mV in amplitude (unpublished observations). As the delay is shortened, EPSP amplitude increases: 25 msec delay = 9.5 mV EPSP, 20 msec = 10 mV, 15 msec = 11 mV, 10 msec = 12 mV, and 5 msec = 15 mV. Spatial summation is also shown in the *in vivo* preparation where simultaneous subthreshold stimulation of thalamic and subicular afferents, which have partially separated termination zones, produces higher-amplitude suprathreshold EPSPs. It is also presumed that one of the primary reasons for higher-amplitude EPSPs in anterior than posterior cingulate cortex (*vide infra*) is spatial summation of the activity of a large number of afferent axons over a larger area of the dendritic tree.

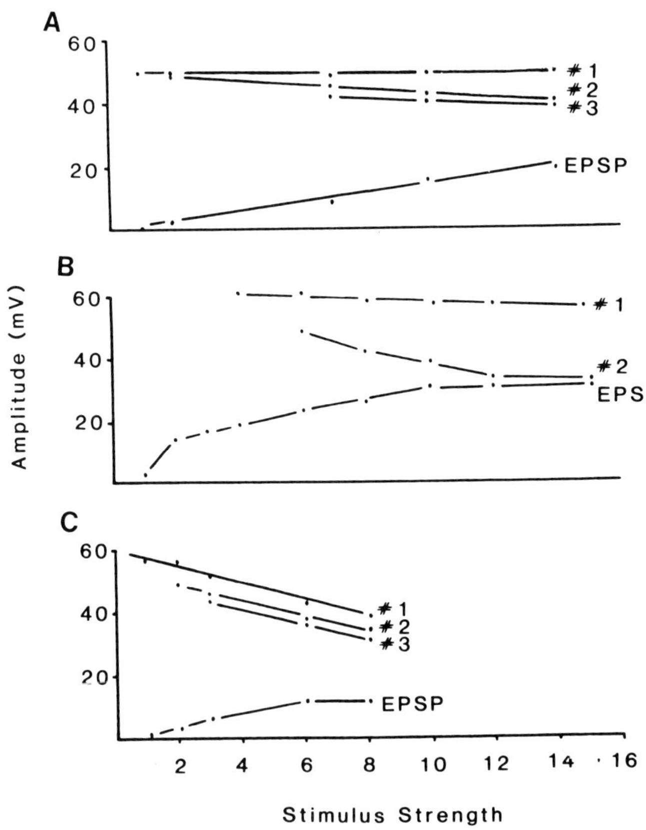

Figure 22. Alterations in spike amplitudes that are associated with increasing-amplitude callosal-evoked EPSPs. (A) Little change in spike amplitudes. (B) Change primarily in spike #2. (C) Significant changes in all three callosal-evoked EPSPs.

11.2. Neurons in Areas 24 and 29

Although the membrane properties of anterior and posterior cingulate neurons are very similar (i.e., resting V_m, time constant, electrotonic length), comparison of callosal-evoked EPSPs in anterior and posterior cingulate cortex has suggested differences between these populations of neurons (Vogt and Gorman, 1982). The amplitude and duration of these EPSPs in layer V of area 24 are almost twice those in area 29 at close to maximal CC stimulation. In addition, the shapes of subthreshold EPSPs are different, for in area 24 time-to-peak amplitude and the duration of half-amplitude EPSPs the greater. EPSP amplitudes do not appear to be related to resting V_m, input resistance, or variation in conduction velocities. Figure 23 presents a sample of four subthreshold, biphasic EPSPs. Although resting V_m is virtually the same for neurons in both regions, mean time-to-peak in the anterior cells is 9.5 versus 6.2 msec in area 29 and half-widths are 19 versus 16 msec in areas 24 and 29, respectively. These higher-amplitude EPSPs could result in part from higher spike initiation thresholds in anterior cingulate cells (Fig. 21). Finally, higher-amplitude EPSPs in area 24 are associated with significant decreases in spike amplitude. In area 24, high-intensity CC stimulation reduces spike amplitude produced by low-intensity CC stimulation by 14, 51, and 25% for spikes, #1, 2, and 3, while in area 29 the reduction is only 9, 12, and 10% for the first, second, and third spikes.

The high-amplitude EPSP may be a consequence of the density and distribution of callosal afferents as well as thresholds for spike initiation. In anterior

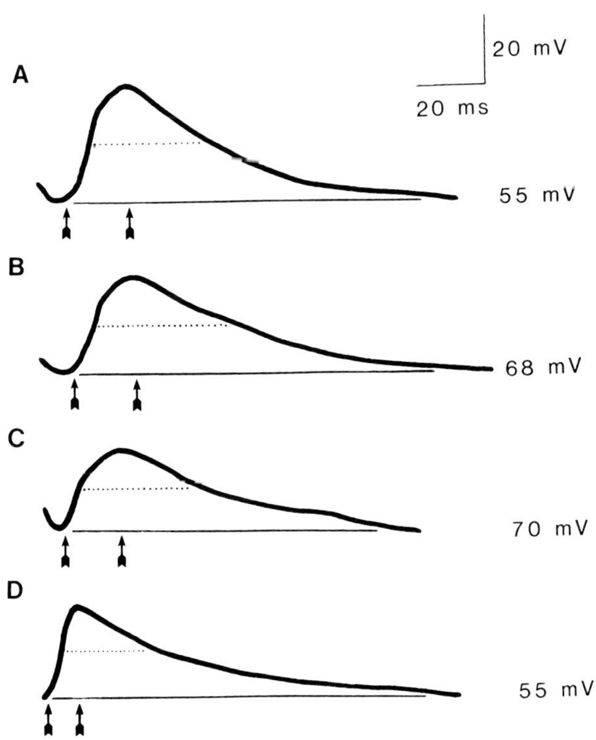

Figure 23. Examples of subthreshold, biphasic EPSPs in areas 24 (A and B) and 29 (C and D). Arrows indicate time-to-peak, dotted lines the half-width, and resting V_m is presented on the right.

cortex, callosal afferents form five to six times more synapses than in posterior cortex and terminate mainly in layers I–III, above the somata of layer V neurons, so they are further from the axon hillock of layer V neurons. In area 29, callosal afferents form fewer synapses and most of these afferents terminate in layers V and VI, close to the somata of layer V neurons.

12. Structure and Function of Thalamic and Subicular Afferents: A Summary

Anterior thalamic afferents provide the most extensive input to areas 29a–c. Understanding the organization and functions of this afferent system is essential to elucidating the role of posterior cingulate cortex in the behavior of lower mammals.

The axons of neurons in the AD and AV nuclei terminate mainly in layer Ia and to a lesser extent in deeper cortical layers. These axons form asymmetric synapses with cells in areas 29a–c and electrical stimulation of this projection produces EPSPs in layer V neurons. Furthermore, this input plays an important role in producing discriminative responses to auditory stimuli during active avoidance conditioning (Foster *et al.*, 1980; Gabriel *et al.*, 1980a, 1983). In agreement with anatomical and physiological studies, ablation of the ATN abolishes excitatory, discriminative multiple-unit discharges that normally develop to auditory conditional stimuli. Finally, since muscarinic AChR are associated with the axons of ATN terminals in layers Ia and IV, it is to be expected that cholinergic connections of the DBB and/or bipolar neurons have a direct influence over the discharge of neurotransmitter from ATN afferents. It is possible, therefore, that one of the primary influences of basal forebrain afferents in memory function is exerted via presynaptic control over neurons which are involved in mediating the retention of certain behavioral responses.

Subicular cortex is the origin of another major input to posterior cingulate areas 29a–c. In addition to the density of termination of this input, subicular afferents are unique because they are the only ones which terminate directly in layer II–III. Electrical stimulation of the subicular complex evokes both EPSPs and IPSPs in cingulate neurons, while ablation of this complex enhances inhibitory neuronal responses produced by conditional auditory stimuli during the acquisition of active avoidance behavior (M. Gabriel, personal communication). Thus, the hippocampal system has two different routes by which it controls the action of cingulate cortex neurons. There is a direct pathway via subicular afferents and an indirect route through the mammillary bodies and the ATN.

The perspective of this review has suggested a number of problem areas and broad issues that will need to be addressed before structure–function relationships are fully understood in this part of the brain. First, the functions of many connections of cingulate cortex have not been analyzed at either the cellular or behavioral level. These projections originate in the claustrum, visual cortex, the intralaminar thalamic nuclei, the raphé and coeruleal nuclei as well as lesser inputs from the lateral hypothalamus and ventral tegmental area. Another level of complexity arises when "multisynaptic" influences are evaluated as those which are exerted by the mammillary bodies. Also, there are numerous projections into various parts of the motor system, but little is known about the role of cingulate cortex in releasing "motor programs" associated with specific behaviors. Second, ablation and multiunit techniques are important approaches for determining the general state of neuronal excitability throughout the acquisition of avoidance behavior. However, intracellular analyses during acquisition and retention training will be required in order to determine the membrane and synaptic dynamics which are the basis for establishing and maintaining these behaviors. Third, there is an elaborate network of neurotransmitter and neuromodulator systems associated with the extrinsic and intrinsic circuits of cingulate cortex. It will be necessary to determine the action of each of these systems in reducing the influence of "behaviorally unnecessary" inputs while enhancing those of others which are required to produce new behaviors.

ACKNOWLEDGMENTS. I would like to thank Drs. Alan Peters, Robert W. Sikes, and Karl Zilles for their careful reading of the manuscript. Research in the author's laboratory is supported by NIH Grants NS 18745 and 16841.

13. References

Abraham, W. C., Delanoy, R. L., Dunn, A. J., and Zornetzer, S. F., 1979, Locus coeruleus stimulation decreases deoxyglucose uptake in ipsilateral mouse cerebral cortex, *Brain Res.* **172:**387–392.

Aghajanian, G. K., and Vandermaelen, C. P., 1982, Intracellular identification of central noradrenergic and serotonergic neurons by a new double labeling procedure, *J. Neurosci.* **2:**1786–1792.

Aguilar, J. S., Jerusalinsky, D., Stockert, M., Medina, J. H., and DeRobertis, E., 1982, Localization of hippocampal muscarinic receptors after kainic acid lesion of CA3 and fimbria–fornix transection, *Brain Res.* **247:**335–340.

Amromin, G. D., Crue, B. L., Felsöory, A., and Todd, E. M., 1975, Bilateral stereotaxic cingulotomy following thoracic rhizotomy, cervical cordotomy and thalamotomy in a patient with intractable pain: A clinical-pathological study, in: *Pain Research and Treatment* (B. L. Crue, ed.), Academic Press, New York, pp. 227–241.

Anand, B. K., and Dua, S., 1954, Circulatory and respiratory changes induced by electrical stimulation of limbic system (visceral brain), *J. Neurophysiol.* **19:**393–400.

Arimatsu, Y., Seto, A., and Amano, T., 1981, An atlas of α-bungarotoxin binding sites and structures containing acetylcholinesterase in the mouse central nervous system, *J. Comp. Neurol.* **198:**603–631.

Armstrong, D. M., Kress, Y., and Terry, R. D., 1983, Electron microscopic immunocytochemical characterization of cholinergic neurons in the rat brain, *Soc. Neurosci. Abstr.* **9:**80.

Baleydier, C., and Mauguiere, F., 1980, The duality of the cingulate gyrus in monkey: Neuroanatomical study and functional hypothesis, *Brain* **103:**525–554.

Bassett, J. L., and Berger, T. W., 1982, Associational connections between the anterior and posterior cingulate gyrus in rabbit, *Brain Res.* **248:**371–376.

Beckstead, R. M., 1976, Convergent thalamic and mesencephalic projections to the anterior medial cortex in the rat, *J. Comp. Neurol.* **166:**403–416.

Beckstead, R. M., 1979, An autoradiographic examination of corticocortical and subcortical projections of the mediodorsal-projection (prefrontal cortex) in the rat, *J. Comp. Neurol.* **184:**43–62.

Beinfeld, M. C., Meyer, D. K., Eskay, R. L., Jensen, R. T., and Brownstein, M. J., 1981, The distribution of cholecystokinin immunoreactivity in the central nervous system of the rat as determined by radioimmunoassay, *Brain Res.* **212:**51–57.

Benjamin, R. M., Jackson, J. C., and Gordon, G. T., 1978, Cortical projections of the thalamic mediodorsal nucleus in the rabbit, *Brain Res.* **141:**251–265.

Bennett-Clarke, C., Romagnano, M. A., and Joseph, S. A., 1980, Distribution of somatostatin in the rat brain: Telencephalon and diencephalon, *Brain Res.* **188:**473–486.

Berger, T. W., Milner, T. A., Swanson, G. W., Lynch, G. S., and Thompson, R. F., 1980, Reciprocal anatomical connections between anterior thalamus and cingulate–retrosplenial cortex in the rabbit, *Brain Res.* **201:**411–417.

Bigl, V., Woolf, N. J., and Butcher, L. L., 1982, Cholinergic projections from the basal forebrain to frontal, parietal, temporal occipital and cingulate cortices: A combined fluorescent tracer and acetylcholinesterase analysis, *Brain Res. Bull.* **8:**727–749.

Bobillier, P., Seguin, S., Degueurce, A., Lewis, B. D., and Phjol, J. F., 1979, The efferent connections of the nucleus raphé centralis superior in the rat revealed by radioautography, *Brain Res.* **166:**1–8.

Boivie, J., 1979, An anatomical reinvestigation of the termination of the spinothalamic tract in the monkey, *J. Comp. Neurol.* **186:**343–370.

Braak, H., 1979, Pigment architecture of the human telencephalic cortex. IV. Regio retrosplenialis, *Cell Tissue Res.* **204:**431–440.

Broca, P., 1878, Anatomie comparée circonvolutions cérébrales: Le grand lobe limbique et la scissure limbique dans la série des mammifères, *Rev. Anthropol. Ser. 2* **1:**384–498.

Brodmann, K., 1909, *Vergleichende Lokalistionslehre der Grosshirnrinde in ihren Prinzipien dargestellt auf Grund des Zellenbaues,* Barth, Leipzig.

Buchanan, S. L., and Powell, D. A., 1982, Cingulate cortex: Its role in Pavlovian conditioning, *J. Comp. Physiol. Psychol.* **96:**755–774.

Bunney, B. S., and Aghajanian, G. K., 1976, Dopamine and norepinephrine innervated cells in the rat prefrontal cortex: Pharmacological differentiation using microiontophoretic techniques, *Life Sci.* **19:**1783–1792.

Casey, K. L., 1966, Unit analysis of nociceptive mechanisms in the thalamus of the awake squirrel monkey, *J. Neurophysiol.* **29:**727–750.

Caviness, V. S., 1975, Architectonic map of neocortex of the normal mouse, *J. Comp. Neurol.* **164:**247–264.

Cespuglio, R., Gomez, M. E., Faradji, H., and Jovet, M., 1982, Alterations in the sleep–waking cycle induced by cooling of the locus coeruleus area, *Electroencephalogr. Clin. Neurophysiol.* **54:**570–578.

Chow, K. L., Masland, R. H., and Stewart, D. L., 1971, Receptive field characteristics of striate cortical neurons in the rabbit, *Brain Res.* **33:**337–352.

Clark, W., and Boggon, R. H., 1933, On the connections of the anterior nucleus of the thalamus, *J. Anat.* **67:**215–231.

Clarke, P. B. S., Pert, C. B., and Pert, A., 1984, Autoradiographic distribution of nicotine receptors in rat brain, *Brain Res.* **323:**390–395.

Connors, B. W., Gutnick, M. J., and Prince, D. A., 1982, Electrophysiological properties of neocortical neurons in vitro, *J. Neurophysiol.* **48:**1302–1320.

Conrad, C. D., and Stumpf, W. E., 1975, Direct visual input to the limbic system: Crossed retinal projections to the nucleus anterodorsalis thalami in the tree shrew, *Exp. Brain Res.* **23:**141–149.

Cowan, W. M., and Powell, T. P. S., 1954, An experimental study of the relation between the medial mammillary nucleus and the cingulate cortex, *Proc. R. Soc. London Ser. B* **143:**114–125.

Cruce, J. A. F., 1975, An autoradiographic study of the projections of the mammillothalamic tract in the rat, *Brain Res.* **85:**211–219.

Cuénod, M., Casey, K. L., and MacLean, P. D., 1965, Unit analysis of visual input to posterior limbic cortex. I. Photic stimulation, *J. Neurophysiol.* **28:**1101–1117.

Dam, M., Wamsley, J. K., Rapoport, S. I., and London, E. D., 1982, Effects of oxotremorine on local glucose utilization in the rat cerebral cortex, *J. Neurosci.* **2:**1072–1078.

Dekker, J. J., and Kuypers, H. G. J. M., 1976a, Morphology of rat's AV thalamic nucleus in light and electron microscopy, *Brain Res.* **117:**387–398.

Dekker, J. J., and Kuypers, H. G. J. M., 1976b, Quantitative EM study of projection terminals in the rat's AV thalamic nucleus: Autoradiographic and degeneration techniques compared, *Brain Res.* **117:**399–422.

Diao, Y.-C., Wang, Y.-K., and Pu, M.-L., 1983, Binocular responses of cortical cells and the callosal projection in the albino rat, *Exp. Brain Res.* **49:**410–418.

Domesick, V. B., 1969, Projections from the cingulate cortex in the rat, *Brain Res.* **12:**296–320.

Domesick, V. B., 1972, Thalamic relationships of the medial cortex in the rat, *Brain Behav. Evol.* **6:**457–483.

Dong, W. K., Ryu, H., and Wagman, I. H., 1978, Nociceptive responses of neurons in medial thalamus and their relationship to spinothalamic pathways, *J. Neurophysiol.* **41:**1592–1613.

Donoghue, J. P., and Wise, S. P., 1982, The motor cortex of the rat: Cytoarchitecture and microstimulation mapping, *J. Comp. Neurol.* **212:**76–88.

Donovan, M. K., and Wyss, J. M., 1983, Evidence for some collateralization between cortical and diencephalic efferent axons of the rat subicular cortex, *Brain Res.* **259:**181–192.

Eckenstein, F., and Thoenen, H., 1983, Cholinergic neurons in the rat cerebral cortex demonstrated by immunohistochemical localization of choline acetyltransferase, *Neurosci. Lett.* **36:**211–215.

Espinoza, S. G., and Thomas, H. C., 1983, Retinotopic organization of striate and extrastriate visual cortex in the hooded rat, *Brain Res.* **272:**137–144.

Fibiger, H. C., 1982, The organization and some projections of cholinergic neurons of the mammalian forebrain, *Brain Res. Rev.* **4:**327–388.

Finch, D. M., Derian, E. L., and Babb, T. L., 1984a, Excitatory projection of the rat subicular complex to the cingulate cortex and synaptic integration with thalamic afferents, *Brain Res.* **301:**25–37.

Finch, D. M., Derian, E. L., and Babb, T. L., 1984b, Afferent fibers to rat cingulate cortex, *Exp. Neurol.* **83:**468–485.

Flynn, J. H., Vanegas, H., Foote, W., and Edwards, S., 1970, Neural mechanisms involved in a cat's attack on a rat, in: *The Neural Control of Behavior* (R. E. Whalen, R. F. Thompson, M. Verzeano, and N. M. Weinberger, eds.), Academic Press, New York.

Foltz, E. L., and White, L. E., 1962, Pain "relief" by frontal cingulumotomy, *J. Neurosurg.* **19:**89–100.

Foster, K., Orona, E., Lambert, R. W., and Gabriel, M., 1980, Early and late acquisition of discriminative neuronal activity during differential conditioning in rabbits: Specificity within the laminae of cingulate cortex and the anteroventral thalamus, *J. Comp. Physiol. Psychol.* **94:**1069–1086.

Gabriel, M., Foster, K., and Orona, E., 1980a, Interaction of laminae of the cingulate cortex with the anteroventral thalamus during behavioral learning, *Science* **203:**1050–1052.

Gabriel, M., Foster, K., Orona, E., Stalwick, S. E., and Stanton, M., 1980b, Neuronal activity of cingulate cortex, anteroventral thalamus, and hippocampal formation in discriminative conditioning: Encoding and extraction of the significance of conditioned stimuli, in: *Progress in Psychobiology and Physiological Psychology*, Volume 9 (J. M. Sprague and A. N. Epstein, eds.), Academic Press, New York, pp. 125–231.

Gabriel, M., Orona, E., Foster, K., and Lambert, R. W., 1982, Mechanism and generality of stimulus significance coding in a mammalian model system, *Adv. Behav. Biol.* **26:**535–567.

Gabriel, M., Lambert, R. W., Foster, K., Orona, E., Sparenborg, S., and Majorca, R. R., 1983, Anterior thalamic lesions and neuronal activity in the cingulate and retrosplenial cortices during discriminative avoidance behavior in rabbits, *Behav. Neurosci.* **97:**675–696.

Galaburda, A. M., and Pandya, D. N., 1983, The intrinsic architectonic and connectional organization of superior temporal regions of the rhesus monkey, *J. Comp. Neurol.* **221:**169–184.

Giesler, G. J., and Liebeskind, J. C., 1976, Inhibition of visceral pain by electrical stimulation of the periaqueductal gray matter, *Pain* **2:**43–48.

Giolli, R. A., Towns, L. C., Takahashi, T. T., Karamanlidis, A. N., and Williams, D. D., 1978, An autoradiographic study of the projections of visual cortical area 1 of the thalamus, pretectum and superior colliculus of the rabbit, *J. Comp. Neurol.* **180:**743–752.

Globus, A., and Scheibel, A. B., 1967, Pattern and field in cortical structure: The rabbit, *J. Comp. Neurol.* **131:**155–172.

Goldman-Rakic, P. S., Selemon, L. D., and Schwartz, M. L., 1984, Dual pathways connecting the dorsolateral prefrontal cortex with the hippocampal formation and parahippocampal cortex in the rhesus monkey, *Neuroscience* **12:**719–743.

Hall, R. D., and Lindholm, E. P., 1974, Organization of motor and somatosensory neocortex in the albino rat, *Brain Res.* **66:**23–38.

Hammer, R., Berrie, C. P., Birdsall, N. J. M., Burgen, A. S. V., and Hulme, E. C., 1980, Pirenzepine distinguishes between different subclasses of muscarinic receptors, *Nature* **283:**90–92.

Hara, Y., Shiosaka, S., Senba, E., Sakanaka, M., Inagaki, S., Taagi, H., Kawai, Y., Takatsuki, K., Matsuzaki, T., and Tohyama, M., 1982, Ontogeny of the neurotensin-containing neuron system of the rat: Immunohistochemical analysis. I. Forebrain and diencephalon, *J. Comp. Neurol.* **208:**177–195.

Hardy, S. G. P., and Leichnetz, G. R., 1981, Frontal cortical projections to the periaqueductal gray in the rat: A retrograde and orthograde horseradish peroxidase study, *Neurosci. Lett.* **23:**13–17.

Hattori, T., and Fibiger, H. C., 1982, On the use of lesions of afferents to localize neurotransmitter receptor sites in the striatum, *Brain Res.* **238:**245–250.

Herkenham, M., 1978, The connections of the nucleus reuniens thalami: Evidence for a direct thalamo-hippocampal pathway in the rat, *J. Comp. Neurol.* **177:**589–610.

Herrmann, C., and Schulz, E., 1978, Quantitative Untersuchungen an Sternzellen in Bereich der cingulären Rinde der Ratte, *J. Hirnforsch.* **19:**519–531.

Hiley, C. R., and Burgen, A. S. V., 1974, The distribution of muscarinic receptor sites in the nervous system of the dog, *J. Neurochem.* **22:**159–162.

Hiller, J. M., Pearson, J., and Simon, E. J., 1973, Distribution of stereospecific binding of the potent narcotic analgesic etorphine in the human brain: Predominance in the limbic system, *Res. Commun. Chem. Pathol. Pharmacol.* **6:**1052–1062.

Hoover, D. B., and Baisden, R. H., 1980, Localization of putative cholinergic neurons innervating the anteroventral thalamus, *Brain Res. Bull.* **5:**519–524.

Hoover, D. B., Muth, E. A., and Jacobowitz, D. M., 1978, A mapping of the distribution of acetylcholine, choline acetyltransferase and acetylcholinesterase in discrete areas of rat brain, *Brain Res.* **153:**295–306.

Houser, C. R., Crawford, G. D., Salvaterra, P. M., and Vaughn, J. E., 1983, Cholinergic neurons in rat cerebral cortex indentified by immunocytochemical localization of choline acetyltransferase, *Soc. Neurosci. Abstr.* **9:**80.

Hunt, S., and Schmidt, J., 1978, Some observations on the binding patterns of α-bungarotoxin in the central nervous system of the rat, *Brain Res.* **157:**213–232.

Innis, R. B., Corrêa, F. M. A., Uhl, G. R., Schneider, B., and Snyder, S. H., 1979, Cholecystokinin octapeptide-like immunoreactivity: Histochemical localization in rat brain, *Proc. Natl. Acad. Sci. USA* **76:**521–525.

Itaya, S. K., Van Hoesen, G. W., and Jenq, C.-B., 1981, Direct retinal input to the limbic system of the rat, *Brain Res.* **226:**33–42.

Iwahori, N., and Mizuno, N., 1981a, A Golgi study on the neuronal organization of the interhemispheric cortex in the mouse. I. Projection neurons, *Anat. Embryol.* **161:**465–481.

Iwahori, N., and Mizuno, N., 1981b, A Golgi study on the neuronal organization of the interhemispheric cortex in the mouse. II. Intrinsic neurons, *Anat. Embryol.* **161:**483–498.

Jacobowitz, D. M., and Goldberg, A. M., 1977, Determination of acetylcholine in discrete regions of the rat brain, *Brain Res.* **122:**575–577.

Jacobson, S., 1970, Distribution of commissural axon terminals in the rat neocortex, *Exp. Neurol.* **28:**193–205.

Johnston, M. V., McKinney, M., and Coyle, J. T., 1981, Neocortical cholinergic innervation: A description of extrinsic and intrinsic components in the rat, *Exp. Brain Res.* **43:**159–172.

Jones, B. E., and Moore, R. Y., 1977, Ascending projections of the locus coeruleus in the rat. II. Autoradiographic study, *Brain Res.* **127:**23–53.

Jones, E. G., and Leavitt, R. Y., 1974, Retrograde axonal transport and the demonstration of nonspecific projections to the cerebral cortex and striatum from thalamic intralaminar nuclei in the rat, cat and monkey, *J. Comp. Neurol.* **154:**349–378.

Jürgens, U., and Pratt, R., 1979, Role of the periaqueductal gray in vocal expression of emotion, *Brain Res.* **167:**367–378.

Kaada, B. R., 1951, Somato-motor, autonomic and electrocorticographic responses to electrical stimulation of 'rhinencephalic' and other structures in primates, cat and dog, *Acta Physiol. Scand.* **24**(Suppl. 83)**:**1–285.

Kaelber, W. W., Mitchell, C. L., Yarmat, A. J., Afifi, A. K., and Lorens, S. A., 1975, Centrum medianum–parafascicularis lesions and reactivity to noxious and non-noxious stimuli, *Exp. Neurol.* **46:**282–290.

Kaitz, S. S., and Robertson, R. T., 1981, Thalamic connections with limbic cortex. II. Corticothalamic projections, *J. Comp. Neurol.* **195:**527–545.

Kamiya, H., Rotter, A., and Jacobowitz, D. M., 1981, Muscarinic receptor binding following cholinergic nerve lesions of the cingulate cortex and hippocampus of the rat, *Brain Res.* **209:**432–439.

Kandel, E. R., and Spencer, W. A., 1961, Electrophysiology of hippocampal neurons. II. Afterpotentials and repetitive firing, *J. Neurophysiol.* **24:**243–259.

Kappers, C. U. A., Huber, G. C., and Crosby, E. C., 1967, *The Comparative Anatomy of the Nervous System of Vertebrates Including Man*, Volume 3, Hafner, New York.

Keefer, D. A., Chung, J.-W., and Heimer, L., 1980, Morphology and ultrastructure of anterior thalamic neurons solidly labeled with horseradish peroxidase, *Brain Res.* **191**:301–311.

Kemper, T. L., Caviness, W. F., and Yakovlev, P. I., 1973, The neuronographic and metric study of the dendritic arbours of neurons in the motor cortex of *Macaca mulatta* at birth and at 24 months of age, *Brain* **96**:765–782.

Kerr, F. W. L., 1975, The ventral spinothalamic tract and other ascending systems of the ventral funiculus of the spinal cord, *J. Comp. Neurol.* **159**:335–356.

Khachaturian, H., Lewis, M. E., Hollt, V., and Watson, S. J., 1983, Telencephalic enkephalinergic systems in the rat, *J. Neurosci.* **3**:844–855.

Kimura, H., McGeer, P. L., Peng, J. H., and McGeer, E. G., 1981, The central cholinergic system studied by choline acetyltransferase immunohistochemistry in the cat, *J. Comp. Neurol.* **200**:151–201.

Kobayashi, R. M., Palkovits, M., Hruska, R. E., Rothschild, R., and Yamamura, H. I., 1978, Regional distribution of muscarinic cholinergic receptors in rat brain, *Brain Res.* **154**:13–23.

Köhler, C., and Schwarcz, R., 1983, Comparison of ibotenate and kainate neurotoxicity in rat brain: A histological study, *Neuroscience* **8**:819–835.

Köhler, C., Schwarcz, R., and Fuxe, H., 1979, Intrahippocampal injections of ibotenic acid provide histological evidence for a neurotoxic mechnaims different from kainic acid, *Neurosci. Lett.* **15**:223–228.

Kremer, W. F., 1947, Autonomic and somatic reactions induced by stimulation of the cingular gyrus in dogs, *J. Neurophysiol.* **10**:371–379.

Krieg, W. J. S., 1946, Connections of the cerebral cortex. I. The albino rat. B. Structure of the cortical areas, *J. Comp. Neurol.* **84**:277–323.

Kruger, L., and Albe-Fessard, D., 1960, Distribution of responses to somatic afferent stimuli in the diencephalon of the cat under choralose anesthesia, *Exp. Neurol.* **2**:442–467.

Kuhar, M. J., and Yamamura, H. I., 1976, Localization of cholinergic muscarinic receptors in rat brain by light microscopic radioautography, *Brain Res.* **110**:229–243.

Kuhar, M. J., Pert, C. B., and Snyder, S. H., 1973, Regional distribution of opiate receptor binding in monkey and human brain, *Nature* **245**:447–450.

Künzle, H., 1978, An autoradiographic analysis of the efferent connections from premotor and adjacent prefrontal regions (area 6 and 9) in *Macaca fascicularis*, *Brain Behav. Evol.* **15**:185–234.

Kusicka, C., and Schulz, E., 1981, Quantitative Untersuchungen von Sternzelltypen der Regio cingularis mesoneocorticalis in Vergleich, zu den Nachbarregionen (Meso-archicortex und Neocortex) und nach Afferenzaus schaltung bei der Ratte, *J. Hirnforsch.* **22**:591–619.

Lang, W., and Henke, H., 1983, Cholinergic receptor binding and autoradiography in brains of non-neurological and senile dementia of Alzheimer-type patients, *Brain Res.* **267**:271–280.

Lent, R., 1982, The organization of subcortical projections of the hamster's visual cortex, *J. Comp. Neurol.* **206**:227–242.

Lewis, M. E., Pert, A., Pert, C. B., and Herkenham, M., 1983, Opiate receptor localization in rat cerebral cortex, *J. Comp. Neurol.* **216**:339–358.

Lewis, M. S., Molliver, M. E., Morrison, J. H., and Lidov, H. G. W., 1979, Complementarity of dopaminergic and noradrenergic innervation in anterior cingulate cortex of the rat, *Brain Res.* **164**:328–333.

Lewis, P. R., and Shute, C. C. D., 1967, The cholinergic limbic system: Projections to hippocampal formation, medial cortex, nuclei of the ascending cholinergic reticular system, and the subfornical organ and supra-optic crest, *Brain* **90**:521–542.

Lidov, H. G. W., Grzanna, R., and Molliver, M. E., 1980, The serotonin innervation of the cerebral cortex in the rat—An immunohistochemical analysis, *Neuroscience* **5**:207–227.

Lindvall, O., Björklund, A., Moore, R. Y., and Stenevi, U., 1974, Mesencephalic dopamine neurons projecting to neocortex, *Brain Res.* **81**:325–331.

Locke, S., Angevine, J. B., and Yakovlev, P. I., 1964, Limbic nuclei of thalamus and connections of limbic cortex. VI. Thalamocortical projection of lateral dorsal nucleus in cat and monkey, *Arch. Neurol.* **11**:1–12.

Lorén, I., Emson, P. C., Fahrenkrug, J., Björklund, A., Alumets, J., Håkanson, R., and Sundler, F., 1979, Distribution of vasoactive intestinal polypeptide in the rat and mouse brain, *Neuroscience* **4**:1953–1976.

Lorente de Nó, R., 1933, Studies on the structure of the cerebral cortex. I. The area entorhinalis, *J. Psychol. Neurol.* **45:**381–438.

Lorente de Nó, R., 1938, Cerebral cortex: Architecture, intracortical connections, motor projections, in: *Physiology of the Nervous System* (J. F. Fulton, ed.), Oxford University Press, London, pp. 291–325.

Lubar, J. F., 1964, Effect of medial cortical lesions on the avoidance behavior of the cat, *J. Comp. Physiol. Psychol.* **58:**38–46.

Lubar, J. F., and Perachio, A. A., 1965, One-way and two-way learning and transfer of an active avoidance response in normal and cingulectomized cats, *J. Comp. Physiol. Psychol.* **60:**46–52.

Luthin, G. R., and Wolfe, B. B., 1984, Comparison of ^{3}H-pirenzepine and ^{3}H-quinuclidinylbenzilate binding to muscarinic cholinergic receptors in rat brain, *J Pharmacol. Exp. Ther.* **228:**648–655.

Lux, H. D., and Pollen, D. A., 1966, Electrical constants of neurons in the motor cortex of the cat, *J. Neurophysiol.* **29:**207–220.

Macchi, G., Bentivoglio, M., D'Atena, C., Rossini, P., and Tempesta, E., 1977, The cortical projections of the thalamic intralaminar nuclei restudied by means of the HRP retrograde axonal transport, *Neurosci. Lett.* **4:**121–126.

McKinney, M., and Coyle, J. T., 1982, Regulation of neocortical muscarinic receptors: Effects of drug treatment and lesions, *J. Neurosci.* **2:**97–105.

Mangini, N. J., and Pearlman, A. L., 1980, Laminar distribution of receptive field properties in the primary visual cortex of the mouse, *J. Comp. Neurol.* **193:**203–222.

Mantyh, P. W., 1982, Forebrain projections to the periaqueductal gray in the monkey, with observations in the cat and rat, *J. Comp. Neurol.* **206:**146–158.

Marin-Padilla, M., 1967, Number and distribution of the apical dendritic spines of the layer V pyramidal cells in man, *J. Comp. Neurol.* **131:**475–490.

Marin-Padilla, M., and Stibitz, G. R., 1968, Distribution of the apical dendritic spines of the layer V pyramidal cells of the hamster neocortex, *Brain Res.* **11:**580–592.

Mark, V. H., Ervin, F. R., and Yakovlev, P. I., 1963, Stereotactic thalamotomy. III. The verification of anatomical lesion sites in the human thalamus, *Arch. Neurol.* **8:**528–538.

Mash, D. C., Flynn, D. D., and Potter, L. T., 1985, Loss of M2 muscarinic receptors in the cerebral cortex in Alzheimer's disease and experimental cholinergic denervation, *Science* **228:**1115–1117.

Massari, V. J., Gottesfeld, Z., and Jacobowitz, D. M., 1976, Distribution of glutamic acid decarboxylase in certain rhombencephalic and thalamic nuclei of the rat, *Brain Res.* **118:**147–151.

Mayer, D. J., and Liebeskind, J. C., 1974, Pain reduction by focal electrical stimulation of the brain: An anatomical and behavioral analysis, *Brain Res.* **68:**73–93.

Mehler, W. R., Feferman, M. E., and Nauta, W. J. H., 1960, Ascending axon degeneration following anterolateral cordotomy: An experimental study in the monkey, *Brain* **83:**718–752.

Meibach, R. C., and Siegel, A., 1977, Subicular projections to the posterior cingulate cortex in rats, *Exp. Neurol.* **57:**264–274.

Messing, R. B., and Lytle, L. D., 1977, Serotonin-containing neurons: Their possible role in pain and analgesia, *Pain* **4:**1–21.

Miller, M. W., and Vogt, B. A., 1984a, Direct connections of rat visual cortex with sensory, motor, and association cortices, *J. Comp. Neurol.* **226:**184–202.

Miller, M. W., and Vogt, B. A., 1984b, Heterotopic and homotopic callosal connections in rat visual cortex, *Brain Res.* **297:**75–89.

Montero, V. M., 1981, Comparative studies on the visual cortex, in: *Cortical Sensory Organization*, Volume 2 (C. N. Woolsey, ed.), Humana Press, Clifton, N. J.

Montero, V. M., Rojas, A., and Torrealba, F., 1973, Retinotopic organization of striate and peristriate visual cortex in the albino rat, *Brain Res.* **53:**197–201.

Morley, B. J., and Kemp, G. E., 1981, Characterization of a putative nicotinic acetylcholine receptor in mammalian brain, *Brain Res. Rev.* **3:**81–104.

Moroni, F., Bianchi, C., Monet, G., Tanganelli, S., Spidalieri, G., Guandalini, P., and Beani, L., 1982, Release of GABA from the guinea pig neocortex induced by electrical stimulation of the locus coeruleus or by norepinephrine, *Brain Res.* **232:**216–221.

Morrell, J. I., Greenberger, L. M., and Pfaff, D. W., 1981, Hypothalamic, other diencephalic, and telencephalic neurons that project to the dorsal midbrain, *J. Comp. Neurol.* **201:**589–620.

Morrison, J. H., Grzanna, R., Molliver, M. E., and Coyle, J. T., 1978, The distribution and orientation of noradrenergic fibers in neocortex of the rat: An immunofluorescence study, *J. Comp. Neurol.* **181:**17–40.

Müller-Preuss, P., and Jürgens, U., 1976, Projections from the "cingular" vocalization area in the squirrel monkey, *Brain Res.* **103:**29–43.

Murphy, E. H., and Berman, N., 1979, The rabbit and the cat: A comparison of some features of response properties of single cells in the primary visual cortex, *J. Comp. Neurol.* **188:**401–428.

Nagai, T., McGeer, P. L., and McGeer, E. G., 1983, Distribution of GABA-T-intensive neurons in the rat forebrain and midbrain, *J. Comp. Neurol.* **218:**220–238.

Nakamura, S., 1977, Some electrophysiological properties of neurons of rat locus coeruleus, *J. Physiol. (London)* **267:**641–658.

Ogawa, T., Ito, S., and Kato, H., 1981, Membrane characteristics of visual cortical neurons in in vitro slices, *Brain Res.* **226:**315–319.

Olpe, H.-R., 1981, The cortical projection of the dorsal raphe nucleus: Some electrophysiological and pharmacological properties, *Brain Res.* **216:**61–71.

Olpe, H.-R., Glatt, A., Laszlo, J., and Schellenberg, A., 1980, Some electrophysiological and pharmacological properties of the cortical, noradrenergic projection of the locus coeruleus in the rat, *Brain Res.* **186:**9–19.

Olszewski, J., 1952, *The Thalamus of Macaca mulatta,* Karger, Basel.

Palacios, J. M., Wamsley, J. K., and Kuhar, M. J., 1981, High affinity GABA receptors—Autoradiographic localization, *Brain Res.* **222:**285–307.

Pandya, D. N., Dye, P., and Butters, N., 1971, Efferent cortico-cortical projections of the prefrontal cortex in the rhesus monkey, *Brain Res.* **31:**35–46.

Pandya, D. N., Van Hoesen, G. W., and Mesulam, M.-M., 1981, Efferent connections of the cingulate gyrus in the rhesus monkey, *Exp. Brain Res.* **42:**319–330.

Papez, J. W., 1937, A proposed mechanism of emotion, *Arch. Neurol. Psychiatry* **38:**725–744.

Parent, A., and Butcher, L. L., 1976, Organization and morphologies of acetylcholinesterase-containing neurons in the thalamus and hypothalamus of the rat, *J. Comp. Neurol.* **170:**205–226.

Parnavelas, J. G., Globus, A., and Kaups, P., 1973, Continuous illumination from birth affects spine density of neurons in the visual cortex of the rat, *Exp. Neurol.* **40:**742–747.

Parnavelas, J. G., Burne, R. A., and Lin, C.-S., 1983, Distribution and morphology of functionally identified neurons in the visual cortex of the rat, *Brain Res.* **261:**21–29.

Penny, G. R., Fitzpatrick, D., Schmechel, D. E., and Diamond, I. T., 1983, Glutamic acid decarboxylase-immunoreactive neurons and horseradish peroxidase-labeled projection neurons in the ventral posterior nucleus of the cat and *Galago senegalensis, J. Neurosci.* **3:**1868–1887.

Peretz, E., 1960, The effects of lesions of the anterior cingulate cortex on the behavior of the rat, *J. Comp. Physiol. Psychol.* **53:**540–548.

Perl, E. R., and Whitlock, D. G., 1961, Somatic stimuli exciting spinothalamic projections to thalamic neurons in cat and monkey, *Exp. Neurol.* **3:**256–296.

Pert, A., and Yaksh, T., 1974, Sites of morphine induced analgesia in the primate brain: Relation to pain pathways, *Brain Res.* **80:**135–140.

Peters, A., and Proskauer, C. C., 1980, Synaptic relationships between a multipolar stellate cell and a pyramidal neuron in the rat visual cortex: A combined Golgi–electron microscopic study, *J. Neurocytol.* **9:**163–183.

Peters, A., Proskauer, C. C., and Ribak, C. E., 1982, Chandelier cells in rat visual cortex, *J. Comp. Neurol.* **206:**397–416.

Peters, A., Miller, M., and Kimerer, L. M., 1983, Cholecystokinin-like immunoreactive neurons in rat cerebral cortex, *Neuroscience* **8:**431–448.

Petras, J. M., 1971, Connections of the parietal lobe, *J. Psychiat. Res.* **8:**189–201.

Pool, J. L., and Ransohoff, J., 1949, Autonomic effects on stimulating rostral portions of cingulate gyri in man, *J. Neurophysiol.* **12:**385–392.

Porrino, L. J., and Goldman-Rakic, P. S., 1982, Brainstem innervation of prefrontal and anterior cingulate cortex in the rhesus monkey revealed by retrograde transport of HRP, *J. Comp. Neurol.* **205:**63–76.

Powell, E. W., 1978, The cingulate bridge between allocortex, isocortex, and thalamus, *Anat. Rec.* **190:**783–794.

Price, D. D., and Dubner, R., 1977, Neurons that subserve the sensory–discriminative aspects of pain, *Pain* **3:**307–338.

Raiteri, M., Leardi, R., and Marchi, M., 1984, Heterogeneity of presynaptic muscarinic receptors regulating neurotransmitter release in the rat brain, *J. Pharmacol. Exp. Ther.* **228:**209–214.

Ramón y Cajal, S., 1911, *Histologie du Système Nerveux de l'Homme et des Vertébrés*, Volume II, Maloine, Paris.

Ramón y Cajal, S., 1922, Estrudios sobre la Fina Estructura de la Corteza Regional de los Roedores. 1. Corteza suboccipital (Retrosplenial de Brodmann), *Trab. Lab. Biol. Univ. Madrid* **20:**1–30.

Ribak, C. E., and Peters, A., 1975, An autoradiographic study of the projections from the lateral geniculate body of the rat, *Brain Res.* **92:**349–368.

Richardson, D. E., and Akil, H., 1977, Pain reduction by electrical brain stimulation in man, *J. Neurosurg.* **47:**184–194.

Roberts, G. W., Woodhams, P. L., Polak, J. M., and Crow, T. J., 1984, Distribution of neuropeptides in the limbic system of the rat: The hippocampus, *Neuroscience* **11:**35–77.

Robertson, R. T., and Kaitz, S. S., 1981, Thalamic connections with limbic cortex. I. Thalamocortical projections, *J. Comp. Neurol.* **195:**501–525.

Robertson, R. T., Kaitz, S. S., and Robards, M. J., 1980, A subcortical pathway links sensory and limbic systems of the forebrain, *Neurosci. Lett.* **17:**161–165.

Robertson, R. T., Thompson, S. M., and Kaitz, S. S., 1983, Projections from the pretectal complex to the thalamic lateral dorsal nucleus of the cat, *Exp. Brain Res.* **51:**157–171.

Robinson, B. W., 1967, Vocalization evoked from forebrain in *Macaca mulatta, Physiol. Behav.* **2:**345–354.

Rose, J. E., and Woolsey, C. N., 1948, Structure and relations of limbic cortex and anterior thalamic nuclei in rabbit and cat, *J. Comp. Neurol.* **89:**279–340.

Rose, M., 1912, Histologische Lokalisation der Grosshirnrinde bei kleinen Saugetieren (Rodentia, Insectivora, Chiroptera), *J. Psychol. Neurol.* **19:**125–207.

Rose, M., 1927, Gyrus limbicus anterior und Regio retrosplenialis (Cortex holoprotoptychos quinquestratificatus) Vergleichende Architektonik bei Tier und mensch, *J. Psychol. Neurol.* **43:**65–173.

Rose, M., 1931, Cytoarchitektonischer Atlas der Grosshirnrinde des Kaninchens, *J. Psychol. Neurol.* **43:**353–440.

Rosene, D. L., and Van Hoesen, G. W., 1977, Hippocampal efferents reach widespread areas of cerebral cortex and amygdala in the rhesus monkey, *Science* **198:**315–317.

Rosenfeld, J. P., and Holzman, B. S., 1978, Effects of morphine on medial thalamic and medial bulboreticular aversive stimulation thresholds, *Brain Res.* **150:**436–440.

Rotter, A., Birdsall, N. J. M., Burgen, A. S. V., Field, P. M., Hulme, F. C., and Raisman, G., 1979, Muscarinic receptors in the central nervous system of the rat. I. Techniques for autoradiographic localization of the binding of ^{3}H-propylbenzilylcholine mustard and its distribution in the forebrain, *Brain Res. Rev.* **1:**141–165.

Royce, G. J., 1982, Laminar origin of cortical neurons which project upon the caudate nucleus: A horseradish peroxidase investigation in the cat, *J. Comp. Neurol.* **205:**8–29.

Royce, G. J., 1983, Cells of origin of corticothalamic projections upon the centromedian and parafascicular nuclei in the cat, *Brain Res.* **258:**11–21.

Sanides, F., 1970, Functional architecture of motor and sensory cortices in primates in the light of a new concept of neocortex evolution, in: *The Primate Brain* (C. R. Noback and W. Montagna, eds.), Appleton–Century–Crofts, New York, pp. 137–208.

Sanides, F., and Sanides, D., 1972, The "extraverted neurons" of the mammalian cerebral cortex, *Z. Anat. Entwicklungsgesch.* **136:**272–293.

Sar, M., Stumpf, W. E., Miller, R. J., Chang, K.-J., and Cuatrecasas, P., 1978, Immunohistochemical localization of enkephalin in rat brain and spinal cord, *J. Comp. Neurol.* **182:**17–38.

Satoh, K., Armstrong, D. M., and Fibiger, H. C., 1983, A comparison of the distribution of central cholinergic neurons as demonstrated by AChE-pharmacohistochemistry and ChAT-immunohistochemistry, *Soc. Neurosci. Abstr.* **9:**80.

Schierhorn, H., Doedens, K., and Nagel, I., 1972, Über die laminäre Zuordnung der apikalen Dendritenspines der lamina-V-Pyramidenzellen in der sensomotorischen Hirnrinde der Albinoratte, *Gegenbaurs Morph. Jahrb.* **118:**465–487.

Scholfield, C. N., 1978, Electrical properties of neurons in the olfactory cortex slice in vitro, *J. Physiol. (London)* **275:**535–546.

Schönheit, B., and Schulz, E., 1982, Korrelation neuronaler Parameter von cingulären Lamina V- und Lamina III-Pyramidenneuronen der Ratte, *J. Hirnforsch.* **23:**203–209.

Schulz, E., and Schönheit, B., 1974, Neurohistologische Untersuchungen zur Neuronstruktur der Regio limbica anterior der Ratte, *J. Hirnforsch.* **15:**469–490.

Schwab, M., Agid, Y., Glowinski, J., and Thoenen, H., 1977, Retrograde axonal transport of ^{125}I-tetanus toxin as a tool for tracing fiber connections in the central nervous system: Connections of the rostral part of the rat neostriatum, *Brain Res.* **126:**211–224.

Schwarcz, R., Hökfelt, T., Fuxe, K., Jonsson, G., Goldstein, M., and Terenius, L., 1979, Ibotenic acid-induced neuronal degeneration: A morphological and neurochemical study, *Exp. Brain Res.* **37:**199–216.

Schwartzkroin, P. A., 1975, Characteristics of CA1 neurons recorded intracellularly in the hippocampal *in vitro* slice preparation, *Brain Res.* **85:**423–436.

Schwartzkroin, P. A., and Knowles, W. D., 1984, Intracellular study of human epileptic cortex: In vitro maintenance of epileptiform activity? *Science* **223:**709–712.

Segal, M., 1981, The action of norepinephrine in the rat hippocampus: Intracellular studies in the slice preparation, *Brain Res.* **206:**107–128.

Segal, M., Dudai, Y., and Amsterdam, A., 1978, Distribution of an α-bungarotoxin-binding cholinergic nicotinic receptor in rat brain, *Brain Res.* **148:**105–119.

Sherk, H., and LeVay, S., 1983, Contribution of the corticoclaustral loop to receptive field properties in area 17 of the cat, *J. Neurosci.* **3:**2121–2127.

Showers, M. J. C., 1959, The cingulate gyrus: Additional motor areas and cortical autonomic regulator, *J. Comp. Neurol.* **112:**231–287.

Sikes, R. W., and Vogt, B. A., 1985, Anatomical and physiological characterization of the cingulate–visual border and connections between these areas in rabbits, *Anat. Rec.* **211:**177A.

Sikes, R. W., Chronister, R. B., and White, L. E., Jr., 1977, Origin of the direct hippocampus–anterior thalamic bundle in the rat: A combined horseradish peroxidase–Golgi analysis, *Exp. Neurol.* **57:**379–395.

Skultety, F. M., 1963, Stimulation of periaqueductal gray and hypothalamus, *Arch. Neurol.* **8:**609–620.

Smith, W. K., 1945, The functional significance of the rostral cingular cortex as revealed by its responses to electrical excitation, *J. Neurophysiol.* **8:**241–255.

Somogyi, P., 1977, A specific axon-axonal neuron in the visual cortex of the rat, *Brain Res.* **136:**345–350.

Somogyi, G., Tömböl, T., and Hajdu, F., 1973, Golgi architecture and connections of the nucleus anterodorsalis thalami, *Acta Anat.* **85:**466–476.

Somogyi, G., Hajdu, F., and Hassler, R., 1977, Electron microscopic study of terminal degeneration in the anterodorsal thalamic nucleus of the cat, *Cell Tissue Res.* **182:**455–467.

Somogyi, G., Hajdu, F., Tömböl, T., and Madarász, M., 1979, Types of thalamocortical relay neurons in the anteroventral nucleus of the cat, *Cell Tissue Res.* **196:**175–179.

Somogyi, P., Freund, T. F., Halász, N., and Kisvárday, Z. F., 1981, Selectivity of neuronal [^{3}H] GABA accumulation in the visual cortex as revealed by Golgi staining of the labeled neurons, *Brain Res.* **225:**431–436.

Somogyi, P., Kisvárday, Z. F., Freund, T. F., and Cowey, A., 1984, Characterization by Golgi impregnation of neurons that accumulate ^{3}H-GABA in the visual cortex of monkey, *Exp. Brain Res.* **53:**295–303.

Sørensen, K. E., 1980, Ipsilateral projection from the subiculum to the retrosplenial cortex in the guinea pig, *J. Comp. Neurol.* **193:**893–911.

Spector, I., Hassmannova, J., and Albe-Fessard, D., 1974, Sensory properties of single neurons of cat's claustrum, *Brain Res.* **66:**39–65.

Stephan, H., 1975, *Handbuch der mikroskopischen Anatomie des Menschen,* Volume 9, Springer-Verlag, Berlin.

Sutton, D., Larson, C., and Lindeman, R. C., 1974, Neocortical and limbic lesion effects on primate phonation, *Brain Res.* **71:**61–75.

Swanson, L. W., and Hartman, B. K., 1975, The central adrenergic system: An immunofluorescence study of the location of cell bodies and their efferent connections in the rat utilizing dopamine-β-hydroxylase as a marker, *J. Comp. Neurol.* **163:**467–506.

Talairach, J., Bancaud, J., Geier, S., Bordas-Ferrer, M., Bonis, A., Szikla, G., and Rusu, M., 1973, The cingulate gyrus and human behaviour, *Electroencephalogr. Clin. Neurophysiol.* **34:**45–52.

Tappaz, M. L., Brownstein, M. J., and Palkovitz, M., 1976, Distribution of glutamate decarboxylase in discrete brain nuclei, *Brain Res.* **108:**371–379.

Taylor, D. A., and Stone, T. W., 1980, The action of adenosine on noradrenergic neuronal inhibition induced by stimulation of locus coeruleus, *Brain Res.* **183:**367–376.

Thomas, G. J., and Slotnick, B. M., 1963, Impairment of avoidance responding by lesions in cingulate cortex in rats depends on food drive, *J. Comp. Physiol. Psychol.* **56:**959–964.

Thomas, G. J., Hostetter, G., and Barker, D. J., 1968, Behavioral function of the limbic system, in: *Progress in Physiological Psychology,* Volume 2 (E. Stellar and J. M. Sprague, eds.), Academic Press, New York.

Towns, L. C., Burton, S. L., Kimberly, C. J., and Fetterman, M. R., 1982, Projections of the dorsal lateral geniculate and lateral posterior nuclei to visual cortex in the rabbit, *J. Comp. Neurol.* **210:**87–98.

Trojanowski, J. Q., and Jacobson, S., 1976, Areal and laminar distribution of some pulvinar cortical efferents in rhesus monkey, *J. Comp. Neurol.* **169:**371–392.

Trulson, M. E., and Trulson, V. M., 1982, Differential effects of phasic auditory and visual stimuli on serotonergic neurons in the nucleus raphe dorsalis and nucleus raphe pallidus in freely moving cats, *Neurosci. Lett.* **32:**137–142.

Tsumoto, T., and Suda, K., 1982, Effects of stimulation of the dorsocaudal claustrum on activities of striate cortex neurons in the cat, *Brain Res.* **240:**345–349.

Turnbull, I. M., 1972, Bilateral cingulumotomy combined with thalamotomy or mesencephalic tractotomy for pain, *Surg. Gynecol. Obstet.* **134:**958–962.

Uematsu, S., Konigsmark, B., and Walker, A. E., 1974, Thalamotomy for alleviation of intractable pain, *Confin. Neurol.* **36:**88–96.

U'Prichard, D. C., Greenberg, D. A., Sheehan, P., and Snyder, S. H., 1977, Regional distribution of α-noradrenergic receptor binding in calf brain, *Brain Res.* **138:**151–158.

Valverde, F., 1967, Apical dendritic spines of the visual cortex and light deprivation in the mouse, *Exp. Brain Res.* **3:**337–352.

Vastola, E. F., 1982, Electrical signs of an oligo-synaptic visual projection to rat hippocampus, *Brain Behav. Evol.* **20:**1–18.

Vilensky, J. A., and Van Hoesen, G. W., 1981, Corticopontine projections from the cingulate cortex in the rhesus monkey, *Brain Res.* **205:**391–395.

Vinogradova, O. S., 1975, Functional organization of the limbic system in the process of registration of information: Facts and hypotheses, in: *The Hippocampus*, Volume 2 (R. L. Isaacson and K. H. Pribram, eds.), Plenum Press, New York, pp. 3–69.

Vogt, B. A., 1976, Retrosplenial cortex in the rhesus monkey: A cytoarchitectonic and Golgi study, *J. Comp. Neurol.* **169:**63–98.

Vogt, B. A., 1984, Afferent specific localization of muscarinic acetylcholine receptors in cingulate cortex, *J. Neurosci.* **9:**2191–2199.

Vogt, B. A., and Gorman, A. L. F., 1982, Responses of cortical neurons to stimulation of corpus callosum in vitro, *J. Neurophysiol.* **48:**1257–1273.

Vogt, B. A., and Miller, M., 1982, Cortical connections of cingulate area 29d: A limbic sensorimotor association cortex, *Soc. Neurosci. Abstr.* **8:**952.

Vogt, B. A., and Miller, M. W., 1983, Cortical connections between rat cingulate cortex and visual, motor and postsubicular cortices, *J. Comp. Neurol.* **216:**192–210.

Vogt, B. A., and Pandya, D. N., 1986, Primate cingulate cortex. II. Corticocortical connections (in preparation).

Vogt, B. A., and Peters, A., 1981, Form and distribution of neurons in rat cingulate cortex: Areas 32, 24, and 29, *J. Comp. Neurol.* **195:**603–625, **200:**461.

Vogt, B. A., Rosene, D. L., and Pandya, D. N., 1979, Thalamic and cortical afferents differentiate anterior from posterior cingulate cortex in the monkey, *Science* **204:**205–207.

Vogt, B. A., Rosene, D. L., and Peters, A., 1981, Synaptic termination of thalamic and callosal afferents in cingulate cortex of the rat, *J. Comp. Neurol.* **201:**265–283.

Vogt, B. A., Townes-Anderson, E., and Burns, D. L., 1985, Distribution of M_1 and M_2 muscarinic acetylcholine receptors on cortical neurons and thalamic afferents, *Trends Pharmacol. Sci.* (in press).

Vogt, B. A., Sikes, R. W., Swadlow, H. A., and Weyand, T. G., 1986, Rabbit cingulate cortex: Cytoarchitecture, physiological border with visual cortex and afferent cortical connections including those of visual, motor, postsubicular, and transcingulate origin, *J. Comp. Neurol.* (in press).

von Economo, C., 1929, *The Cytoarchitectonics of the Human Cerebral Cortex*, Oxford University Press, London.

Waller, W. H., 1934, Topographical relations of cortical lesions to thalamic nuclei in the albino rat, *J. Comp. Neurol.* **60:**237–269.

Wamsley, J. K., Gehlert, D. R., Roeske, W. R., and Yamamura, H. I., 1984, Muscarinic antagonist binding site heterogeneity as evidenced by autoradiography after direct labeling with ^{3}H-QNB and ^{3}H-pirenzepine, *Life Sci.* **34:**1395–1402.

Watabe, K., Nakai, K., and Kasamatsu, T., 1982, Visual afferents to norepinephrine-containing neurons in cat locus coeruleus, *Exp. Brain Res.* **48:**66–80.

Ward, A. A., 1948, The cingular gyrus: Area 24, *J. Neurophysiol.* **11:**13–23.

Weber, J. T., and Harting, J. K., 1980, The efferent projections of the pretectal complex: An autoradiographic and horseradish peroxidase analysis, *Brain Res.* **194:**1–28.

White, J. C., and Sweet, W. H., 1969, *Pain and the Neurosurgeon,* Thomas, Springfield, Ill.

Wiesendanger, R., and Wiesendanger, M., 1982a, The corticopontine system in the rat. I. Mapping of corticopontine neurons, *J. Comp. Neurol.* **208:**215–226.

Wiesendanger, R., and Wiesendanger, M., 1982b, The corticopontine system in the rat. II. The projection pattern, *J. Comp. Neurol.* **208:**227–238.

Willis, W. D., Kenshalo, D. R., and Leonard, R. B., 1979, The cells of origin of the primate spinothalamic tract, *J. Comp. Neurol.* **188:**543–574.

Wong, R. K. S., and Prince, D. A., 1981, Afterpotential generation in hippocampal pyramidal cells, *J. Neurophysiol.* **45:**86–97.

Woodward, D. J., Moises, H. C., Waterhouse, B. D., Hoffer, B. J., and Freedman, R., 1979, Modulatory actions of norepinephrine in the central nervous system, *Fed. Proc.* **38:**2109–2116.

Wyss, J. M., and Sripanidkulchai, K., 1984, The topography of the mesencephalic and pontine projections from the cingulate cortex of the rat, *Brain Res.* **293:**1–15.

Yaksh, T. L., 1979, Direct evidence that spinal serotonin and noradrenaline terminals mediate the spinal antinociceptive effects of morphine in the periaqueductal gray. *Brain Res.* **160:**180–185.

Yamamura, H. I., Kuhar, M. J., Greenberg, D., and Snyder, S. H., 1974, Muscarinic cholinergic receptor binding: Regional distribution in monkey brain, *Brain Res.* **66:**541–546.

Yeterian, E. H., and Van Hoesen, G. W., 1978, Cortico-striate projections in the rhesus monkey: The organization of certain cortico-caudate connections, *Brain Res.* **139:**43–63.

Yeung, J. C., Yaksh, T. L., and Rudy, T. A., 1977, Concurrent mapping of brain sites for sensitivity to the direct application of morphine and focal electrical stimulation in the production of antinociception in the rat, *Pain* **4:**23–40.

Yorke, C. H., Jr., and Caviness, V. S., Jr., 1975, Interhemispheric neocortical connections of the corpus callosum in the normal mouse: A study based on anterograde and retrograde methods, *J. Comp. Neurol.* **164:**233–246.

Zarbin, M. A., Wamsley, J. K., and Kuhar, M. J., 1982, Axonal transport of muscarinic cholinergic receptors in rat vagus nerve: High and low affinity agonist receptors move in opposite directions and differ in nucleotide sensitivity, *J. Neurosci.* **2:**934–941.

Zarbin, M. A., Innis, R. B., Wamsley, J. K., Snyder, S. H., and Kuhar, M. J., 1983, Autoradiographic localization of cholecystokinin receptors in rodent brain, *J. Neurosci.* **3:**877–906.

Zilles, K., Zilles, B., and Schleicher, A., 1980, A quantitative approach to cytoarchitectonics, VI. The areal pattern of the cortex of the albino rat, *Anat. Embryol.* **159:**335–360.

4

The Prefrontal Cortex and Temporal Integration

JOAQUIN M. FUSTER

1. Introduction

The associative cortex of the frontal lobe, commonly termed the prefrontal cortex, undergoes considerable expansion in the course of evolution. It reaches its greatest development in the human brain, constituting nearly one-third of the neocortex. Its phylogenetic expansion outstrips that of other cortical regions and also that of the mediodorsal (MD) nucleus of the thalamus, a nucleus so heavily and distinctly projecting to the prefrontal cortex that the latter is conventionally defined as the cortical territory of MD projection.

For a number of reasons, including its remarkable evolution, the prefrontal cortex has long been thought to play an important role in higher integrative functions. The evidence that this is indeed the case is now plentiful (Fuster, 1980). Yet the precise definition of those functions remains elusive, in part, no doubt, because of the complex and diversified nature of the neural signals that are apparently integrated in the prefrontal cortex, one of the most richly innervated of all cortical regions. In its location, anterior to the central sulcus and neighboring motor areas, many have naturally seen a clear indication that the prefrontal cortex plays primarily a motor role. That opinion is strengthened by the presence within that cortex of areas that are evidently engaged in at least two forms of motility: eye movements and speech. Nonetheless, even the most

---

JOAQUIN M. FUSTER • Department of Psychiatry and Brain Research Institute, University of California School of Medicine, Los Angeles, California 90024.

cursory analysis of the data from those areas, or from any other prefrontal area, reveals how simplistic the assignment would be of merely motor functions to the prefrontal cortex or any of its parts.

In the light of current knowledge, however, it is appropriate to postulate a critical role of the prefrontal cortex in certain complex forms of motor behavior or in certain aspects of motor control. This chapter will deal with one such postulate, namely, that the principal and most characteristic function of the prefrontal cortex is the temporal organization of behavior. Below, the evidence is reviewed that bears on this hypothesis. Data from three main sources will be considered: connectivity, electrophysiology, and neuropsychology. As we shall see, the evidence supporting the hypothesis is far more substantial than that conflicting with it and, of the latter, none invalidates it. We shall also see, however, that for it to retain its validity, the concept of a role of the prefrontal cortex in temporal organization needs to be qualified. Furthermore, it should become clear in this review that a broad and parsimonious postulate such as we propose may not be construed as implying the functional homogeneity of the prefrontal cortex at any level of analysis. In fact, the results of many studies point to the functional specialization of various areas of prefrontal cortex, although those various areas may be seen to subserve in one way or another the supraordinate function of temporal organization.

2. Connectivity

An overview of its extrinsic connections clearly suggests that the prefrontal cortex is the substrate for neural integrations at a very high level, where internal as well as external streams of information converge and interact and where signals or commands to a variety of effector mechanisms originate. Even allowing for the relatively large size of the prefrontal region in some species and for the differences in the connectivity of its parts, the fact remains that the prefrontal cortex is endowed with an exceptional profusion of afferent and efferent connections. In the light of the published reports of those connections, the prefrontal cortex, at least in primates, appears as a giant hub of neural communications. It is not only connected with neocortex, paleocortex, and archicortex, but with assorted structures at all other levels of the neuraxis. That diversity of anatomical links is the more remarkable for its functional implications, that is, for the multiplicity of functional domains that the prefrontal cortex evidently enters by virtue of those links. Now we have experimental and clinical evidence indicating that the prefrontal cortex is somehow involved in certain aspects of sensation and movement, in skeletal motility as well as autonomic function, in cognition as well as affect and emotion. However, even in the absence of that evidence, we might infer those involvements merely on the basis of the anatomical evidence and of what we know about the functions of the cerebral structures with which the prefrontal cortex is associated.

The most important thalamic projection to the prefrontal cortex is, by definition, that from the MD nucleus. However, other nuclei, especially those in the anterior and intralaminar sectors of the thalamus, also send projection fibers to the prefrontal cortex. For a long time it has been known that the MD–prefrontal connections are topologically organized and that different parts of the nucleus are connected with different parts of the cortex (Akert, 1964). However, in recent years a pattern of topological relationship has been established that applies to all the thalamic projections to the prefrontal cortex, including those of the MD nucleus (Kievit and Kuypers, 1977). That pattern overrides nuclear boundaries and the internal medullary lamina, as well as the sulcal pattern on the surface of the frontal lobe. Sagittally oriented slabs of thalamus—each comprising a portion of the MD nucleus and cutting across boundaries and laminae—project to transverse strips of prefrontal cortex. The most medial slab of thalamus, adjacent to the midplane, projects to the orbitofrontal cortex, and progressively more lateral slabs of thalamus project to progressively more posterior strips of the dorsolateral prefrontal cortex.

By virtue of its afferents from the lower brain stem and the limbic system (Nauta, 1972), the medial thalamus, including the medial (magnocellular) portion of the MD nucleus and anterior and intralaminar nuclei, most probably provides the prefrontal cortex with input related to the internal state and motivations of the organism. However, information of that kind, from the internal environment, arrives at the prefrontal cortex not only through the thalamus but also by direct projections from the hypothalamus (Kievit and Kuypers, 1975), the amygdala (Jacobson and Trojanowski, 1975), and others parts of the limbic system (Jacobson and Trojanowski, 1977; Rosene and Van Hoesen, 1977).

The large lateral (parvocellular) portion of the MD nucleus does not receive extrinsic afferents other than from the dorsolateral prefrontal cortex, to which it projects. Thus, that part of the MD nucleus seems to be extrinsically connected only with the prefrontal cortex. Accordingly, the reciprocal connections between parvocellular MD and prefrontal cortex may be suspected to serve as the substrate for reverberating circuits, perhaps involved in some form of memory. That suspicion is supported by data indicating that the cooling of the dorsolateral prefrontal cortex results in a short-term memory deficit and, concomitantly, a disruption of the sustained patterns of MD cell discharge associated with the retention of stimuli in short-term memory (Alexander and Fuster, 1973). Furthermore, MD lesions have also been noted to impair performance of short-term memory tasks (Schulman, 1964; Isseroff *et al.*, 1982).

Probably no other neocortical region is connected to so many others as the prefrontal cortex (Kuypers *et al.*, 1965; Pandya and Kuypers, 1969; Jones, 1969; Jones and Powell, 1970; Pandya *et al.*, 1969, 1971; Benevento and Fallon, 1975; Chavis and Pandya, 1976; Jacobson and Trojanowski, 1977; Mesulam *et al.*, 1977; Künzle, 1978; Goldman-Rakic and Schwartz, 1982). That connectivity, linking the prefrontal cortex to various and distant areas of the cerebral mantle, seems to be, for the most part, reciprocal. Its functions are not yet clear. However, some general inferences can be drawn from the pattern of corticocortical connections that has emerged in the past 20 years of extensive studies in primates and carnivores.

The prefrontal cortex is not directly connected to the primary sensory areas but to associative areas, in occipital, temporal, and parietal regions, which are themselves the recipients of afferents from those primary sensory areas (Jones and Powell, 1970). Thus, although the prefrontal cortex is not known to receive direct input from any cortical or subcortical component of any primary sensory pathway, it apparently receives, transcortically, indirect sensory input of the visual, auditory, and somesthetic modalities. Some of that input is most likely the result of elaborate integrations in the three major sensory association domains and possibly dependent, in part at least, on the perceptual and cognitive aspects of the sensory information that is processed in those domains.

The analysis of the known corticocortical connections of the prefrontal cortex reveals a degree of topological order and mutual correspondence, especially in the primate, between parts of the prefrontal cortex and other cortical areas. Different prefrontal areas receive projection fibers from different parasensory areas (Jones and Powell, 1970; Chavis and Pandya, 1976; Jacobson and Trojanowski, 1977). Thus, groups of projections from those distant and widely separated cortical sources converge on neighboring or overlapping parcels of the prefrontal cortex. In the light of that anatomical evidence, the prefrontal cortex may be viewed as a polysensory region, and therefore a region for cross-modal association. Electrophysiological evidence supports that view (Bignall and Imbert, 1969; Mohler *et al.*, 1973; Schechter and Murphy, 1975; Tanabe *et al.*, 1975; Benevento *et al.*, 1977).

The mechanisms of interaction of inputs from different sources within the prefrontal cortex are not known, but it is worthwhile in this regard to take note of two significant anatomical findings, one pointing to the possibility of interaction within certain cortical layers and the other between cortical columns. The first is the evidence that in the prefrontal cortex, as in other cortical regions, afferent fibers of various cortical sources terminate preponderantly in supragranular layers (Luttenberg, 1974; Jacobson and Trojanowski, 1977). As it pertains to afferents from the inferotemporal cortex, which presumably convey visual information, the results of a cooling study (Fuster *et al.*, 1982b) are in agreement with the anatomical evidence: cooling the inferotemporal cortex alters the reactions of prefrontal cells to visual stimuli, especially in supragranular layers. The second is the recent finding that, within certain zones of the prefrontal cortex, fibers of at least two cortical sources—contralateral frontal cortex and homolateral parietal cortex—innervate spatially alternating columns, suggesting a pattern of interdigitation of afferent terminals (Goldman-Rakic and Schwartz, 1982).

Practically all the cortical and subcortical structures known to send projections to the prefrontal cortex are the targets of efferent fibers from it. However, an important contingent of prefrontal efferents flow into a group of subcortical nuclei involved in motor functions, which are not themselves the source of direct fiber connections to the prefrontal cortex. Most prominent among these nuclei, for their size as well as the abundance of their prefrontal afferents, are the basal ganglia, notably the caudate nucleus (Nauta, 1964; Webster, 1965; Kemp and Powell, 1970). Nuclei of the hypothalamus and lower brain stem which participate in various aspects of movement also receive direct projections from the prefrontal cortex (DeVito and Smith, 1964; Leichnetz and Astruc, 1977). Subcortical structures implicated in eye movements receive projection from prefrontal area 8, the frontal eye fields (Künzle *et al.*, 1976; Leichnetz *et al.*, 1981). Apparently, whatever influences the prefrontal cortex has upon the primary motor cortex are exerted, like those of the primary sensory cortex upon the prefrontal cortex, through intermediaries. The available evidence indicates that prefrontal control of motor cortex is not direct but through the basal ganglia, the lateral thalamus, and premotor cortex (Fuster, 1981).

In conclusion, the connectivity of the prefrontal cortex is compatible with several theoretical constructs of its functions. It is not only fully congruent with the hypothesis of a prefrontal role in the temporal organization of behavior but also provides that hypothesis with at least indirect support. Indeed, any cerebral structure that plays a critical role in the concerted organization of goal-directed sequences of behavior must have access to information from a wide variety of sources, as well as to a wide variety of effectors. By reason of its connections, the prefrontal cortex is exceptionally well suited to perform that role. From limbic and diencephalic structures, it probably receives impulses related to goal and motivation, as well as the motivational significance of environmental events. From associative cortices, it probably receives sensory and cognitive information about the environment, as well as the state of proprioceptors. Through its output connections, the prefrontal cortex can modulate the many internal and external effectors that are engaged, concomitantly or in sequence, in order to attain the behavioral goal. On the basis of connectivity alone, only the reticular core of the brain stem is conceivably suited for performing the behavior-organizing role that we postulate for the prefrontal cortex. However, data from other sources, reviewed below, suggest that no cortical or subcortical formation has a role that is as selective and at the same time all-encompassing, in the temporal structuring of behavior, as the prefrontal cortex seems to have.

3. Electrophysiology

Electrical stimulation and recording techniques have been extensively and successfully used to trace neural pathways to and from the prefrontal cortex. Also useful in this regard have been the studies of single-cell and field-potential responses to sensory stimuli. Essentially, the electrical data on prefrontal connections are in accord with and confirm the anatomical data (Fuster, 1981).

The stimulation of certain areas of the prefrontal cortex provides evidence of its involvement in some important aspects of behavior. For instance, mild electrical stimulation of area 8, the frontal eye field, elicits movements of the eyes and the head (Robinson and Fuchs, 1969; Wagman and Mehler, 1972). Maps of that area have been made on the basis of the ocular and head motility resulting from stimulation of its parts. Because some of the elicited motility and concomitant pupillary changes are reminiscent of those accompanying attentive reactions in normal behavior, it has been suggested that the frontal eye field is a component of a so-called "teleokinetic system" (Hess, 1943; Fuster, 1980), a system of neural structures ensuring the attentive guidance of gaze and the coordination of related goal-directed acts. The superior colliculus, the pulvinar, and certain portions of the basal ganglia would also be part of that system. Furthermore, there is evidence that stimulation of the prefrontal cortex, particularly its orbital aspect, induces a variety of inhibitory phenomena in the motor and autonomic spheres (Fuster, 1981). However, the results of direct electrical stimulation of the prefrontal cortex on somatic and visceral functions do not provide but faint indications of the role of the areas concerned in the temporal organization of behavior. In that respect, recording studies are more explicit.

3.1. Field Potentials

The electrophysiological evidence of the participation of the prefrontal cortex in the temporal organization of behavior derives fundamentally from observations, in that cortex, of changes of cell discharge and field potentials during performance of behavioral tasks that require the integration of temporally discontiguous items of sensory information. The first such observation was made by Walter and his colleagues (Walter *et al.,* 1964; Walter, 1973) in humans. The behavioral paradigm they used is as follows: the subject is presented with a first (conditional) stimulus—a click—and, after a delay, a second (imperative) stimulus—a flickering light—which in turn demands a motor act (pressing a button). A slow negative potential can be recorded from anterior regions of the scalp during the delay that separates the two stimuli. The potential vanishes upon execution of the motor response. It is only after the subject has become accustomed to the relation of contingency between the two stimuli that the potential can be seen. For that reason the potential was termed the "contingent negative variation" (CNV). Since it appears immediately after the first stimulus, in anticipation of the second—and the subsequent response—it was also termed the "expectancy wave." Both designations imply relationships of significance, even casuality, that have not yet been made clear. What is clear is that a slow surface negative potential develops on the frontal region in the interval between events that are mutually dependent and, though temporally discontiguous, are part of the same behavioral structure. As we shall see, that field potential has a counterpart, which is probably its bioelectrical foundation, at the cellular level.

Subsequent experiments showed that the CNV is not restricted to humans but can also be seen in other primates (Borda, 1970; Rebert, 1972). Other experiments showed that its source is prefrontal and that it should be distinguished from at least one other negative wave that precedes motor acts and develops in a slightly more posterior region (Borda, 1970; Jaervilehto and Fruhstorfer, 1970). Sasaki and Gemba (1982), in the monkey, have demonstrated close relationship between the development of sensory-motor associations, as manifested behaviorally, and that of slow potentials in the frontal cortex. Others (Pirch *et al.,* 1983), in the rat, have demonstrated similar relationships at the single-cell and slow-potential levels.

3.2. Unit Discharge

In the late 1960s and early 1970s, investigations were undertaken (Fuster and Alexander, 1971; Kubota and Niki, 1971; Fuster, 1973; Niki, 1974a) with the aim of elucidating the relationship between the neuronal activity in the prefrontal cortex of the monkey and the behavioral performance of delay tasks—e.g., delayed response, delayed alternation. These are tasks that essentially re-

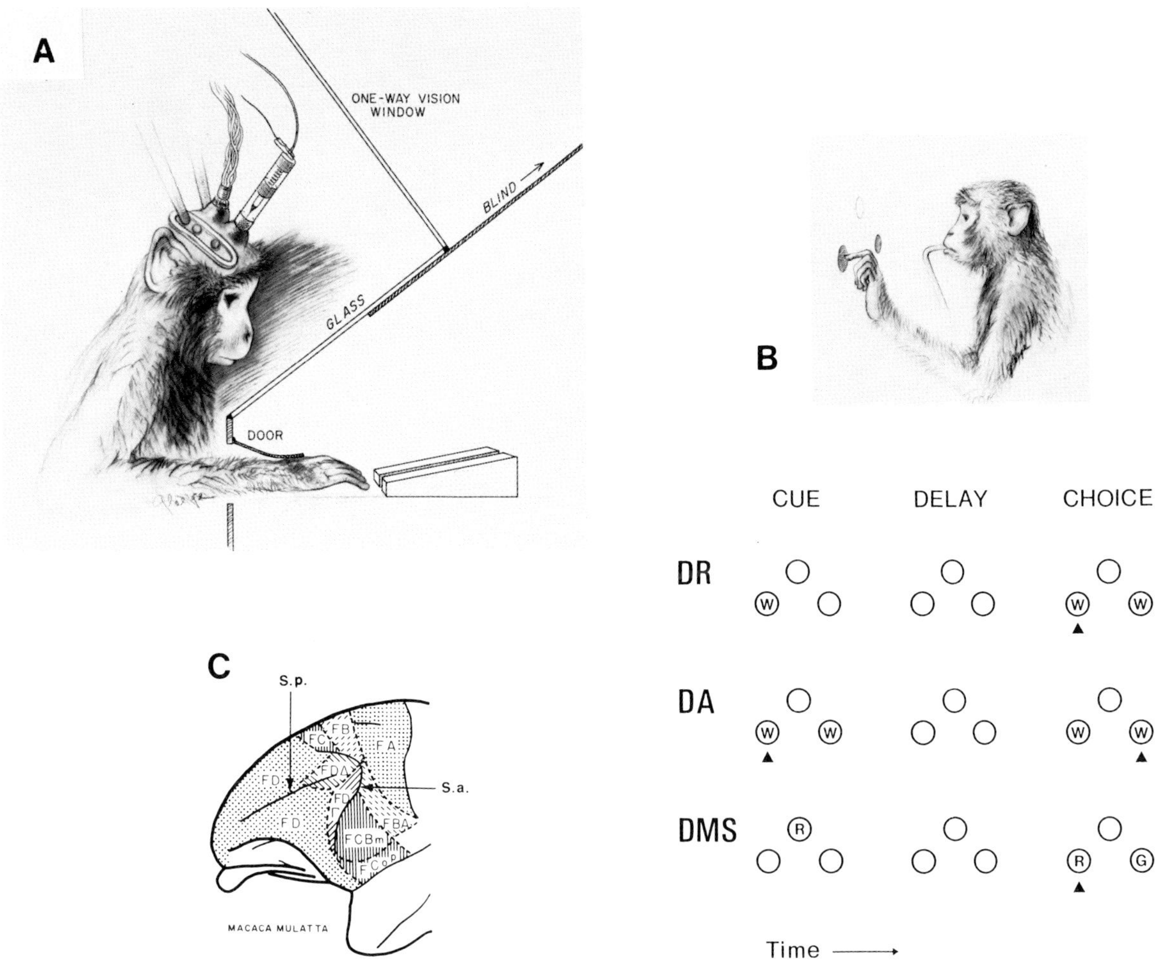

Figure 1. (A) Monkey performing a delayed-response task (direct method). A trial begins with elevation of the opaque screen ("blind") and, in full view of the animal, the placement of a piece of food under one of the two identical objects behind the glass. After that, the screen is lowered, blocking the view of the objects for a period of a few seconds or minutes. At the end of that delay, the screen is again raised and the monkey chooses one of the objects by introducing a hand through one of two corresponding trap doors. If the chosen object is the correct (baited) one, the monkey is allowed to retrieve the food from under it. Otherwise, the trial is terminated without reward. The position of the bait is changed randomly between trials. (The animal in this diagram carries a microelectrode positioner for single-cell recording.)

(B) At top, monkey facing a panel with stimulus–response buttons for performance of short-term memory tasks. Below, sequence of events in three such tasks: delayed response (DR), delayed alternation (DA), and delayed matching-to-sample (DMS). In a DR trial, the cue to be remembered is the brief white-light (W) illumination of a button (right or left, randomly changing); after a delay, two buttons are simultaneously lit and the animal must press the button that contained the cue. In DA, the animal must alternately press the right and the left of a pair of buttons that are simultaneously lighted between delays. In DMS, the cue (sample) is a colored light, red (R) or green (G), on the upper button; after a delay, both colors appear in the lower buttons and the animal must press the one with the sample color; both that color and the position of the colors in the lower buttons are changed randomly between trials. In the three sequence diagrams, black triangles mark the site of correct response, which is rewarded with a squirt of fruit juice automatically delivered to the monkey's mouth.

(C) Cytoarchitectonic map of the monkey's frontal cortex according to von Bonin and Bailey. The prefrontal cortex is the area designated FD (S.p., principal sulcus; S.a., arcuate sulcus).

quire that the animal retain an item of information, a cue, for execution of the appropriate motor act after a period of delay (which differs in duration from one study to another; it is usually between 10 and 20 sec in our studies). Thus, from an operational viewpoint, they are short-term memory tasks (Fig. 1). The rationale for neuronal-activity studies was firmly based on the old and extensive evidence that primates with lesions of the prefrontal cortex are severely impaired in the performance of those tasks (Jacobsen, 1935, 1936; Pribram *et al.*, 1952; Fuster, 1981).

Here, in broad outline, are the single-cell observations made in the prefrontal cortex of performing monkeys:

1. A large proportion of cells, a clear majority in certain prefrontal areas, react with increased firing frequency to sensory stimuli that serve the animal as cues for the performance of its task (Fuster, 1973; Niki, 1974a–c; Fuster *et al.*, 1982a). Some of the reactions are rather nonspecific, in that they are not related in either magnitude or duration to the specific cue for each trial. They appear to be related to attention or to certain features of the cue that are invariant, common to all cues, and constant for all trials. Such reactions are difficult to distinguish from those observable in many parts of the neocortex in a variety of behavioral tasks. The reactions of some prefrontal cells, however, are closely related to the specific cue for each trial, whether it be the location of a food-baited object, the color of a light, or the site of a just-completed motor act. Thus, those cells appear to distinguish visual, proprioceptive, or kinesthetic features.

2. The reactions of some prefrontal cells to visual stimuli are closely dependent on the behavioral implications of the stimuli—more specifically, on whether or not they call for a motor response (Kubota *et al.*, 1974; Sakai, 1974; Suzuki and Azuma, 1977). In other words, the cells are only responsive to a given stimulus if that stimulus, by appropriate manipulation of reinforcement, has become a signal for an instrumental response of the animal. For that reason, such cells, clearly sensitive to the relationship of contingency between stimulus and motor act, have been called "visuokinetic" (Kubota *et al.*, 1974). More recent evidence indicates that the phenomenon is not restricted to visual events; some prefrontal units appear to be sensitive to the relationship between auditory stimuli and movement (Kubota *et al.*, 1980).

3. Some cells, particularly common in the region of the sulcus principalis, are attuned to the spatial relationship of stimuli and response. They differentiate right from left, up from down. This was made clear in spatial delay tasks (Niki, 1974a–c; Fuster, 1975; Fuster *et al.*, 1982a; Kojima and Goldman-Rakic, 1982). A number of cells do not so much respond to different stimuli depending on their position or on that of the required response in the testing field as on the relative position of stimuli or responses with respect to one another (Niki, 1974c). Thus, they appear to code relative spatial location in the context of the testing situation and not in absolute terms.

4. Especially in the orbital prefrontal cortex, units can be found which show changes of discharge in relation to various aspects of reinforcement, the reward

provided to the animal after each correct response (Niki *et al.*, 1972; Rosenkilde *et al.*, 1981a). Some units seem to be influenced by the presence of appetitive reward under any conditions, others by the absence of that reward (when expected), and still others by the termination of the trial sequence, whether the response is rewarded or not.

5. One of the most relevant findings, in terms of our hypothesis, is that of prefrontal cells whose activity is altered in a sustained manner during the delay period, the time interposed between a given event and the appropriate response of the animal. Such cells are not found exclusively in the prefrontal cortex. They have also been observed in structures connected to that cortex, such as the MD nucleus of the thalamus (Fuster and Alexander, 1973) and the inferotemporal cortex (Fuster and Jervey, 1982), during performance of visual delay tasks.

3.2.1. Properties of Delay Discharge

The study of sustained changes in the discharge of prefrontal cells during the delay of delay tasks sheds some light on the functional characteristics of those cells and their possible role in the temporal structuring of behavior. Those changes, unfortunately, have not been examined in a consistent manner and by uniform methods. Furthermore, it is clear that the direction and course of the changes differ greatly between cells. Nevertheless, an overview of the available data justifies the assignment of a temporal function to the prefrontal cortex and provides some insights on how that function may be subserved by various contingents of prefrontal cells. Even though the analysis of the delay-discharge has been to a large extent qualitative and limited in terms of cell samples, the properties of that discharge are highly revealing. The following paragraphs summarize those properties:

1. Relation to learning. In delayed-response and delayed-matching tasks, where the delay follows a sensory cue which is to determine the animal's response at the end of the trial, the changes of cell firing during the delay are by and large attributable to the learned behavioral significance of that cue and cannot be interpreted as sensory reactions exclusively dependent on the physical characteristics of the stimulus. The principal reason for this assertion is the evidence that cortical cell responses to temporally discrete, unconditioned, sensory stimuli are shorter than those observed in response to the cues in delay tasks; indeed, the duration of the responses of prefrontal cells to those cues, persisting through a good portion of the ensuing delay, is generally much greater than that of the cell responses which can be expected in the untrained animal to the same stimuli (Fuster, 1973).

2. Relation to performance. A positive relation has been noted between the degree of discharge that some cells manifest during the delay and the level of performance of the animal in terms of correct behavioral responses (Fuster, 1973; Batuev *et al.*, 1979). That relationship suggests that the cells are involved in the neural processes that mediate performance of the task and determine the

accuracy of instrumental response; in other words, that the elevated discharge exhibited by some cells between a particular cue and the appropriate motor response is functionally significant with regard to the animal's capacity to successfully bridge the temporal gap between one and the other.

3. Relation to arousal. The activity of prefrontal cells during the delay, as during any other phase of a delayed-response trial, may be to some extent dependent on the level of arousal, a variable related to the general state of the animal and its nervous system. Although the precise role of such a variable in that activity cannot be easily determined, it is quite apparent that arousal *per se* does not determine or significantly affect the discharge of at least some of the cells during the delay. The arousal factor can be ruled out in the many cells in which the peak of activation of discharge does not occur at the time of the cue, when arousal is maximal (as evident from EEG analysis), but during the subsequent delay. It can also be ruled out in those cells whose discharge is only elevated in the delay and either inhibited or not changed at the cue (Fuster, 1973).

4. Relation to cross-temporal contingencies. Firing during the delay can be directly attributed to the behavioral contingency between the events that precede the delay and those that succeed it. That relation of delay discharge to the contingency between the temporally separate events that bracket the delay is made evident when the contingency is experimentally abolished, while preserving the basic format of the behavioral delayed-response paradigm. This has been achieved by depriving the cue of its value as the determinant of a specific motor act, while maintaining the paradigm basically intact, that is, with a cue (though incomplete), a delay, and a second stimulus (or stimuli) prompting the animal's response. Under such circumstances, the normal firing elevation (or inhibition) during the delay disappears (Fuster, 1973). It seems compelling to conclude that the firing change during the delay, as the CNV, depends heavily on the contingency between cue and response, much as that of some cells (e.g., "visuokinetic" cells) in tasks with minimal or no delay (Kubota *et al.,* 1974).

5. Cue-specificity. In several studies, cells have been found whose discharge during the delay is different depending on the specific cue for the trial (Niki, 1974a; Fuster, 1975; Niki and Watanabe, 1976a,b; Fuster *et al.,* 1982a; Kojima and Goldman-Rakic, 1982). Presumably, such cells are not only subject to some of the factors noted thus far, but to specific physical traits of the cue (e.g., color, position). Since that kind of differential firing takes place during the delay period, in some cases well after the cue has disappeared, it is reasonable to suppose that the cells in question are involved in the memory of the cue. Obviously, cells showing nondifferential changes of delay discharge cannot be excluded from that memory process, because in any given task the different cues share a number of traits (e.g., shape, brightness, contrast). Indeed, such cells might be involved in coding and retaining those common traits. But cue-specific delay discharge provides a more convincing reason for the attribution of short-term memory to some of the neuronal populations of the prefrontal cortex.

6. Response-specificity. Since in some delay tasks the site of the animal's response is predetermined by the cue from the very moment that cue appears, it is natural to suspect that some of the apparent cue-specificity of delay discharge may actually be response-specificity. In other words, the differential delay discharge may not be due to differences between cues but to differences between impending, predetermined responses. That possibility is obvious in the delayed-response and delayed-alternation tasks. In the second of those tasks, where the response of one trial (right or left) is the "cue" for the next, cue-specificity and response-specificity cannot be practically distinguished from each other in the firing of cells during the delay. In delayed matching to sample, on the other hand, where the cue (usually a colored light) does not allow the animal to predict the site of response, response-specificity plays presumably no role during the delay, whereas cue-specificity may be ascertained in the delay discharge of some cells (Kubota *et al.*, 1980; Fuster *et al.*, 1982a). Niki and Watanabe (1976b), testing monkeys with a clever combination of delayed-response and conditional-position discrimination tasks, were able to determine that some prefrontal units are attuned to the cue, while others are attuned to the response; some show different delay discharge depending on the cue and others depending on the site of the impending motor response.

7. Temporal course. The duration and time course of delay discharge vary greatly between cells. Both vary also, to some degree, from one trial to the next. Moreover, differences between the procedural parameters (e.g., length of the delay) utilized in separate studies make it difficult to determine general and systematic patterns of that discharge. Some such patterns, however, have been identified (Fuster, 1973; Batuev *et al.*, 1979; Fuster *et al.*, 1982a; Kojima and Goldman-Rakic, 1982). In some cells, discharge is maximal at the beginning of the delay and tends to diminish, more or less gradually, in the course of it. Others show the opposite trend: a gradual acceleration of firing as the time for response grows near. These cells seem to time the delay and anticipate the response. Not only during the delay of delay tasks, but also in a variety of situations in which the animal can time and anticipate events, prefrontal cells have been found that fire faster in anticipation of those events (Niki, 1974a,b; Sakai, 1974; Niki and Watanabe, 1979; Kojima *et al.*, 1981). It is reasonable to suppose that the cells whose discharge peaks at the cue or early in the delay, and then diminishes as the delay progresses, are engaged in the process of retaining the cue. Conversely, those cells that appear to anticipate the motor response might be engaged in preparation for that response. The two types of cells would participate in two complementary functions at the service of temporal integration: a short-term memory function, transferring information across time, and a preparatory function for the consequent response. Because of the presence of cells that show relatively brief trains or bursts of firing at various times in the course of long delays, it has been suggested that the temporal transfer function may be carried out sequentially by different cell groups, as in a kind of relay race (Batuev *et al.*, 1979). In any event, the organization and circuitry of the prefrontal cortex are not known sufficiently to provide such functional assumptions with a plausible morphological base.

3.2.2. Topographic Distribution of Unit Types

As noted above, certain types of cells (e.g., space-differential cells, reward-related cells) predominate in certain prefrontal areas, but a clear-cut anatomical separation of cell types is not possible, for there is considerable overlap between the areas populated by the various cell types. In very general terms, however, it is worthwhile to point out that units whose activity during the delay is different from intertrial baseline are much more heavily represented in the cortex of the dorsolateral prefrontal convexity than in orbital or medial prefrontal cortex (Fuster *et al.*, 1982a).

At the microscopic level, the distribution of cell types is even more difficult to establish than at the gross anatomical level. Nonetheless, three important points can be made: (1) Units of like properties tend to cluster together, although the configuration of the clusters cannot yet be ascertained (Fuster, 1973; Fuster *et al.*, 1982a). (2) Similarities are more apparent among units recorded along microelectrode paths orthogonal to the cortical surface than among those along nonorthogonal paths with respect to that surface; thus, a semblance of columnar distribution has been observed, particularly as it pertains to delay-discharge types (Fuster *et al.*, 1982a). (3) Units whose reactions to sensory (visual) stimuli are subject to modulation from another cortical area (inferotemporal cortex) are more common in supragranular than infragranular layers of the prefrontal cortex (Fuster *et al.*, 1985).

4. Neuropsychology

In this section we will briefly review the neuropsychological facts bearing on the postulated role of the prefrontal cortex in the temporal organization of behavior. We will examine the literature dealing with the effects of lesions of the prefrontal cortex in animals and humans, with special emphasis on the behavioral manifestations of the cognitive impairments resulting from such lesions. We will also examine the data from blood-flow and metabolic-rate studies in humans.

4.1. Animals

The best known and most characteristic manifestation of experimental prefrontal lesion in animals is the deficit in performance of so-called delay tasks. As noted above, those are behavioral tasks that essentially require from the animal the execution of motor acts in accord with events in the recent past. Each act is to some extent predicated on a specific event or stimulus (ordinarily visual) that has occurred a few seconds or minutes before the animal is called to act. Thus, the tasks in question, among which delayed response, delayed alternation, and delayed matching-to-sample are the best known, are considered tests of recent or short-term memory. They are tests of the animal's capacity to remember sensory items after prescribed periods of time.

Animals that have undergone ablations of the prefrontal cortex, especially its dorsolateral aspects (i.e., in primates, the granular cortex of the convexity of the frontal lobe), have an inordinate difficulty in performing delay tasks correctly (Jacobsen, 1935, 1936; Mishkin and Pribram, 1955; Lawicka, 1972). Many attempts have been made, especially in primates, to identify precisely the part of the prefrontal cortex that is essential for delay-task performance, the cerebral structures that in addition to the prefrontal cortex are involved in that form of behavior, and the factors underlying the delay-task deficit. As a result, different prefrontal subareas have been found critical for different delay tasks, suggesting areal specialization with regard to the modality of the cues used by the animal in the task or with regard to other (e.g., spatial) aspects of the information they convey (Goldman *et al.*, 1971; Passingham, 1975; Mishkin and Manning, 1978; Rosenkilde, 1979). Delay-task deficits have also been obtained by lesions of structures directly connected to the prefrontal cortex, such as the MD nucleus (Schulman, 1964; Isseroff *et al.*, 1982), the caudate nucleus (Divac, 1972), and the hippocampus (Mishkin and Pribram, 1954; Karmos and Grastyan, 1962).

On reviewing the literature from lesion work in monkeys (Fuster, 1980), it becomes evident that the delay, that is, the time interposed between a cue-stimulus and the consequent response, is the single most important factor in the delay-task deficit from lesions of the prefrontal convexity. Neither the perception of the cue, regardless of its attributes, nor the ability to respond to it is impaired in the animal with prefrontal lesion. Thus, that animal can perform its task without trouble if little or no time is interposed by the experimenter between cue and response. A time delay is the prerequisite for the deficit to surface. Furthermore, the use of reversible lesions—by cooling—has shown that the magnitude of that deficit is directly related to the length of the delay (Fuster and Bauer, 1974; Bauer and Fuster, 1976). That relationship suggests that the function most affected in the deficit is one with a temporal decay, thus lending plausibility to the long-held—and debated—view that the delay-task deficit is fundamentally one of short-term memory.

At a more general level, however, the delay-task deficit can be rightfully ascribed to an impairment in the animal's ability to connect events *across time*; in other words, a diminished ability to integrate information from temporally separate events for the execution of motor action congruent with them. It is equally appropriate, therefore, to speak of lesser ability to organize behavior on the basis of temporally discontinuous information.

Inasmuch as the prefrontal cortex, or a part thereof, plays a critical role in short-term memory, it plays a no less critical role in the temporal organization of behavior. A temporally extended structure of behavior requires a memory function that will carry the trace of cues across time and thus allow subsequent and fully consequent responses. An animal deprived of the functional substrate for that form of memory, as the dorsolateral prefrontal cortex appears to be, will and does manifest difficulties organizing behavior along the time axis. It is mainly on the basis of such reasoning and the results of cortical lesions on delay tasks that Jacobsen and his colleagues (Jacobsen, 1935; Jacobsen *et al.*, 1935) first proposed the idea of a function of frontal association cortex in the temporal organization and synthesis of behavior.

The difficulty that an animal with prefrontal lesion shows in performing delay tasks can be partly overcome by better definition of the structure and components of each individual trial. This can be accomplished, in the classical delayed-response test, for instance, by making the cues more salient (Finan, 1942; Blake *et al.*, 1966). In delayed alternation, it can be accomplished by parsing the trials in such a way that each one becomes more distinct as a separate temporal structure or "gestalt." The finding that the latter procedure alleviated the animal's lesion-induced deficit led Pribram and Tubbs to ascribe a function of temporal organization to the prefrontal cortex (Pribram and Tubbs, 1967; Tubbs, 1969), thus adding support to the concept previously advanced by Jacobsen. Other support, from use of behavioral paradigms other than delay tasks, was provided by the evidence that prefrontal animals had difficulty ordering items in a sequence (Pinto-Hamuy and Linck, 1965).

So far I have referred mainly to the results of lesions of the prefrontal cortical convexity. In support of the temporal-organization argument, I have adduced, as most important, the evidence of delay-task deficits, which *prima facie* and operationally can be defined as short-term memory deficits and, as such, are detrimental to the integration of temporally discontinuous information. Now I will add that ventromedial portions of the prefrontal cortex also contribute to the temporal organization of behavior, but probably do so by their involvement in a function different from short-term memory. That function seems to be, simply stated, the suppression of internal and external interference capable of disrupting the behavioral structure at hand. Impairment of such a function has been invoked as the main reason why animals with lesions of ventromedial areas of the prefrontal cortex show a deficit in successive discrimination and reversal tasks (Stepien *et al.*, 1963; Brutkowski, 1964; Mishkin, 1964; Butter, 1969; Passingham and Ettlinger, 1972; Inversen and Mishkin, 1973). For lack of that postulated inhibitory control of interference, the behavior of the animals in those tasks appears to be abnormally subjected to distraction from extraneous sensory stimuli as well as to obsolete internal tendencies or memory traces (Mishkin, 1964).

Successive discrimination tasks, such as *go/no go,* and discrimination reversal are behavioral tasks that essentially require that kind of control. In *go/no go,* a stimulus demands a given action and another stimulus no action at all. The animal with ventral or medial lesion cannot inhibit the tendency to respond to one as well as to the other, as if unable to suppress the response which is not appropriate. In reversal, two different stimuli are presented simultaneously and repeatedly for choice, one of them associated with reward; from time to time the reward is transferred to the other stimulus and thus the animal must reverse the choice. The animal with ventromedial lesion persists in its choice of the wrong stimulus well after it has become obsolete, as if unable to suppress the old association. Here again, the behavioral structure (the trial) is disrupted by an interfering influence from within.

Nonetheless, even in ventromedial lesions a memory factor cannot be excluded, for it is possible that, at least in reversal tasks, the animal fails to retain the last change in the stimulus–reward association. Furthermore, the ventromedial prefrontal cortex has also been implicated in short-term memory by the demonstration that lesions of that cortex, in the monkey, lead to a severe deficit in delayed nonmatching-to-sample, a visual recognition task (Mishkin and Bachevalier, 1983).

To sum up, a temporal structure of behavior, such as a trial in a delay task, in a successive discrimination task, or in a reversal task, can be faultily formed or lead to a faulty outcome for two basic reasons: defective memory of cues or defective control of interference. Dorsolateral prefrontal lesions appear to impair the organization of that behavioral structure by reason of their detrimental effect on short-term memory, wheras ventromedial lesions do so, primarily if not exclusively, by depriving the animal of the inhibitory control of interference.

Are the estimation and discrimation of time disrupted by prefrontal lesions? The experimental evidence in this respect is inconclusive. Some studies have shown a deficit of timing behavior as a result of prefrontal lesion (Glickstein *et al.*, 1964; Nonneman *et al.*, 1974; Rosenkilde and Divac, 1975, 1976; Rosenkilde and Lawicka, 1977), whereas others have shown an absence of such deficit (Stamm, 1963; Manning, 1973; Rosenkilde *et al.*, 1981b). In any event, whatever deficits have been observed seem to have been, for the most part, minor and transient. That would preclude a substantial role of the cortex removed in timing or estimation of time. Nevertheless, the lack of clear and permanent deficits in timing–behavior tests may be less due to the sparing of a timing function than to the nature of some of those tests. Indeed, timing may not be necessary for their performance. The adoption of different bodily positions to signal different responses, for example, may obviate the need to time intervals, allowing the animal to perform the task up to par with the aid of kinesthetic cues.

At all events, even if in the prefrontal animal timing were intact, that would not in itself be a conclusive argument against the temporal organization postulate. It is important to differentiate between the mere estimation of time intervals and the temporal synthesis of behavioral structures, that is, the organization of behavior in the temporal domain. The latter function is not seriously challenged or tested in timing tasks. Yet two of its components, short-term memory and control of interference, appear to be severely impaired, as we have seen, in animals with lesions of the prefrontal cortex.

4.2. Humans

The difficulties that humans with prefrontal pathology experience in organizing their behavior have long been recognized, even though in many cases those difficulties are rather subtle. Clinical descriptions often refer to the patient's lack of spontaneity, poor concentration, and absence of proper planning, to cite but three of the commonly encountered symptoms of prefrontal disorder. These symptoms defy a parsimonious explanation, although several have been attempted. One such explanation was advanced long ago by Brickner (1934), who attributed the prefrontal syndrome to a deficit of integrative synthesis in thinking and behavior.

That the prefrontal cortex plays a role in the goal-directed synthesis of behavioral structures may always have been surmised on cursory consideration of the disorders that prefrontal damage entails, especially the impairments of a form of behavior which is distinctly human, language. But it is only in the past 30 years that, with objective and rigorous neuropsychological tests, the role of the human prefrontal cortex in behavioral synthesis has been subjected to in-depth analysis. The results of that analysis are in remarkable agreement with animal data. For one thing, in the human as in the animal, the role of that cortex in behavior has been shown to be critically dependent on temporal factors.

Humans with prefrontal lesions, as monkeys with such lesions (Pinto-Hamuy and Linck, 1965; Brody and Pribram, 1978), show characteristic impairments in performance of self-ordering tasks, that is, tasks that require the serial organization of items in the temporal axis (Milner, 1971; Petrides and Milner, 1982). They also have difficulties sorting items according to a temporally changing principle of categorization, such as required in card-sorting tasks (Milner, 1963, 1964, 1971; Drewe, 1974). Both self-ordering and card-sorting tasks impose considerable demands on organizing strategy (planning) as well as on short-term memory.

Frontal-lobe lesions lead to impairments of speech and language, which are the highest forms of temporal structuring activity (Lashley, 1951). The best known impairment is Broca's aphasia (Geschwind, 1970), a profound disorder of articulation and grammar that results from lesions of the left inferior frontal gyrus, a prefrontal area by definition (MD projection). More subtle and less known is the disorder of speech resulting from more anterior prefrontal lesions sparing Broca's area. That disorder was called by Goldstein (1948) "central motor aphasia" and by Luria (1970) "frontal dynamic aphasia." Its principal features are the diminished spontaneity of speech and the general impoverishment of verbal expression. The capacity to "propositionize" (Jackson, 1915) is poor, the patient speaking in short and simple sentences with little or no use of dependent clauses (Ramier and Hecaen, 1970, 1977; Lhermitte *et al.*, 1972; Barbizet *et al.*, 1975). In other words, there is a diminution of the so-called recursiveness of language (Chomsky, 1975). The disorder is more prominent after left than after right frontal lesion. Conversely, "fluency" in tasks of nonverbal praxis (e.g., production or self-ordering of abstract designs) is more impaired by right than by left lesions (Jones-Gotman and Milner, 1977; Petrides and Milner, 1982).

In summary, all the disorders of language produced by prefrontal damage can be characterized as syntagmatic disorders, that is, disorders of the capacity to form linguistic structures. Whereas posterior prefrontal lesions (Broca) impair the organization of elementary morphemes, more anterior lesions impair the organization of more complex language. With quantitative, not qualitative, differences, all prefrontal lesions seem to compromise a (syntagmatic) function of sequential ordering of sounds, words, and sentences in the making of semantic structures, in which meaning is both the goal and the organizing principle.

Measurements of cerebral blood flow and metabolic rate, recently facilitated by combined use of radioisotopes and computer tomography, have revealed the participation of the prefrontal cortex in the formation and planning of a variety of behavioral structures, including linguistic structures. During performance of motor sequences, the supplementary motor area (SMA), the mesial premotor cortex, is markedly activated in addition to rolandic motor cortex (Orgogozo and Larsen, 1979; Roland *et al.*, 1980a,b, 1982). In well-trained subjects, the SMA has been noted to be selectively activated also during the conceptualization or programming of movements, without their execution (Roland *et al.*, 1980b). In nontrained subjects, however, motor ideation also activates the more anterior prefrontal region (Ingvar and Philipson, 1977).

During speech production there is increased metabolic activity in the prefrontal cortex, in addition to premotor cortex and the mouth–larynx motor cortex (Ingvar and Schwartz, 1974; Larsen *et al.*, 1978). In the course of inner speech, patches of activation are seen only in prefrontal areas and the SMA (Larsen *et al.*, 1978; Lassen and Larsen, 1980). Thus, a certain parallelism is apparent in the activations of frontal cortex that accompany the structuring of speech and skeletal motility. In both kinds of performance, the areas most heavily activated, and therefore presumably involved, are motor and premotor areas in the posterior part of the frontal lobe, just in front of the central fissure. However, in the programming or conceptualization of speech, as in that of skeletal movements, more anterior areas in the prefrontal cortex itself seem to be preponderantly involved, especially if the programmed or conceptualized performance is unusually elaborate or taxes cognitive functions such as short-term memory.

With regard to short-term memory, the data from use of auditory stimuli are particularly revealing. Although the mere listening to sequences of words has been noted to activate foci in upper and orbital prefrontal cortex (Nishizawa *et al.*, 1982), marked activation of large areas of the prefrontal convexity is reported to occur in tasks that require the effortful retention of auditory information for the short term. In one especially demonstrative study (Risberg and Ingvar, 1973), the subjects had to remember series of digits and reproduce them backwards (digit-span-backward test). In two others (Roland *et al.*, 1981; Mazziotta *et al.*, 1982), the subjects' task was to retain tones for later comparison with other tones, an auditory short-term memory task analogous to the visual delayed matching-to-sample task.

As may have been noted above, in the human, as in the animal, there are indications that the prefrontal cortex is not a functionally homogeneous part of the cerebral mantle and that its temporal-integration role, the evidence for which we have been reviewing, is the composite of several subordinate functions probably represented in different prefrontal territories. At least three can be inferred from human data, as we have seen they can be from animal data: a temporally prospective function of anticipatory preparation, a temporally retrospective function of short term memory, and an interference-control function. Let us briefly examine some of the clinical evidence of them.

Several investigators have attempted to elucidate the neuropsychological foundation of the well-known difficulty that many frontal-lobe patients have in planning their actions and anticipating their consequences. On the basis of extensive testing, Luria (1966) concludes that the symptom implies a disorder of what he considers the most important function of the prefrontal cortex: the formation of preliminary syntheses of action (the metabolic evidence reviewed above would seem to support that view). Those "presyntheses" would be internalized schemata of prospective action, at a linguistic or quasi-linguistic level, that exert regulatory control over behavior. In the prefrontal case, where presumably those inner directives are weaker or absent, the planning of action and the orderly carrying out of intentions succumb to circumstantial influences.

Because of its mentalistic aspects, Luria's proposition is hard to test. Nonetheless, its generality has been found to be somewhat limited (Drewe, 1975a) and, of course, it cannot be easily applied to animals. Moreover, the lack of "presyntheses" is not sufficient to explain other elements of prefrontal pathology, such as the difficulty to benefit from errors, which Luria himself and others have emphasized (Milner, 1964; Luria, 1966) and which can be better understood as the result of memory or interference-control deficits (below). However, Luria's concept provides at least a partially adequate explanation and has the merit of pointing out a general preparatory, anticipatory, function of the prefrontal cortex which most probably transcends any given species.

Teuber (1966) also hypothesized such a preparatory function with his corollary-discharge theory. It postulates a critical role of the prefrontal cortex in the intimate relationship between perception and movement. According to the theory, the prefrontal cortex allows the flow of information (corollary discharge) from motor to sensory systems, whereby the organism can adjust the sensory apparatus in anticipation of its own movements. In this manner, action is balanced against intent. In the prefrontal defect, corollary discharge is missing or inadequate and, as a result, perception becomes unstable. Deliberate action is difficult, since its consequences cannot be properly anticipated. As with Luria's hypothesis, the corollary-discharge hypothesis remains to be proven, especially with regard to its generality, but it clearly offers another plausible explanation for some aspects of the prefrontal syndrome and it also emphasizes the need to postulate a temporally anticipatory prefrontal function.

A memory function is also evidently disturbed in a substantial number of prefrontal patients (Hecaen, 1964; Luria, 1966; Barbizet, 1970). Clearly, the disorder is not the incapacity to memorize (as in the Korsakoff syndrome), but rather a defect in voluntary memorizing, in the active will to remember. That is one reason why the frontal memory deficit has at least the appearance of a short-term memory deficit. Is that deficit equivalent to the memory deficit of the frontal animal? In two studies, in which use was made of the conventional delayed-response and delayed-alternation tasks, frontal-lobe patients showed either no deficit (Ghent *et al.*, 1962) or one no greater than that produced by lesions outside the prefrontal cortex (Chorover and Cole, 1966). Maybe the tasks utilized were too simple or the delays too short for the defect to appear in human subjects. In any event, other studies, using more complex delay tasks, such as delayed matching-to-sample, have clearly demonstrated a short-term memory deficit in the human (Milner, 1964; Lewinsohn *et al.*, 1972). Furthermore, these studies have shown that the deficit is supramodal with respect to the sensory modality of the cues to be remembered and is not restricted to cues of spatial nature.

In humans, the temporally prospective (anticipatory preparation) and retrospective (short-term memory) functions of the prefrontal cortex seem to be primarily represented, as in the monkey, in the dorsolateral prefrontal convexity, for the difficulties with anticipation and memory have been almost exclusively reported in cases with dorsolateral lesions. However, there is little evidence, one way or the other, from the generally less common cases with medial or orbital prefrontal lesions.

Another set of similarities with the animal can be found in the nature and localization of a third prefrontal function: the control of interference. The derangement of such a function may be inferred from the inordinate distractibility of prefrontal patients and their inability to shield their attention from external interference (Luria, 1966). It may also be inferred from the difficulty they have in suppressing interference from within, that it, from obsolete memories and response tendencies; as in the animals with prefrontal lesions, the behavior of humans with such lesions becomes perseverative, subject to what Mishkin has called "perseverative interference" (Milner, 1964; Mishkin, 1964). In line with that is the finding that, on formal testing, they fail to perform correctly a successive discrimination task, such as *go/no go,* that requires inhibition of inappropriate responses (Drewe, 1975b). Both distractibility and poor control of internal interference have been observed in patients with lesions of the dorsolateral convexity, but seem to be particularly prevalent in patients with orbito-medial lesions (Drewe, 1975b; Stuss *et al.,* 1982).

5. Comment and Conclusions

The extrinsic connectivity of the prefrontal cortex appears to be a necessary, though not sufficient, condition for a comprehensive role of that cortex in the organization of behavior. Because of the large variety of its afferent connections, the prefrontal cortex, unlike the primary sensory and motor cortex, and unlike the posterior associative cortex, is in the position to integrate vastly diverse kinds of information on both the external and the internal milieu of the organism (see also Pandya and Yeterian, this volume). Its elaborate efferent connections, on the other hand, theoretically allow the prefrontal cortex to gate or modulate the innumerable sources of that information and to control, directly or indirectly, the effector apparatus of behavior in the broadest sense. At the very least, heterogeneity of inputs and outputs allows for heterogeneity, i.e., complexity, of both integrated information and behavior.

Nonetheless, despite the richness of its connectivity, the prefrontal cortex may not be construed as the supreme integrative or controlling structure in a hierarchy of neural structures, for the attribution of pontifical properties to any of them inevitably leads to an infinite regress. Rather, whatever integrative or motor-control properties the prefrontal cortex has are probably inherent in that extrinsic connectivity and to a large extent the result of it. Those properties of the prefrontal cortex, which emerge with its development and the development of its connections, would be essentially the extension or sum, in one dimension or another, of the basic properties that other structures have which are connected to it. One such dimension might be complexity, implicit in the diversity of prefrontal inputs and outputs. But complexity *per se* does not seem to be the critical dimension, for the organism deprived of prefrontal cortex is capable of both complex sensory integration and complex behavior. We contend that the critical dimension is time and that the prefrontal cortex mediates the formation of temporally extended structures of behavior on the basis of temporally discontiguous items of information, whether they come from inside or outside the organism.

The evidence for a temporally integrative, synthetic function of the prefrontal cortex can be summarized as follows:

1. A slow field potential (CNV) occurring in the prefrontal region between the temporally separate events of a behavioral structure. The potential develops as the events become parts of the structure through the establishment of relationships of mutual contingency (dependency) between them.

2. Prefrontal cell activity changes in temporal relation with the various events of a behavioral structure, such as a delay-task trial: the cue (or sample), the choice stimuli, the motor response, and the reinforcement or reward. Cells reacting to one or more of those events coexist within certain areas of the prefrontal cortex.

3. Sustained changes of prefrontal cell discharge during the delay of delay-task trials. Those changes, which in many cells are seen to bridge the temporal distance between cue (sample) and choice, are a product of learning, related to performance, and related to mutual contingency between the temporally separate cue and choice response.

4. Deleterious effects of prefrontal lesions on an animal's performance of tasks that require one of the following: (a) the integration of temporally separate items of information (delay tasks: e.g., delayed response, delayed matching, delayed alternation); (b) the adjustment of behavior to the temporally changing association between external stimuli and reinforcement (reversal); (c) the inhibition of untimely responses (successive discrimination, *go/no go*); or (d) the temporal ordering of items.

5. Difficulties experienced by human subjects with prefrontal pathology in the performance of the tasks enumerated above and, in addition, in the formation of temporally extended structures of behavior, such as required by language.

6. Increased metabolic activity in prefrontal areas during the planning, conceptualization, or execution of certain complex and temporally extended structures of behavior, including structures of speech.

Of course, *all* behavior involves some degree of temporal integration, yet not all behavior is affected by ablation of prefrontal cortex. Thus, there must be some limitations to the kinds of temporal structuring that the prefrontal cortex helps integrate and there must be qualification of its temporal integrative function. I have argued elsewhere (Fuster, 1980) that novelty and complexity are two necessary qualifications; in other words, that behavioral structures need to be novel and to have a degree of complexity in order to come under the purview of the prefrontal cortex. However, novelty and complexity are probably only necessary if, and inasmuch as, they tax certain functions which make temporal integration possible and in which, as we have seen, the prefrontal cortex is implicated. They are the following:

1. Anticipatory preparation. This is a temporally prospective function readying or presetting the organism for anticipated action. Evidence of prefrontal involvement in it lies in unit records (cells anticipating movement), the disruptive effects of prefrontal lesion on the formation and execution of plans, and the increased metabolic rate in prefrontal cortex during the programming of speech and motor sequences.

2. Short-term memory. This is a temporally retrospective function of provisional mnemonic retention. Evidence of prefrontal participation in it lies in the effects of lesions of prefrontal cortex on performance of delay tasks, the sustained unit activity changes in that cortex during the retention periods of those tasks, and the prefrontal metabolic-rate increases during memorization of digits or tones.

3. Interference control. This function would allow the suppression of internal or external interference which can disrupt the formation of temporally extended behavioral structures. Evidence of prefrontal involvement may be found in the effects of prefrontal lesions on successive discrimination and reversal tasks, and the inordinate distractibility that such lesions produce.

It is not possible, on the basis of available evidence, to determine the precise cortical localization of the three functions mentioned. Nevertheless, as discussed above, the evidence from some lesion and single-unit studies suggests that the preparatory and short-term memory functions are mainly represented in the dorsolateral prefrontal cortex, whereas the interference-control function is mainly represented in the ventromedial prefrontal cortex.

With regard to the mechanisms by which the prefrontal cortex performs its apparently essential role in temporal integration, we can at present only speculate. Prefrontal neurons evidently participate in the neuronal populations representing sensory and motor components of behavioral structures. Such structures are epitomized by the single trial of a delay task, during which prefrontal unit-discharge changes are seen to both succeed sensory stimuli and precede motor actions. Conceivably, the critical role of prefrontal cell assemblies is to expand in the temporal dimension the cortical representation of stimuli inasmuch as they are associated with prospective action and, at the same time, the representation of actions before they are executed. Thus, by extending retention of stimuli and anticipation of actions, those assemblies would permit the bridging of temporal discontinuities between them and the integration of a consistent structure. Prefrontal neurons may accomplish that in intimate cooperation with the rest of the cortical network and by sustained excitation of local and corticocortical reverberating pathways. Also having access to motor control mechanisms in the basal ganglia and the brain stem, the prefrontal assemblies would facilitate the setting or preparation of those mechanisms for the execution of action consistent with events in the recent past.

Because of the participation of prefrontal cell assemblies in the cortical network, any input reaching the network from external or internal origin could be matched to the temporally separate elements of the behavioral structure at hand. Any action and its results could be matched to recent events and to the goal of the structure. Thus, each motor action would be consistent with past and future, and mismatches would be corrected. Extraneous stimuli and inappropriate or untimely actions would be suppressed or inhibited (interference control). In sum, because of the intervention of the prefrontal cortex, each event in a sequence, whether sensory or motor, could be put in the proper temporal perspective, with respect to both what has happened and what is to happen in the formation of the structure. The prefrontal cortex would not superimpose a steering or directing function on the remainder of the nervous system, but rather, by expanding the temporal perspectives of the system, it would allow it to integrate longer, newer, and more complex structures of behavior.

6. Summary

It is postulated that the main function of the prefrontal cortex is the temporal organization of behavior. That hypothesis is examined here in the light of the results of anatomical, electrophysiological, and neuropsychological studies.

The abundant extrinsic connectivity of the prefrontal cortex points to a complex integrative function. Prefrontal cell discharge and field potentials are found related to diverse and temporally separate events in a behavioral sequence. Sustained patterns of cell discharge and slow potentials develop in prefrontal cortex in the time between events and as a function of the establishment of behavioral contingencies between them. The effects of prefrontal lesion, in animals as well as in humans, point to an important role of the prefrontal cortex in the organization of behavioral structures in the time domain. That role applies to a wide range of structures, from the individual trial in a delay task to the more complex structures of language. In humans, prefrontal metabolic activity appears to increase during the formation of such structures.

The available evidence from the various methodologies affirms the plausibility of the hypothesis and suggests, further, that the temporal integration of behavior is subserved by three mutually supportive functions with somewhat different cortical topography: anticipatory preparation, short-term memory, and interference control. The first two seem mainly represented in dorsolateral and the third in ventromedial prefrontal cortex. It is concluded that the prefrontal cortex, by coordinating those functions and by mechanisms probably involving its connections with other parts of cortex and with effector systems, enables the organism to form goal-directed and temporally extended structures of behavior.

7. References

Akert, K., 1964, Comparative anatomy of frontal cortex and thalamo-frontal connections, in: *The Frontal Granular Cortex and Behavior* (J. M. Warren and K. Akert, eds.), McGraw–Hill, New York, pp. 372–396.

Alexander, G. E., and Fuster, J. M., 1973, Effects of cooling prefrontal cortex on cell firing in the nucleus medialis dorsalis, *Brain Res.* **61**:93–105.

Barbizet, J., 1970, *Human Memory and Its Pathology*, Freeman, San Francisco.

Barbizet, J., Diuzabo, P., and Flavigny, R., 1975, Role des lobes frontaux dans le langage, *Rev. Neurol.* **131**:525–544.

Batuev, A. S., Pirogov, A. A., and Orlov, A. A., 1979, Unit activity of the prefrontal cortex during delayed alternation performance in monkey, *Acta Physiol. Hung.* **53**:345–353.

Bauer, R. H., and Fuster, J. M., 1976, Delayed-matching and delayed-response deficit from cooling dorsolateral prefrontal cortex in monkeys, *J. Comp. Physiol. Psychol.* **90**:293–302.

Benevento, L. A., and Fallon, J. H., 1975, The projection of occipital cortex to orbital cortex in the rhesus monkey (*Macaca mulatta*), *Exp. Neurol.* **46**:402–408.

Benevento, L. A., Fallon, J., Davis, B. J., and Rezak, M., 1977, Auditory–visual interaction in single cells in the cortex of the superior temporal sulcus and the orbital frontal cortex of the macaque monkey, *Exp. Neurol.* **57**:849–872.

Bignall, K. E., and Imbert, M., 1969, Polysensory and cortico-cortical projections to frontal lobe of squirrel and rhesus monkey, *Electroencephalogr. Clin. Neurophysiol.* **26**:206–215.

Blake, M., Meyer, D. R., and Meyer, P. M., 1966, Enforced observation in delayed response learning by frontal monkeys, *J. Comp. Physiol. Psychol.* **61**:374–379.

Borda, R. P., 1970, The effect of altered drive states on the contingent negative variation (CNV) in rhesus monkeys, *Electroencepalogr. Clin. Neurophysiol.* **29**:173–180.

Brickner, R., 1934, An interpretation of frontal lobe function based upon the study of a case of partial bilateral frontal lobectomy, *Res. Publ. Assoc. Res. Nerv. Ment. Dis.* **13**:259–351.

Brody, B. A., and Pribram, K. H., 1978, The role of frontal and parietal cortex in cognitive processing: Tests of spatial and sequence functions, *Brain* **101**:607–633.

Brutkowski, S., 1964, Prefrontal cortex and drive inhibition, in: *The Frontal Granular Cortex and Behavior* (J. M. Warren and K. Akert, eds.), McGraw–Hill, New York, pp. 242–270.

Butter, C. M., 1969, Preservation in extinction and in discrimination reversal tasks following selective frontal ablations in *Macaca mulatta, Physiol. Behav.* **4**:163–171.

Chavis, D. A., and Pandya, D. N., 1976, Further observations on corticofrontal connections in the rhesus monkey, *Brain. Res.* **117**:369–386.

Chomsky, N., 1975, *Reflections on Language,* Pantheon Books, New York.

Chorover, S., and Cole, M., 1966, Delayed alternation performance in patients with cerebral lesions, *Neuropsychologia* **4**:1–7.

DeVito, J. L., and Smith, O. A., 1964, Subcortical projections of the prefrontal lobe of the monkey, *J. Comp. Neurol.* **123**:413–424.

Divac, I., 1972, Neostriatum and functions of prefrontal cortex, *Acta Neurobiol. Exp.* **32**:461–477.

Drewe, E. A., 1974, The effect of type and area of brain lesion on Wisconsin card sorting test performance, *Cortex* **10**:159–170.

Drewe, E. A., 1975a, An experimental investigation of Luria's theory on the effects of frontal lobe lesions in man, *Neuropsychologia* **13**:421–429.

Drewe, E. A., 1975b, Go–no go learning after frontal lobe lesions in humans, *Cortex* **11**:8–16.

Finan, J. L., 1942, Delayed response with predelay reinforcement in monkeys after removal of the frontal lobes, *Am. J. Psychol.* **55**:202–214.

Fuster, J. M., 1973, Unit activity in prefrontal cortex during delayed-response performance: Neuronal correlates of transient memory, *J. Neurophysiol.* **36**:61–78.

Fuster, J. M., 1975, Cryogenic and microelectrode studies of the prefrontal cortex, in: *Symp. 5th Congr. Int. Primatol. Soc.* (S. Kondo, M. Kawai, A. Ehara, and S. Kawamura, eds.), Japan Science Press, Tokyo, pp. 445–458.

Fuster, J. M., 1980, *The Prefrontal Cortex,* Raven Press, New York.

Fuster, J. M., 1981, Prefrontal cortex in motor control, in: *Handbook of Physiology,* Volume II (V. B. Brooks, ed.), American Physiological Society, Washington, D. C., pp. 1149–1178.

Fuster, J. M., and Alexander, G. E., 1971, Neuron activity related to short-term memory, *Science* **173**:652–654.

Fuster, J. M., and Alexander, G. E., 1973, Firing changes in cells of the nucleus medialis dorsalis associated with delayed response behavior, *Brain Res.* **61**:79–91.

Fuster, J. M., and Bauer, R. H., 1974, Visual short-term memory deficit from hypothermia of frontal cortex, *Brain Res.* **81**:393–400.

Fuster, J. M., and Jervey, J. P., 1982, Neuronal firing in the inferotemporal cortex of the monkey in a visual memory task, *J. Neurosci.* **2**:361–375.

Fuster, J. M., Bauer, R. H., and Jervey, J. P., 1982a, Cellular discharge in the dorsolateral prefrontal cortex of the monkey in cognitive tasks, *Exp. Neurol.* **77**:679–694.

Fuster, J. M., Jervey, J. P., and Bauer, R. H., 1982b, Indications of functional relationship between prefrontal and inferotemporal cortex, *Soc. Neurosci. Abstr.* **8:**681.

Fuster, J. M., Bauer, R. H., and Jervey, J. P., 1985, Functional interactions between inferotemporal and prefrontal cortex in a cognitive task, *Brain Res.* **330:**299–307.

Geschwind, N., 1970, The organization of language and the brain, *Science* **170:**940–944.

Ghent, L., Mishkin, M., and Teuber, H. L., 1962, Short-term memory after frontal lobe injury in man, *J. Comp. Physiol. Psychol.* **55:**705–709.

Glickstein, M., Quickley, D. E., and Glassman, H. N., 1964, Effect of frontal and parietal lesions on timing behavior in monkeys, *Psychon. Sci.* **1:**265–266.

Goldman, P. S., Rosvold, H. E., Vest, B., and Galkin, T. W., 1971, Analysis of the delayed-alternation deficit produced by dorsolateral prefrontal lesions in the rhesus monkey, *J. Comp. Physiol. Psychol.* **77:**212–220.

Goldman-Rakic, P. S., and Schwartz, M. L., 1982, Interdigitation of contralateral and ipsilateral columnar projections to frontal association cortex in primates, *Science* **216:**755–757.

Goldstein, K., 1948, *Language and Language Disturbances,* Grune & Stratton, New York.

Hecaen, H., 1964, Mental symptoms associated with tumors of the frontal lobe, in: *The Frontal Granular Cortex and Behavior* (J. M. Warren and K. Akert, eds), McGra·v–Hill, New York, pp. 335–352.

Hess, W. R., 1943, Teleokinetisches und ereismatisches Kraeftesystem in der Biomotorik, *Helv. Physiol. Pharmacol. Acta* **1:**c62–c63.

Ingvar, D. H., and Philipson, L., 1977, Distribution of cerebral blood flow in the dominant hemisphere during motor ideation and motor performance, *Ann. Neurol.* **2:**230–237.

Ingvar, D. H., and Schwartz, M. S., 1974, Blood flow patterns induced in the dominant hemisphere by speech and reading, *Brain* **97:**273–288.

Isseroff, A., Rosvold, H. E., Galkin, T. W., and Goldman-Rakic, P. S., 1982, Spatial impairments following lesions of the mediodorsal nucleus of the thalamus in rhesus monkeys, *Brain Res.* **232:**97–113.

Iversen, S. D., and Mishkin, M., 1973, Comparison of superior temporal and inferior prefrontal lesions on auditory and non-auditory tasks in rhesus monkeys, *Brain Res.* **55:**355–367.

Jackson, J. H., 1915, On affections of speech from disease of the brain, *Brain* **38:**107–174.

Jacobsen, C. F., 1935, Functions of the frontal association area in primates, *Arch. Neurol. Psychiatry* **33:**558–569.

Jacobsen, C. F., 1936, Studies of cerebral function in primates. I. The functions of the frontal association areas in monkeys, *Comp. Psychol. Monogr.* **13:**3–60.

Jacobsen, C. F., Wolfe, J. B., and Jackson, T. A., 1935, An experimental analysis of the functions of the frontal association areas in primates, *J. Nerv. Ment. Dis.* **82:**1–14.

Jacobson, S., and Trojanowski, J. Q., 1975, Amygdaloid projections to prefrontal granular cortex in rhesus monkey demonstrated with horseradish peroxidase, *Brain Res.* **100:**132–139.

Jacobson, S., and Trojanowski, J. Q., 1977, Prefrontal granular cortex of the rhesus monkey. I. Intrahemispheric cortical afferents, *Brain Res.* **132:**209–233.

Jaervilehto, T., and Fruhstorfer, H., 1970, Differentiation between slow cortical potentials associated with motor and mental acts in man, *Exp. Brain Res.* **11:**309–317.

Jones, E. G., 1969, Interrelationships of parieto-temporal and frontal cortex in the rhesus monkey, *Brain Res.* **13:**412–415.

Jones, E. G., and Powell, T. P. S., 1970, An anatomical study of converging sensory pathways within the cerebral cortex of the monkey, *Brain* **93:**793–820.

Jones-Gotman, M., and Milner, B., 1977, Design fluency: The invention of nonsense drawings after focal cortical lesions, *Neuropsychologia* **15:**653–674.

Karmos, G., and Grastyan, E., 1962, Influence of hippocampal lesions on simple and delayed conditional reflexes, *Acta Physiol. Acad. Sci. Hung.* **21:**215–224.

Kemp, J. M., and Powell, T. P. S., 1970, The cortico-striate projection in the monkey, *Brain* **93:**525–546.

Kievit, J., and Kuypers, H. G. J. M., 1975, Basal forebrain and hypothalamic connections to frontal and parietal cortex in the rhesus monkey, *Science* **187:**660–662.

Kievit, J., and Kuypers, H. G. J. M., 1977, Organization of the thalamo-cortical connections to frontal and parietal cortex in the rhesus monkey, *Science* **187:**660–662.

Kojima, S., and Goldman-Rakic, P. S., 1982, Delayed-related activity of prefrontal neurons in rhesus monkeys performing delayed response, *Brain Res.* **248:**43–49.

Kojima, S., Matsumura, M., and Kubota, K., 1981, Prefrontal neuron activity during delayed-response performance without imperative GO signals in the monkey, *Exp. Neurol.* **74:**396–407.

Kubota, K., and Niki, H., 1971, Prefrontal cortical unit activity and delayed alternation performance in monkeys, *J. Neurophysiol.* **34:**337–347.

Kubota, K., Iwamoto, T., and Suzuki, H., 1974, Visuokinetic activities of primate prefrontal neurons during delayed-response performance, *J. Neurophysiol.* **37:**1197–1212.

Kubota, K., Tonoike, M., and Mikami, A., 1980, Neuronal activity in the monkey dorsolateral prefrontal cortex during a discrimination task with delay, *Brain Res.* **183:**29–42.

Künzle, H., Akert, K., and Wurtz, R. H., 1976, Projections of area 8 (frontal eye field) to superior colliculus in the monkey: An autoradiographic study, *Brain Res.* **117:**487–492.

Kuypers, H. G. J. M., Szwarcbart, M. K., Mishkin, M., and Rosvold, H. E., 1965, Occipitotemporal corticocortical connections in the rhesus monkey, *Exp. Neurol.* **11:**245–262.

Larsen, B., Skinhoj, E., and Lassen, N. A., 1978, Variation in regional cortical blood flow in the right and left hemispheres during automatic speech, *Brain* **101:**193–209.

Lashley, K. S., 1951, The problem of serial order in behavior, in: *Cerebral Mechanisms in Behavior* (L. A. Jeffress, ed.), Wiley, New York, pp. 112–146.

Lassen, N. A., and Larsen, B., 1980, Cortical activity in the left and right hemispheres during language-related brain functions, *Phonetica* **37:**27–37.

Lawicka, W., 1972, Proreal syndrome in dogs, *Acta Neurobiol. Exp.* **32:**261–276.

Leichnetz, G. R., and Astruc, J., 1977, The course of some prefrontal corticofugals to the pallidum, substantia innominata, and amygdaloid complex in monkeys, *Exp. Neurol.* **54:**104–109.

Leichnetz, G. R., Spencer, R. F., Hardy, S. G. P., and Astruc, J., 1981, The prefrontal corticotectal projection in the monkey: An anterograde and retrograde horseradish peroxidase study, *Neuroscience* **6:**1023–1041.

Lewinsohn, P., Zieler, R., Libet, J., Eyeberg, S., and Nielson, G., 1972, Short-term memory—a comparison between frontal and nonfrontal right- and left-hemisphere brain-damaged patients, *J. Comp. Physiol. Psychol.* **81:**248–255.

Lhermitte, F., Deroulsne, J., and Signoret, J. L., 1972, Analyse neuropsychologique du syndrome frontal, *Rev. Neurol.* **127:**415–440.

Luria, A. R., 1966, *Higher Cortical Functions in Man*, Tavistock Publications, London.

Luria, A. R., 1970, *Traumatic Aphasia*, Mouton, The Hague.

Luttenberg, J., 1974, Heterotopic contralateral projection of neocortical spheres of the cat brain. I. Frontal cortex, *Acta Univ. Carol. Med.* **20:**251–276.

Manning, F. J., 1973, Performance under temporal schedules by monkeys with partial ablations of prefrontal cortex, *Physiol. Behav.* **11:**563–569.

Mazziotta, J. C., Phelps, M. E., Carson, R. E., and Kuhl, D. E., 1982, Tomographic mapping of human cerebral metabolism: Auditory stimulation, *Neurology* **32:**921–937.

Mesulam, M.-M., Van Hoesen, G. W., Pandya, D. N., and Geschwind, N., 1977, Limbic and sensory connections of the inferior parietal lobule (area PG) in the rhesus monkey: A study with a new method for HRP histochemistry, *Brain Res.* **136:**393–414.

Milner, B., 1963, Effects of different brain lesions on card sorting, *Arch. Neurol.* **9:**90–100.

Milner, B., 1964, Some effects of frontal lobectomy in man, in: *The Frontal Granular Cortex and Behavior* (J. M. Warren and K. Akert, eds.), McGraw–Hill, New York, pp. 313–334.

Milner, B., 1971, Interhemispheric differences in the localization of psychological processes in man, *Br. Med. Bull.* **27:**272–277.

Mishkin, M., 1964, Perservation of central sets after frontal lesions in monkeys, in: *The Frontal Granular Cortex and Behavior* (J. M. Warren and K. Akert, eds.), McGraw–Hill, New York, pp. 219–241.

Mishkin, M., and Bachevalier, J., 1983, Object recognition impaired by ventromedial but not dorsolateral prefrontal cortical lesions in monkeys, *Soc. Neurosci. Abstr.* **9:**29.

Mishkin, M., and Manning, F. J., 1978, Nonspatial memory after selective prefrontal lesions in monkeys, *Brain Res.* **143:**313–323.

Mishkin, M., and Pribram, K. H., 1954, Visual discrimination performance following partial ablations of the temporal lobe. I. Ventral vs. lateral, *J. Comp. Physiol. Psychol.* **47:**14–20.

Mishkin, M., and Pribram, K. H., 1955, Analysis of the effects of frontal lesion in monkeys. I. Variations of delayed alterations, *J. Comp. Physiol. Psychol.* **48:**492–495.

Mohler, C. W., Goldberg, M. E., and Wurtz, R. H., 1973, Visual receptive fields of frontal eye field neurons, *Brain Res.* **61:**385–389.

Nauta, W. J. H., 1964, Some efferent connections of the prefrontal cortex in the monkey, in: *The Frontal Granular Cortex and Behavior* (J. M. Warren and K. Akert, eds.), McGraw–Hill, New York, pp. 397–407.

Nauta, W. J. H., 1972, Neural associations of the frontal cortex, *Acta Neurobiol. Exp.* **32:**125–140.
Niki, H., 1974a, Prefrontal unit activity during delayed alternation in the monkey. I. Relation to direction of response, *Brain Res.* **68:**185–196.
Niki, H., 1974b, Prefrontal unit activity during delayed alternation in the monkey. II. Relation to absolute versus relative direction of response, *Brain Res.* **68:**197–204.
Niki, H., 1974c, Differential activity of prefrontal units during right and left delayed response trials, *Brain Res.* **70:**346–349.
Niki, H., and Watanabe, M., 1976a, Prefrontal unit activity and delayed response: Relation to cue location versus direction of response, *Brain Res.* **105:**79–88.
Niki, H., and Watanabe, M., 1976b, Cingulate unit activity and delayed response, *Brain Res.* **110:**381–386.
Niki, H., and Watanabe, M., 1979, Prefrontal and cingulate unit activity during timing behavior in the monkey, *Brain Res.* **171:**213–224.
Niki, H., Sakai, M., and Kubota, K., 1972, Delayed alternation performance and unit activity of the caudate head and medial orbitofrontal gyrus in the monkey, *Brain Res.* **38:**343–353.
Nishizawa, Y., Olsen, T. S., Larsen, B., and Lassen, N. A., 1982, Left–right cortical asymmetries of regional cerebral blood flow during listening to words, *J. Neurophysiol.* **48:**458–466.
Nonneman, A. J., Voigt, J., and Kolb, B. E., 1974, Comparisons of behavioral effects of hippocampal and prefrontal cortex lesions in the rat, *J. Comp. Physiol. Psychol.* **87:**249–260.
Orgogozo, J. M., and Larsen, B., 1979, Activation of the supplementary motor area during voluntary movement in man suggests it works as a supra-motor area, *Science* **206:**847–850.
Pandya, D. N., and Kuypers, H. G. J. M., 1969, Cortico-cortical connections in the rhesus monkey, *Brain Res.* **13:**13–36.
Pandya, D. N., Hallett, M., and Mukherjee, S. K., 1969, Intra- and interhemispheric connections of the neocortical auditory system in the rhesus monkey, *Brain Res.* **14:**49–65.
Pandya, D. N., Dye, P., and Butters, N., 1971, Efferent cortico-cortical projections of the prefrontal cortex in the rhesus monkey, *Brain Res.* **31:**35–46.
Passingham, R., 1975, Delayed matching after selective prefrontal lesions in monkeys (*Macaca mulatta*), *Brain Res.* **92:**89–102.
Passingham, R. E., and Ettlinger, G., 1972, Tactile discrimination learning after selective prefrontal ablations in monkeys (*Macaca mulatta*), *Neuropsychologia* **10:**17–26.
Petrides, M., and Milner, B., 1982, Deficits on subject-ordered tasks after frontal- and temporal-lobe lesions in man, *Neuropsychologia* **20:**249–262.
Pinto-Hamuy, T., and Linck, P., 1965, Effect of frontal lesions on performance of sequential tasks by monkeys, *Exp. Neurol.* **12:**96–107.
Pirch, J. H., Corbus, M. J., and Rigdon, G. C., 1983, Single-unit and slow potential responses from rat frontal cortex during associative conditioning, *Exp. Neurol.* **82:**118–130.
Pribram, K. H., and Tubbs, W. E., 1967, Short-term memory, parsing, and the primate frontal cortex, *Science* **156:**1765–1767.
Pribram, K. H., Mishkin, M., Rosvold, H. E., and Kaplan, S. J., 1952, Effects on delayed-response performance of lesions of dorsolateral and ventromedial frontal cortex of baboons, *J. Comp. Physiol. Psychol.* **45:**565–575.
Ramier, A., and Hecaen, H., 1970, Role respectif des attaines frontales et de la lateralisation lesionalle dans les deficits de la "fluence verbale," *Rev. Neurol.* **123:**17–22.
Ramier, A. M., and Hecaen, H., 1977, Les deficits au test de "fluence verbale" chez les sujets gauchers avec lesions hemispheriques unilaterales, *Rev. Neurol.* **133:**571–574.
Rebert, C. S., 1972, Cortical and subcortical slow potentials in the monkey's brain during preparatory intervals, *Electroencephalogr. Clin. Neurophysiol.* **33:**389–402.
Riseberg, J., and Ingvar, D. H., 1973, Patterns of activation in the gray matter of the dominant hemisphere during memorization and reasoning, *Brain* **96:**737–756.
Robinson, D. A., and Fuchs, A. F., 1969, Eye movements evoked by stimulation of frontal eye fields, *J. Neurophysiol.* **32:**637–648.
Roland, P. E., Larsen, B., Lassen, N. A., and Skinhoj, E., 1980a, Supplementary motor area and other cortical areas in organization of voluntary movements in man, *J. Neurophysiol.* **43:**118–136.
Roland, P. E., Vaernet, K., and Lassen, N. A., 1980b, Cortical activations in man during verbal report from visual memory, *Neurosci. Lett.* **5:**478.
Roland, P. E., Skinhoj, E., and Lassen, N. A., 1981, Focal activations of human cerebral cortex during auditory discrimination, *J. Neurophysiol.* **45:**1139–1151.
Roland, P., Meyer, E., Shibasahi, T., Yamamoto, Y., and Thompson, C., 1982, Regional cerebral

blood flow changes in cortex and basal ganglia during voluntary movements in normal human volunteers, *J. Neurophysiol.* **48:**467–480.

Rosene, D. L., and Van Hoesen, G. W., 1977, Hippocampal efferents reach widespread areas of cerebral cortex and amygdala in the rhesus monkey, *Science* **198:**315–317.

Rosenkilde, C. E., 1979, Functional heterogeneity of the prefrontal cortex in the monkey: A review, *Behav. Neural Biol.* **25:**301–345.

Rosenkilde, C. E., and Divac, I., 1975, DRL performance following anteromedial cortical ablations in rats, *Brain Res.* **95:**142–146.

Rosenkilde, C. E., and Divac, I., 1976, Time-discrimination performance in cats with lesions in prefrontal cortex and caudate nucleus, *J. Comp. Physiol. Psychol.* **90:**343–352.

Rosenkilde, C. E., and Lawicka, W., 1977, Effects of medial and dorsal prefrontal ablations on a go left–go right time discrimination task in dogs, *Acta Neurobiol. Exp.* **37:**209–221.

Rosenkilde, C. E., Bauer, R. H., and Fuster, J. M., 1981a, Single cell activity in ventral prefrontal cortex of behaving monkeys, *Brain Res.* **209:**375–394.

Rosenkilde, C. E., Rosvolk, H. E., and Mishkin, M., 1981b, Time discrimination with positional responses after selective prefrontal lesions in monkeys, *Brain Res.* **210:**129–144.

Sakai, M., 1974, Prefrontal unit activity during visually guided lever pressing reaction in the monkey, *Brain Res.* **81:**297–309.

Sasaki, K., and Gemba, H., 1982, Development and change of cortical field potentials during learning processes of visually initiated hand movements in the monkey, *Exp. Brain Res.* **48:**429–437.

Schechter, P. B., and Murphy, E. H., 1975, Response characteristics of single cells in squirrel monkey frontal cortex, *Brain Res.* **96:**66–70.

Schulman, S., 1964, Impaired delayed response from thalamic lesions, *Arch. Neurol.* **11:**477–499.

Stamm, J. S., 1963, Function of prefrontal cortex in timing behavior of monkeys, *Exp. Neurol.* **7:**87–97.

Stepien, I., Stepien, L., and Kreiner, J., 1963, The effects of total and partial ablation of the premotor cortex on the instrumental conditioned reflexes in dogs, *Acta Biol. Exp. (Warsaw)* **23:**45–59.

Stuss, D., Kaplan, E., Benson, D., Weir, W., Chiulli, S., and Sarazin, F., 1982, Evidence for the involvement of orbitofrontal cortex in memory functions: An interference effect, *J. Comp. Physiol. Psychol.* **96:**913–925.

Suzuki, H., and Azuma, M., 1977, Prefrontal neuronal activity during gazing at a light spot in the monkey, *Brain Res.* **126:**497–508.

Tanabe, T., Yarita, H., Iino, M., Ooshima, Y., and Takagi, S. F., 1975, An olfactory projection area in orbitofrontal cortex of the monkey, *J. Neurophysiol.* **38:**1269–1283.

Teuber, H. L., 1966, The frontal lobes and their functions: Further observations on rodents, carnivores, subhuman primates, and man, *Int. J. Neurol.* **5:**282–300.

Tubbs, W. E., 1969, Primate frontal lesions and the temporal structure of behavior, *Behav. Sci.* **14:**347–356.

Wagman, I. H., and Mehler, W. R., 1972, Physiology and anatomy of the cortico-oculomotor mechanism, *Prog. Brain Res.* **37:**619–635.

Walter, W., 1973, Human frontal lobe function in sensory-motor association, in: *Psychophysiology of the Frontal Lobes* (K. H. Pribram and A. R. Luria, eds.), Academic Press, New York, pp. 109–122.

Walter, W., Cooper, R., Aldridge, V., McCallum, W., and Winter, A., 1964, Contingent negative variation: An electric sign of sensori-motor association and expectancy in the human brain, *Nature* **203:**380–384.

Webster, K. E., 1965, The cortico-striatal projection in the cat, *J. Anat.* **99:**329–337.

5

The Insula of Reil in Man and Monkey

Architectonics, Connectivity, and Function

M.-MARSEL MESULAM and ELLIOTT J. MUFSON

1. Introduction

During the fetal stages of primate development, the insula is situated on the surface of the cerebral hemisphere (Fig. 1A). However, in the more advanced primates, the adjacent neocortical areas develop much more extensively than the insula. This leads to massive frontal, parietal, and temporal opercularization and to the formation of the sylvian fissure within which the insula remains buried from the time of birth onwards. In subprimate mammalian species where the neocortex does not develop as extensively, the homolog of the insula remains exposed on the surface of the brain throughout adult life (Rose, 1928).

While its growth may not keep pace with the massive development of surrounding neocortex, the insula is by no means a vestigial structure. In fact, a gradual increase in the complexity of its organization may be discerned in the course of primate evolution. For example, the insula of the New World monkey

A list of abbreviations for structures discussed in the text is given in Section 5, pages 221–222.

M.-MARSEL MESULAM and ELLIOTT J. MUFSON • Bullard and Denny-Brown Laboratories and Behavioral Neurology Section, Neurology Department, Harvard University, and Charles A. Dana Research Institute, Beth Israel Hospital, Boston, Massachusetts 02215. This chapter is dedicated to Paul I. Yakovlev (1894–1983).

Saimiri sciureus is entirely smooth with no evidence of sulcation whereas in the more highly developed brain of the Old World monkey, a shallow and inconsistent orbitoinsular groove can be identified (Fig. 1B). The complexity of organization is still greater in the insula of apes and humans where several radially oriented sulci divide the insula into the gyri breves and the gyri longus. In the entire animal kingdom, the insula of cetaceans is clearly the most complex and

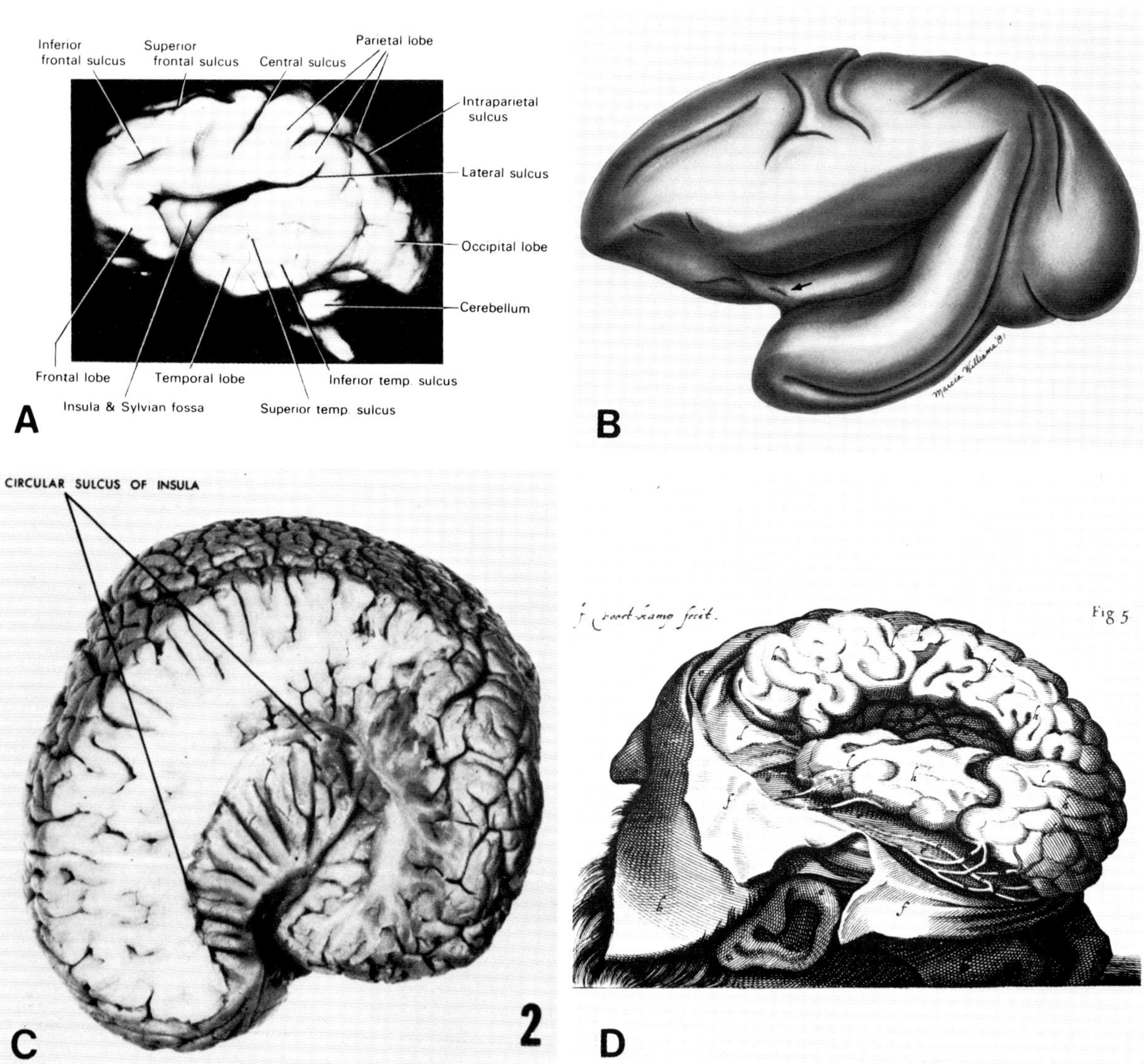

Figure 1. (A) Photograph of the brain of a 7-month human fetus. Note the exposed insula. From Carpenter (1976) with permission. (B) Drawing of an adult rhesus monkey brain. The opercula had to be separated in order to expose the insula. Note the orbitoinsular sulcus (arrow). (C) Photograph of a lateral view of the brain of the dolphin (*Tursiops truncatus*). The opercula have been removed to expose the insula. Note the numerous gyri. From Morgane *et al.* (1980) with permission. (D) The first detailed illustration of the adult human insula. Note the position deep in the sylvian fossa. From Bartholin (1641).

may contain up to 20 sulci (Fig. 1C) whereas the human insula contains no more than 5–7 (Morgane *et al.*, 1980). It is unclear how this extensive development of the insula relates to the behavioral specializations of this species.

Although its position within the sylvian fissure hides it from external inspection, awareness of the insula's presence has existed at least since 1543 when Vesalius included what appears to be insular cortex in Fig. 7 of his *De Humani Corporis Fabrica Libri Septem.* However, the first unmistakable illustration of the insula was not published until a century later, in 1641, in the *Institutiones Anatomicae* of Caspar Bartholin (Fig. 1D). The earliest influential treatise focusing on this region was written by the eminent anatomist Johann Christian Reil (1809) who introduced the term insula (Insel) and whose name is now permanently associated with this part of the brain.

The human insula displays consistent asymmetries in development. For example, sulcal markings appear consistently earlier in the right insula even though the left insula is destined to reach a larger size by adulthood (Clark, 1896). Several complex functional specializations have been attributed to the insula. Aphasia and apraxia were associated with insular damage but definitive evidence was never obtained (see Clark, 1896, for review). Waldschmidt is quoted as having stated in 1887 that the insula tends to have a greater size in more highly educated individuals (Clark, 1896). The insula was even implicated in criminology when it was pointed out that several renown criminals had unusually extensive hemispheric asymmetries in the insula region (Clark, 1896). Although its location deep inside the sylvian fissure initially erected an obstacle to experimental explorations, recently a great deal of information has been gathered on the structure, connections, and physiology of the insula in the monkey brain. This information, much of which is likely to prove relevant to the investigation of the human insula, will constitute the focus of this chapter.

Two considerations have guided our analysis of the insula. First, the insula is intimately related to adjacent cortical areas of the orbitofrontal and temporopolar regions. The structural and functional similarities among these three regions are so strong that it becomes highly desirable to consider the insula, not in isolation, but as one component of an insulo-orbito-temporopolar complex. Second, the insula as well as the adjacent orbitofrontal and temporopolar areas are characterized by a remarkable heterogeneity in cortical architecture, connectivity, and physiology. This heterogeneity which is a hallmark of paralimbic (mesocortical) parts of the brain, indicates neither a chaotic organization nor a haphazard distribution of behavioral specialization. Instead, evidence indicates that the insula, like other components of the paralimbic brain, acts as a multipurpose region of cortex where a remarkably wide range of neural processes modulate behaviors which primarily depend on interactions between the extrapersonal world and the internal milieu.

2. Structure of the Insula

The insula of *Macaca mulatta* has an elliptical shape and covers approximately 160 mm^2 of surface area. The superior limb of the circular sulcus (of Reil), also known as the superior limiting sulcus (SLS in Fig. 2), provides the boundary between the insula and the adjacent frontoparietal operculum. The ventral limb of the circular sulcus, also known as the inferior limiting sulcus, demarcates the insula from the supratemporal plane. The fusion of these two limiting sulci within the fundus of the sylvian fissure (at a level slightly posterior to the ventral tip of the central sulcus) marks the caudal end of the insula. Since the inferior limiting sulcus does not extend rostral to the limen insulae, there is no definite topographic boundary between the anterior insula and the adjacent orbitofrontal cortex (Figs. 2 and 3).

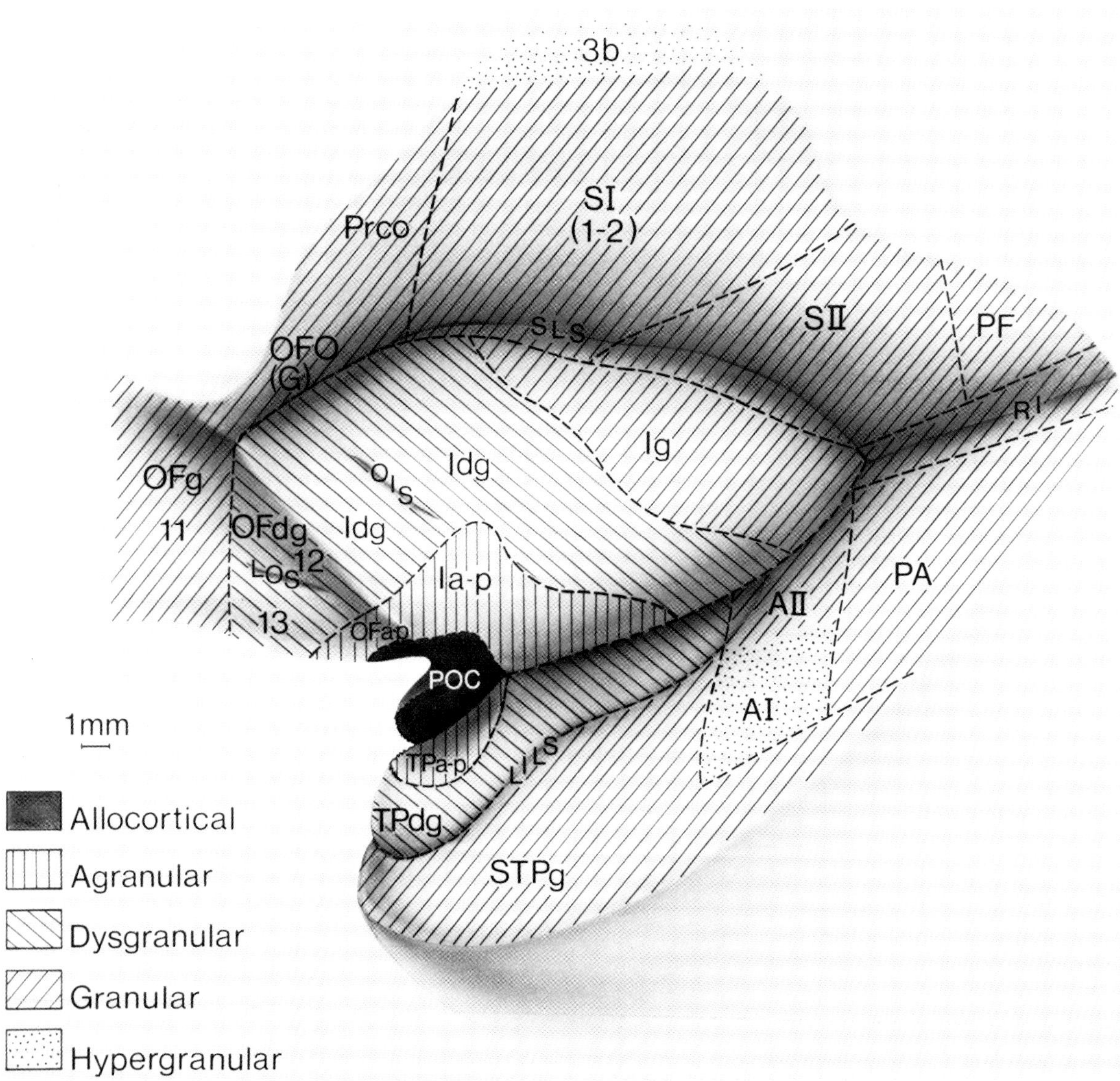

Figure 2. This planar reconstruction of the insula and surrounding regions is based on 35-μm-thick, celloidin-embedded, coronally cut sections from the macaque brain stained with cresyl violet. Dashed lines demarcate architectonic boundaries. Cortical areas which belong to similar architectonic types are indicated by an identical hatching pattern. The boundary between TPdg-AII and the insula is formed by the medial limb of the inferior limiting sulcus (not labeled on this map).

The insula contains several cortical types arranged along a gradient of increasingly more complex architecture (Rose, 1928; Roberts and Akert, 1963; Sanides, 1968, 1970; Jones and Burton, 1976; Mesulam and Mufson, 1982a). Contrary to widespread opinion, the fundamental vector of organization is not directed anteroposteriorly since the most anterior part of the insula is not necessarily its least well-developed component. Instead, the principal vector has a radial orientation emanating outward from the allocortical focus provided by the piriform olfactory cortex (POC in Figs. 2 and 3). Centrifugal waves of cortical development surround this allocortical core both in the insula and in the surrounding temporopolar and orbitofrontal regions (Mesulam and Mufson, 1982a). Traditionally, the cortex of the insula has been subdivided into three belts: (1) an innermost (*periallocortical*) agranular belt which lacks identifiable collections of granule cells (star cells of Lorente de Nó, 1949); (2) an intermediate belt (*pro-* or *periisocortical*) which is called dysgranular because the granule cells in layers 4 and 2 do not display complete laminar differentiation; (3) an outer granular

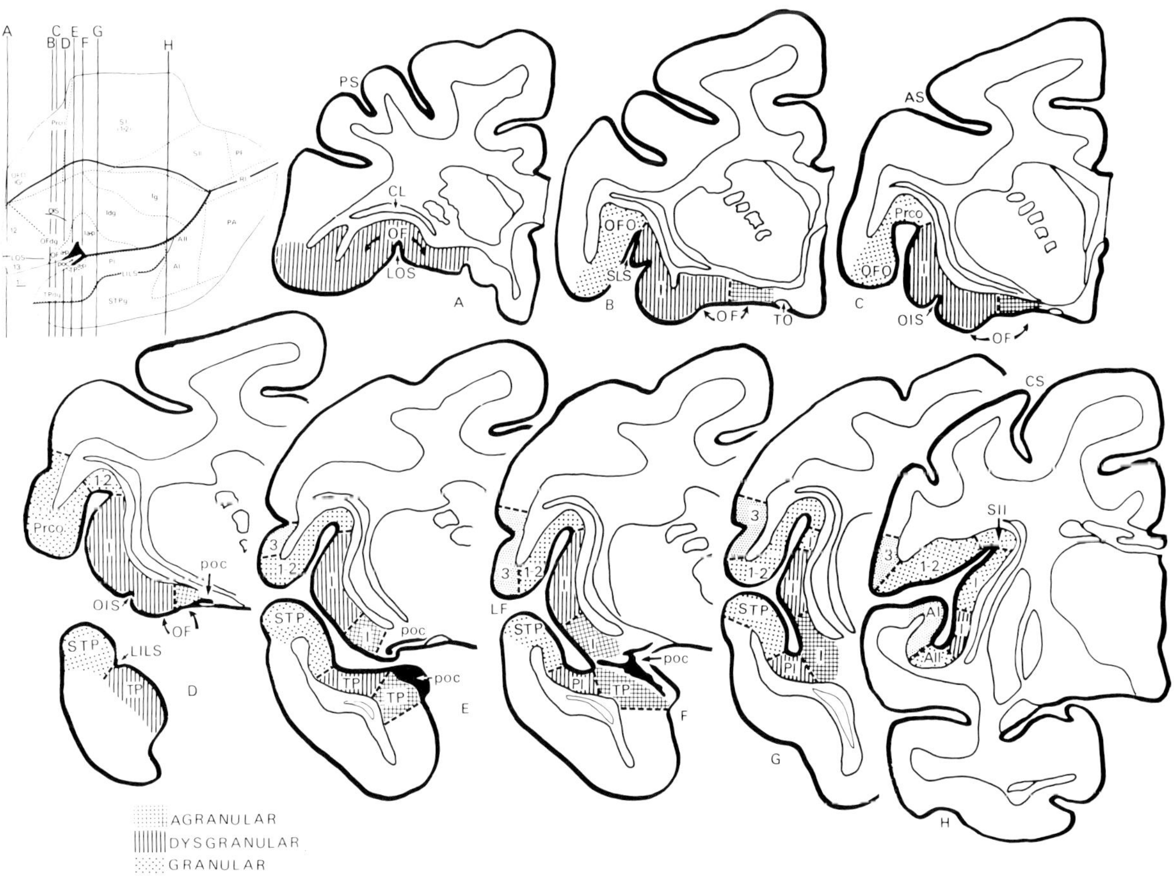

Figure 3. Coronal cross sections showing the cytoarchitectonic subtypes in the insula and adjacent areas in the macaque. This is the type of information that was used to construct the map in Fig. 2. From Mesulam and Mufson (1982a) with permission.

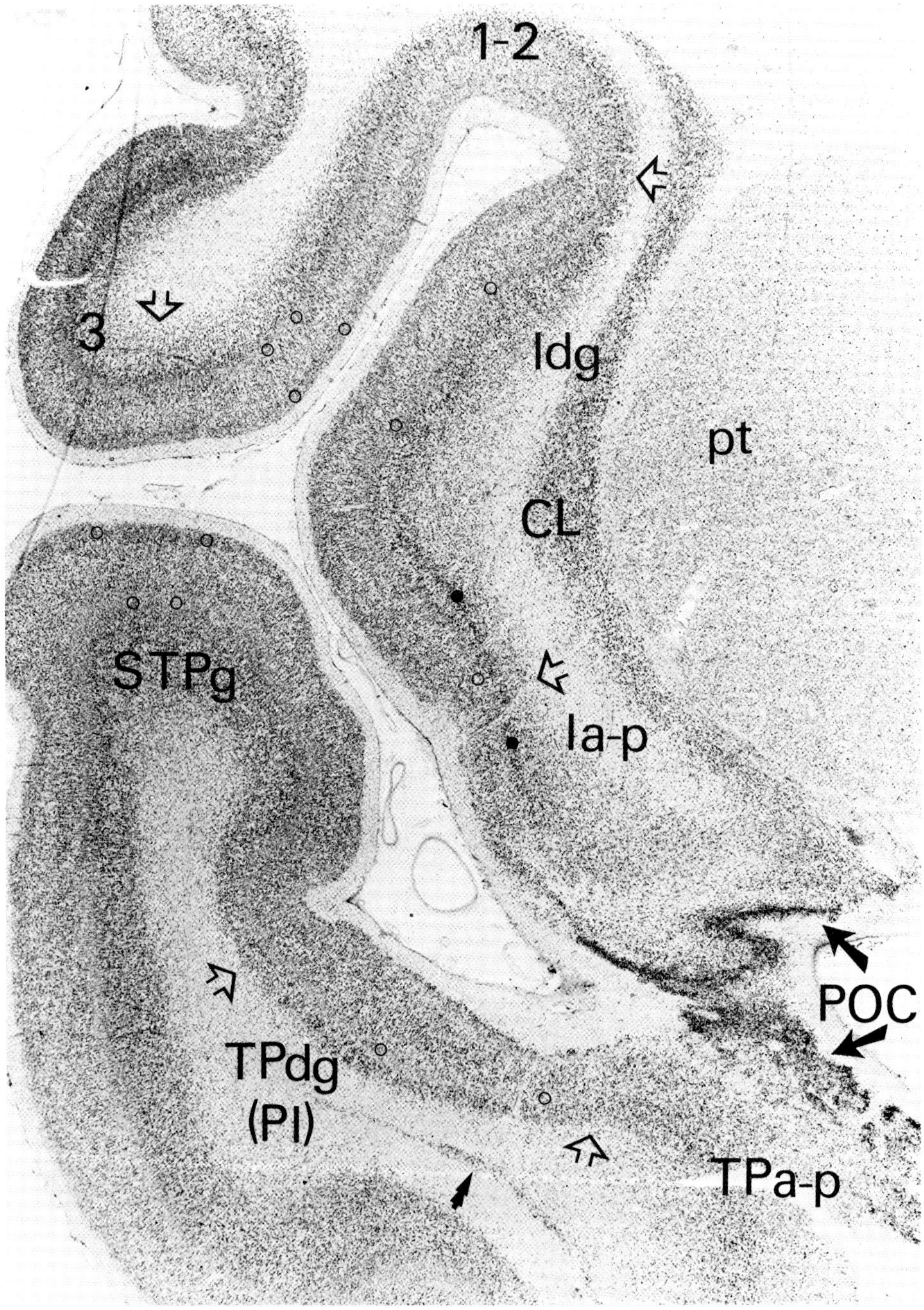

Figure 4. Macaque brain. Open arrows point toward architectonic boundaries. Open circles indicate the position of inner (L4) and outer (L2) granular layers. This section corresponds to Fig. 3F and is at the level of the limen insulae. The trifurcation of POC occurs at this level. The insular extension of POC is continuous with Ia-p. A brief lamina dissecans can be seen in this part of Ia-p. More dorsally, an intermediate stratum (solid circles) inserts itself into Ia-p. Idg begins with the appearance of granule cell clusters in L4 (open circles). In Idg L2 is not demarcated; L5 is not differentiated from L6; and the infragranular layers stand out in Nissl stains. Dorsally in Idg, L4 becomes more

belt which essentially has an *isocortical,* homotypical structure with fully demarcated granule cell layers *both* in layer 4 and in layer 2. This classification will be followed in our discussion of insular architectonics.

2.1. The Piriform Allocortex and the Agranular-Periallocortical Sector of the Insula

At the level of the limen insulae, the POC trifurcates and one of its limbs extends into the insula where it becomes the focus for concentrically arranged insular cortical fields (Fig. 4). At this level, the part of the insula which is coextensive with piriform cortex forms the innermost agranular-periallocortical belt (Ia-p) of the insula. Although a few scattered granule cells can be identified within this sector, they fail to form identifiable clusters or layers. The Ia-p is essentially constructed of two strata: a superficial one in direct continuity with the pyramidal layer of piriform cortex and an inner stratum which is coextensive with the claustrum and perhaps also with the plexiform plate of piriform cortex (Fig. 4). A brief lamina dissecans can be identified between these two strata. With increasing distance from the piriform cortex, an intermediate stratum of deeply staining pyramidal neurons inserts itself into Ia-p (Fig. 4). This intermediate stratum extends into the more dorsal dysgranular insula.

2.2. The Dysgranular-Periisocortical Insula

The boundary between Ia-p and the dysgranular belt (Idg) of the insula is not sharp. Most investigators would place this boundary at the point where clusters of granule cells begin to form individual islets in the region of L4 (Fig. 4). Several centrifugal trends, such as increased granularization and laminar differentiation, can be identified within Idg along a gradient of increasing radial distance from Ia-p. For example, the L4 granules which initially form islets in Idg gradually increase in density and become organized into a continuous layer while granule cell clusters begin to appear in L2. Furthermore, there is also a gradual differentiation of the cortical plate from the underlying white matter so that fewer and fewer neurons can be seen straddling the gap between claustrum and insula. However, Idg does not contain a full demarcation of L5 from L6. Moreover, a true sublamination of L3 is lacking although a trend toward the columnar organization of pyramidal neurons can be identified in this layer.

---

fully demarcated. The temporopolar extension of POC overlies TPa-p. The outer stratum of TPa-p is continuous with POC and the inner stratum is continuous with a subcortical cell cluster, which may correspond to a temporal extension of the claustrum (double arrow). TPa-p does not have an intermediate stratum. TPdg begins with the appearance of granule cell clusters in L4 (open circles). At this level TPdg lacks a granular L2 but contains small deeply staining pyramids in that layer. STPg contains a differentiated granular layer not only in L4 but also in L2 (open circles). Areas 1–2 contain granular layers 2 and 4 (open circles) and also a sublamination in L3. Area 3 is hypergranular so that L4, 3, and 2 tend to merge. ×26. From Mesulam and Mufson (1982a) with permission.

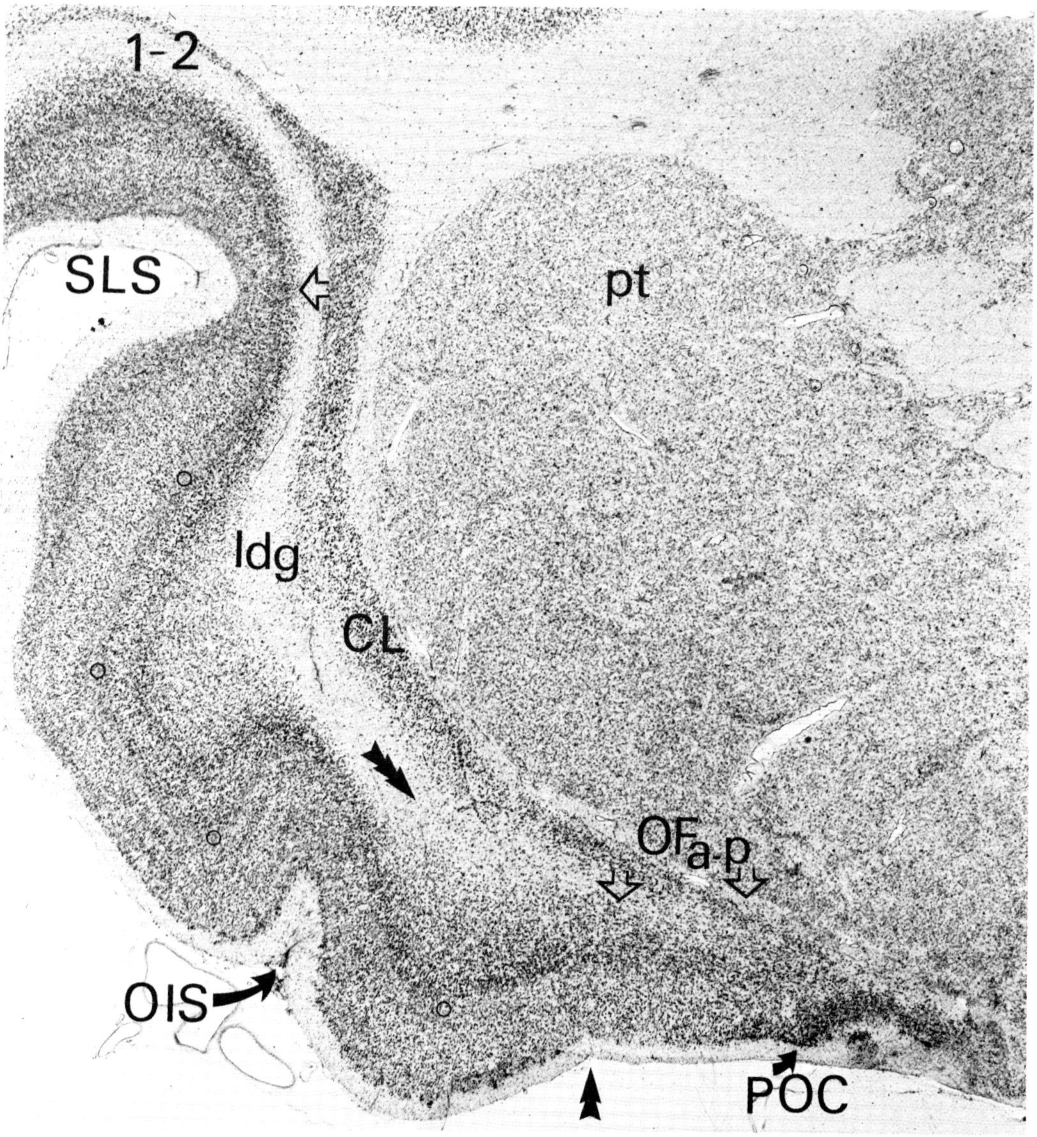

Figure 5. Macaque brain. Photomicrograph of the section shown in Fig. 3D. The open arrows point to architectonic boundaries. In OFa-p, the outer stratum is continuous with the pyramidal layer of POC; the inner stratum merges with the claustrum. Virtually no granule cells are present. More laterally the cortex takes on a dysgranular form as indicated by the appearance of clusters of granule cells (open circles). The double arrowhead points to the arbitrary division between orbitofrontal cortex medially and the insular region laterally. At this level there is only a thin sliver of OFdg, (unlabeled in this photomicrograph) just lateral to OFa-p; the rest of the dysgranular cortex is Idg. In the ventral part of Idg, L2 contains clusters of deeply staining small pyramids rather than granule cells. The triple arrowhead points to neurons in the extreme capsule which form a bridge between the claustrum and Idg. The open circles in Idg indicate the position of a moderately well-granularized L4. The absence of a granular L2 makes this cortex dysgranular. Area 1–2 appears dorsally and can be identified by the presence of a more granular L2 and more distinct sublamination in L3. ×18. From Mesulam and Mufson (1982a) with permission.

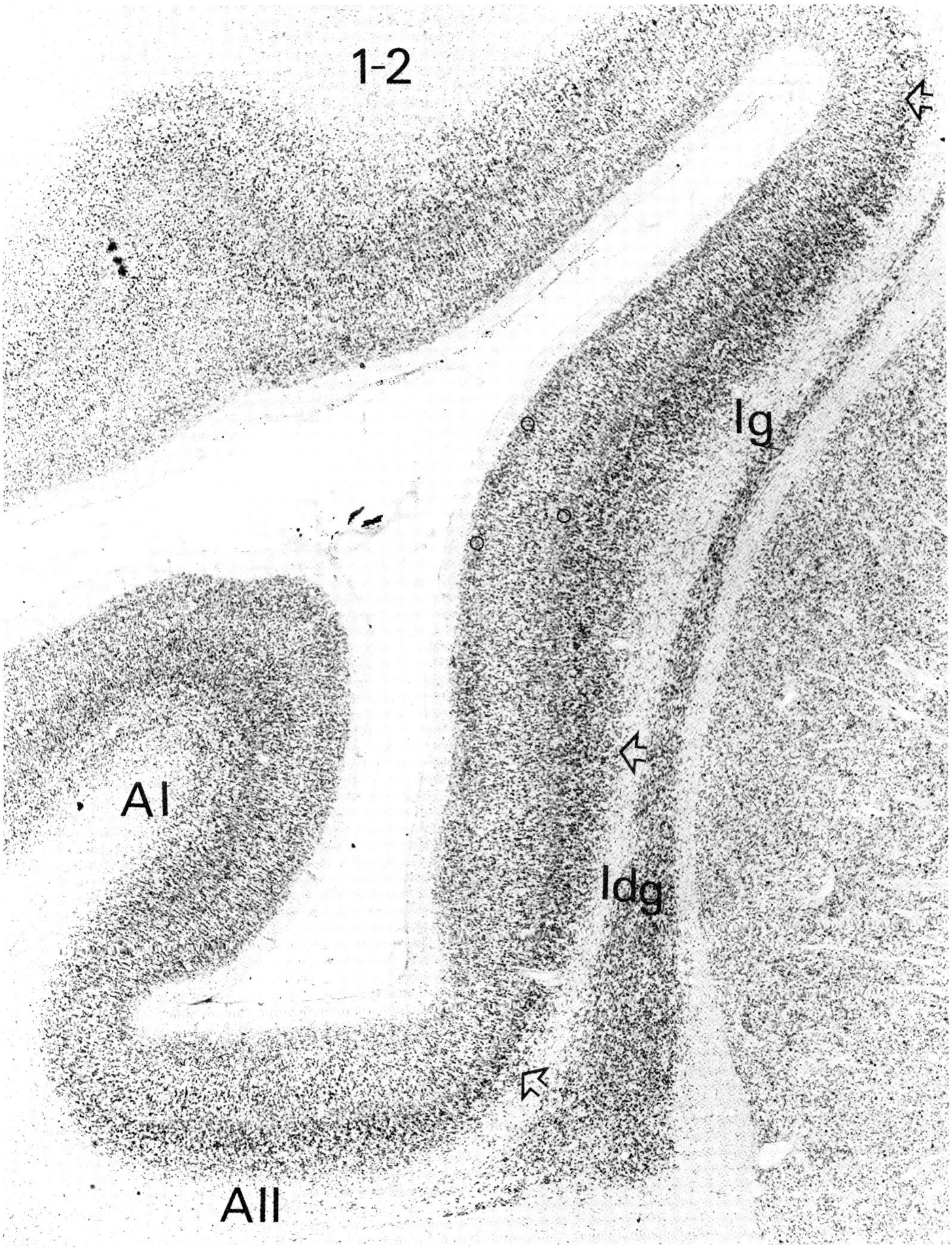

Figure 6. Macaque brain. This photomicrograph corresponds to Fig. 3H. The granular Ig sector is now well established with a differentiated granular L2 and L4 (open circles). There is incipient sublamination in L3 but not as clearly as in area 1–2. AII is characterized by an increased granularity of L2 when compared to Idg. AI is hypergranular cortex where the large number of granules cause layers 4, 3, and 2 to merge with each other. ×20. From Mesulam and Mufson (1982a) with permission.

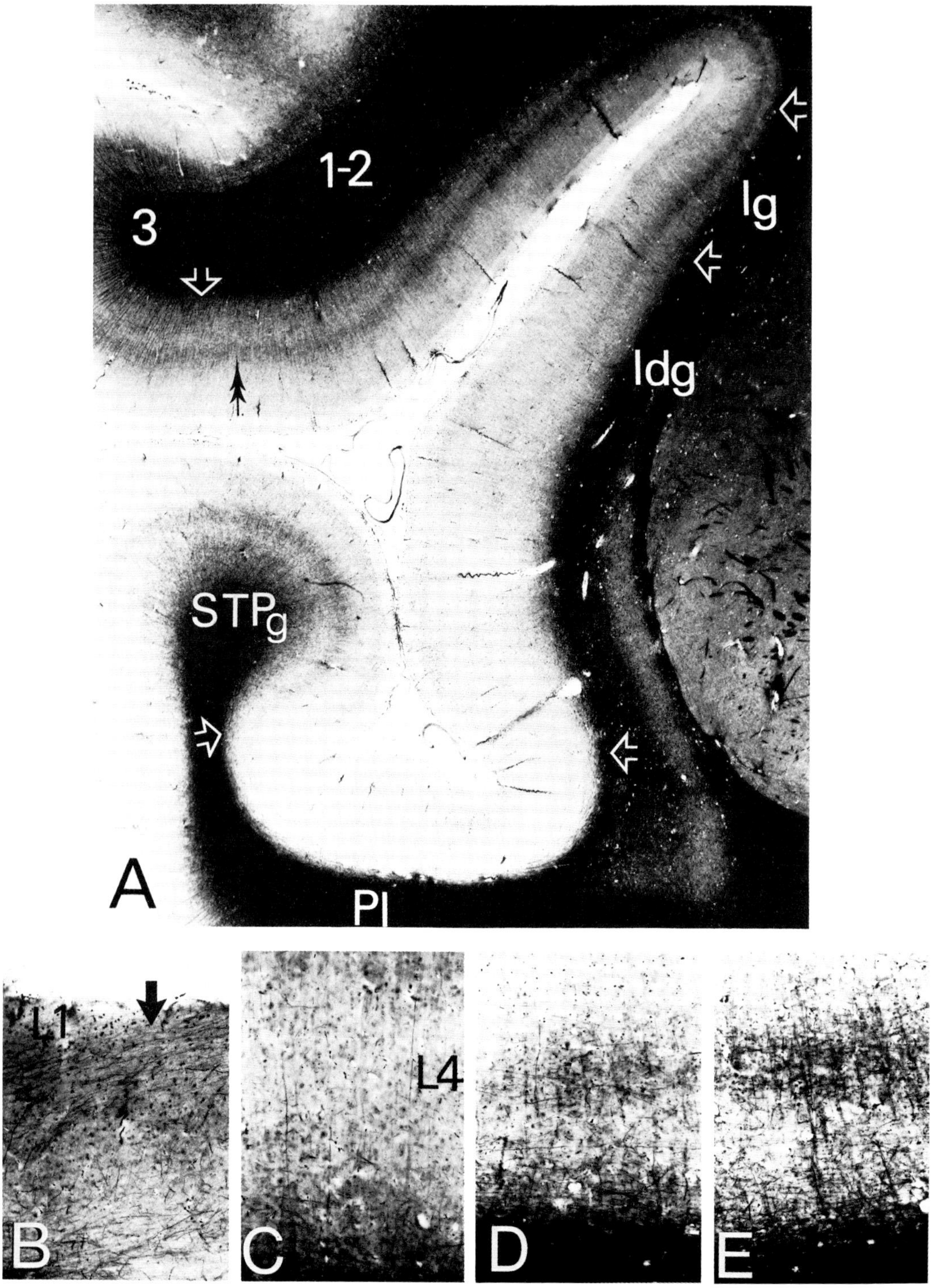

Figure 7. Macaque brain. Loyez stain for myelin. (A) The open arrows point toward lines of cytoarchitectonic demarcation. These were determined by examining an immediately adjacent section stained with cresyl violet. The pattern of intracortical myelin shows regional changes which reflect cytoarchitectonic demarcations. Areas 3, 1, and 2 have the highest level of

A distinctive cytoarchitectonic feature of Idg is the tinctorial prominence of the infragranular pyramidal neurons. The most anterior part of the insula contains the Idg type of cortex rather than Ia-p (Figs. 2, 3, and 5). This configuration highlights the radial organization of insular cortical fields around piriform cortex.

2.3. The Granular-Isocortical Insula and Beyond

The presence of fully demarcated granule cell layers both in L2 and in L4 marks the beginning of the granular-isocortical insula (Ig). In addition, a sublamination appears in L3 so that an L3A and L3B can be identified on the basis of differences in cell density. The tinctorial prominence of infragranular layers gradually disappears in Ig (Fig. 6). It is also possible to discern an incipient differentiation of L5 from L6. This sector is the most heavily myelinated one in the insula, especially with respect to the outer layer of Baillarger (Fig. 7).

The centrifugal trends toward increased granularization, laminar differentiation, and myelination continue beyond the topographical boundary of the insula and into the adjacent opercular cortex. For example, in SI (areas 1–2) the granularization and differentiation of L2 and L4 are even more advanced; the demarcation of L5 from L6 becomes complete; a more conspicuous sublamination, based on differences of cell size, becomes established in L3; and cortical myelination is even more advanced than in Ig (Figs. 6 and 7).

2.4. The Insulo-Orbito-Temporopolar Component of the Paralimbic Brain

As the POC trifurcates at the limen insulae, one limb extends into the insula and provides the focus for the insular fields described above. The other two limbs extend into the orbitofrontal and temporopolar regions where architectonic developments analogous to those in the insula can be identified (Figs. 2–4).

The frontal component of piriform cortex abuts on the most caudal part of areas 12 and 13 of Walker (1940) in the lateral orbitofrontal cortex (Figs. 2, 3, and 5). This portion of orbitofrontal cortex has an agranular-periallocortical structure virtually identical to that of Ia-p. This sector has been designated "Allo" by Roberts and Akert (1963), "J" cortex by Jones and Burton (1976), and OFa-p by us (Mesulam and Mufson, 1982a) (see Fig. 2). With increasing rostral

---

intracortical myelin. The outer line of Baillarger (double arrowhead) is quite conspicuous. There is less intracortical myelin both in radial fibers and in the outer line of Baillarger in Ig. In Idg, myelin is confined to a few radial fibers in deep layers. Area PI (TPdg) contains virtually no intracortical myelin whereas STPg is characterized by myelin in radial fibers as well as in the outer line of Baillarger. ×15. (B) The Ia-p showing the zonal layer of myelin in L1 and the low overall myelin. (C) Myelin in the Idg sector. (D) Myelin in Ig. An incipient outer layer of Baillarger appears in L4. (E) Myelin in area 1–2. Note the intense intracortical myelin in radial fibers and especially in the outer line of Baillarger. (B–E) ×200. From Mesulam and Mufson (1982a) with permission.

distance from piriform cortex, this orbital region assumes first a dysgranular (OFdg) and then a granular (OFg) architecture (Figs. 2 and 5). The architectonic organization of these three orbitofrontal sectors and their relationship to piriform cortex and claustrum are virtually identical to that of the analogous insular sectors. In view of these parallel developments and since no distinct landmarks separate the anterior insula from orbitofrontal cortex, insulo-orbital cortex may represent an indivisible unit of cortical organization (Sanides, 1962; Mesulam and Mufson, 1982a).

The third limb of the POC gives rise to concentric waves of cortical fields in two different directions within the temporal pole. One of these extends ventrally into the inferotemporal isocortex, whereas the second one extends dorsally toward the superior temporal isocortex (Gower, 1981; Gower and Mesulam, 1982). As in the insula and the orbitofrontal region, an agranular-periallocortical type of cortex (TPa-p) abuts on the temporopolar extension of piriform cortex. With increasing distance from TPa-p, first a dysgranular-periisocortical (TPdg) and then a granular-isocortical (STPg) type of cortex can be identified in the supratemporal plane (Figs. 2–4). The superior dysgranular temporopolar belt (TPdg) extends beyond the limen insulae both rostrally and caudally (Figs. 2 and 3). Caudally, this dysgranular type of cortex has been called parainsular cortex (Sanides, 1972; Jones and Burton, 1976; see Fig. 2). The granular supratemporal cortex (STPg in Fig. 2) contains the Ts3 and paAlt fields of Sanides (1972) and the T1 and T2 fields of Jones and Burton (1976).

The insula, caudal orbital cortex, and the temporal pole each contain architectonic transitions between simple allocortex and more fully developed isocortex. Regions which contain this type of transition may be designated by the generic terms "paralimbic" or "mesocortical." In addition to the insula, the orbitofrontal cortex, and the temporal pole, this type of architectonic transition also occurs in the parahippocampal region and in the gretrosplenial-cingulate-parolfactory areas. However, in contrast to the insulo-orbito-temporopolar contingent of the paralimbic brain where the piriform cortex provides the allocortical focus, it is the hippocampal formation and its induseal rudiment which form the allocortical focus for this second group of paralimbic areas (Abbie, 1942; Filimonoff, 1947; Yakovlev, 1959; Sanides, 1968, 1970; Mesulam and Mufson, 1982a). Thus, the medial and basal parts of the hemispheric mantle are surrounded by a continuous girdle of paralimbic areas (Fig. 8A).

2.5. The Olfactocentric Plan of Cortical Organization

In addition to the orbitofrontal and temporal paralimbic regions, the insula is surrounded by a number of additional cortical fields to which it is closely related (Fig. 2). Anteriorly, the inner face of the frontoparietal operculum is covered by granular cortical areas which have been designated OFO and Prco by Roberts and Akert (1963). According to Sanides (1968, 1970) and Jones and Burton (1976), OFO corresponds to a primary gustatory cortex (G) in the monkey brain. Posterior to OFO–Prco, the inner face of the frontoparietal operculum contains areas 1–2 (SI), SII, PF, and RI each exhibiting well-developed granular

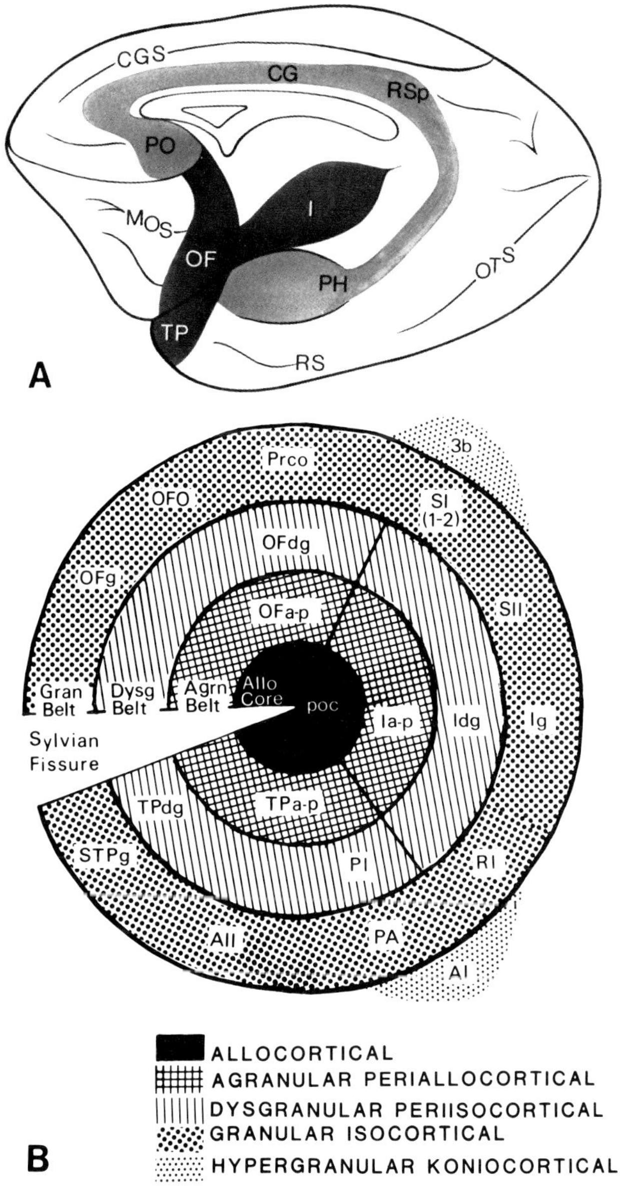

Figure 8. (A) Diagram showing the two divisions of the paralimbic brain. The parahippocampal (PH), retrosplenial (RSp), cingulate (CG), and parolfactory (PO) areas make up the hippocampocentric division; the insula (I), orbitofrontal (OF), and temporopolar (TP) regions make up the olfactocentric component. (B) Diagrammatic representation of the concentric arrangements of granular (Gran), dysgranular (Dysg), and agranular (Agrn) belts around the olfactory (poc) allocortical (Allo) focus.

type of cortex (Fig. 2). The part of SI which is adjacent to areas 1–2, area 3b, has a hypergranular (koniocortical) architecture.

In addition to TPdg, PI, and STPg, which have been described above, the supratemporal plane also contains the auditory association area PA (Jones and Burton, 1976) which contains a homotypical, granular type of cortex (Fig. 2). The PA region includes the relt, paAc, and Tpt fields of Sanides (1972). The principal auditory areas AII and AI are adjacent to PA and STPg. The AII zone has a granular isocortical architecture whereas AI is hypergranular in appearance. The architectonic details of these juxtainsular areas and the features which help to delineate them from insular cortex have been described elsewhere (Roberts and Akert, 1963; Jones and Burton, 1976; Mesulam and Mufson, 1982a).

When the insula and the surrounding fields are considered collectively, the concentric plan of cortical organization around piriform cortex can be discerned with greater clarity (Fig. 8B). In a figurative sense, the piriform cortex appears to provide an allocortical source from which emanate, in three principal directions, concentric waves of gradually more highly developed cortical fields. The first ring consists of three agranular-periallocortical fields: OFa-p, Ia-p, TPa-p. The second ring consists of the three dysgranular-periisocortical fields: OFdg, Idg, PI–TPdg. The third ring consists of granular cortical fields which include OFg, OFO–Prco, SI (areas 1–2), SII, Ig, and also RI, PA, AII, and STPg. It is even possible to identify a fourth, albeit incomplete, hypergranular belt consisting of the koniocortical regions 3b and AI. It is conceivable as suggested by Yakovlev (1959) and Sanides (1963, 1970) that this concentric plan of organization may reflect stages of cortical development in phylogeny and ontogeny.

2.6. The Human Insula

Despite its considerably greater size, the general architectonic plan of the human insula is very similar to that observed in the macaque brain (Fig. 9). For example, at the origin of the limen insulae it is possible to identify an agranular Ia-p sector which is coextensive with the piriform cortex and the claustrum. The distribution of myelin in this sector is similar to that of the monkey Ia-p (Fig. 9). The gradual transition from Ia-p to Idg and from Idg into Ig show the same trends which have been described in the monkey brain. The concentric organization depicted in Fig. 8B can also be identified in the human brain.

Figure 9. Human brain. (A) Cresyl violet stain of the agranular sector at thc level of the limen insulae. There are no recognizable collections of granule cells. Cell bridges from the claustrum are continuous with the inner stratum of Ia-p while superficial POC cells are continuous with the outer stratum. ×10. (B) Immediately adjacent section stained for myelin shows the zonal band of myelin in Ia-p (arrow). ×7.5. (C) Photomicrograph of the insula at a level slightly posterior to the limen. A stratum of deeply staining pyramids (black circles) is conspicuous in Idg and forms the intermediate stratum of Ia-p. The cell bridges with the claustrum are still present. Idg shows the tinctorial prominence of infragranular layers, an incipient granularity in L4, and very little granularity or differentiation in L2. ×9. From Mesulam and Mufson (1982a) with permission.

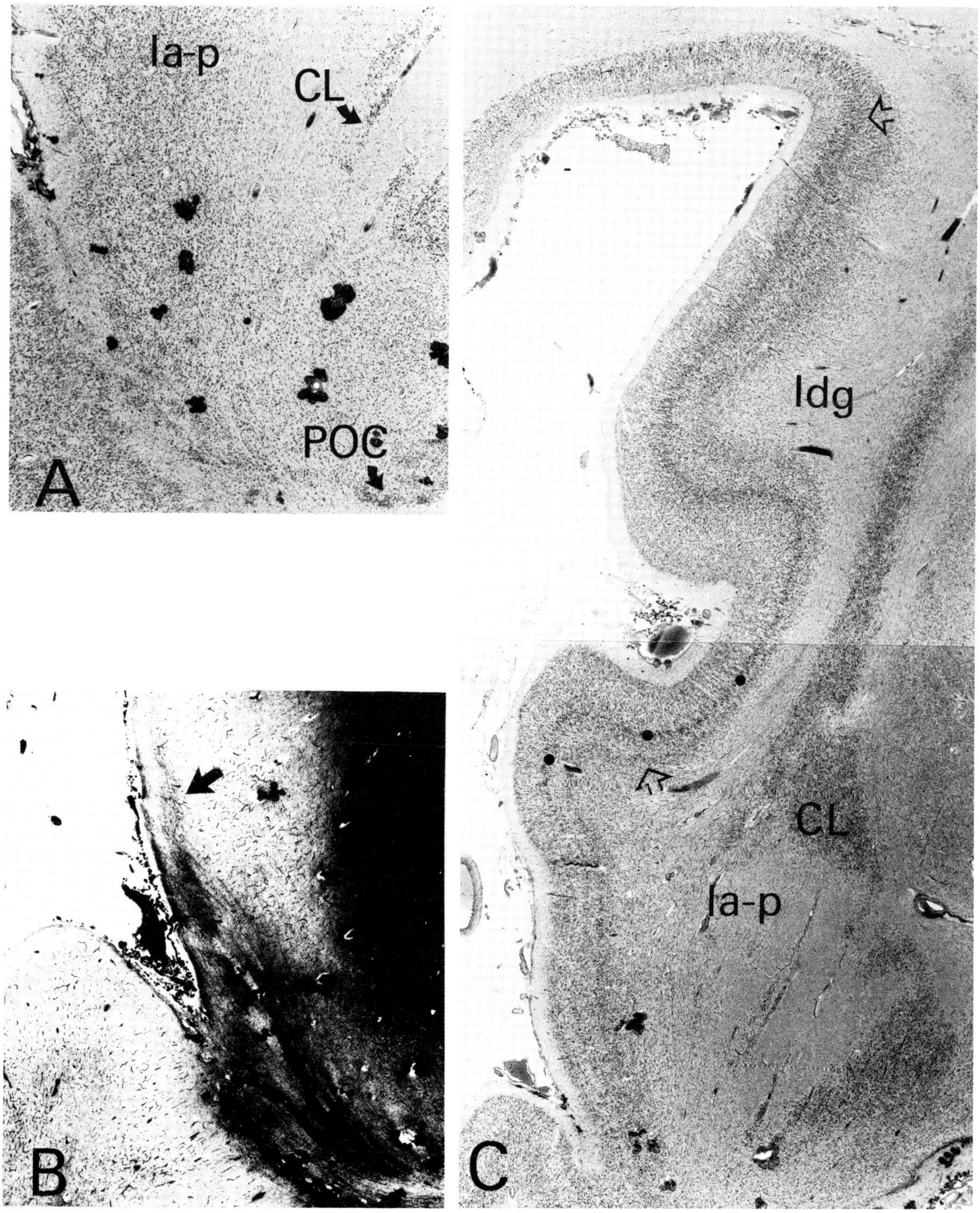
Ia-p
CL
POC
A
B
Idg
CL
Ia-p
C

3. Connections of the Insula

The architectonic heterogeneity of the insula is reflected in the remarkable spectrum of its connections. In this section we will review the cortical, amygdaloid, and thalamic connectivity of the insula in the Old World monkey. The information in this section is based on 44 macaque cerebra, each with an injection of tritiated amino acids (TAA) or of horseradish peroxidase (HRP) within a specified region of the brain. Cases with TAA injections were used to determine the anterograde projections of the injection site whereas the HRP cases were used to determine the distribution of neurons which project into the injection site. In many cases, the HRP tracer was conjugated to the lectin wheat germ agglutinin. The technical considerations which are relevant to the preparation of this material and the caveats which guided the interpretations of the results have been discussed elsewhere (Mufson *et al.*, 1981; Mufson and Mesulam, 1982, 1984; Mesulam and Mufson, 1982b). Our previous observations had indicated that the connections of the insula with other cortical areas and with most thalamic and amygdaloid nuclei are essentially reciprocal. Therefore, in the following section, the cases with TAA and with HRP injections will be considered together in establishing the essential outlines of insular connectivity.

Three types of cases have been used to analyze insular connectivity. Some cases had relatively extensive insular injections (Fig. 10). Although the injection sites in these cases include minor extensions into the underlying white matter and claustrum, the resultant distribution of labeling provided a useful overview of putative insular connections. This information was supplemented with additional confirmatory cases. In the first type of confirmatory experiment, the injection site was entirely confined to the insula (Fig. 11). While the small injection site in this type of preparation revealed only a subset of all the insular connections, these could be interpreted with confidence. In 26 additional experiments, 19 of which are illustrated in Fig. 12, regions which appeared to have insular connections on the basis of the first two types of cases were themselves injected with TAA or HRP. The presence of HRP or TAA labeling within the insula in these 26 cases provided definitive confirmation for the existence of the connection under investigation and also revealed its architectonic and laminar distribution within the insula.

3.1. Somatosensory Connections

The insula is related to a surprisingly large number of somatosensory areas, including SI–SII, area 5 (PE), and area 7b (PF) (Fig. 10). Cases with large tracer

Figure 10. Schematic summary diagram of the cortical and amygdaloid connections of the insula, summarizing the observations from 17 macaque monkeys each with an injection of TAA or HRP in the insula. A composite of the anterior injections is shown in solid black while a composite of the posterior cases is indicated by the cross-hatching in (C). The black circles indicate the sites of transported label in cases with the more anterior injections; the open circles indicate the location of transported label in cases with the more posterior injections. In each of these animals, the injection site involved areas outside of insular cortex. While these cases cannot provide definitive information, they were very useful in delineating the entire spectrum of potential insular connections. The solid lines in A–C indicate sulci. The area between the solid and interrupted lines designates the cortex lining sulcal banks.

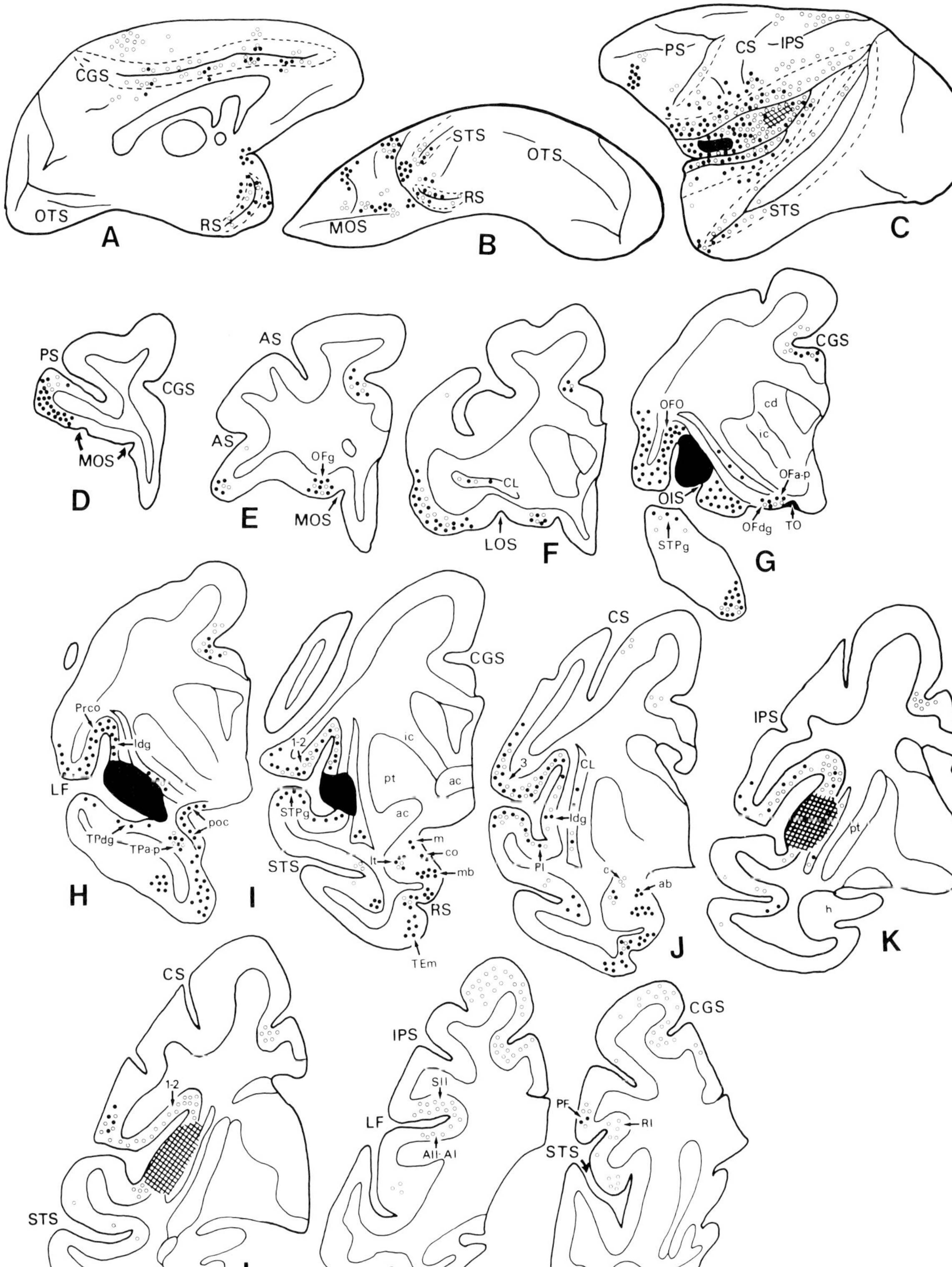

CGS
OTS
RS
A
STS
OTS
RS
MOS
B
PS
CS
IPS
STS
C
PS
CGS
MOS
D
AS
AS
OFg
MOS
E
CL
LOS
F
CGS
OFO
cd
ic
OFa-p
OIS
OFdg
TO
STPg
G
Prco
Idg
LF
TPdg
TPa-p
poc
H
CGS
1-2
ic
pt
ac
ac
STPg
STS
m
co
lt
mb
RS
TEm
I
CS
3
CL
Idg
PI
c
ab
J
IPS
pt
h
K
CS
1-2
STS
L
IPS
SII
LF
AII-AI
OTS
M
CGS
PF
RI
STS
N

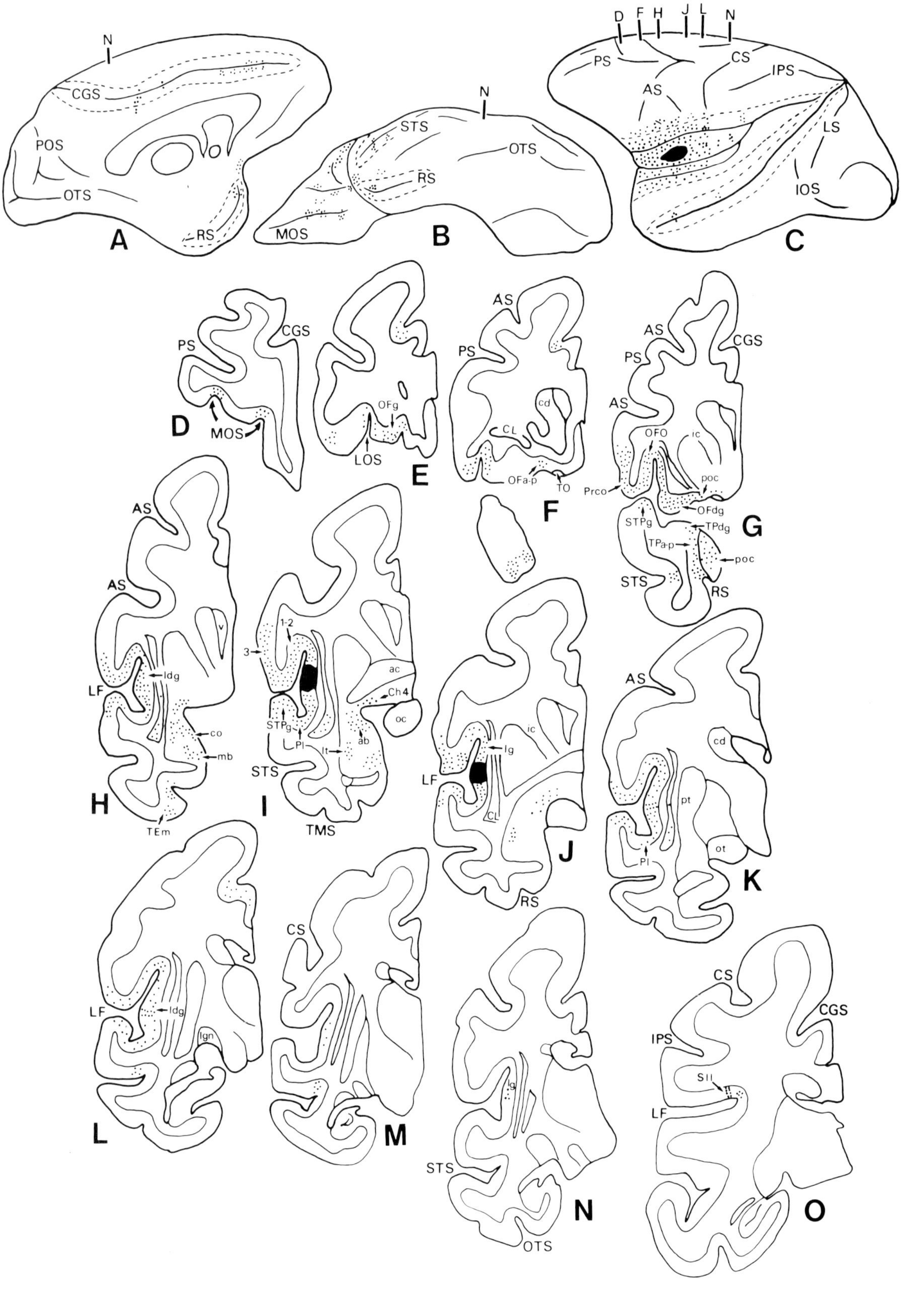

A
B
C
N
CGS
POS
OTS
RS
STS
MOS
D F H J L N
PS
CS
IPS
AS
LS
IOS
D
E
F
G
PS
CGS
MOS
OFg
LOS
AS
cd
CL
OFa-p
TO
OFO
ic
Prco
poc
OFdg
STPg
TPdg
TPa-p
STS
RS
H
I
J
K
LF
Idg
co
mb
TEm
v
1-2
3
ac
Ch4
oc
STPg
Pl
lt
ab
STS
TMS
Ig
CL
pt
ot
L
M
N
O
lgn
SII
IPS
OTS

injections suggest that the insula may have two foci of SI connectivity, one at the foot of the central sulcus, in a region which contains the representation of the face and mouth, and a second one at the dorsal crest of the postcentral gyrus in a region that contains the representation of the leg and foot (Fig. 10C,J). Confirmatory evidence has been obtained for both of these connections. For example, a small TAA injection in the ventral part of SI (mostly within area 3b, with a minor extension into area 1–2) results in terminal labeling within the dorsal part of Idg (Fig. 12A). In another animal, a TAA injection in the dorsal part of SI (mostly within area 1–2 with a minor extension into area 3b) shows the presence of a projection virtually confined to the anterior portion of Ig (Fig. 12B). These cases indicate that the insula definitely receives SI projections, and that these may be arranged in a way which could be related to a body map. It is unclear if parts of SI not included in these two small injection sites also project to the insula. An experiment in which a TAA injection was confined to the insula reveals definitive evidence for the existence of connectivity with SII as well (Fig. 11O).

Area 5 (PE) in the superior parietal lobule constitutes a somatosensory association area since it receives massive neural input from area 2 of SI (Jones *et al.*, 1978). Tracer injections into the dorsal and medial aspects of area 5 demonstrate the presence of connections between area 5 and dorsal Ig (Fig. 12C,D). A third stage in the processing of somatosensory information takes place in area 7b (PF). This part of the inferior parietal lobule also sends neural projections to caudal Idg and especially to Ig (Fig. 12E). These cases show that the dorsocaudal portion of the insula has access to a wide spectrum of somatosensory inputs. In fact, these may provide the predominant sensory afferents into parts of Ig. This component of the insula could therefore constitute a somatosensory association area.

3.2. Auditory Connections

Information derived from cases with large tracer injections suggests a widespread connectivity between the insula and the auditory regions of the supratemporal plane, including PI, STPg, PA, RI, and perhaps even AI–AII (Fig. 10C,G–J). The presence of projections from the Idg sector to STPg and PI is confirmed following a TAA injection within the dysgranular insula (Fig. 11C,G–K). An additional experiment in which a small TAA injection is placed within area PA shows the presence of auditory association projections to the posterior insula, especially to Ig (Fig. 12F).

3.3. Visual Connections

Even experiments with large insular injections fail to show connections with most of the traditional visual areas in cortex (Fig. 10). The one exception appears to be the anteromedial part of the inferotemporal region (area TEm, Figs. 10I

Figure 11. This case had a TAA injection site entirely confined to the insula. The sites of anterograde TAA transport are indicated by dots.

and 11H). Further support for the existence of this connection is provided by an experiment where a small TAA injection is made almost entirely within TEm, but with some minor extension into the adjacent perirhinal cortex (Fig. 12G). This case shows two foci of insular labeling, one ventral (Ia-p, Idg) and the other more dorsal (Idg, Ig). The dorsal extent of this projection raises the possibility of a potential somesthetic–auditory–visual convergence in this part of the insula.

3.4. Gustatory Connections

A connection between Idg and the gustatory cortex (OFO) is demonstrated by a TAA injection limited to the insula (Fig. 11G). We do not know if other regions of the insula also have gustatory connectivity.

3.5. Motor Connections

The upper bank of the anterior cingulate sulcus and the cortex dorsal to it contain the MII region of Woolsey (1965) which is also referred to as the supplementary motor cortex. Although some of our material raises the possibility of a projection between MII and the posterior insula (Fig. 10), confirmatory evidence is not yet available. Another motor connection of the insula is directed to the ventral portion of area 6 (area Prco). This portion of premotor cortex contains the representation mostly of the face region (McGuiness *et al.*, 1980). An experiment in which an HRP injection is made into area Prco shows that this connection arises predominantly from the mid-insula, at the junction of Idg with Ig (Fig. 12H). When the cases shown in Figs. 11 and 12A,H are considered collectively, it appears that the mid-insula may provide a site where gustatory input could partially overlap with somesthetic input from the face–mouth region. Furthermore, the same region may have access to the premotor face area. It is conceivable that this arrangement is related to the organization of complex behaviors related to the face and mouth.

3.6. High-Order Association Connections

The insula has connections with three high-order association areas: prefrontal cortex (areas 45–46), rostral orbital cortex [area 11 (OFg)], and the banks of the caudal superior temporal cortex (Fig. 10B–E,I–K). The case illustrated in Fig. 11B–E confirms the presence of a projection from the insula to area 11 (OFg). An additional case with a small HRP injection mostly in area 11 shows

Figure 12. In these 19 cases, HRP or TAA was injected into cerebral areas with potential insular connections in order to obtain more definitive confirmation. On the left, the distribution of label within the insula is shown within the planar map which corresponds to that of Fig. 2. Two coronal cross sections through the insula give another view of the labeling and its architectonic distribution. On the right, the injection site is shown in black and the type of tracer is indicated as TAA or HRP. White arrows on the injection site indicate that sulcal cortex was also included in the injection site.

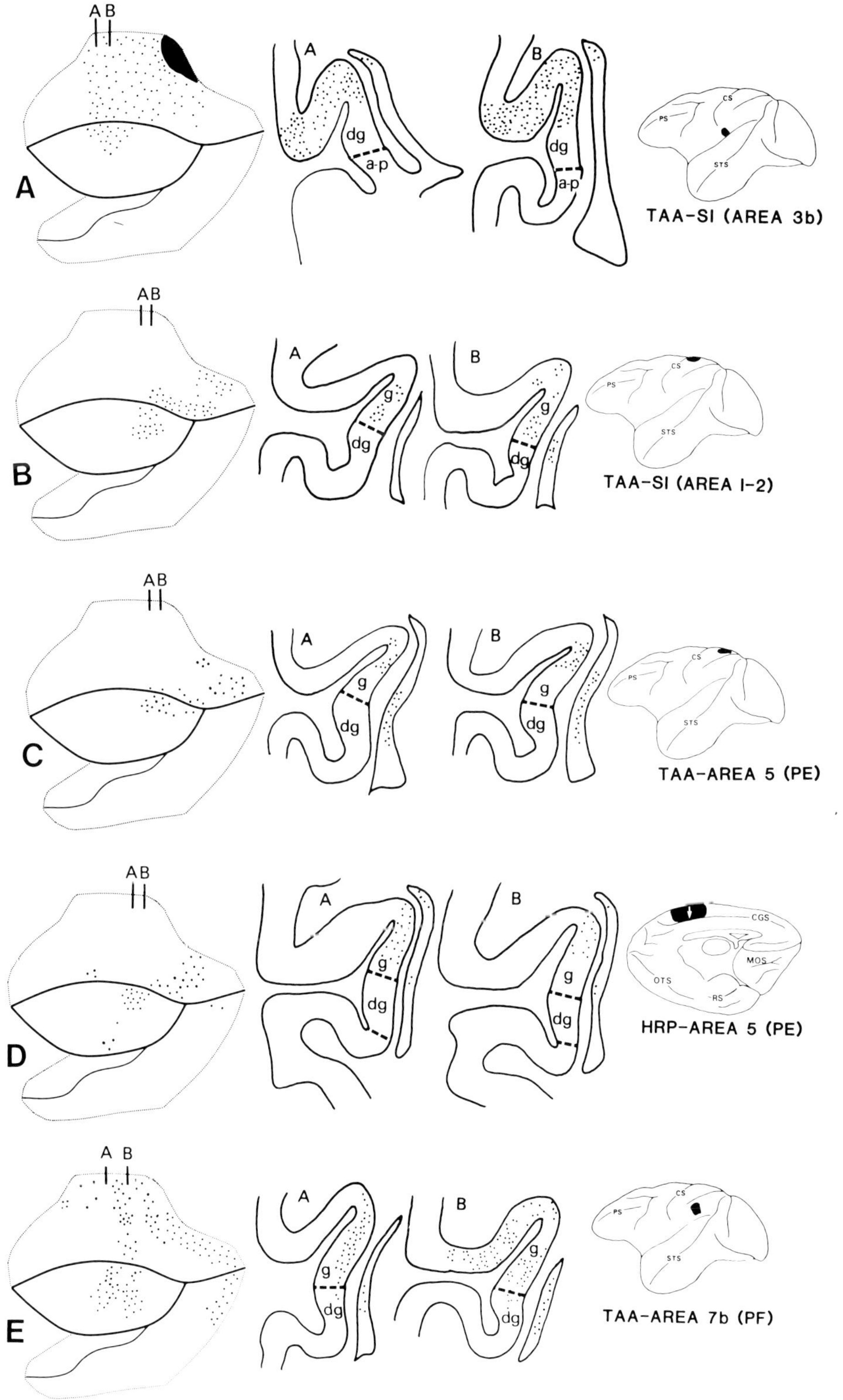
A
B
C
D
E
dg
a-p
g
CS
PS
STS
CGS
MOS
OTS
RS
TAA-SI (AREA 3b)
TAA-SI (AREA I-2)
TAA-AREA 5 (PE)
HRP-AREA 5 (PE)
TAA-AREA 7b (PF)

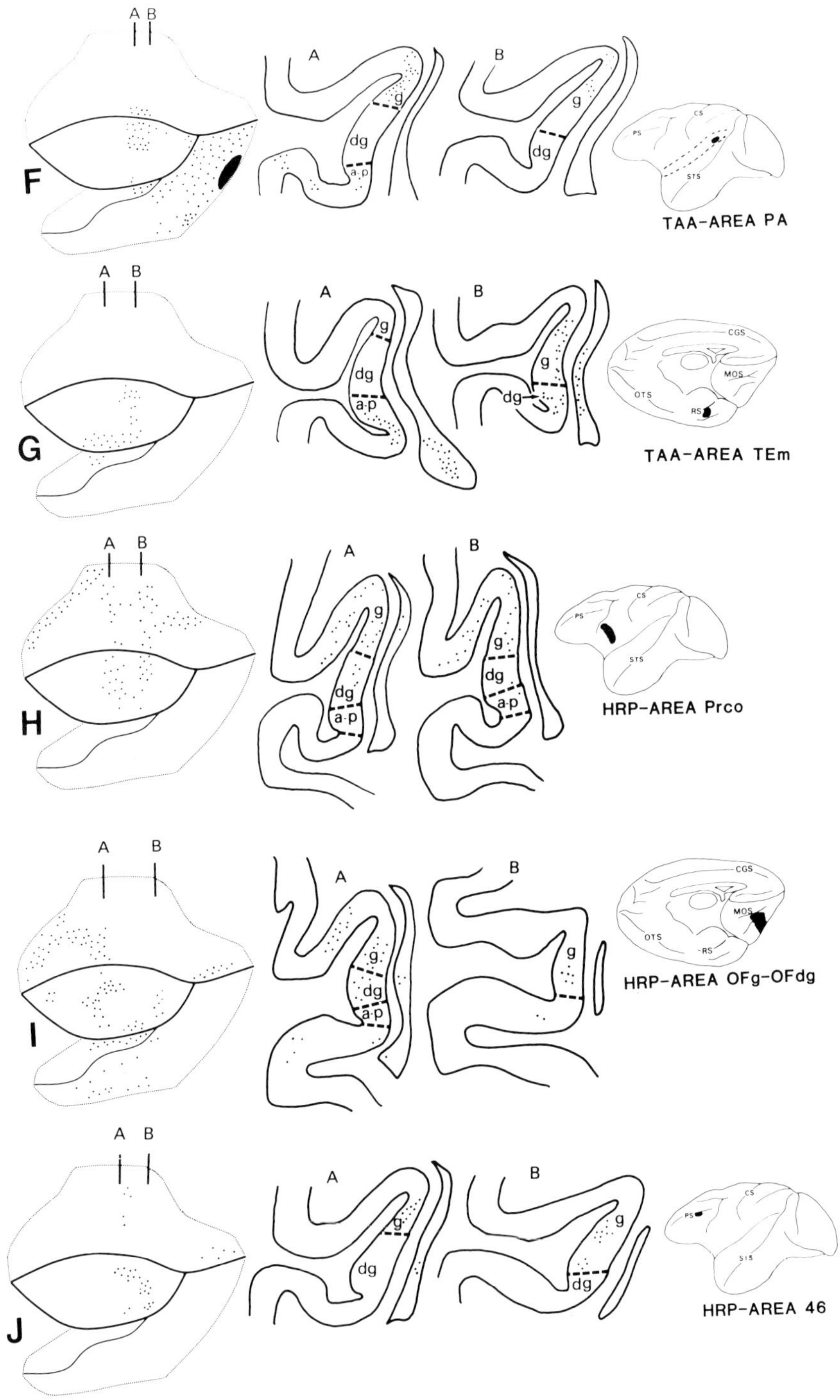

Figure 12. *(continued)*

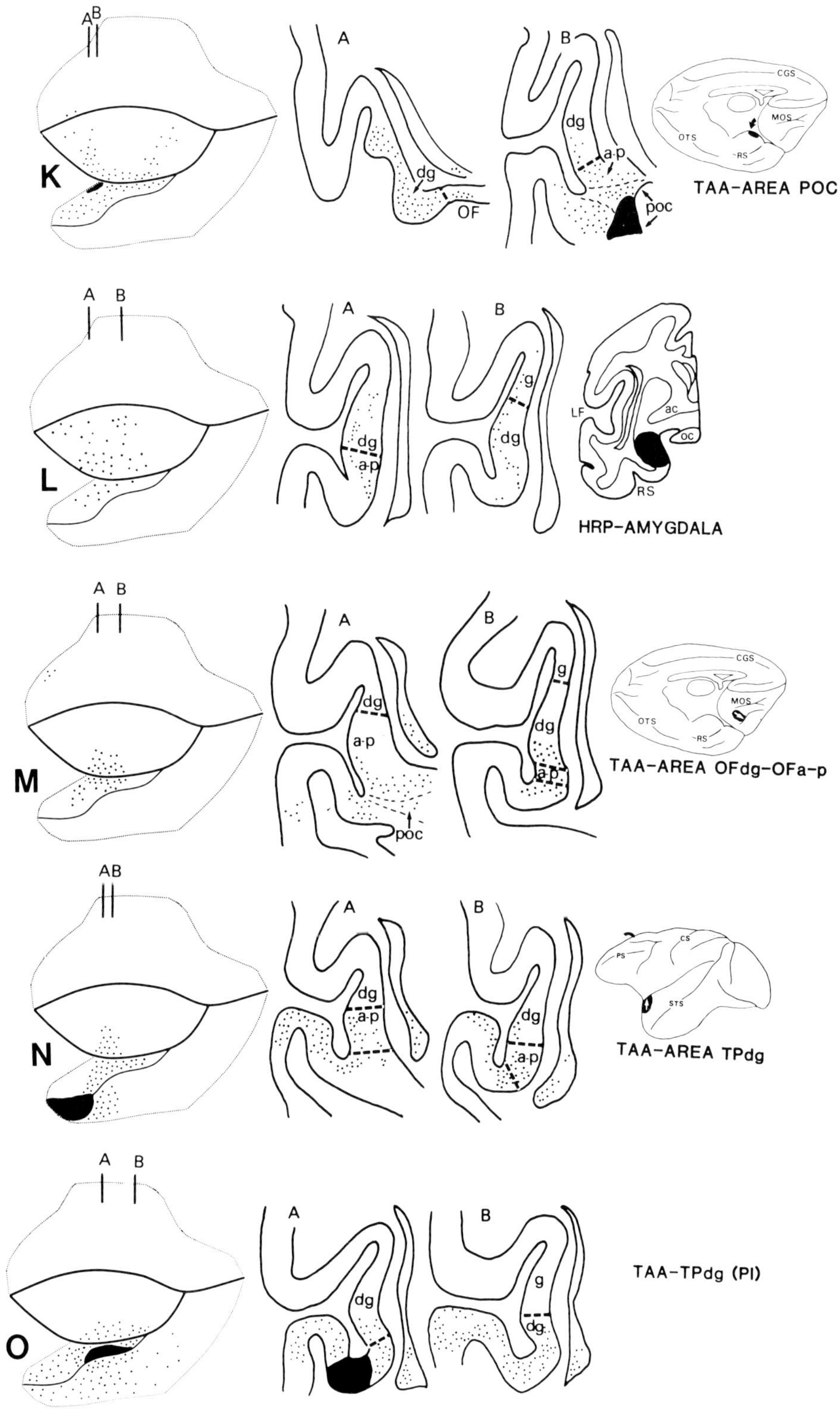

Figure 12. *(continued)*

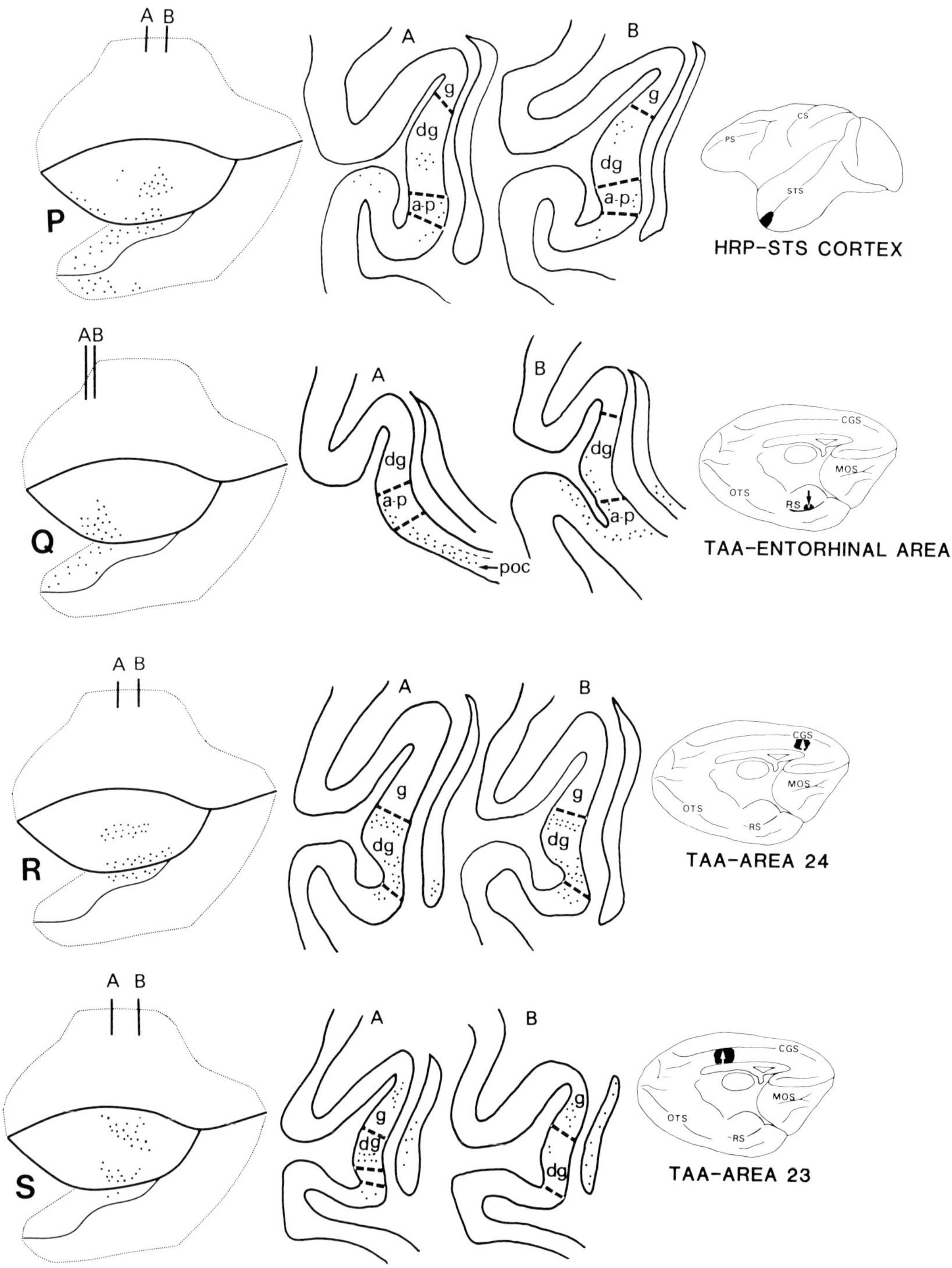

Figure 12. *(continued)*

that this projection originates mostly from mid-Idg and Ig (Fig. 12I). Confirmation for the projection to prefrontal association cortex is obtained in a case with an HRP injection confined to the lower bank of the principal sulcus (Fig. 12J). This projection appears to arise almost exclusively from Ig in the caudal insula.

3.7. Olfactory and Amygdaloid Connections

The sensory and association connections described above are mostly related to the middle, caudal, and dorsal regions of the insula (Ig and the adjacent Idg). In contrast, the ofactory, limbic, and paralimbic connections tend to concentrate in the more ventral and anterior parts of the insula (Ia-p and the adjacent Idg). The case illustrated in Fig. 11 demonstrates the presence of insular connections with all three (temporal, frontal, and insular) limbs of the POC (Fig. 11G). Further confirmation for this connection is offered by an experiment in which a TAA injection is confined almost entirely to the temporal limb of the POC (Fig. 12K). This case shows the presence of a projection from POC to Ia-p and to the anterior Idg. The existence of such a connection suggests that the insula may provide a site for the integration of gustatory with olfactory information.

Amygdaloid nuclei have reciprocal connections with orbitofrontal, temporal, and cingulate regions (Van Hoesen, 1981). However, of all the cortical regions studied thus far, it appears that the insula is the one which has the most intense and widespread reciprocal connections with the amygdala (Mufson *et al.*, 1981). Information obtained from several types of experiments suggests that the anterior insula is predominantly connected with the anterior amygdaloid area, as well as with the cortical, medial, accessory basal, lateral, laterobasal, and basomedial nuclei (Figs. 10I,J and 11H,I). The caudal insula, on the other hand, is interconnected mostly with the central nucleus and with the dorsal part of the lateral nucleus. The anterior insula (Ia-p and adjacent Idg) appears to have a much more intense connectivity with the insula. Of all the amygdaloid nuclei which have insular connectivity, those of the basomedial and lateral nuclei are particularly intense while those of the laterobasal nucleus are among the most sparse (Mufson *et al.*, 1981).

In order to obtain further information on the connectivity between the insula and the amygdala, an animal was prepared with an HRP injection which covered most of the amygdala with the possible exception of the laeral, medial, and central nuclei (Fig. 12L). The distribution of retrograde labeling shows that the projection to this part of the amygdala arises mostly from the Ia-p and the adjacent parts of the anterior Idg. Additional cases with tracer injections confined to the central or lateral amygdaloid nuclei would be required to investigate the organization of their connections with the insula.

3.8. Paralimbic Connections

The insula has widespread connections with other paralimbic regions of the brain, especially with those that belong to the olfactocentric group. For example, the case in Fig. 11 confirms the presence of insular connections with TPa-p, TPdg (PI), STPg, and also with OFa-p, OFdg, and OFg (Fig. 11D–K). The isocortical OFg sector appears to have most of its insular connections with Ig and the adjacent parts of Idg (Fig. 12I). On the other hand, the connections of the nonisocortical orbitofrontal cortex (OFa-p and the adjacent OFdg) appear to be directed preferentially to Ia-p and the adjacent Idg (Fig. 12M). These observations suggest that there may be a preferential interconnectivity between those sectors of paralimbic areas which share a similar cortical architecture.

Additional information on the connections between the insula and the temporopolar paralimbic region was obtained from two autoradiographic cases, one with an injection confined to TPdg (Fig. 12N) and another with an injection within the PI region (Fig. 12O). In both cases, insular labeling was concentrated ventrally in Ia-p and the adjacent Idg. The neural projection from PI may provide another source of auditory input for the insula. The anterior tip of the superior temporal sulcus can also be included as a component of the temporopolar paralimbic region since it has a transitional dysgranular cytoarchitecture. This region of the temporal lobe is connected with Ia-p and with Idg (Fig. 12P).

The insula is also connected with the hippocampocentric paralimbic fields. Projections to prorhinal and perirhinal areas along the banks of the rhinal sulcus (RS) originate from Idg and perhaps also from additional regions of the insula (Figs. 10I,J and 11G,H). The entorhinal cortex also has insular connections, mostly with the Ia-p and adjacent Idg sectors (Fig. 12Q). Another paralimbic connection of the insula occurs with the cingulate cortex. It appears that these connections are predominantly directed to the ventral bank of the cingulate sulcus and to the immediately adjacent cortical areas (Figs. 10A and 11A). The cingulate connections appear to have two targets within the insula. One is ventral, mostly in Idg, the other is more dorsal and includes Ig (Fig. 12R,S). There is also widespread intrainsular connections which interconnect the various sectors of this paralimbic region (Fig. 11).

3.9. Some Regional and Architectonic Patterns in the Cortical Connections of the Insula

The distribution of insular connectivity shows regional and architectonic variations (Table I). For example, the caudal-dorsal insula (Ig and the adjacent parts of Idg) have extensive somatosensory, auditory, motor, and association connections whereas the olfactory, amygdaloid, and paralimbic connections are more concentrated within the anterior-ventral insula (Ia-p and adjacent parts of Idg). These are not rigid patterns and considerable overlap exists, especially since some of the insular connections appear to be distinctly multifocal. For example, the connection with the cingulate gyrus seems to have a consistent bifocal distribution, perhaps reflecting the cytoarchitectonic heterogeneity of the cingulate cortex itself (Fig. 12R,S).

Some connections of the insula are almost exclusively concentrated within the Ig sector. These include the connections with area 1–2, area 5, area PA, and

Table I. Confirmed Cortical Connections of the Insula

	Ia-p	Idg	Ig
Somatosensory			
SI		+	+ +
Area 5			+ +
Area 7b			+ +
Auditory			
STPg (area TA)		+	
PA (area TA)		+	+ +
PI	+	+	
Visual			
TEm	+	+	+
Gustatory			
OFO		+ +	
Motor			
Prco (area 6)		+ +	+ +
High-order association			
OFg (area 11)		+ +	+ +
Area 46			+ +
Olfactory–limbic			
Piriform cortex	+ +	+	
Amygdala	+ +	+ +	
Paralimbic			
TPa-p		+ +	
TPdg	+ +	+	
OFa-p	+ +	+	
OFdg	+ +	+	
Anterior STS cortex	+	+ +	
Pro- and perirhinal		+	
Entorhinal	+ +	+	
Cingulate		+ +	+ +

area 46 (Fig. 12B–F,J). It is interesting to note that these areas are themselves characterized by an isocortical, granular architecture. On the other hand, Ia-p becomes a major focus in the connectivity with regions such as the piriform cortex, amygdala, TPdg, and entorhinal cortex (Fig. 12K,L,N,Q). None of these areas has a granular architecture. It seems, therefore, that there is a concordance between the architecture of regions that are connected to the insula and the architecture of the insular sector which acts as a focus for that connection. For example, Ia-p has virtually no connection with isocortical regions outside of the insula whereas Ig has no substantial connections with distant allocortical or agranular regions. The Idg sector has the most heterogeneous set of connectivities and this may reflect the transitional character of its cytoarchitecture.

Some of the insular connections demonstrated in Figs. 10–12 have been shown previously. For example, experiments based on strychnine neuronography showed reciprocal insular connections with adjacent orbital, temporopolar, opercular, and supratemporal regions (Pribram *et al.*, 1950; Pribram and MacLean, 1953). Other investigations based on ablation–degeneration methods have also described insular connections with the amygdala, prefrontal cortex, orbitofrontal cortex, motor–premotor cortex, the postcentral gyrus, the inferior parietal lobule, and the frontoparietal operculum (Showers, 1958; Showers and Lauer, 1961; Johnson *et al.*, 1968; Jones and Powell, 1970; Pandya *et al.*, 1971; Pandya and Vignolo, 1971; Van Hoesen *et al.*, 1975; Turner *et al.*, 1980; Aggleton *et al.*, 1980; Pandya and Seltzer, 1982).

3.10. Lamination of Cortical Connections

Information on the lamination of insular connections is far from being complete. When cases with tracer injections outside of the insula were examined, very few (mostly projections from sensory and association areas) showed a columnar arrangement of labeling within the insula (Figs. 13 and 14). A laminar predilection for the distribution of neural inputs into the insula could be detected only in a minority of cases. It appears, for example, that the projection from SI and from area 5 are directed predominantly to L4 of the insula (Fig. 14). On the other hand, there is a tendency for the limbic–paralimbic inputs from the piriform, temporopolar, entorhinal, and amygdaloid areas to concentrate in the more superficial layers.

Additional information on laminar organization was obtained by quantitating the distribution of retrogradely labeled perikarya in cases with HRP injections outside the insula. Such an analysis suggests that most of the insulopetal projections to limbic, paralimbic, and high-order association areas originate from superficial layers (L2–L3) whereas most of the projections to somatosensory regions originate in the deeper, infragranular layers. However, this is by no means an absolute distinction (Fig. 14).

3.11. Thalamic Connectivity

The insula of the macaque has widespread thalamic connections (Roberts and Akert, 1963; Wirth, 1973; Burton and Jones, 1976). Burton and Jones (1976) had shown that the suprageniculate and limitans nuclei of the thalamus project to Ig whereas the ventroposterior inferior nucleus projects to Idg. In a case with a TAA injection confined to the insula (mostly Idg) we found evidence for connections with the principal and parvicellular components of the ventroposterior medical nucleus, the ventroposterior inferior nucleus, the oral and medial pulvinar nuclei, the nucleus reuniens, the parvicellular and magnocellular com-

Figure 13. (A) Brightfield photomicrograph of perikaryal retrograde labeling in the depth of the rhinal sulcus after an HRP injection into the anterior insula of a macaque. ×200. (B) Darkfield photomicrograph of a column of label in the insula after a TAA injection in granular prefrontal cortex (area 46) of a macaque brain. ×200. From Mufson and Mesulam (1982b) with permission.

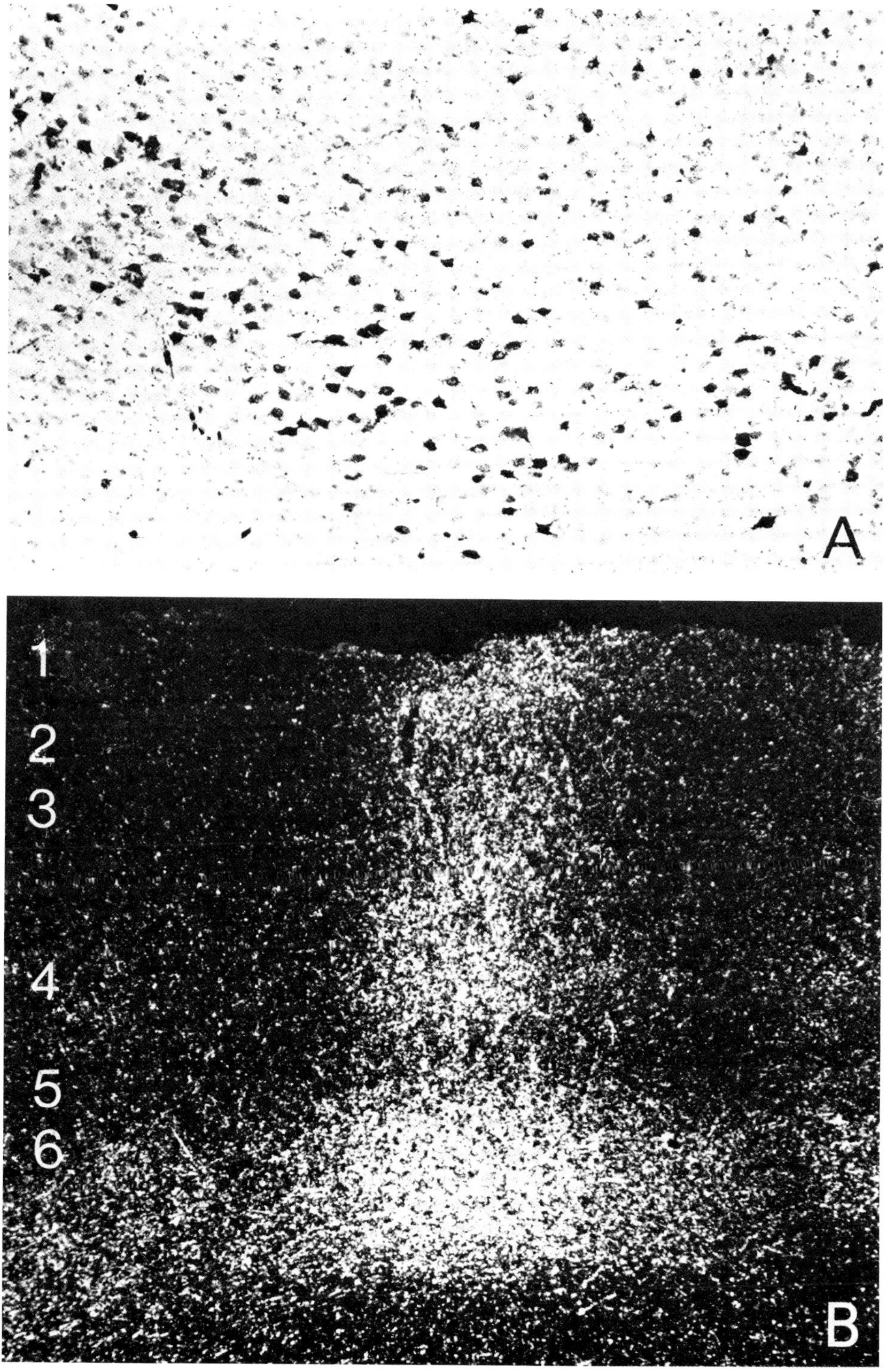
A
1
2
3
4
5
6
B

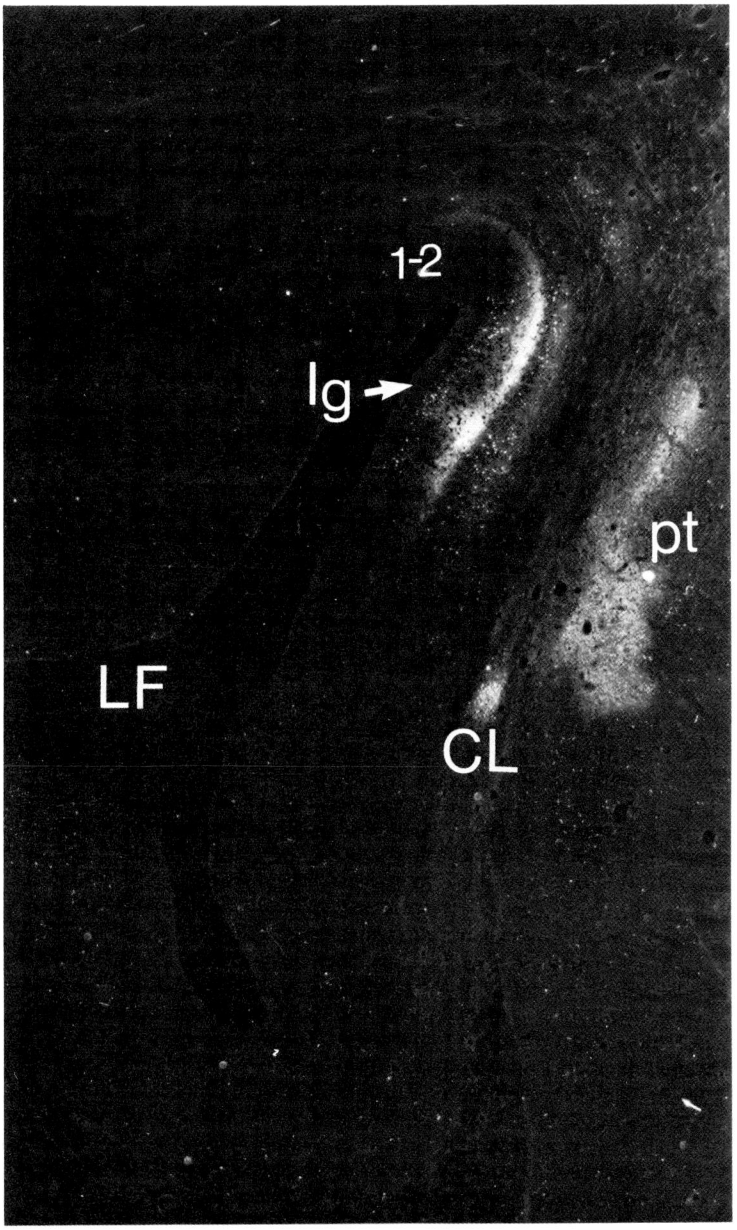

Figure 14. Darkfield photomicrograph of HRP labeling in the case shown in Fig. 12D (injection site in medial area 5). The intense labeling in L4 of Ig reflects the anterograde transport from the injection site. Retrogradely labeled neurons are seen in the supra- and infragranular layers. Cell counting shows that such labeled neurons are more numerous in the infragranular layers. There is a suggestion that the intensity of labeling in Ig shows columnar variations. This case shows that there are closely reciprocal connections between area 5 and the insula. Note the anterograde labeling in the putamen and the mixed anterograde–retrograde labeling in the claustrum. ×15. From Mesulam and Mufson (1982b) with permission.

ponents of the medial dorsal nucleus, the centromedian–parafascicular complex, and the reticular nucleus (Mufson and Mesulam, 1984).

In order to obtain more information on the organization of these thalamic connections, additional cases were examined in which TAA or HRP injections had been placed into different parts of the insula. The injection sites in these cases did have minor extension into the underlying white matter and claustrum so that these cases cannot yield definitive results. However, the information obtained from these cases, when considered in conjunction with the case in which the TAA was confined to the insula, reveals the presence of rostrocaudal variations in the organization of connections between the insula and thalamic nuclei (Fig. 15).

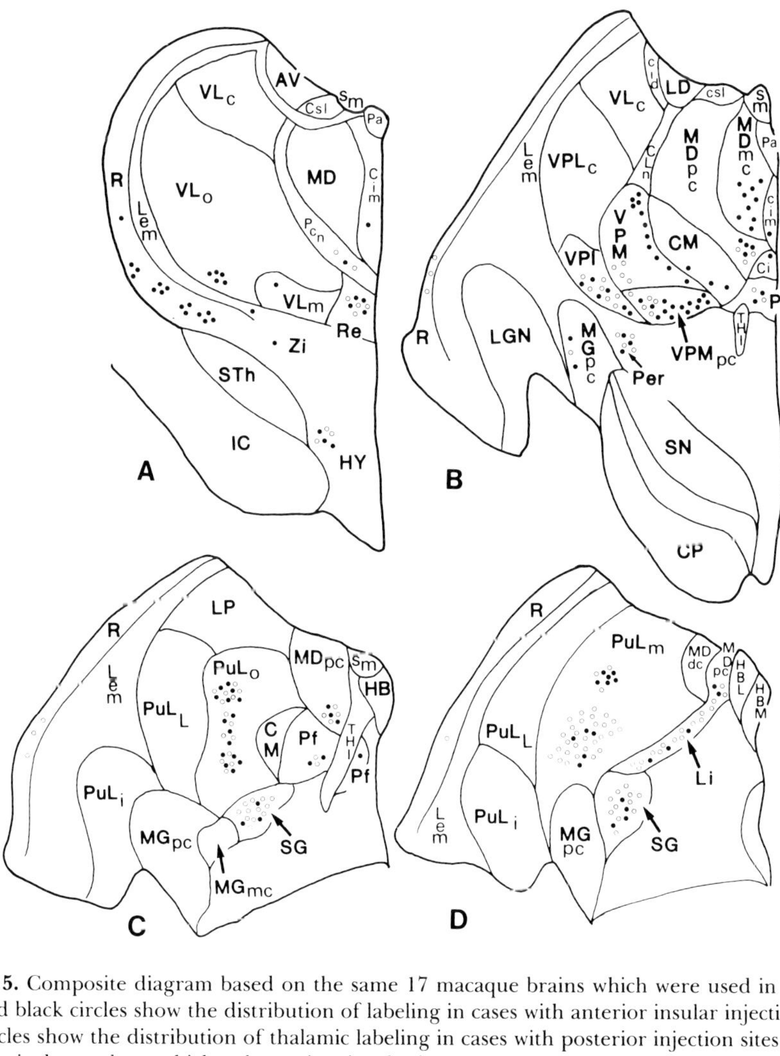

Figure 15. Composite diagram based on the same 17 macaque brains which were used in Fig. 10. The solid black circles show the distribution of labeling in cases with anterior insular injections; the open circles show the distribution of thalamic labeling in cases with posterior injection sites. Except for the reticular nucleus, which only receives insular input, all other thalamic nuclei have reciprocal connections with the insula.

These cases suggest that the anterior insula (Ia-p and adjacent Idg) has particularly extensive interconnections with the ventroposterior medial complex (especially VPM_{pc} in Figs. 15B and 16), the medial dorsal nucleus (especially its magnocellular component MD_{mc}, in Fig. 15B), the centromedian–parafascicular complex (CM and Pf in Fig. 15B), and midline nuclei such as the nucleus reuniens and the centralis inferior complex (Re and Ci, Cim in Fig. 15A,B). In contrast, the more posterior insula (Ig and adjacent Idg) appears to have more extensive connections with the ventroposterior inferior nucleus (VPI in Fig. 15B), the oral and medial pulvinar nuclei (PuL_o and PuL_m in Fig. 15C,D), and the suprageniculate nucleus (SG in Fig. 15D). Both anterior and posterior parts of the insula appeared to have comparable connections with the parvicellular medial geniculate, limitans, and reticular nuclei (MG_{pc}, Li, and R in Fig. 15A–D). Except for the reticular nucleus, all other thalamic nuclei have reciprocal connections with the insula. In keeping with all its other known cortical connections, the reticular nucleus receives insular projections but does not send reciprocal projections back into the insula.

These regional variations in thalamic connectivity seem to parallel the variations in the cortical connections of the insula. For example, the oral pulvinar, the ventroposterior inferior, and the suprageniculate nuclei which are involved in processing multimodal and especially somesthetic information project preferentially to the posterior insula whose cortical connectivity also emphasizes somatosensory and sensory association information. On the other hand, the anterior insula which has the better developed olfactory, limbic, and paralimbic connections is preferentially connected to the medial dorsal and midline nuclei which are known to have strong olfactory and limbic affiliations (Nauta, 1962; Herkenham, 1978; Krayniak *et al.*, 1980). The parvicellular component of the ventroposterior medial nucleus (VPM_{pc}) is the major thalamic target of monosynaptic inputs from the nucleus of the tractus solitarius which is, itself, the first relay nucleus of visceral and gustatory afferents (Beckstead *et al.*, 1980). Thus, in addition to its olfactory and limbic affiliations, its pattern of preferential connectivity with VPM_{pc} suggests that the anterior insula also has an intimate role in gustatory and autonomic functions.

In comparative neuroanatomy, the projection field of the medial dorsal nucleus has been used to define the boundaries of prefrontal cortex. In particular, the orbitofrontal cortex is often defined on the basis of its connectivity with the magnocellular part of the medial dorsal nucleus (Rose and Woolsey, 1948). The presence of a connection between the anterior insula and this thalamic nucleus may reflect the topographic continuity of the anterior insula with orbitofrontal cortex (Fig. 2).

Figure 16. Darkfield photomicrograph showing the distribution of autoradiographic label in the principal (VPM) and parvicellular (VPM_{pc}) divisions of the ventroposterior medial thalamic nucleus in a case with an anterior insular injection of TAA. (B) Brightfield photomicrograph of the same area. For purposes of orientation, the double arrowhead points to the same vascular marking in the two photomicrographs. ×25. From Mufson and Mesulam (1984) with permission.

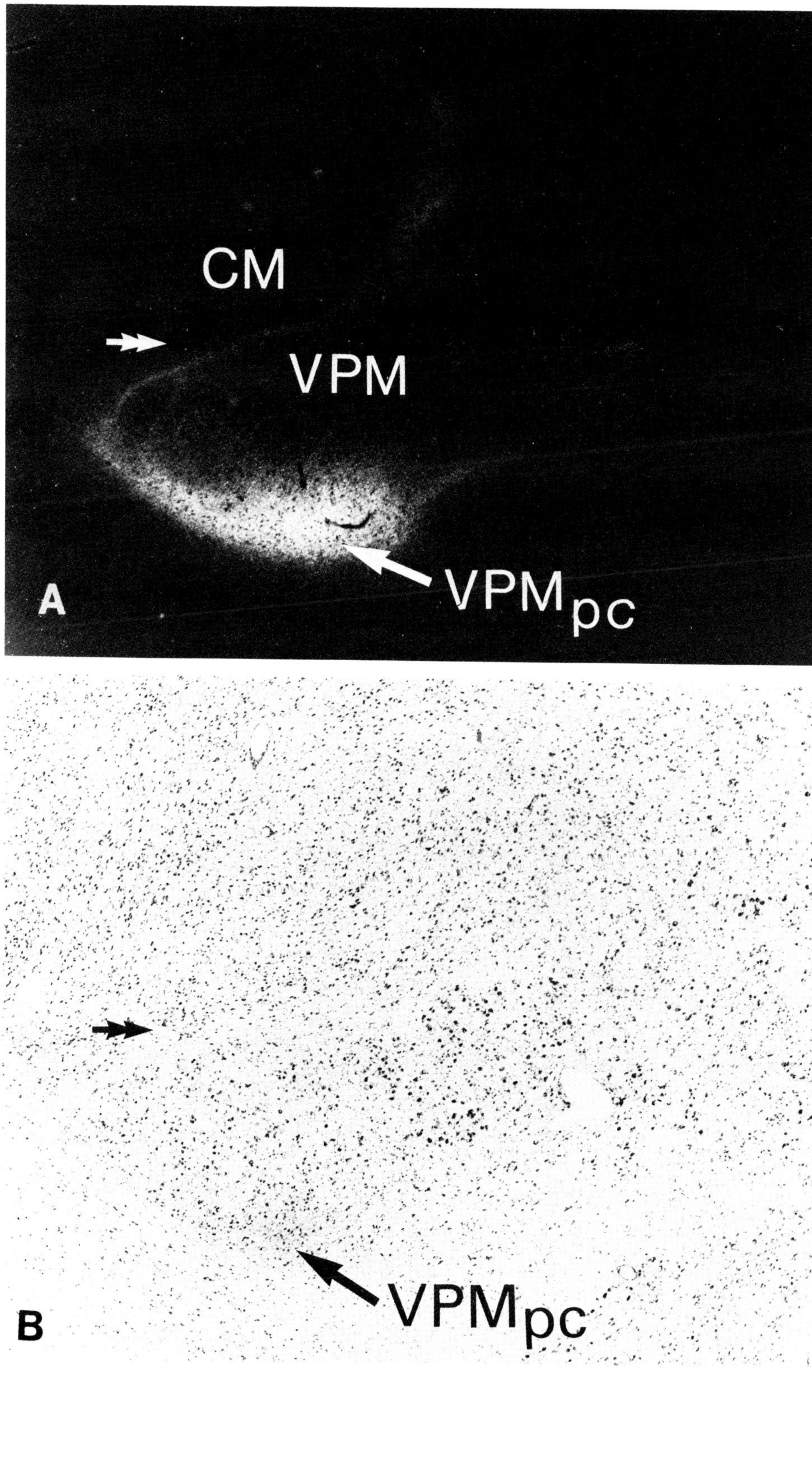
CM
VPM
VPMpc
A
VPMpc
B

3.12. Additional Miscellaneous Connections of the Insula

The posterior insula contributes fibers to corticobulbar and corticospinal pathways (Catsman-Berrevoets and Kuypers, 1976). This is in keeping with the somatosensory and motor affiliations of the posterior insula. We have also observed projections from the insula to the striatum and claustrum but these have not been analyzed in detail yet. Additional insular connections with the hypothalamus, the zona incerta, the peripeduncular nucleus, the nucleus basalis (Ch4), the pontomesencephalic reticular formation (Ch5–Ch6), the brain-stem raphe nuclei, and the nucleus locus coeruleus are almost certainly present but require further delineation (Mufson *et al.,* 1982; Mesulam *et al.,* 1983; Mesulam and Mufson, 1984; Mufson and Mesulam, 1984). In the rodent, a direct projection from the insula to the nucleus of the solitary tract has been described (Shipley, 1982; Saper, 1982). Such a projection has not yet been reported in the monkey.

4. Behavioral Affiliations of the Insula with Special Reference to Cortical Autonomic Control and Temporal Lobe Epilepsy

In neurological practice, the insula is often the site of stroke, tumor, and epileptic discharges. In virtually all such cases, however, there is also involvement of adjacent white matter, basal ganglia, and fronto-parieto-temporal cortex. It is the involvement of these additional areas which is usually held responsible for the resulting neurological picture. In fact, there is still no clinical deficit which can specifically be attributed to the involvement of the insula. Insular damage has been thought to contribute to aphasia and apraxia but the evidence is scanty at best and overshadowed by the role of other adjacent perisylvian regions in praxis and language. The insula has also been implicated in more complex aspects of human behavior ranging from overall intelligence to criminal tendencies (Clark, 1896). For example, unusually profound asymmetries in insular structure have been described in the brains of notorious criminals such as Guiteau, the assassin of President Garfield, whose insula was found to have seven fissures on the left but only five on the right (Clark, 1896). While the behavioral affiliations of the human insula need further clarification, relevant data have been obtained during investigations of the monkey brain. In this section, the limited information available on the behavioral specializations of the human brain will be considered together with the experimental observations in the monkey.

The primate brain contains many different types of cortical areas. Some among these (e.g., the striate cortex) are highly specialized in that they have a homogeneous set of cortical connections, most of which can be identified as belonging to a single category or modality. Others, such as prefrontal or posterior parietal association cortex, appear to have a much more varied set of connections but even these can be classified into four or five major categories (Mesulam *et al.,* 1977; Barbas and Mesulam, 1981). The insula, on the other hand, seems to be characterized by the greatest conceivable spectrum of cortical connections, ranging from those with limbic allocortex to those with primary sensory and high-order association cortices. This heterogeneity may well be a hallmark of all paralimbic areas rather than a feature unique to the insula. This large spectrum of connections is in keeping with the panoply of behavioral affiliations associated with the insula. It is possible to detect a topographic gradient in the distribution of behavioral specializations within the insula so that the functions of the anteroventral insula can be compared to those of the dorsocaudal insula (Mesulam and Mufson, 1982b). Although this anteroposterior gradient is relative rather than absolute, it will form the basis for much of the following discussion.

4.1. Behavioral Affiliations of the Anterior-Ventral Insula

The connections of the insula which were described in Section 3 show that the anterior-ventral insula (Ia-p and adjacent parts of Idg) has more extensive connections with limbic, paralimbic, olfactory, gustatory, and autonomic structures whereas the posterior-dorsal insular (Ig and adjacent parts of Idg) is more closely connected to somesthetic, auditory, motor, and high-order association areas (Fig. 17). Several independent lines of investigation show that this pattern

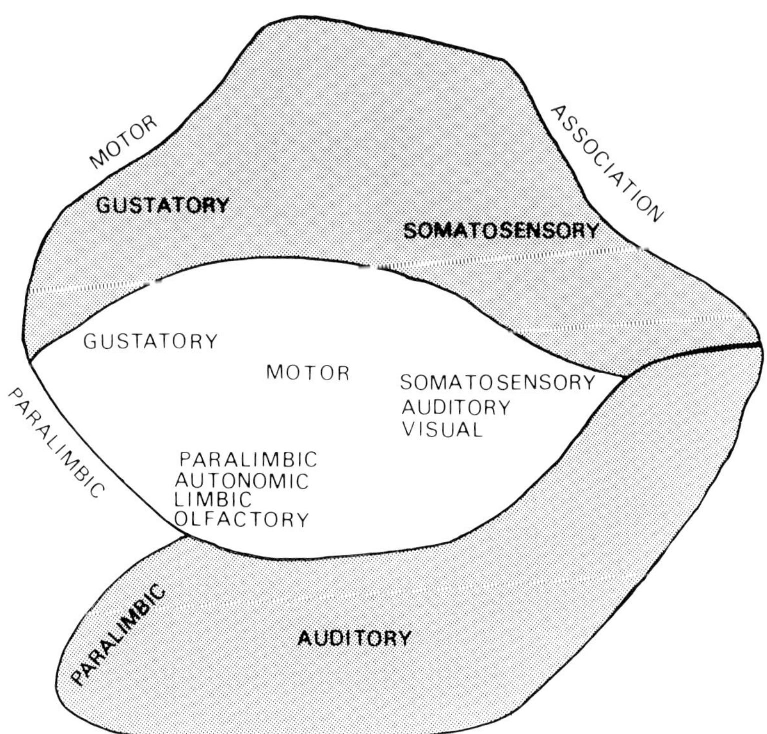

Figure 17. Schematic diagram of the distribution of various behavioral specializations in the insula and adjacent regions.

of connectivity is consistent with the regional distribution of behavioral specializations.

For example, in a patient who underwent surgical exploration, olfactory sensations were reported only when the anterior insular cortex was stimulated (Penfield and Faulk, 1955). Gustatory function is also well represented within the anterior insula. In the rhesus monkey, for example, deficits in gustatory behavior emerge after ablations of the anterior insula (Bagshaw and Pribram, 1953). Furthermore, electrical stimulation of taste nerves as well as direct gustatory stimulation of the tongue elicited responses that were most readily obtained in the anterior insula (Benjamin and Burton, 1968; Sudakov *et al.*, 1971).

In Section 3 we showed that VPM_{pc}, which is the major thalamic target for projections from the nucleus of the solitary tract (nTS), has much stronger connections with the anterior insula. Since the nTS is the first central relay for gustatory and visceral input, this organization is consistent with a preferential involvement of the anterior insula not only in gustatory sensation but also in autonomic function. Indeed, Penfield and Faulk (1955) found that stimulation of the anterior insula in the human elicited marked alterations of gastrointestinal motility as well as a variety of sensations associated with the digestive tract. In the rhesus monkey, stimulation of the anterior but not the posterior insula results in a large spectrum of cardiovascular, respiratory, salivary, and gastrointestinal responses (Hoffman and Rasmussen, 1953; Showers and Lauer, 1961). This overlap of autonomic, gustatory, and olfactory mechanisms within the anterior insula suggests that this region may participate in the organization of complex alimentary behaviors.

During neurological exploration of epileptic patients, it has been shown that the activation of medial temporal regions, especially the amygdala, is necessary for imparting affective coloring to mental and perceptual phenomena (Gloor *et al.*, 1982). Since it is so extensively interconnected with the amygdala, the anterior insula may also participate in the process of determining the affective tone of experience and behavior. This would be consistent with the role of the anterior insula in the modulation of autonomic responses since autonomic activation is such a prominent feature of many emotional states. Its close affiliation with autonomic control, alimentary behavior, and affect (drive) suggests that the anterior insula may be particularly sensitive to shifts in the internal milieu.

4.2. Behavioral Affiliations of the Posterior-Dorsal Insula

The major connections of the posterior-dorsal insula (Ig and adjacent parts of Idg) include those with somatosensory areas (SI, SII, area 5, area 7b), auditory areas (STPg, PI, PA, and perhaps AI–AII), visual areas (TEm), paramotor cortex (area 6 and perhaps MII), and high-order association regions (banks of the STS, prefrontal cortex, granular orbitofrontal cortex). Furthermore, neurons in the posterior rather than the anterior insula seem to be the source of direct descending projections to somatosensory nuclei in the brain stem. The somatosensory and auditory connections of the posterior insula are particularly well developed, almost giving the impression that the adjacent somatosensory and auditory fields simply extend into the insula (or have arisen from the insula, as the case may be). Physiological observations are consistent with this pattern of connectivity. For example, Robinson and Burton (1980) showed that 76% of all sampled units in the Ig of rhesus monkeys responded to somesthetic stimulation while Juliano *et al.* (1983) showed extensive metabolic activation in the posterior insula in response to somatosensory stimulation. In the squirrel monkey, the units responsive to auditory and somesthetic stimulation appear to be more concentrated in the posterior insula (Sudakov *et al.*, 1971). Furthermore, electrical stimulation of the posterior but not the anterior insula is more likely to elicit skeletal movement (Showers and Lauer, 1961). Thus, the posterior insula appears more involved in auditory–somatosensory–skeletomotor functions which are closely related to the extrapersonal world and its manipulation. This contrasts with the specialization of the anterior insula which places a stronger emphasis on the internal milieu and its relationship to behavior.

Several authors have implicated the human insula, especially its posterior aspects, in language function (see Clark, 1896; Rose, 1928). The pattern of insular connections is not inconsistent with this possibility. In the left hemisphere of the human brain, Wernicke's area (posterior temporoparietal cortex) and Broca's area (premotor frontal operculum) are the two most essential components of an anatomical network subserving language. Fibers interconnecting these two areas are thought to run in the arcuate fasciculus (Dejerine, 1895) which is located within the white matter of the parietal operculum. Damage to the arcuate fasciculus is thought to disconnect Wernicke's area from Broca's area in a way which leads to the development of ideomotor apraxia and conduction aphasia (Benson and Geschwind, 1985). However, its anatomical connectivity indicates that the insula may also participate in the interactions between Wernicke's area and Broca's area. For example, the evidence in Section 3 shows that the posterior insula not only receives projections from the posterior auditory association areas in the superior temporal gyrus (area PA) but that it also sends projections to opercular paramotor cortex (area Prco). Therefore, the posterior insula may provide an alternative neural relay for connecting Wernicke's area to Broca's area. This could provide one plausible anatomical substrate for the putative association between insular damage and disturbances of language and praxis.

4.3. Sensory–Limbic and Sensory–Sensory Interactions in the Insula

The insula may also provide a neural relay for conveying sensory information into the limbic system. Connections between sensory areas and limbic structures are important for at least two reasons. First they provide a means for interrelating events of the extrapersonal world with relevant motivational states. This may provide a fundamental anatomical substrate for learning and memory functions. Second, sensory–limbic interactions may be necessary for determining the affective coloring and hedonic valence of perceptual experience. The insula may participate in both types of processes.

Several multisynaptic routes have been identified for relaying auditory and visual information from association areas of the temporal lobe into the limbic system (Van Hoesen and Pandya, 1975; Herzog and Van Hoesen, 1976; Turner *et al.*, 1980; Van Hoesen, 1981). However, analogous pathways for the somesthetic modality have been difficult to specify. The case material in Section 3 shows that the insula has massive connections with somatosensory areas, on one hand, and with the amygdala, on the other. This pattern suggests that the insula may act as a major neural relay for conveying somatosensory impulses into limbic areas. In fact, anatomical (see Section 3.1) and physiological (Schneider *et al.*, 1984) evidence suggests that parts of the posterior granular insula may act as a "downstream" modality-specific somatosensory association area which acts as a principal source of somatosensory information for limbic areas. In this respect, the role of Ig in the somatosensory system is analogous to that of medial inferotemporal cortex (TEm) in the visual modality. Perhaps SII is the major source of the somatosensory information which is relayed via the insula toward limbic targets (Mishkin, 1979). However, the insula also receives other somatosensory inputs and their contribution needs to be considered as well. The pivotal role of the insula in somesthetic–limbic interactions suggests that damage to this part of the brain, especially to the posterior insula, may impair learning based on somesthetic discriminations. This expectation receives support from the experiments of Murray *et al.* (1980) which showed that learning based on tactual discrimination is disrupted after damage involving the posterior insula.

Its role as a somatosensory–limbic relay suggests that another consequence of insular damage may be to alter the hedonic aspects of somatosensory experience. It is therefore interesting that at least one human case has been reported where a lesion involving the insula and SII was associated with a dramatic loss of pain perception (Biemond, 1956). In the clinical literature, this loss of pain sensation is often attributed to the involvement of SII which is considered to provide a primary sensory area for pain perception. However, the involvement of the insula in Biemond's (1956) case would also have produced a somesthetic–limbic disconnection. Since the experience of pain undoubtedly has a strong limbic component, this neural disconnection may explain, at least in part, the loss of pain sensation that results from this type of brain damage.

In addition to somesthetic information, the insula may also participate in relaying auditory, gustatory, and visual input into the limbic system. Furthermore, in addition to the amygdala, the insula may also relay sensory information into the hippocampal system throughout its projections into prorhinal, perirhinal, and entorhinal regions (see Section 3).

The insula has connections with piriform cortex, primary gustatory cortex (OFO), SI–SII, and probably AI–AII. This suggests that the insula may be the site of multimodal convergence among inputs from the olfactory, gustatory, somesthetic, and auditory modalities. Although multimodal convergence occurs in many other parts of the brain, notably in high-order association cortex, many intervening relays are usually interposed between the primary sensory area and the site of convergence. This means that the sensory information in a given modality is extensively processed before being integrated with information from another modality (Mesulam, 1981a). In the insula, however, it appears that the putative multimodal convergence may occur without such intervening relays, directly among inputs from primary sensory areas. The behavioral implications of this arrangement and whether such multimodal convergence at a relatively early stage in the processing of sensory processing is also present in other paralimbic areas remain to be determined.

4.4. Behavioral Affiliations of the Insula in the Context of Other Paralimbic Areas

In Section 2 we showed that the POC is the cortical focus not only for the insula but also for orbitofrontal and temporopolar paralimbic areas. In Section 3 we showed that components of these three paralimbic areas are extensively interconnected. In addition to these anatomical relations, these three areas also share common functional specializations in at least three realms of behavior: visceral–autonomic function, gustatory–olfactory sensation, and affiliative–motivational behavior (Mesulam, 1985).

With respect to visceral control, electrical stimulation not only in the anterior insula but also in caudal orbitofrontal cortex as well as in the temporal pole yields consistent and marked autonomic responses (Sugar *et al.*, 1948; Kaada *et al.*, 1949; Wall and Davis, 1951; Hoffman and Rasmussen, 1953; Kaada, 1960; Showers and Lauer, 1961). Some of these effects can be quite dramatic and include inhibition of gastric peristalsis, respiratory arrest, and blood pressure changes up to 100 mm Hg. Even multifocal cardiac necrosis can be obtained when monkeys with no prior cardiovascular disease receive electrical stimulation to orbitofrontal cortex (Hall and Cornish, 1977). Marked autonomic responses have also been obtained in the human upon stimulation of analogous sites (Chapman *et al.*, 1950; Penfield and Faulk, 1955).

It is well known that the pacemakers and final common pathways for autonomic regulation are located in diencephalic and other brain-stem nuclei. Nonetheless, the insulo-orbito-temporopolar paralimbic regions may provide a matrix for the higher control of autonomic activity, especially according to the prevailing mental state. Such interactions between autonomic responses and mental states exist in many realms of behavior. For example, cognitive tasks as well as emotional states are associated with specific patterns of autonomic activity which are determined by the difficulty of the task, the type of ongoing information processing, the nature of the emotion, and even the magnitude of its impact on the individual (Ax, 1953; Fenz and Epstein, 1967; Lacey, 1967; Kahneman *et al.*, 1969; Mesulam and Perry, 1972). In medical practice, mental stress is known to increase blood pressure and even to induce potentially lethal cardiac arrhythmias in the absence of any cardiovascular predisposing factors (Lown *et al.*, 1976). Furthermore, manometrically proven abnormalities of esophageal motility are known to be more common in individuals with psychiatric problems (Clause and Lustman, 1983). In general, a disturbance in the relationship between mental events and autonomic activity is thought to be a crucial factor in the pathogenesis of all psychosomatic diseases (Alexander, 1950). It is conceivable that the integration of mental state with patterns of autonomic activation in these types of normal and abnormal conditions is modulated by the insulo-orbito-temporal sector of paralimbic areas. These cerebral sites may therefore provide a potential anatomical substrate for psychosomatic disease, essential hypertension, and perhaps even some types of heart disease. One important outcome of investigating the insula and other paralimbic areas might be to gain further insight into the pathogenesis and treatment of such conditions. MacLean (1949) emphasized the involvement of paralimbic regions in autonomic control and proposed the term "visceral brain" for these areas.

A second specialization common to these three areas is olfactory–gustatory sensation. In addition to the participation of the insula in these sensory modalities, it has been shown by Tanabe *et al.* (1975a,b) that a region in the caudal orbitofrontal cortex, probably corresponding to our OFdg and OFa-p, yielded powerful evoked potentials to olfactory stimuli and that this area was essential for odor discrimination. Bagshaw and Pribram (1953) also suggested that the temporopolar parts of the brain may be essential for determining gustatory preference patterns. Recently, deficits in odor discrimination and olfactory memory have been described in patients with anterior temporal lobectomy (Eichenbaum *et al.*, 1983). It is conceivable that the loss of temporopolar paralimbic cortex contributed to the emergence of this deficit.

A third group of functional specializations common to this group of paralimbic areas occurs in motivational, mnemonic, and affiliative behavior. For example, the caudal parts of area 13 (corresponding to OFdg and OFa-p) sustain self-stimulation behavior in the rhesus monkey (Mora *et al.*, 1980). While posterior insular lesions interfere with somatosensory learning, lesions of orbitofrontal cortex are known to cause severe and multimodal learning deficits (Iversen and Mishkin, 1970). Orbitofrontal and temporopolar areas are also important for the regulation of social interactions among conspecifics. Lesions in these areas decrease the effectiveness of aggressive encounters, result in social isolation, and lead to a reduction of adaptive affiliative behaviors in monkeys (Kling and Steklis, 1976; Raleigh *et al.*, 1979; Franzen and Myers, 1981; Raleigh and Steklis, 1981).

This information shows that components of the olfactocentric paralimbic zone have many overlapping behavioral affiliations. It is conceivable that the components of the hippocampocentric paralimbic areas will also share many anatomical and behavioral properties among themselves and that these common properties will have some overlap with the specializations of the olfactocentric paralimbic areas.

We would like to suggest that all paralimbic areas, the olfactocentric as well as the hippocampocentric, are of particular importance for those classes of complex behavior where it is essential to coordinate the internal milieu with the extrapersonal space. In providing a matrix for this type of interaction, paralimbic areas could take part in at least two types of neural integration. One could be to use information about the extrapersonal space in order to guide or inhibit

Figure 18. Darkfield photography of intercortical AChE which appears white against a dark background. Open arrows point to cytoarchitectonic subdivisions which were determined in matching sections stained with neutral red. Ia-p has the greatest density of AChE, mostly in the intermediate and deep strata. Idg has less AChE, mostly in the infragranular layers. The solid arrows point to the position of L4. Areas 1–2 have the least AChE. PI (TPdg) has a high level of AChE and most of it is in the superficial layers. STPg has less AChE than PI. The most intense AChE in STPg is confined to a narrow band in layers 1 and 2. ×16. From Mesulam and Mufson (1982a) with permission.

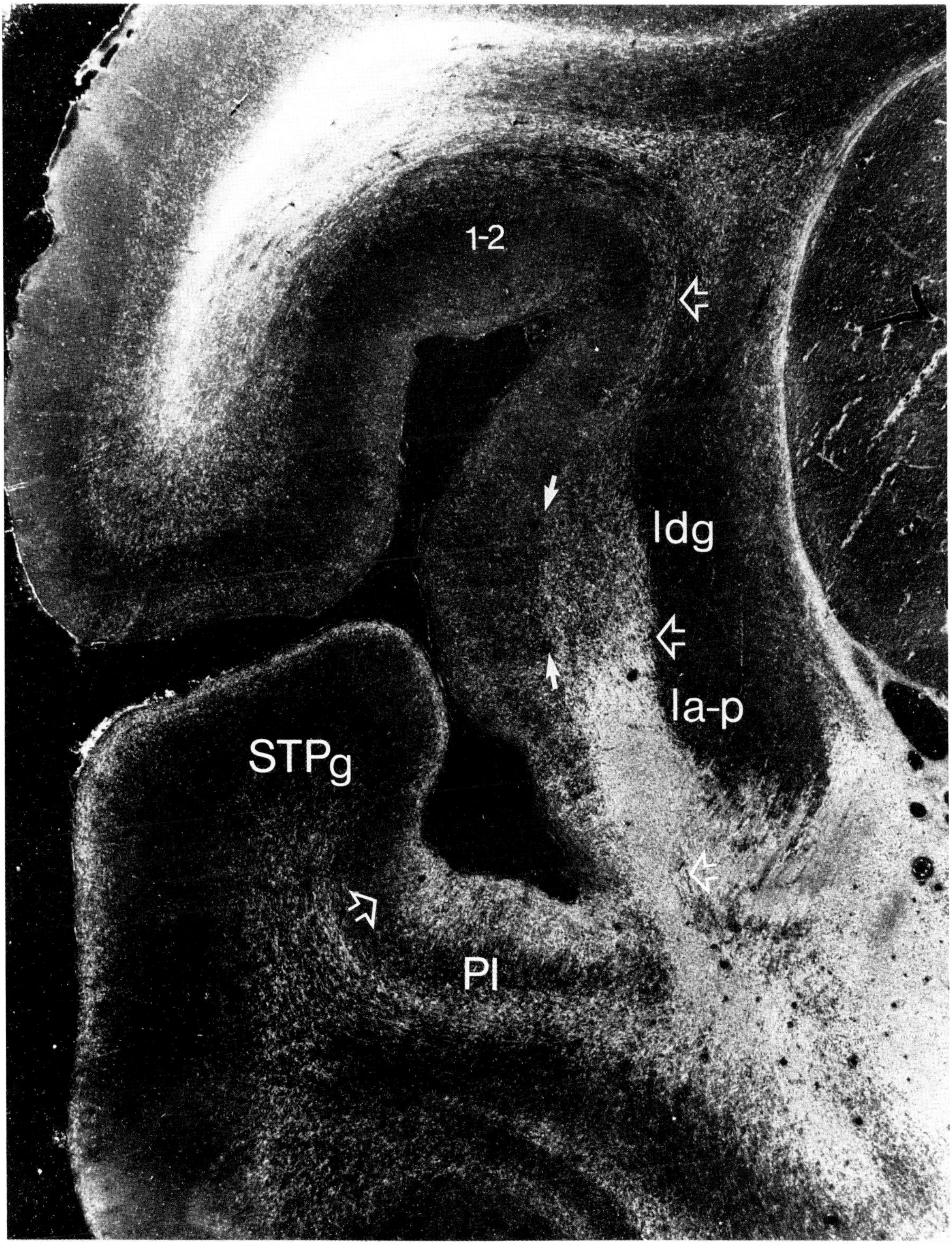

1-2
Idg
Ia-p
STPg
Pl

behaviors which aim to satisfy the requirements of the internal milieu. Another could be to impart affective tone and hedonic valence to experience and behavior. Appetitive and affiliative activities, the emotional coloring of experience, its association with the proper visceral tone, and many aspects of learning and memory are behaviors which depend on these two kinds of neural processing and which are vulnerable to disruption when paralimbic areas are damaged.

It is also interesting to note that paralimbic areas throughout the brain appear to share some similarities in neurotransmitter distribution. For example, the density of naloxone-binding opiate receptors and of cholinergic innervation (Fig. 18) is higher in paralimbic regions than in other sensory and association areas of cortex (Lewis *et al.,* 1981; Mesulam *et al.,* 1984, 1985). This selective regional distribution may explain, at least in part, the powerful effect of opiates on pain perception and of cholinergic agents on mood and memory.

4.5. Epilepsy and the Paralimbic Brain

Limbic and paralimbic components of the brain frequently become the focus of complex partial epileptic discharges. Although there is a tendency to use the term "temporal lobe epilepsy" in describing these conditions, it is important to realize that nontemporal regions in insular, orbitofrontal, and cingulate areas are also frequently involved in the generation and spread of these seizures. Perhaps the high level of cholinergic innervation in these paralimbic regions is partially responsible for this predilection (Mesulam *et al.,* 1984). Patients with this type of epilepsy report a most remarkable set of experiences which range from brief ictal events to prolonged chronic changes in behavior. The ictal events include olfactory and gustatory hallucinations, autonomic discharges, visceral sensations, episodic amnesias, and sudden experiential phenomena ranging from isolated affective discharges to déjà vu and depersonalization. The evidence reviewed in this chapter shows that the paralimbic areas contain the anatomical substrate for each of these ictal experiences.

It appears reasonable to assume that a healthy mental life requires a proper balance between experience, affect, and behavior. The paralimbic parts of the brain are likely to play a prominent role in establishing this balance. Under normal conditions, the affective coloring of perception and thought is expected to reflect the individual's past experience, the present internal state, and the intrinsic characteristics of the ongoing mental experience. In patients who have epileptogenic foci and seizure activity within limbic–paralimbic regions, this relationship may be severely disrupted even in the interictal periods. This may lead to an unpredictable and incongruous affective coloring of mental activity in a way that may disrupt the entire fabric of experience. This disruption may account for the remarkable spectrum of chronic psychiatric syndromes including manic-depressive illness, schizophreniform states, and multiple personalities which have been described in conjunction with this type of epilepsy (see Mesulam, 1981b, and Spiers *et al.,* 1985, for review). These observations on patients with temporolimbic epilepsy suggest that research on the limbic and paralimbic parts of the brain is likely to provide a productive avenue for elucidating the neurological bases of many psychiatric conditions.

5. Abbreviations

ab — Accessory basal nucleus of the amygdala
ac — Anterior commissure
AS — Arcuate sulcus
AV — Nucleus anterior ventral
AI — First auditory area
AII — Second auditory area
c — Central nucleus of the amygdala
cd — Caudate
CG — Cingulate cortex
CGS — Cingulate sulcus
Ch4 — Cholinergic cell group (Mesulam *et al.*, 1983)
Ci — Nucleus centralis inferior
Cim — Nucleus centralis intermedialis
CL — Claustrum
Cld — Capsule of the nucleus lateralis dorsalis
CLn — Nucleus centralis lateralis
CM — Centromedian nucleus
co — Cortical nucleus of the amygdala
CP — Cerebral peduncle
CS — Central sulcus
Csl — Nucleus centralis superior lateralis
dg — Insula, dysgranular (Mesulam and Mufson, 1982a)
G — Gustatory cortex (Sanides, 1968)
g — Insula, granular (Mesulam and Mufson, 1982a)
h — Hippocampus
HB — Habenula
HBL — Lateral habenular nucleus
HBM — Medial habenular nucleus
Hy — Hypothalamus
I — Insula
Ia-p — Insula, agranular-periallocortical (Mesulam and Mufson, 1982a)
ic — Internal capsule
Idg — Insula, dysgranular (Mesulam and Mufson, 1982a)
Ig — Insula, granular (Mesulam and Mufson, 1982a)
IPS — Intraparietal sulcus
LD — Nucleus lateralis dorsalis
Lem — External medullary lamina
LF — Lateral fissure
LGN — Lateral geniculate nucleus
Li — Nucleus limitans
LILS — Lateral limb of inferior limiting sulcus
LOS — Lateral orbitofrontal sulcus
LP — Nucleus lateralis posterior
LS — Lunate sulcus
lt — Lateral nucleus of the amygdala
m — Medial nucleus of the amygdala
mb — Medial basal nucleus of the amygdala
MD — Nucleus medial dorsal
MD_{dc} — Densocellular division of the medial dorsal nucleus
MD_{mc} — Magnocellular division of the medial dorsal nucleus
MD_{pc} — Parvicellular division of the medial dorsal nucleus
MG_{mc} — Magnocellular division of the medial geniculate nucleus
MG_{pc} — Parvicellular division of the medial geniculate nucleus
MII — Secondary motor area (Woolsey, 1965)
MOS — Medial orbital sulcus
OC — Optic chiasma
OF — Orbitofrontal cortex
OFa-p — Orbitofrontal cortex, agranular-periallocortical (Mesulam and Mufson, 1982)
OFdg — Orbitofrontal cortex, dysgranular (Mesulam and Mufson, 1982a)
OFg — Orbitofrontal cortex, granular (Mesulam and Mufson, 1982a)
OFO — Opercular cortex (Roberts and Akert, 1963)
OIS — Orbitoinsular sulcus
Ot — Optic tract
OTS — Occipitotemporal sulcus
PA — Postauditory cortex (Jones and Burton, 1976)
Pa — Paraventricular nucleus

Pcn	Paracentral nucleus
Per	Peripeduncular nucleus
PF	Anterior inferior parietal cortex (von Bonin and Bailey, 1947)
Pf	Parafascicularis nucleus
PH	Parahippocampal area
PI	Parainsular cortex (Jones and Burton, 1976)
PO	Parolfactory area
POC	Piriform olfactory cortex
Prco	Precentral operculum (Roberts and Akert, 1963)
PS	Principal sulcus
pt	Putamen
PuL_i	Inferior division of the pulvinar nucleus
PuL_L	Lateral division of the pulvinar nucleus
PuL_m	Medial division of the pulvinar nucleus
PuL_o	Oral division of the pulvinar nucleus
R	Reticular thalamic nucleus
RE	Nucleus reuniens
RI	Retroinsular cortex (Jones and Burton, 1976)
RS	Rhinal sulcus
SG	Suprageniculate nucleus
SI	First somatosensory cortex
SII	Second somatosensory cortex
SLS	Superior limiting sulcus
SM	Stria medullaris
SN	Substantia nigra
STh	Subthalamic nucleus
STP	Superior temporal plane
STPg	Superior temporal cortex, granular (Mesulam and Mufson, 1982a)
STS	Superior temporal sulcus
TEm	Medial portion of inferior temporal gyrus (Turner *et al.*, 1980)
THI	Habenular interpeduncular tract
TO	Olfactory tract
TP	Temporopolar cortex
TPa-p	Temporopolar cortex, agranular-periallocortical (Mesulam and Mufson, 1982a)
TPdg	Temporopolar cortex, dysgranular (Mesulam and Mufson, 1982a)
VL_c	Caudal division of the ventral lateral nucleus
VL_m	Medial division of the ventral lateral nucleus
VL_o	Oral division of the ventral lateral nucleus
VPI	Ventral posterior inferior nucleus
VPL_c	Caudal division of the ventral posterior lateral nucleus
VPM	Ventral posterior medial nucleus
VPM_{pc}	Parvicellular division of the ventral posterior medial nucleus
Zi	Zona incerta
3,1,2	Areas 3, 1, 2 (Brodmann, 1905)
11,12,13	Orbitofrontal cortex (Walker, 1940)

ACKNOWLEDGMENTS. We are grateful to Leah Christie, Terry Martin, Rick Plourde, and Marcia Williams for secretarial, technical, and artistic assistance. Supported in part by NS-07011, NS-09211, NS-20285, The McKnight Foundation, and a Javits Neuroscience Investigator Award.

6. References

Abbie, A. A., 1942, Cortical lamination in a polyprotodont marsupial, *Perameles nasuta, J. Comp. Neurol.* **76:**509–535.

Aggleton, J. P., Burton, M. J., and Passingham, R. E., 1980, Cortical and subcortical afferents to the amygdala of the rhesus monkey (*Macaca mulatta*), *Brain Res.* **190:**347–368.

Alexander, F., 1950, *Psychosomatic Medicine: Its Principles and Applications,* Norton, New York.
Ax, A. F., 1953, The physiological differentiation between fear and anger in humans, *Psychosom. Med.* **15:**433–442.
Bagshaw, M. H., and Pribram, K. H., 1953, Cortical organization in gustation (*Macaca mulatta*), *J. Neurophysiol.* **16:**499–508.
Barbas, H., and Mesulam, M.-M., 1981, Organization of afferent input to subdivisions of area 8 in the rhesus monkey, *J. Comp. Neurol.* **200:**407–431.
Bartholin, C., 1641, *Institutiones Anatomicae ab auctoris filio Thoma Bartholino,* Hack, Leiden.
Beckstead, R. M., Morse, J. R., and Norgren, R., 1980, The nucleus of the solitary tract in the monkey: Projections to the thalamus and brain stem nuclei, *J. Comp. Neurol.* **190:**259–282.
Benjamin, R. M., and Burton, H., 1968, Projection of taste nerve afferents to anterior opercular–insular cortex in squirrel monkey (*Saimiri sciureus*), *Brain Res.* **7:**221–231.
Benson, D. F., and Geschwind, N., 1985, Aphasia and related disorders, in: *Principles of Behavioral Neurology* (M.-M. Mesulam, ed.), Davis, Philadelphia.
Biemond, A., 1956, The conduction of pain above the level of the thalamus opticus, *Arch. Neurol. Psychiatry* **75:**221–231.
Burton, H., and Jones, E. G., 1976, The posterior thalamic region and its cortical projection in New World and Old World monkeys, *J. Comp. Neurol.* **168:**249–301.
Carpenter, M. B., 1976, Development and histogenesis of the nervous system, in: *Human Neuroanatomy* (M. B. Carpenter, ed.), Williams & Wilkins, Baltimore, pp. 49–70.
Catsman-Berrevoets, C. E., and Kuypers, H. G. J. M., 1976, Cells of origin of cortical projections to dorsal column nuclei, spinal cord and bulbar medial reticular formation in the rhesus monkey, *Neurosci. Lett.* **3:**245–252.
Chapman, W. P., Livingston, K. E., and Poppen, J. L., 1950, Effect upon blood pressure of electrical stimulation of tips of temporal lobes in man, *J. Neurophysiol.* **13:**65–71.
Clark, T. E., 1896, The comparative anatomy of the insula, *J. Comp. Neurol.* **6:**59–100.
Clause, R. E., and Lustman, P. J., 1983, Psychiatric illness and contraction abnormalities of the esophagus, *N. Engl. J. Med.* **309:**1337–1342.
Dejerine, J., 1895, *Anatomie des Centres Nerveux,* Rueff, Paris.
Eichenbaum, H., Morton, T. H., Potter, H., and Corkin, S., 1983, Selective olfactory deficits in case H. M., *Brain* **106:**459–472.
Fenz, W. D., and Epstein, S., 1967, Gradients of physiological arousal in parachutists as a function of an approaching jump, *Psychosom. Med.* **29:**33–51.
Filimonoff, I. N., 1947, A rational subdivision of the cerebral cortex, *Arch. Neurol. Psychiatry* **58:**296–311.
Franzen, E. A., and Myers, R. E., 1981, Neural control of social behavior: Prefrontal and anterior temporal cortex, *Neuropsychologia* **11:**141–157.
Gloor, P., Oliver, A., Quesney, L. F., Andermann, F., and Horowitz, S., 1982, The role of the limbic system in experiential phenomena of the temporal pole, *Ann. Neurol.* **12:**131–144.
Gower, E. C., 1981, Architectonic and Thalamic Connections of Macaque Temporal Polar Cortex, Ph.D. thesis, Massachusetts Institute of Technology.
Gower, E. C., and Mesulam, M.-M., 1982, Cytoarchitecture correlated with distribution of AChE in monkey temporopolar cortex, *Anat. Rec.* **202:**67–68 (abstr.).
Hall, R. E., and Cornish, K., 1977, Role of the orbital cortex in cardiac dysfunction in unanesthetized rhesus monkey, *Exp. Neurol.* **56:**289–297.
Herkenham, M. A., 1978, The connections of the nucleus reuniens thalami: Evidence for a direct thalamo-hippocampal pathway in the rat, *J. Comp. Neurol.* **177:**589–610.
Herzog, A. G., and Van Hoesen, G. W., 1976, Temporal neocortical afferent connections to the amygdala in the rhesus monkey, *Brain Res.* **115:**57–69.
Hoffman, B. L., and Rasmussen, T., 1953, Stimulation studies of insular cortex of *Macaca mulatta, J. Neurophysiol.* **16:**343–351.
Iversen, S. D., and Mishkin, M., 1970, Perseverative interference in monkeys following selective lesions of the inferior prefrontal convexity, *Exp. Brain Res.* **11:**376–386.
Johnson, T. N., Rosvold, H. E., and Mishkin, M., 1968, Projections from behaviorally-defined sectors of the prefrontal cortex to the basal ganglia, septum and diencephalon of the monkey, *Exp. Neurol.* **12:**20–34.
Jones, E. G., and Burton, H., 1976, Areal differences in the laminar distribution of thalamic afferents in cortical fields of the insular, parietal and temporal regions of primates, *J. Comp. Neurol.* **168:**197–248.

Jones, E. G., and Powell, T. P. S., 1970, An anatomical study of converging sensory pathways within the cerebral cortex of the monkey, *Brain* **93:**793–820.

Jones, E. G., Coulter, J. D., and Hendry, S. H. C., 1978, Intracortical connectivity of architectonic fields in the somatic sensory motor and parietal cortex of monkeys, *J. Comp. Neurol.* **181:**291–348.

Juliano, S. L., Hand, P. J., and Whitsel, B. L., 1983, Patterns of metabolic activity in cytoarchitectural area SII and surrounding cortical fields of the monkey, *J. Neurophysiol.* **50:**961–980.

Kaada, B. R., 1960, Cingulate, posterior orbital, anterior insular and temporal pole cortex, in: *Neurophysiology* (H. W. Magoun, ed.), Waverly Press, Baltimore, pp. 1345–1372.

Kaada, B. R., Pribram, K. H., and Epstein, J. A., 1949, Respiratory and vascular responses in monkeys from temporal pole, insula, orbital surface and cingulate gyrus, *J. Neurophysiol.* **12:**348–356.

Kahneman, D. B., Tursky, B., Shapiro, D., and Crider, A., 1969, Pupillary, heart rate, and skin resistance changes during a mental task, *J. Exp. Psychol.* **79:**164–167.

Kling, A., and Steklis, H. D., 1976, A neural substrate for affiliative behavior in nonhuman primates, *Brain Behav. Evol.* **13:**216–238.

Krayniak, P. F., Weiner, S., and Siegel, A., 1980, An analysis of the efferent connections of the septal area in the cat, *Brain Res.* **189:**15–29.

Lacey, J. I., 1967, Somatic response patterning and stress: Some revisions of activations theory, in: *Psychological Stress* (M. H. Appley and R. Trumbull, eds.), Appleton–Century–Crofts, New York, pp. 14–37.

Lewis, M. E., Mishkin, M., Bragin, E., Brown, R. M., Pert, C. B., and Pert, A., 1981, Opiate receptor gradients in monkey cerebral cortex: Correspondence with sensory processing hierarchies, *Science* **211:**1166–1169.

Lorente de Nó, R., 1949, Cerebral cortex: Architecture, intracortical connections, motor projections, in: *Physiology of the Nervous System* (J. F. Fulton, ed.), Oxford University Press, London, pp. 288–330.

Lown, B., Temte, J. V., Reich, P., Gaughan, C., Regestein, Q., and Hai, H., 1976, Basis for recurring ventricular fibrillation in the absence of coronary heart disease and its management, *N. Engl. J. Med.* **294:**628–629.

McGuiness, E., Sivertsen, D., and Allman, J. M., 1980, Organization of the face representation in macaque motor cortex, *J. Comp. Neurol.* **193:**591–608.

MacLean, P. D., 1949, Psychosomatic disease and the "visceral brain": Recent developments bearing on the Papez theory of emotion, *Psychosom. Med.* **11:**338–353.

Mesulam, M.-M., 1981a, A cortical network for directed attention and unilateral neglect, *Ann. Neurol.* **10:**309–325.

Mesulam, M.-M., 1981b, Dissociative states with abnormal temporal lobe EEG, *Arch. Neurol.* **38:**176–181.

Mesulam, M.-M., 1985, Patterns in behavioral neuroanatomy: Association areas, the limbic system and hemispheric specialization, in: *Principles of Behavioral Neurology* (M.-M. Mesulam, ed.), Davis, Philadelphia.

Mesulam, M.-M., and Mufson, E. J., 1982a, Insula of the Old World monkey. I. Architectonics in the insulo-orbito-temporal component of the paralimbic brain, *J. Comp. Neurol.* **212:**1–22.

Mesulam, M.-M., and Mufson, E. J., 1982b, Insula of the Old World monkey. Part III. Efferent cortical output, *J. Comp. Neurol.* **212:**38–52.

Mesulam, M.-M., and Mufson, E. J., 1984, Neural inputs into the nucleus basalis of the substantia innominata (Ch4) in the rhesus monkey, *Brain* **107:**253–274.

Mesulam, M.-M., and Perry, J., 1972, The diagnosis of love sickness: Experimental psychophysiology without the polygraph, *Psychophysiolgy* **9:**546–551.

Mesulam, M.-M., Van Hoesen, G. W., Pandya, D. N., and Geschwind, N., 1977, Limbic and sensory connections of the inferior parietal lobule (area PG) in the rhesus monkey: A study with a new method for horseradish peroxidase histochemistry, *Brain Res.* **136:**393–414.

Mesulam, M.-M., Mufson, E. J., Levey, A. I., and Wainer, B. H., 1983, Cholinergic innervation of cortex by the basal forebrain: Cytochemistry and cortical connections of the septal area, diagonal band nuclei, nucleus basalis (substantia innominata) and hypothalamus in the rhesus monkey, *J. Comp. Neurol.* **214:**170–197.

Mesulam, M.-M., Rosen, A. D., and Mufson, E. J., 1984, Regional variations in cortical cholinergic innervation: Chemoarchitectonics of acetylcholinesterase-containing fibers in the macaque brain, *Brain Res.* **311:**245–258.

Mesulam, M.-M., Volicer, L., Marquis, J., Mufson, E. J., and Green, R., 1985, Regional variations in the cholinergic innervation of the primate cortical surface, *Soc. Neurosci. Abstr.* (in press).

Mishkin, M., 1979, Analogous neural models for tactual and visual learning, *Neuropsychologia* **17:**139–151.
Mora, F., Avrith, D. B., and Rolls, E. T., 1980, An electrophysiological and behavioral study of self-timulation in the orbitofrontal cortex of the rhesus monkey, *Brain Res. Bull.* **2:**111–115.
Morgane, P. J., Jacobs, M. S., and McFarland, W. L., 1980, The anatomy of the brain of the bottlenose dolphin (*Tursiops truncatus*): Surface configurations of the telencephalon of the bottlenose dolphin with comparative anatomical observations in four other cetacean species, *Brain Res. Bull.* **5**(Suppl. 5):1–107.
Mufson, E. J., and Mesulam, M.-M., 1982, Insula of the Old World monkey. Part II. Afferent cortical input, *J. Comp. Neurol.* **212:**23–37.
Mufson, E. J., and Mesulam, M.-M., 1984, Thalamic connections of the insula in the rhesus monkey and comments on the paralimbic connectivity of the medial pulvinar nucleus, *J. Comp. Neurol.* **227:**109–120.
Mufson, E. J., Mesulam, M.-M., and Pandya, D. N., 1981, Insular interconnections with the amygdala in the rhesus monkey, *Neuroscience* **6:**1231–1248.
Mufson, E. J., Levey, A. I., Wainer, B. H., and Mesulam, M.-M., 1982, Cholinergic projections from the mesencephalic tegmentum to neocortex in rhesus monkey, *Soc. Neurosci. Abstr.* **8:**202.
Murray, E. A., Nakamura, R. K., and Mishkin, M., 1980, A possible cortical pathway for somatosensory processing in monkeys, *Soc. Neurosci. Abstr.* **6:**654.
Nauta, W. J. H., 1962, Neural associations of the amygdaloid complex in the monkey, *Brain* **85:**505–520.
Pandya, D. N., and Seltzer, B., 1982, Intrinsic connections and architectonics of posterior parietal cortex in the rhesus monkey, *J. Comp. Neurol.* **204:**196–210.
Pandya, D. N., and Vignolo, L. A., 1971, Intra- and interhemispheric projections of the precentral, premotor and arcuate areas in the rhesus monkey, *Brain Res.* **26:**217–233.
Pandya, D. N., Dye, P., and Butters, N., 1971, Efferent cortico-cortical projections of the prefrontal cortex in the rhesus monkey, *Brain Res.* **31:**31–46.
Penfield, W., and Faulk, M. E., 1955, The Insula: Further observations on its function, *Brain* **78:**445–470.
Pribram, K. H., and MacLean, P. D., 1953, Neuronographic analysis of medial and basal cerebral cortex. II. Monkey, *J. Neurophysiol.* **16:**324–340.
Pribram, K. H., Lennox, M. A., and Dunsmore, R. H., 1950, Some connections of the orbito-fronto-temporal, limbic and hippocampal areas of *Macaca mulatta*, *J. Neurophysiol.* **13:**127–135.
Raleigh, M. J., and Steklis, H. D., 1981, Effects of orbitofrontal and temporal neocortical lesions on the affiliative behavior of vervet monkeys (*Cercopithecus aethiops sabaeus*), *Exp. Neurol.* **73:** 378–389.
Raleigh, M. J., Steklis, H. D., Ervin, F. R., Kling, A. S., and McGuire, M. T., 1979, The effects of orbitofrontal lesions on the aggressive behavior of vervet monkeys (*Cercopithecus aethiops sabaeus*), *Exp. Neurol.* **66:**158–168.
Reil, J. C., 1809, Die sylvische Grube, *Arch. Physiol. (Halle)* **9:**195–208.
Roberts, T. S., and Akert, K., 1963, Insular and opercular cortex and its thalamic projection in *Macaca mulatta*, *Schweiz. Arch. Neurol. Neurochir. Psychiat.* **92:**1–43.
Robinson, C. L., and Burton, H., 1980, Somatic submodality distribution within the second somatosensory (SII), 7b, retroinsular, post-auditory and granular insular cortical areas of *M. fascicularis*, *J. Comp. Neurol.* **192:**93–108.
Rose, J. E., and Woolsey, C. N., 1948, The orbitofrontal cortex and its connections with the mediodorsal nucleus in rabbit, sheep and cat, *Res. Publ. Assoc. Res. Nerv. Ment. Dis.* **27:**210–233.
Rose, M., 1928, Die Inselrinde des Menschen und der Tieren, *J. Psychol. Neurol.* **37:**467–624.
Sanides, F., 1968, The architecture of the cortical taste nerve areas in squirrel monkey (*Saimiri sciureus*) and their relationships to insular, sensorimotor and prefrontal regions, *Brain Res.* **8:**97–124.
Sanides, F., 1970, Functional architecture of motor and sensory cortices in primates in the light of a new concept of neocortex evolution, in: *The Primate Brain* (C. R. Noback and W. Montagna, eds.), Appleton–Century–Crofts, New York, pp. 137–208.
Sanides, F., 1972, Representation in the cerebral cortex and its areal lamination pattern, in: *The Structure and Function of Nervous Tissue*, Volume 5 (G. H. Bourne, ed.), Academic Press, New York.
Saper, C. B., 1982, Convergence of autonomic and limbic connections in the insular cortex of the rat, *J. Comp. Neurol.* **210:**163–173.
Schneider, R. J., Nelson, R. J., Friedman, D. P., O'Neill, J. B., and Mishkin, M., 1984, The granular insula in the rhesus monkey: Somatic sensory properties, *Soc. Neurosci. Abstr.* **10:**496.

Shipley, M. T., 1982, Insular cortex projection to the nucleus of the solitary tract and brainstem visceromotor regions in the mouse, *Brain Res. Bull.* **8:**139–148.

Showers, M. J. C., 1958, Correlation of medial thalamic nuclear activity with cortical and subcortical neuronal arcs, *J. Comp. Neurol.* **109:**261–315.

Showers, M. J. C., and Lauer, E. W., 1961, Somatovisceral motor patterns in the insula, *J. Comp. Neurol.* **117:**107–116.

Spiers, P., Schomer, D., Blume, H., and Mesulam, M.-M., 1985, Temporolimbic epilepsy and behavior, in: *Principles of Behavioral Neurology* (M.-M. Mesulam, ed.), Davis, Philadelphia.

Sudakov, K., MacLean, P. D., Reeves, A., and Marino, R., 1971, Unit study of exteroceptive inputs to claustrocortex in awake, sitting squirrel monkeys, *Brain Res.* **28:**19–34.

Sugar, O., Chusid, J. G., and French, J. D., 1948, A second motor cortex in the monkey (*Macaca mulatta*), *J. Neuropathol. Exp. Neurol.* **7:**182–189.

Tanabe, T., Iino, M., and Takagi, S. F., 1975a, Discrimination of odors in olfactory bulb, pyriform–amygdaloid areas, and orbitofrontal cortex of the monkey, *J. Neurophysiol.* **38:**1284–1296.

Tanabe, T., Yarita, H., Iino, M., Ooshima, Y., and Takagi, S. F., 1975b, An olfactory projection area in orbitofrontal cortex of the monkey, *J. Neurophysiol.* **38:**1269–1283.

Turner, B. H., Mishkin, M., and Knapp, M., 1980, Organization of the amygdalopetal projections from modality-specific cortical association areas in the monkey, *J. Comp. Neurol.* **191:**515–543.

Van Hoesen, G. W., 1981, The differential distribution, diversity and sprouting of cortical projection to the amygdala in the rhesus monkey, in: *The Amygdaloid Complex* (Y. Ben-Ari, ed.), Elsevier, Amsterdam, pp. 77–90.

Van Hoesen, G. W., and Pandya, D. N., 1975, Some connections of the entorhinal (area 28) and perirhinal (area 35) cortices of the rhesus monkey. I. Temporal lobe afferents, *Brain Res.* **95:**1–24.

Van Hoesen, G. W., Pandya, D. N., and Butters, N., 1975, Some connections of the entorhinal (area 28) and perirhinal (area 35) cortices of the rhesus monkey. II. Frontal lobe afferents, *Brain Res.* **95:**25–38.

von Bonin, G., and Bailey, P., 1974, *The Neocortex of Macaca mulatta,* University of Illinois Press, Urbana.

Walker, A. E., 1940, A cytoarchitectural study of the prefrontal area of the macaque monkey, *J. Comp. Neurol.* **73:**59–86.

Wall, P. D., and Davis, G. D., 1951, Three cerebral cortical systems affecting autonomic function, *J. Neurophysiol.* **14:**508–517.

Wirth, F. P., 1973, Insular–diencephalic connections in the macaque, *J. Comp. Neurol.* **150:**361–392.

Woolsey, C. N., 1965, Organization of somatic sensory and motor areas of the cerebral cortex, in: *Biological and Biochemical Bases of Behavior* (H. F. Harlow and C. N. Woolsey, eds.), University of Wisconsin Press, Madison, pp. 63–81.

Yakovlev, P. I., 1959, Pathoarchitectonic studies of cerebral malformations. III. Arrhinencephalies (holotelencephalies), *J. Neuropathol. Exp. Neurol.* **18:**22–55.

II

Auditory Cortex

6
Auditory Cortex

JOHN F. BRUGGE and RICHARD A. REALE

1. Introduction

Auditory cortex refers, in the classical sense, to the temporal region of cerebral cortex that receives a major ascending afferent input from the medial geniculate body of the thalamus and contains neurons responsive to acoustic stimulation. Anatomical, physiological, and behavioral studies, some dating back to the last century, have shown repeatedly that auditory cortex, so defined, is a complex structure made up of not one field, but several fields that can be distinguished from one another on the bases of cytoarchitecture, connectivity patterns, functional maps, and the coding properties of single neurons. In a more general sense, the definition of auditory cortex may be expanded to include those areas referred to as "polysensory," "nonspecific," or "associational," areas whose major ascending inputs are derived mainly from sources outside of those making up the main auditory lemniscal pathway (Graybiel, 1973, 1974; Irvine and Phillips, 1982; see Pandya and Yeterian, this volume). This so-called "diffuse" or "lemniscal adjunct" system is characterized by poor frequency selectivity, little tonotopic organization, and considerable convergent input from other sensory systems.

Systematic work on the functional organization of auditory cortex began with the classical 1942 study of Woolsey and Walzl in which they mapped on the ectosylvian region of cat cerebral cortex the distribution of slow-wave potentials evoked by focal electrical stimulation of auditory nerve fibers in the osseous spiral lamina of the cochlea. By 1961, Woolsey had developed a scheme

---

JOHN F. BRUGGE and RICHARD A. REALE • Department of Neurophysiology and Waisman Center on Mental Retardation and Human Development, University of Wisconsin, Madison, Wisconsin 53705.

of organization of cat auditory cortex that was based largely on evoked potential studies carried out by him and his colleagues over nearly two decades (Woolsey, 1960, 1961). The scheme was consistent with both the cytoarchitectonic analysis of Rose (1949) and the results of studies of retrograde degeneration patterns following regional auditory cortical lesions (Rose and Woolsey, 1949, 1958). Woolsey (1971; Woolsey and Walzl, 1982) later applied this model to other mammalian cortices, especially that of the monkey. His scheme, in which auditory cortex is parcelled into a primary field (AI) and several surrounding fields, has remained essentially intact although in recent years a number of refinements and modifications have been made based on the results of detailed microelectrode mapping studies (Merzenich and Brugge, 1973; Merzenich *et al.*, 1975; Imig *et al.*, 1977; Knight, 1977; Andersen *et al.*, 1980a; Reale and Imig, 1980).

With the boundaries and frequency organizations of several auditory cortical fields firmly established in cat and monkey and with the recent developments of powerful neuroanatomical tracer methods, attention turned again to the patterns of connections these fields have with one another and with the thalamus, midbrain, and basal ganglia. Some of these connectivity patterns are closely related to the functional maps determined electrophysiologically. Single-unit studies of auditory cortex carried out over the past 30 years have revealed some of the coding properties of AI neurons and, most recently, of cells in surrounding fields. In this chapter we review some of the major developments in these areas. We proceed by first outlining the locations and boundaries of auditory cortical fields. We then go on to describe the coding properties of single auditory cortical neurons, paying particular attention to cells in AI. Next we discuss the functional organization of auditory cortex based on the spatial distribution of neurons having particular coding properties. Finally, we take up the relationships between the functional organizations of these fields, as determined electrophysiologically, and the patterns of connections these areas have established with each other and with subcortical structures. The material is necessarily selective and we have omitted, because of space limitations, a discussion of all work relevant to our understanding of auditory cortex. Some of the gaps can be filled by referring to other reviews (Goldstein, 1968; Goldstein and Abeles, 1975; Brugge, 1975; Imig and Morel, 1983; Aitkin *et al.*, 1984; Masterton and Imig, 1984; Phillips and Brugge, 1985).

2. Parcellation of Auditory Cortex

2.1. Auditory Cortical Fields in the Cat

Woolsey's map of cat auditory cortex is based on evoked potential recordings (Woolsey and Walzl, 1942; Woolsey, 1960, 1961), cytoarchitectonic analysis (Rose, 1949), and studies of the patterns of retrograde degeneration in the medial geniculate body after regional cortical lesions (Rose and Woolsey, 1949, 1958). Four tonotopic (or cochleotopic) fields were recognized by him: the primary auditory field (AI), the second auditory field (AII), the posterior ectosylvian area (EP), and the suprasylvian fringe area (SSF). He also pointed to several other outlying auditory fields, including the temporal field (T), an area in the dog referred to as AIII by Tunturi (1960), the insular field (I), after Goldberg *et al.* (1957), and association cortex, which was studied by Thompson and Sindberg (1960). Auditory evoked potentials have also been recorded under chloralose anesthesia in precentral motor cortex (MI) and in visual area II. The results of fine-grain microelectrode mapping experiments (e.g., Fig. 10A) later added details that required revision of Woolsey's original scheme (Fig. 1).

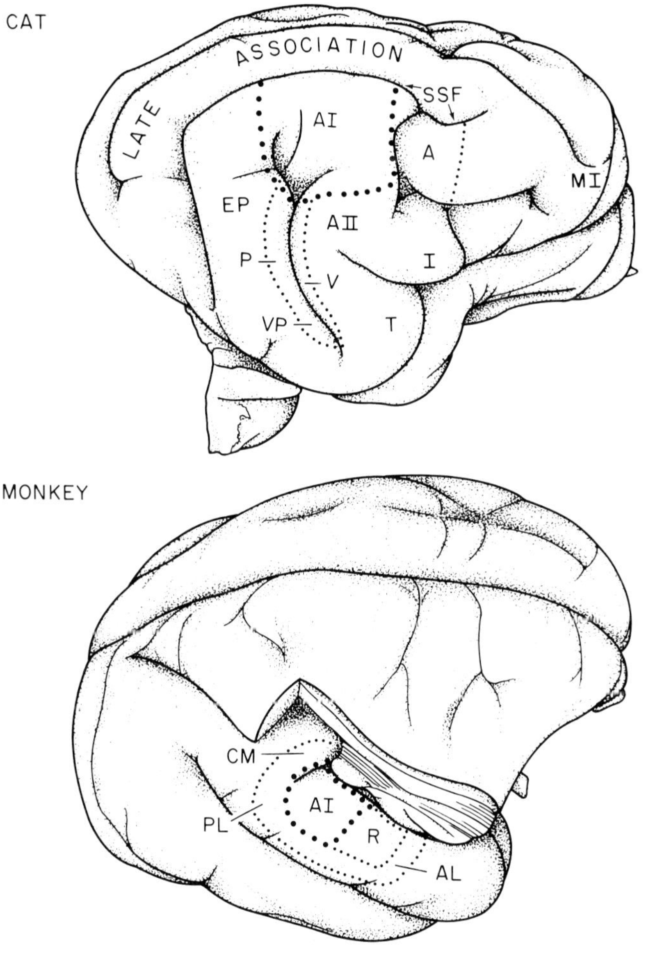

Figure 1. Locations and boundaries of auditory cortical fields in the cat and monkey based on electrophysiological mapping experiments. Shown on the cat brain are the primary (AI), second (AII), anterior (A), posterior (P), ventral posterior (VP), and ventral (V) fields mapped by microelectrode recording of single neurons or neuronal clusters. The presence of the posterior ectosylvian (EP), temporal (T), insular (I), motor (MI), association, visual cortex (late) fields has been determined by evoked potential mapping. On the monkey brain are outlined the core tonotopically organized areas, the primary (AI) and rostral (R) fields. The auditory belt includes the caudomedial (CM), posterolateral (PL), and anterolateral (AL) fields. All of these fields have been studied by fine-grain microelectrode mapping.

The location and frequency map of AI has remained largely as Woolsey and Walzl first described it. Neurons with low characteristic frequencies (CFs) are found caudally within the field whereas the representations of higher CFs are found progressively more rostrally. Rostral to AI, on the anterior ectosylvian gyrus, is found another complete and orderly representation of the audible frequency spectrum first mapped by Woolsey (1961) and included by him in the suprasylvian fringe area. Knight (1977) and Reale and Imig (1980) mapped this area in even greater detail and referred to it as the anterior auditory field (A or AAF). The tonotopically organized posterior (P) and ventral posterior (VP) fields discovered by Reale and Imig (1980) lie mainly on the buried banks of the posterior ectosylvian sulcus, a region not explored by Woolsey in his evoked potential work. A portion of Woolsey's area EP probably corresponds to the posterior portions of P and VP. A belt of cortex containing neurons responsive to acoustic stimulation surrounds the core fields of AI, A, P, and VP. These belt areas have not been studied at the single-unit level. One of these areas, AII, which in the cat is contiguous with AI ventrally, has recently been remapped by Schreiner and Cynader (1984). Their observations of broad tuning of single neurons and blurred tonotopy are in accord with those of others and with the anatomical findings that AII, and perhaps other fields in the auditory belt, receives a substantially different thalamic afferent supply than the four tonotopically organized core areas. Neurons in the "association cortex" have been studied systematically and the results of those and related experiments on the thalamus and reticular formation have provided much of the evidence for a diffuse system of afferents arising at the level of the midbrain and projecting forward along pathways separate from those of the lemniscal system (Irvine and Phillips, 1982).

2.2. Auditory Cortical Fields in Primates

The organizational scheme for auditory cortex in the primate has been derived from evoked-potential and single-unit mapping experiments in New World monkeys (Woolsey, 1971; Imig *et al.*, 1977), Old World monkeys (Merzenich and Brugge, 1973; Woolsey and Walzl, 1982), chimpanzees (Woolsey, 1971), and the prosimian primate, *Galago crassicaudatus* (Brugge, 1982). In no single primate species has auditory cortex been mapped in its entirety. The original electrophysiological mapping data, again from the 40-year-old experiments of Woolsey and Walzl, have only recently been published (Woolsey and Walzl, 1982). Microelectrode mapping work has added a great deal to Woolsey and Walzl's results, which were considerably less extensive in the monkey than in the cat. Figure 1 shows a composite map of auditory cortical fields drawn on the superior temporal gyrus of an owl monkey brain. The map is derived from data pooled from microelectrode mapping studies of several primate species.

Like that of the cat, auditory cortex of primates contains multiple fields that are distinguished from one another anatomically and physiologically. Most of the acoustically responsive cortex lies on the superior bank of the superior temporal gyrus, deep within the lateral fissure. Area AI is a tonotopically organized field that is coextensive with an area of koniocortex. It has been referred to in the rhesus monkey as area TC based on cytoarchitectonic analyses (Pandya and Sanides, 1973). AI has been mapped extensively in several primate species and may be considered the homolog of the field by that name in nonprimate species and of that field of temporal cortex in apes and humans identified cytoarchitectonically as koniocortex (von Economo, 1929; Galaburda and Sanides, 1980; Seldon, 1981, and this volume). The fields surrounding AI include the rostral (R), anterolateral (AL), posterolateral (PL), and caudomedial (CM) fields, so named because of their positions relative to AI. In the rhesus monkey, Pandya and Sanides (1973) recognize areas TA and TB adjacent to TC (AI). The correspondence between these areas and the ones mapped electrophysiologically is not entirely clear.

2.3. Auditory Cortical Fields in Other Mammals

About the time that Woolsey and his colleagues were studying the auditory cortex in the cat, Tunturi was mapping the corresponding area in the dog. Using the evoked potential method, he defined four acoustically responsive cortical fields which he named the posterior ectosylvian (PES), middle ectosylvian (MES), anterior ectosylvian (AES), and third auditory fields (Tunturi, 1960). A complete tonotopic map is shown in MES, the apparent primary field in this animal. Evidence for a tonotopic organization exists for AES, a field that occupies the same position with respect to MES as does field A with respect to AI in the cat. PES occupies a region that corresponds to Woolsey's field EP.

An acoustically responsive region of temporal cortex in the gray squirrel has been subdivided on the basis of its electrophysiologic and cytoarchitectonic properties (Kaas *et al.*, 1972; Merzenich *et al.*, 1976). One of these subdivisions contains a complete tonotopic map derived from the responses of sharply tuned neurons; the map is coextensive with a cytoarchitectonic field of koniocortex. Merenich *et al.* (1976) propose that this field be considered the homolog of AI in cat and monkey. Surrounding this field is a belt of responsive cortex the frequency organization of which has yet to be worked out.

In another rodent, the guinea pig, two tonotopically organized auditory cortical fields have been identified (Hellweg *et al.*, 1977). One of them has within it a rostral-to-caudal, low-to-high-frequency gradient similar to that in the presumed primary field of the squirrel. The high-frequency representation of the second of the two areas joins the high-frequency representation of the presumed AI so that together their maps are in a mirror-image relationship.

The auditory region in the rabbit also appears to contain multiple cortical fields which differ in their cellular architecture and tonotopic organization (Woolsey, 1971; Kraus and Disterhoft, 1982; McMullen and Glaser, 1982). Moreover, the relative positions and orientations of the tonotopic maps of the primary and secondary fields differ from those in other mammals studied this way (McMullen and Glaser, 1982). In the presumed AI, neurons with the highest CFs are located in the dorsal part of the field whereas cells having the lowest CFs are found ventrally. The second auditory cortical field in this animal lies dorsal and rostral to the primary field. Its tonotopic sequence is the reverse of that found in AI. The only other species shown so far to possess a dorsal-to-ventral, high-to-low-frequency representation in the primary field is the marsupial possum (Gates and Aitkin, 1982). The available data on this animal are not complete enough to determine whether there are any surrounding auditory cortical fields.

2.4. Homologies

From the available electrophysiological and anatomical data, there seems to be little doubt that areas designated AI in cat and monkey are homologs and that these likely correspond to areas designated AI in dogs, rodents, and rabbits. In the tree shrew (*Tupaia glis*), a field corresponding to the mammalian AI has been defined on the bases of cytoarchitecture and thalamocortical connections (Casseday *et al.*, 1976; Oliver and Hall, 1978).

The parallels between fields surrounding AI are even more problematic and, except perhaps for the dog, can be considered only in the cat and monkey, animals in which extensive mapping data are available. By considering the evolutionary trends in the development of the temporal lobe, the frequency organization of the auditory fields, and the connections these fields make with the thalamus, Woolsey and Walzl (1982) were able to suggest the homologies between the main auditory fields in the cat and monkey. Woolsey considered area A of the cat (the rostral portion of the suprasylvian fringe) the homolog of PL in the monkey. Woolsey also parcelled the area rostral to AI of the monkey into four areas rather than the two (R and AL) suggested by Imig *et al.* (1977). He further suggested that the two of them adjacent to AI are homologs of P and V of the cat. Rostral to these are two others, one of which he labeled the homolog of cat VP. AII does not appear on the recent monkey maps, although Woolsey would put it at the caudal end of the insula, medial to AI.

At this point we postpone further discussion of the organization of auditory cortical fields until later in the chapter. First it is necessary to consider the coding mechanisms of auditory cortical neurons upon which some of this organization is based.

3. Neural Coding Mechanisms within Field AI

For more than three decades, efforts have been made to understand, at the single-neuron level, the cortical mechanisms involved in a listener's ability to detect and discriminate tones of different frequencies or intensities, to locate the sources of sounds in space, and to process more complex acoustic stimuli, such as noises, frequency- or amplitude-modulated tones, clicks, and communication sounds. With few exceptions, the approach has been one of a detailed analysis of the response properties of single cortical neurons recorded extracellularly in anesthetized animals. Most of this work has been carried out in the middle ectosylvian region of the cat cortex, the region designated on Woolsey's map as AI. Unfortunately, the contours of the ectosylvian sulci differ considerably from one cat to the next and, thus, they are poor landmarks for the accurate localization of AI borders (e.g., Merzenich *et al.*, 1975). A similar situation obtains in the monkey, for there are no good gross landmarks on the superior surface of the superior temporal gyrus of this animal's brain to identify unequivocally the boundaries of any of the auditory cortical fields located there (e.g., Merzenich and Brugge, 1973; Imig *et al.,* 1977). Thus, the properties attributed to AI neurons have been derived both from experiments in which the electrode was inserted into the middle ectosylvian area of the cat or supratemporal plane of the monkey assuming the areas penetrated to be AI and from experiments in which the locations of recorded neurons were verified histologically or by electrophysiological mapping of the AI borders.

3.1. Coding for Stimulus Frequency

AI neurons are responsive only within a restricted domain of stimulus frequencies and intensities referred to as the response area of the cell. The frequency selectivity of a neuron may be described by its threshold tuning curve which is derived by determining systematically the sound pressure level required to evoke a threshold response at various stimulus frequencies (Figs. 2 and 8). The frequency at which threshold intensity is lowest, i.e., the tip of the tuning curve, is called the "characteristic" or "best" frequency of that neuron and is related to the place of resonance along the cochlear partition to which the neuron is ultimately connected. For primary auditory nerve fibers, the shape of the tuning curve rather accurately reflects the mechanical tuning properties of the basilar membrane (e.g., Rhode, 1978; Khanna and Leonard, 1982; Sellick *et al.,* 1982). For a neuron within the central auditory pathways, any departure from the primary-like shape of the tuning curve is interpreted to be the result of activation of convergent excitatory and inhibitory inputs to that neuron derived from different afferent sources. The shapes of tuning curves can vary considerably from one AI neuron to the next even for cells of the same CF (Hind *et al.,* 1960; Evans and Whitfield, 1964; Oonishi and Katsuki, 1965; Goldstein *et al.,* 1968; Phillips and Irvine, 1981a; Phillips *et al.,* 1985). Under general anesthesia, the great majority of AI neurons exhibit tuning curves that are narrow and V-shaped, like those of primary auditory nerve fibers (Fig. 2A–C). For other AI cells the tuning curves may be exceptionally broad or contain multiple tips. Some response areas are circumscribed (Fig. 2D–F). Exceptionally broad or multiply tipped tuning curves reflect convergent input from a number of sharply tuned elements whose CFs span a relatively wide spectral range and, hence, are derived from a relatively wide segment of cochlear partition. The fact that some response areas are circumscribed implies that these neurons are selective not only for stimulus frequency but for stimulus intensity as well. Intensity-selective properties of AI neurons are described later in this chapter.

Various proportions of unusually shaped AI tuning curves have been reported from different laboratories. There are at least two possible explanations for the discrepancies. First, the results attributed to AI neurons may be contaminated with data from adjoining fields where neurons are known to have irregular or broad tuning properties. Second, anesthesia may reduce much of the afferent input to AI cells, other than that arising directly from the thalamus. In this regard, the greatest number of AI cells having nonprimary-like tuning curves are recorded from unanesthetized animals (Evans and Whitfield, 1964; Oonishi and Katsuki, 1965; Goldstein *et al.*, 1968). In the unanesthetized condition, not only are AI neurons more active than under general anesthesia but there is a greater probability of recording from active neurons in cortical layers above and below laminae III and IV which receive the bulk of the thalamic afferent input. The question of proportions aside, it is clear that for a large number of AI neurons, frequency selectivity established at the level of the cochlea is preserved at the level of the cortex (Calford *et al.*, 1983). For other AI neurons, cochlear frequency selectivity is altered by neuronal interactions along the central auditory pathway.

The fact that AI neurons are frequency selective immediately argues for the involvement of this field in a central place mechanism for frequency encoding and, hence, pitch recognition. Although a place principle is clearly supported by many kinds of evidence, it is not possible to explain pitch perception with

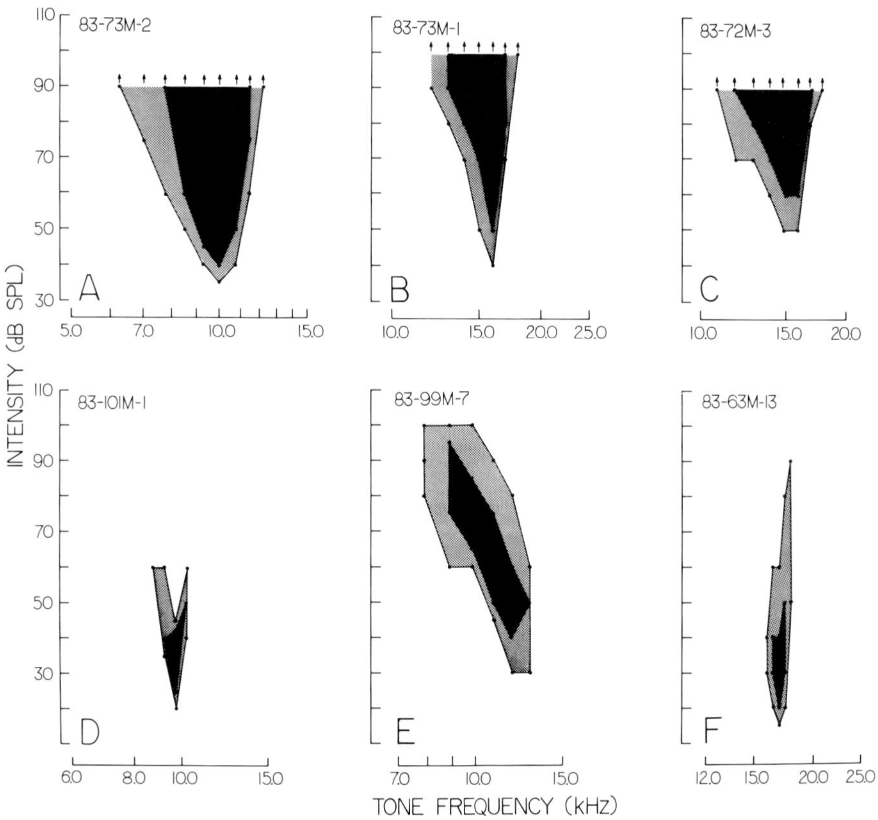

Figure 2. Threshold tuning curves from single AI neurons in the cat. (A–C) V-shaped curves similar to those derived from primary nerve fibers. (D–F) Circumscribed response areas. Stippled area indicates the frequency-intensity domain within which an AI neuron responds to 50% or fewer stimuli. Black area is the region in which fewer than half the stimuli are effective. From Phillips *et al.* (1985).

the simplistic concept that each auditory neuron corresponds to a specific frequency. The shape of even the sharpest tuning curve in cat or monkey AI shows that at moderate suprathreshold sound pressure levels, the neuron associated with that curve would respond over a relatively wide range of frequencies and, moreover, that at any one frequency a sizable number of neurons with different CFs would be included in the active neuronal pool. Hence, place theorists must postulate that the central nervous system, for the purpose of pitch identification, ignores all of the activity except the peak or boundary of the array of active neurons.

3.2. Coding for Stimulus Intensity

The sensitivity and dynamic range of hearing are remarkable: a young listener is able to discriminate tones close in frequency, and even understand speech, over an intensity range of more than 100 db on the scale of sound pressure. A primary auditory nerve fiber increases monotonically its discharge rate over a range of no more than about 40 db. How, then, with that dynamic range characterizing individual cochlear nerve fibers, can normal hearing achieve its effective range of 100–120 db? At the level of the auditory nerve, part of the answer appears to reside in the spread of excitation along the basilar membrane and the recruitment of additional fibers into the active population (see Green, 1976).

The dynamic range for intensity is, as a rule, not extended greatly, if at all, for a single neuron in the central auditory system. Perhaps 75% of the neurons recorded in primary auditory cortex display rate-vs.-intensity functions that are virtually identical to those derived from auditory nerve fibers (Fig. 3A). For them, spike counts increase from threshold over a 10- to 40-db range at which point further increase in stimulus intensity is ineffective in further raising the spike count, i.e., saturation firing is reached. Neurons with V-shaped tuning curves tend to exhibit this property. The responses of other AI neurons to changing intensity may be quite different from this. For these cells, the number of spikes evoked by a tone rises to a maximum as intensity is raised and then, with further increase in sound level, the count falls (Fig. 3B). This property is associated with neurons having the circumscribed response areas previously described. Because cochlear nerve fibers show only monotonic rate-vs.-intensity functions, any nonmonotonic behavior in the rate curve derived from a central auditory cortical neuron may be considered the result of inhibitory processes.

In AI of the anesthetized cat (Brugge *et al.*, 1969; Phillips and Irvine, 1981a) and awake and unanesthetized monkey (Brugge and Merzenich, 1973a,b), the nonmonotonic functions relating spike count to intensity can be sharply peaked and a change of 10 db on either side of the most effective SPL may reduce the count by as much as 90% of its maximum value; other cells have a broad maximum (Fig. 3B). Recent evidence suggests that cells displaying nonmonotonic rate-vs.-intensity functions might represent a class of neurons functionally distinct and spatially separate from those displaying monotonic functions (Phillips *et al.*, 1985). In any given animal, the most effective SPLs differ from one cell to the next. In monkey AI, best SPLs have been recorded between 15 and 94 db whereas in cat the values reported have ranged between 10 and 85 db. The presence of intensity-selective neurons suggests another mechanism, introduced in the central auditory system, whereby the dynamic range of hearing might be accomplished. Namely, regardless of where the stimulus lies within the dynamic range of hearing, there exists a population of cortical neurons that is firing maximally. Other pools of cells will be firing at submaximal rates and still others will be silent or spontaneously active because the sound intensity is too low or too high to evoke from them a vigorous discharge. Changing SPL results in a shift in the active neuronal pool such that many neurons once maximally active become less so and many of those once marginally active or unresponsive are brought to maximal response. Such a finding leads to the hypothesis of a "place" mechanism for intensity coding (Brugge and Merzenich, 1973b). This hypothesis has not yet been tested in either cat or monkey. Evidence for an "amplitopic organization" in the primary auditory cortex of the mustache bat has been obtained, however (see Suga, 1977).

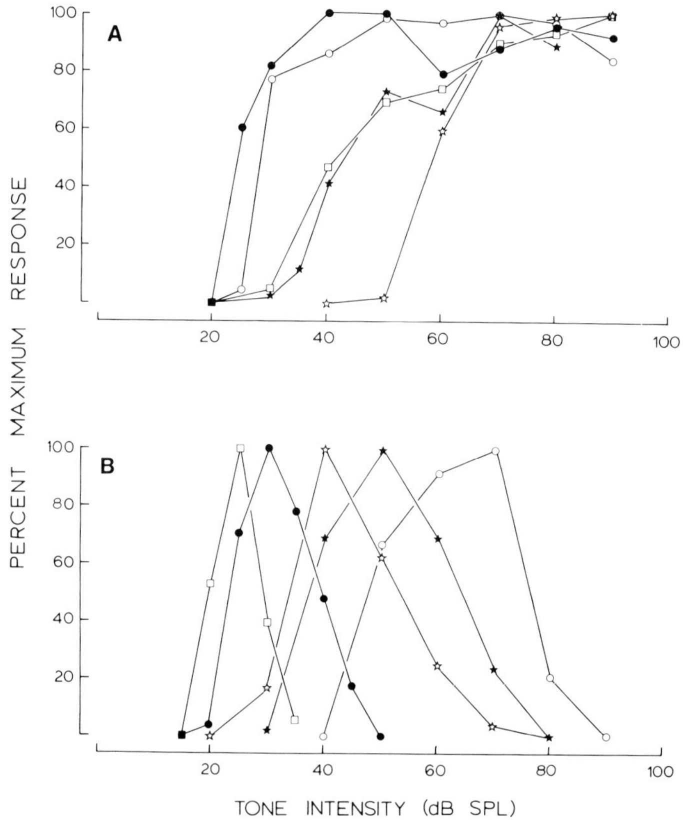

Figure 3. Normalized spike count-vs.-stimulus intensity for repeated tone bursts at CF derived from 10 AI neurons in the cat. Five of the neurons exhibited monotonic (A) and five nonmonotonic (B) behaviors. From Phillips *et al.* (1985).

3.3. Coding for Binaural Stimulus

There is abundant evidence that the integrity of AI is necessary for normal azimuthal sound localization behavior (e.g., Jenkins and Merzenich, 1984), although the mechanisms involved are far from being understood. Much of the information concerning the coding of binaural signals and the neural mechanisms of sound localization has been derived from studies of the brain-stem nuclei where activity originating in the two ears first converges and binaural interactions first occur (for recent reviews see Yin and Kuwada, 1984, and Phillips and Brugge, 1985). Brain-stem neurons receiving input from the two ears relay the results of that interaction to higher auditory centers including the cortex. Thus, many neurons within auditory cortical fields are sensitive to the binaural stimulus parameters of interaural time differences (ITDs) and interaural intensity differences (IIDs), two cues used by human listeners for localizing the source of a sound in space.

In order to understand the mechanisms that underlie binaural sensitivity and the possible roles played by cortical cells in sound localization and other binaural phenomena, the discharge properties of neurons in AI have been studied parametrically both under conditions where the stimuli to the two ears are controlled independently and delivered to the two eardrums via speculums sealed into the external ear canals (Hall and Goldstein, 1968; Brugge *et al.,* 1969; Brugge and Merzenich, 1973b; Benson and Teas, 1976; Phillips and Irvine, 1981b, 1983) and under more natural conditions where sounds are generated from speakers located in space around the head (Evans, 1968; Eisenman, 1974; Sovijarvi and Hyvarinen, 1974; Middlebrooks and Pettigrew, 1981). Most studies of binaural interactions carried out under earphone listening conditions have sought to define relationships between neural discharge patterns and variations in the intensities and the times of arrival of the stimuli delivered to the two ears.

3.3.1. Neurons Responding to ITDs

In the free field, a difference in the time the stimuli arrive at the tympanic membranes occurs when the sound source is located at a point not equidistant from the two ears. If the stimulus is a pure tone, which it often is in experiments of the kind to be described, then two time disparities are realized: an interaural arrival-time disparity and an ongoing phase disparity. Auditory cortical neurons are sensitive to both temporal cues.

Many AI neurons responding to low-frequency tones, below about 4000 Hz, are exquisitely sensitive to small interaural differences in the interaural phase differences of the tonal stimuli at the two ears (Brugge *et al.,* 1969; Brugge and Merzenich, 1973b; Benson and Teas, 1976). The function relating the number of spikes evoked to the interaural time delay is a periodic one, the period of the function being equal to the period of the stimulating tone (Fig. 4). The periodicity of that function indicates that the neuron is responding to the differences in interaural phase of the stimuli. At favorable interaural delays, the response of a binaural AI neuron may be as strong or stronger than the sum of the two monaural responses. At least-favorable delays, the spike count may fall below that evoked by stimulation of either ear alone, indicating that a suppressive mechanism may be involved in the interaural delay sensitivity of these neurons. Studies of interaural phase sensitivity in the lower auditory brain stem reveal that monaural phase-locked inputs each have an excitatory half-cycle and an inhibitory half-cycle (Goldberg and Brown, 1969; Brugge *et al.*, 1970). When the impulses arriving from the two ears arrive at the cell of convergence at the same time, i.e., when the two excitatory half-cycles are in time register, the output

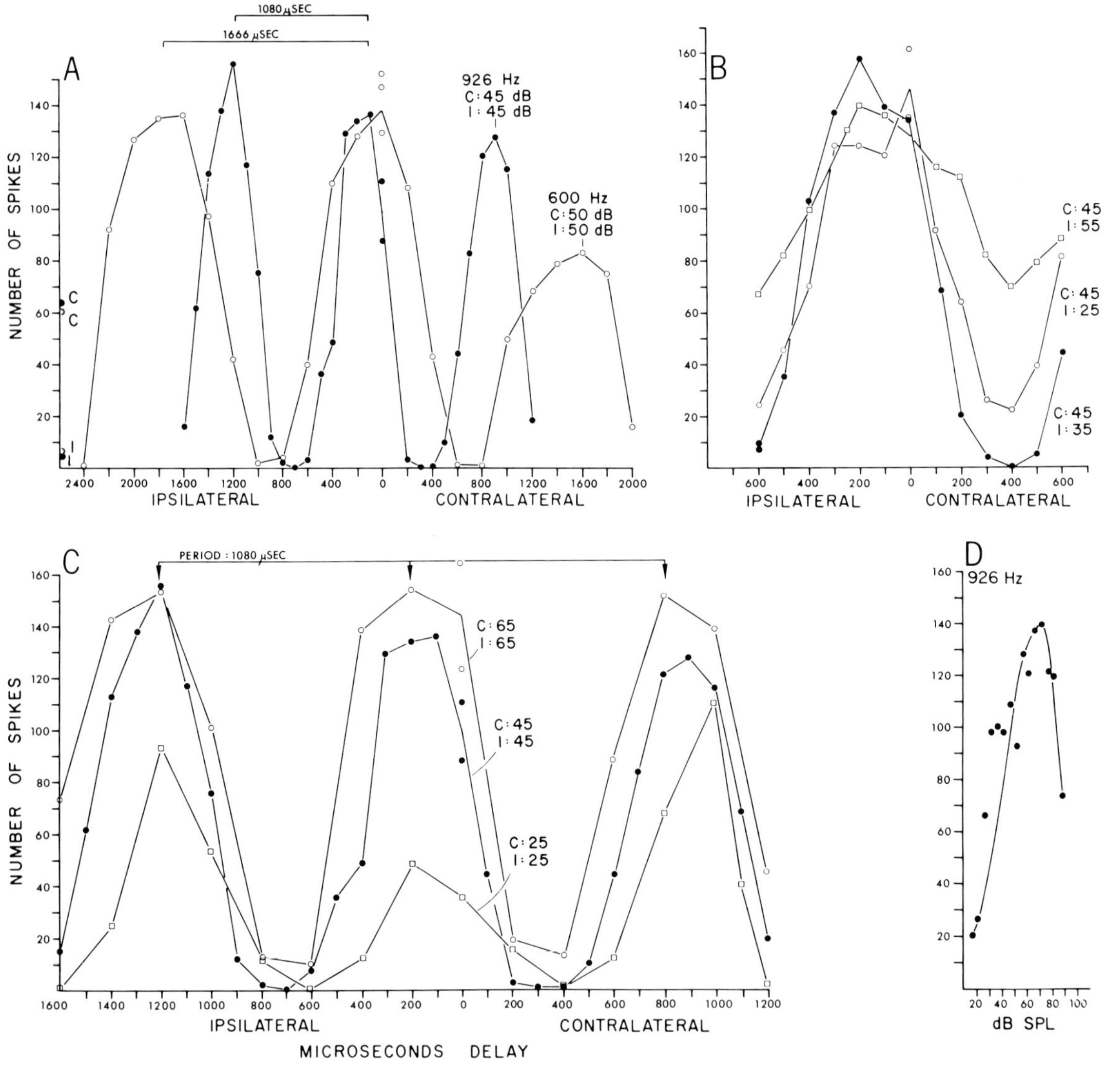

Figure 4. Spike count as a function of interaural time delay of low-frequency tones delivered to the two ears. Data are derived from a single AI neuron in an awake rhesus monkey. (D) Spike count-vs.-intensity function exhibited by this same unit at CF. In (A) the periodic functions obtained at 926 and 600 Hz reach the same maximal values when the tone to the ipsilateral ear lags that to the contralateral ear by around 100 μsec. This preferred time delay is uneffected by changing the interaural (B) or binaural (C) intensity. From Brugge and Merzenich (1973b).

of the binaural neuron is maximal. When the optimal delay is changed, so that the excitatory half-cycle of one input is shifted in a way that makes it coincide in time with the inhibitory half-cycle of the other, the result is cancellation of the monaural response. Thus, the sensitivity to interaural phase delay reflects a temporal interaction of phase-locked excitatory and inhibitory inputs to neurons of the brain stem. The activity that results from this cycle-by-cycle interaction is transmitted faithfully to higher auditory centers including the cortex.

In 1966 Rose *et al.* described a class of binaural neurons in the cat inferior colliculus that detect a specific interaural time delay for all low frequencies to which the neuron is phase sensitive (see also Kuwada and Yin, 1983; Yin and Kuwada, 1983a,b). These cells were said to have a "characteristic delay." This delay would correspond to the interaural phase delay created by a sound lying on the surface of a hyperboloid extending laterally from one ear (the so-called "cone of confusion"). The concept of a "characteristic delay" was attractive and widely accepted despite the fact that until recently there was little experimental evidence to support it. The notion was questioned by some (e.g., Benson and Teas, 1976) largely on the grounds that frequently the characteristic delays recorded were outside of the useful range for animals with small head sizes (e.g., chinchillas and kangaroo rats). The arguments related to this and other issues that take us beyond the scope of this chapter have been discussed recently elsewhere (Phillips and Brugge, 1985). Suffice it to say that many AI neurons in both cat and monkey are phase sensitive and that this interaural phase sensitivity meets the criteria for a characteristic delay as originally put forward by Rose *et al.* (1966). Brugge and Merzenich (1973b) suggested from their studies of AI neurons in the unanesthetized monkey that phase sensitivity may be a column-dependent property of AI. This hypothesis has not been rigorously tested.

The sensitivity of AI neurons to onset-time disparities has received considerably less attention (Brugge *et al.*, 1969; Benson and Teas, 1976; Kitzes *et al.*, 1980). Nonetheless, there is evidence to indicate that this property is not governed by the same phase-locked, cycle-by-cycle mechanisms described previously. Rather, the discharge of such an AI neuron appears determined by the relative times of arrival of the excitatory inputs from one ear and the inhibitory inputs from the other.

Still another group of time-dependent neurons was discovered recently by Kitzes *et al.* (1980). These neurons tend not to be excited by monaural stimulation but respond vigorously when tones are presented to both ears with a zero, or near zero, interaural onset delay (Fig. 5A). This dependence on binaural stimulation prompted Kitzes *et al.* (1980) to refer to these cells as being "primarily binaural" (PB). The PB cells prefer not only an ITD around zero but also a zero, or near zero, IID (Fig. 5B) and hence are good candidates for detecting the location of sound in the vicinity of the median plane.

3.3.2. Neurons Responding to IIDs

IIDs are generated mainly by the acoustic shadow created by the head and pinnae (Moore and Irvine, 1979). They are significant only for high-frequency sound where the wavelengths are smaller than the head diameter or pinna height.

The specification of the IID is most accurately and conveniently determined under earphone listening conditions. The IID may be controlled by changing the intensity of a tone delivered to one ear while holding the intensity of the tone to the other ear constant or by differentially changing the intensities of the tones delivered to both ears. For the most part, the connection between sensitivities to IID and absolute intensity has not been thoroughly analyzed at the cortical level despite the fact that both cues are needed to accurately locate the source of sound in space. Thus, it is often not clear from the published data whether and to what extent the apparent IID sensitivity of an auditory cortical neuron is also affected by absolute intensity.

For our present purposes we consider neurons in the high-frequency region of AI as distributed along a continuum based upon the sensitivity of these neurons to changes in the IID of binaurally presented tones. At one end of this distribution are those neurons whose discharge properties are particularly sensitive to changes in the IID. We presume that these neurons participate in the process of detecting the location of a sound in space. At the other end of the spectrum are the IID-insensitive neurons. The position of a neuron along this IID-sensitivity scale is determined to a large degree by the extent to which IID sensitivity of the cell is influenced by the absolute intensity of the binaural stimulus.

Neurons that occupy one of the extreme positions near the IID-sensitive end of this continuum are those that are excited by stimulation of one ear (usually

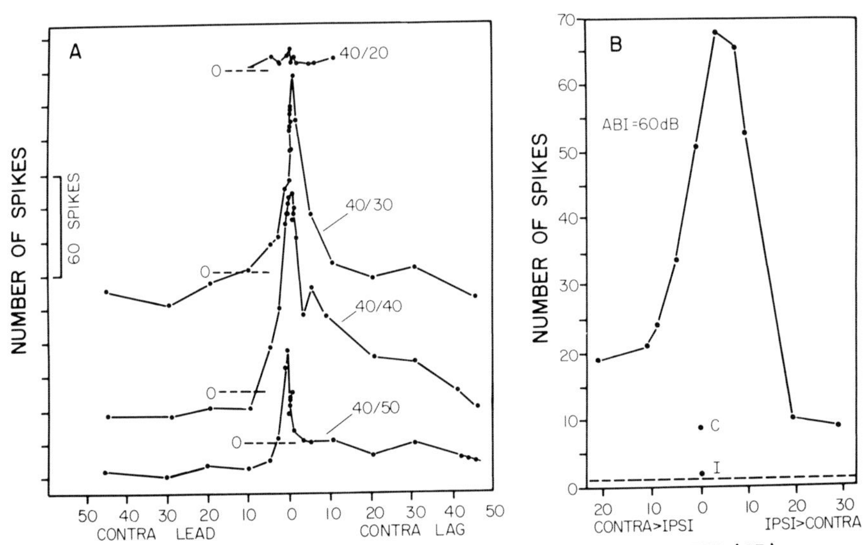

Figure 5. Spike count as a function of interaural time delay (A) or interaural intensity difference (B) for two PB neurons recorded in AI of the cat. (A) Four spike count functions, each at a different IID, are separated along the ordinate for easy comparison. Dashed line on each plot (0) indicates the baseline level of spontaneous activity. Maximal discharge occurs around an interaural delay and interaural intensity difference of zero. Redrawn from Kitzes *et al.* (1980). (B) IID functions obtained while holding the average binaural intensity constant at 60 db SPL. Counts generated by monaural stimulation of the contralateral (C) or ipsilateral (I) ears and the level of spontaneous activity (dashed line) are indicated on the plot. From Phillips and Irvine (1981b).

the one contralateral to the cortex under study) and inhibited by simultaneous stimulation of the opposite ear (the so-called EI neurons after Goldberg and Brown, 1969). Typically, these cells discharge maximally when the IID is adjusted to favor the contralateral ear and may be completely inhibited from discharging when the IID favors the ipsilateral ear. Figures 6A and 7A show the familiar sigmoidal relationship between spike rate and IID exhibited by these kinds of neurons. As shown in Fig. 6A, within the dynamic range of IID sensitivity, the change in discharge rate that accompanies a shift in IID (long arrow) is far greater than the rate change produced by an equivalent decibel shift in absolute intensity (short arrow). Neglecting any front-to-back ambiguity, a neuron with these properties should be able to detect the presence of a sound in the contralateral hemifield and to give information regarding the azimuthal location of that sound, at least in one plane of space (Phillips and Irvine, 1981b). It is also important to note, however, that while such a neuron may provide information about the direction from which a sound emanates, it cannot, because of its relative insensitivity to the overall sound intensity, provide as much information about the distance of the sound source from the head.

A second variant of IID-sensitive AI cell, the PB neuron, exhibits maximal discharge only when the IID and ITD are both around zero (Figs. 5 and 6B).

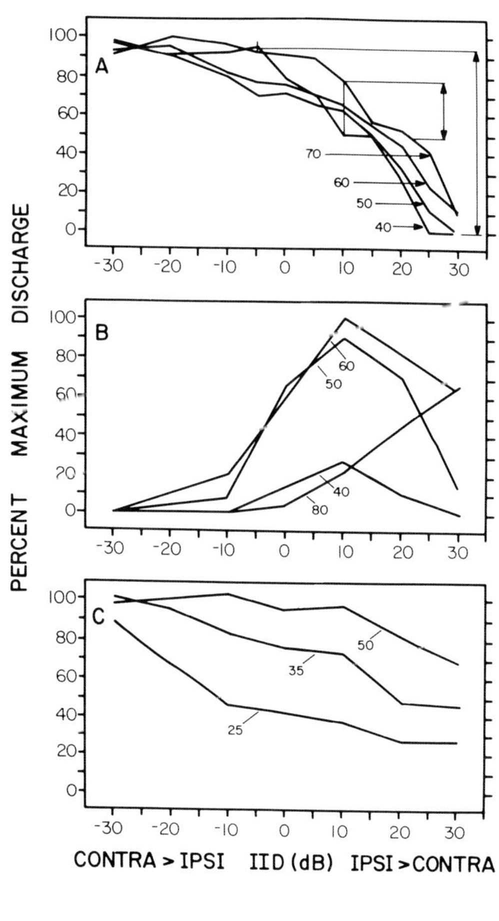

Figure 6. Normalized spike count-vs.-IID functions for three AI neurons in the cat. (A) IID functions derived from an EI cell at the four different average binaural intensities shown on the graph. The sensitivity to change in the IID (long arrow) is far greater than the sensitivity to change in the average level (short arrow) of the CF tone. (B) IID functions derived from a PB cell at four average intensity levels. (C) IID functions derived from a neuron relatively insensitive to IID.

At around zero ITD, there is a gradual reduction in the discharge rate as the IID is increased, until virtually no action potentials are evoked when the IID favors one ear or the other by some 20–30 db (Phillips and Irvine, 1981b). The distribution of IID functions shown in Fig. 6B indicates that the discharge for this PB neuron may also be highly dependent upon change in absolute intensity of the binaural stimulus. The relationship between discharge rate and IID is clearly different when one compares the curves taken at an average binaural intensity of 40 or 80 db with the curves taken with the average binaural intensity

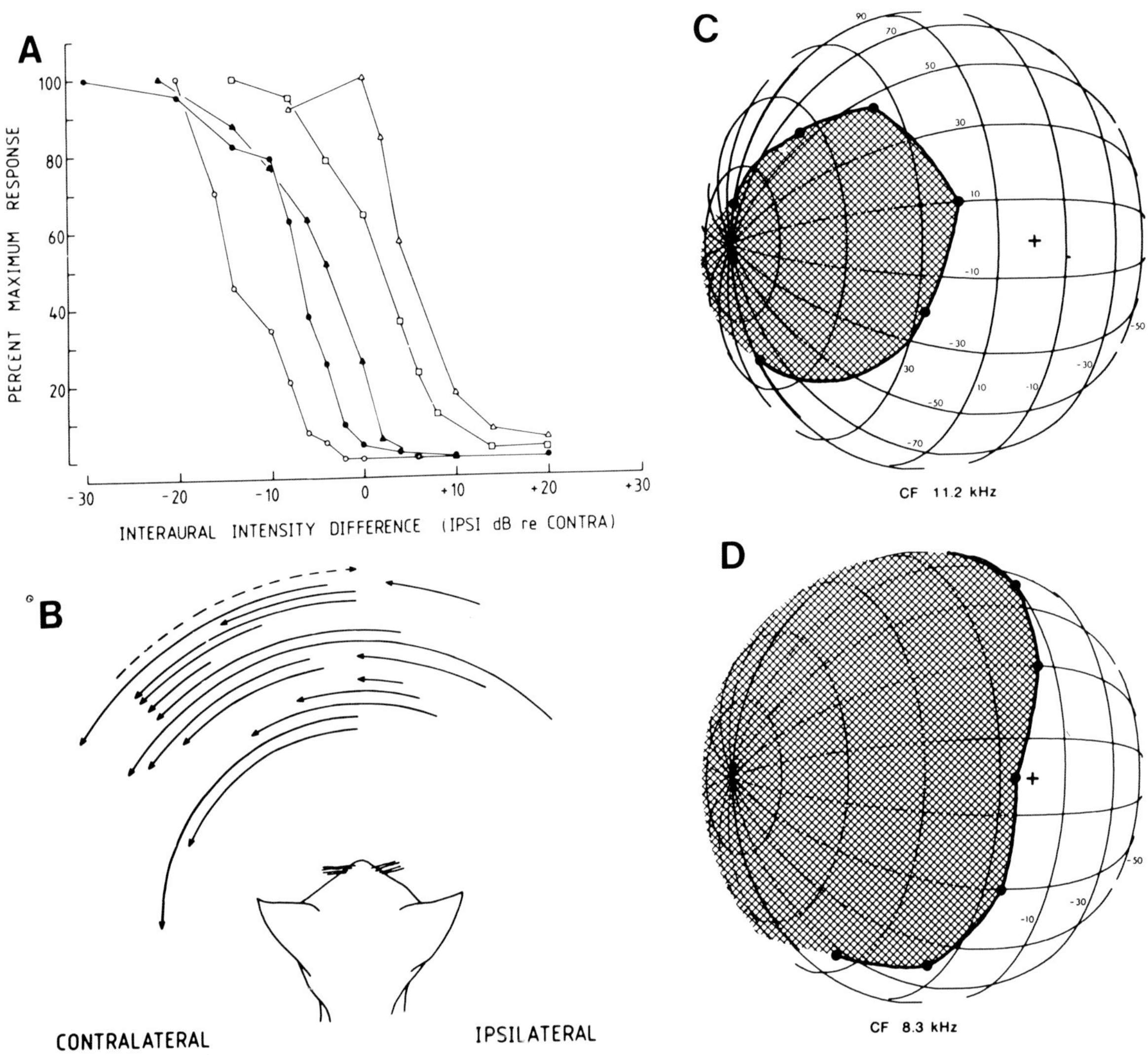

Figure 7. (A) IID functions derived from five AI neurons in cats listening through earphones. (B) Schematic view of a cat's head and sound field viewed from above showing the aximuthal projections of IID functions for 17 cells [including the 5 shown in (A)] isolated in AI. Each arrow represents the azimuthal range corresponding to the IIDs to which an individual unit was most sensitive. (A,B) from Phillips and Irvine (1981b). (C,D) Spatial receptive fields of two AI cells (hemifield units) recorded under free-field listening conditions. (C,D) from Middlebrooks and Pettigrew (1981).

at 50 or 60 db. When the average intensity at the two ears is between 50 and 60 db, small changes in IID around the most effective IID result in substantial decreases in spike counts. In contrast, when the average intensity is reduced to 40 db or elevated to 80 db, the sensitivity to IID is greatly diminished or lost. This dual sensitivity to *both* IID and absolute intensity may render to PB neurons the capability of detecting the actual location of a sound source in space if that source is near the midline.

It is not uncommon to find in AI a third variant of IID-sensitive neuron. Cells in this group are excited by monaural stimulation of both ears (so-called EE neurons). When the IID favors one ear over the other, the rate and timing of the binaural response are similar to, if not indistinguishable from, the rate and timing of the response to monaural stimulation of the favored ear. Hence, such neurons not only can detect the hemifield within which the sound source lies but can provide information about the azimuthal location in *both* hemifields.

At this time there are not enough data available to give the complete input–output functions for each variant of AI neuron sensitive to IID. We can, however, suggest that the rules governing binaural interactions involving IID of the kinds previously described may, for many AI neurons responding under dichotic conditions, elicit two simple outcomes. First, if the intensity at one ear far exceeds that at the other, the resulting discharge pattern is virtually identical to the one evoked by monaural stimulation of the ear receiving the more intense sound. Second, if the intensities of the sounds at the two ears are relatively close to one another, the binaural response is dominated by neither the left ear response nor the right ear response, i.e., the response pattern resembles neither monaural response.

For EI neurons, if the intensity at the excitatory (usually contralateral) ear is much greater than that at the inhibitory (usually ipsilateral) ear, the binaural response pattern is virtually identical to the one evoked by stimulation of the contralateral ear alone. As far as this cell is concerned, the sound is perceived as coming only through the contralateral ear or, extrapolating to the free-field listening condition, from a source in the contralateral hemifield. If, on the other hand, the ipsilateral stimulus is the stronger, then the binaural response may contain no spikes, which is for EI neurons the ipsilateral monaural response.

For EE neurons under conditions of large IIDs, the discharge pattern evoked monaurally by the weaker stimulus does not appear in the joint response. From the neuron's point of view, the relatively weak sound at one ear has been masked by the presence of a more intense sound at the opposite ear and the sound is lateralized to the ear receiving the greater intensity; the sound is perceived as coming through the ear receiving the more intense sound or, extrapolating to the free-field listening condition, from a source in that ear's hemifield. If the IID is reversed, the sound is then lateralized to the opposite ear or localized to the opposite hemifield. PB neurons also obey these rules but, because they produce few or no spikes unless the two ears are stimulated nearly simultaneously and with IIDs near zero, they can only code for stimuli in the vicinity of the midline.

Neurons that are placed near the other extreme of the IID-sensitivity continuum are, because of their relative insensitivity to changes in IID, considered poor candidates for detectors of the locus of a sound in space. Again, stimulation of one ear, usually the contralateral one excites such a neuron whereas monotic stimulation of the opposite ear may or may not evoke spikes. For some of these IID-insensitive neurons, simultaneous stimulation of the two ears results in a rate and temporal pattern of the discharge that are distinct from that recorded from the neuron under monotic conditions regardless of the IID (Brugge *et al.*, 1969; Brugge and Imig, 1978; Phillips and Irvine, 1983). The family of IID curves shown in Fig. 6C illustrates that the discharge of an IID-insensitive neuron changes rather little as a function of IID over a wide range of absolute intensities. Furthermore, the maximal change in discharge rate produced by either small or large changes in IID is virtually identical to the change in discharge rate which can be produced by comparable changes in absolute intensity.

3.3.3. Neurons Reponding under Free-Field Stimulus Conditions

The studies mentioned above using dichotic stimulation did not answer directly the questions of whether binaural neurons of any class actually encode the spatial location of sounds and, if they do so, whether these neurons are distributed in a topographic fashion that could be construed to be a neural map of auditory space. These questions have been addressed to some extent in studies of AI neurons carried out under free-field stimulus conditions in both cat (Evans, 1968; Eisenman, 1974; Sovijarvi and Hyvarinen, 1974; Middlebrooks and Pettigrew, 1981) and monkey (Benson *et al.*, 1979). The results of these studies show that, in general, the response strength of a cortical neuron is a graded function of azimuth for sound in the contralateral hemifield and that some neurons show a spatial selectivity to relatively specific locations of the sound source.

In Middlebrooks and Pettigrew's (1981) experiments, an anesthetized cat was held in such a way that its head was in the center of a hemispheric field around which could be moved a loudspeaker. The speaker locations that were effective in eliciting a discharge from an AI neuron were used to determine the locus and extent of a spatial receptive field. About 50% of the neurons isolated in these experiments were selective for the location of sounds. This population could be divided into two subclasses of cells. One contained those cells that responded only to sounds in the contralateral hemifield (hemifield units). The properties of these neurons resemble most closely those of IID-sensitive EI cells recorded under dichotic listening conditions. Figure 7 illustrates the spatial receptive fields of two hemifield neurons and the corresponding rate-vs.-IID curves that such neurons might generate under earphone listening conditions. A second category of location-selective units (axial units) contained cells with small circumscribed receptive fields whose locations were constant for each experiment and which were determined by the orientation of the pinna. The presence of neurons with these properties points to the importance of the pinnae in sound localization mechanisms (Phillips *et al.*, 1982; Calford and Pettigrew, 1984). The remaining half of the AI cells isolated in these experiments were not sensitive to sound location and, thus, were referred to as omnidirectional units. These latter neurons are likely the IID-insensitive cells recorded under dichotic listening conditions. There was no indication of a space map in AI resembling the one discovered by Knudsen and Konishi (1978) in the midbrain of the barn owl. Rather, neurons with similar receptive field properties tend to aggregate into patches similar to those that make up binaural interaction bands in AI mapped under dichotic listening conditions as described later in this chapter.

Sovijarvi and Hyvarinen (1974) discovered in AI of the cat neurons that responded selectively not only to the spatial location of a sound but to the direction of its angular movement. Movement in one direction in the horizontal or vertical direction produced excitation in these neurons whereas motion in the opposite direction evoked inhibition or no change in the spontaneous activity. Although no systematic mapping of these properties was carried out, it is interesting to note that cells whose responses were determined by the location or movement of the sound source were frequently found along the length of an electrode penetration.

3.4. Coding for Complex Sounds

3.4.1. Coding of Species-Specific Vocalizations

In certain primates the known repertoires of species-specific vocalizations have provided a wealth of natural acoustic stimuli with which to study the possible roles played by auditory cortical neurons in processing information of communicative significance (Wollberg and Newman, 1972; Newman and Wollberg, 1973; Glass and Wollberg, 1983). In general, however, the stimulus–response relationships derived form primate auditory cortical neurons suggest no simple one-to-one mapping between a neuron's response selectivity and a specific vocalization (Newman, 1978; Newman and Symmes, 1979; Glass and Wollberg, 1983). Neither the specific spectral content nor the communicative significance serves as a feature sufficient to evoke a unique neuronal response. Rather, responses of auditory cortical cells appear to depend in complex ways on specific patterns of acoustic transients embedded in time-varying and spectrally diverse stimuli (Wollberg and Newman, 1972; Newman and Symmes, 1979; Creutzfeldt *et al.*, 1980; Steinschneider *et al.*, 1982; Glass and Wollberg, 1983). Thus, a more suitable model of neural responsiveness may involve mechanisms similar to those proposed for the bat in which neurons are specialized to respond more on the basis of the presence or absence of certain transient components embedded in a species-specific vocalization rather than on the basis of a unique vocalization (Glass and Wollberg, 1983). Such a neural mechanism may also be involved in speech waveform processing since transient components in speech sounds, known to be linguistic parameters of perceptual significance in human communication, are clearly reflected in the temporal responses of primate cortical and geniculate neurons (Steinschneider *et al.*, 1982). Quantitative differences in the selectivities for primate vocalizations and human speech sounds have been documented for neurons believed to be located in different auditory cortical areas (Newman and Wollberg, 1973; Glass and Wollberg, 1983). Such findings would tend to support the idea that different cortical areas detect biologically significant sounds by unique, but not mutually exclusive, processing mechanisms. While there are little data to support this notion (see also Manley and Müller-Preuss, 1978), mechanisms like these are apparently operating at the level of the auditory cortex in the bat.

3.4.2. Repetitive Acoustic Pulses

For human listeners, a train of short noise pulses or clicks evokes a pitch sensation at pulse repetition rates that approach several hundred pulses per second. The pitch sensation is not based on spectral cues but, rather, on the temporal waveform, hence its name, periodicity pitch. Monkeys, like humans, can discriminate a train of noise bursts having a rate of 10/sec from one with a rate of 300/sec (Symmes, 1974). This ability is permanently lost in such a trained monkey when the auditory cortex is destroyed bilaterally, suggesting that the cortex is involved in the detection of these temporal cues. De Ribaupierre and Goldstein (1972) discovered in the auditory cortex of the cat several classes of cells responding to click and noise trains. One class contains neurons that could encode the periodicity of clicks or noise bursts that elicit, in human listeners, a strong sensation of periodicity pitch. The single-spike discharge of these cells is precisely time-locked to the click and is perfectly entrained at click rates between 10/sec and 1000/sec. These cells represented 39% of the population of cortical neurons isolated in this experiment. Intracellular recordings of these neurons and neurons in other categories revealed that one mechanism that could limit the rate of locking to repeated acoustic transients is a hyperpolarization that follows a brief depolarization giving rise to the short-latency excitatory onset response. These so-called "locker" neurons resemble functionally certain cells of the cochlear nuclei and, hence, may be part of a neural circuit, originating in the brain stem, that encodes for the periodicity of the acoustic stimulus.

4. Coding Mechanisms in Fields Outside of AI

The publication of fine-grained frequency maps of the organization and boundaries of several auditory cortical fields in the cat (Knight, 1977; Merzenich *et al.*, 1975; Reale and Imig, 1980) was soon followed by several parametric studies in this animal of the coding properties of single neurons in areas outside of AI. These experiments probed the questions of whether and to what extent the different auditory fields divide the labor of encoding the many acoustic features found in the animal's natural environments.

4.1. The Anterior Auditory Field

The anterior auditory field (A) occupies the dorsal part of the anterior ectosylvian gyrus adjacent to (and partially overlapping) the hindquarter representation of the second somatic field. The similarities in the thalamic projections upon this field and AI led to an earlier prediction that the information reaching the two fields would be similar and that the processing of this information by the two fields would be done in parallel (Andersen *et al.*, 1980a). The electrophysiological results of Phillips and Irvine (1982) confirmed these expectations. Like many AI neurons, many cells in field A are sharply tuned to frequency, have dynamic ranges of no more than 30–40 db, and respond to CF tones with latencies as short as 10–12 msec. Their binaural sensitivity to interaural time and intensity differences of tonal stimulation has not been distinguishable from that of AI cells (see also Brugge *et al.*, 1969). Latency data are consistent with the view that the two fields derive parallel inputs directly and mainly from the ventral division of the MGB.

4.2. The Posterior Auditory Field

Field P is located on the caudal bank of the posterior ectosylvian sulcus adjacent to both AI and VP. It receives a widespread distribution of afferent input from its counterpart on the opposite hemisphere, from fields AI, A, and VP on the same side (Imig and Reale, 1980, 1981b), and from at least four subdivisions of the MGB including the ventral and deep dorsal nuclei, the magnocellular division, and those periventral nuclei bordering the ventral division (FitzPatrick *et al.*, 1977; Imig and Reale, 1981a; Morel and Imig, 1984). The physiology of field P cells takes on a character somewhat different from that of neurons in A and AI owing, presumably, to the fact that its thalamic input is weighted in favor of structures other than the ventral division of the MGB. Posterior field neurons have extraordinarily long minimal response latencies, on the order of 50 msec or more (Phillips and Orman, 1984). Such a latency would easily permit input from fields AI and A whose minimal latencies are, on the average, around 12–20 msec (Phillips and Irvine, 1981a). Even longer latencies have been described in field P (Reale and Imig, 1980). More than 85% of the neurons recorded in this area generate spike counts that are nonmonotonic functions of stimulus intensity and for 20% of these the counts fall to nearly zero at high stimulus levels (Phillips and Orman, 1984). This feature is present for all frequencies that excite the neuron and, hence, the response area for this neuron is a circumscribed one (see Fig. 2D–F). Another remarkable property of some of these cells is that the best SPL remains constant across frequency. The mechanisms operating to achieve such intensity selectivity are not known but they must necessarily involve neuronal interactions for the output of the cochlea shows no such selectivity. The high proportion of intensity-selective neurons would suggest that field P may be specialized for encoding intensity shifts by a place mechanism previously suggested for AI neurons. Whether this auditory field, or any other, possesses a map of stimulus intensity, as does one of the cortical fields in the mustache bat (Suga, 1977), is not known.

Field P neurons exhibit binaural interactions of the kind displayed by cells in both A and AI and in many brain-stem auditory nuclei (Orman and Phillips, 1984). Thus, there appear to be striking similarities in the mechanisms of encoding dichotic listening cues between field P cells and cells in other auditory cortical fields. Moreover, the relatively high frequency–intensity selectivity displayed by field P cells has been achieved apparently without cost to the detection of these interaural time and intensity localization cues.

4.3. Second Auditory Field (AII)

Systematic quantitative studies of the discharge properties of single AII neurons have yet to be carried out. Qualitative studies of the responses of neuronal clusters have, however, provided a picture of AII, which is different in several respects from that of AI (Andersen *et al.*, 1980a; Reale and Imig, 1980; Middlebrooks and Zook, 1983; Schreiner and Cynader, 1984). In contrast to AI neurons, AII cells are broadly tuned and have relatively high thresholds to tonal stimulation. These properties may reflect the relatively diffuse afferent input AII cells apparently have from the MGB (Andersen *et al.*, 1980a). Binaural interactions of the kind recorded in AI, A, and P are preserved in AII as well.

4.4. Association Cortex

In unanesthetized cats or in cats under chloralose anesthesia, responses to auditory stimulation are recorded in three regions outside of the ectosylvian auditory area. These fields on the medial suprasylvian gyrus (MSA), anterior lateral gyrus (ALA), and pericruciate gyrus (PCA) receive convergent auditory, visual, and somatic sensory input, hence the name "polysensory" or "nonspecific" cortex. Neurons in these areas are, as a rule, broadly tuned to frequency, respond to tones or noise at each ear with a long-latency (16–54 msec) onset spike which is followed by a 200- to 300-msec suppression of spontaneous activity, and exhibit a binaural interaction described as occlusion (Irvine and Huebner, 1979). The responses to acoustic stimulation of neurons in the medial/intralaminar (M/IL) nuclei of the thalamus are virtually identical to those in association cortical fields, a finding which supports the notion that these areas are reached by a nonspecific projection system that is distinct from the lateral lemniscal pathway reaching ectosylvian fields via the MGB (Irvine, 1980). Input to the intralaminar nuclei comes from the brain-stem reticular formation which contains neurons whose acoustic response properties are similar to those in the M/IL. Another potential source of diffuse input to association cortex is the pulvinar–posterior complex of the thalamus. This work has been reviewed by Irvine and Phillips (1982).

5. Functional Architecture

5.1. Frequency Selectivity and the Cortical Representation of Cochlear Place

Neurons located along a cell column radial to the cortical surface tend to have the same or very similar CFs as illustrated in Fig. 8 (Hind *et al.*, 1960; Abeles and Goldstein, 1970; Merzenich and Brugge, 1973; Merzenich *et al.*, 1975; Phillips and Irvine, 1981a). The frequency organization of a radial cell column has been difficult to determine accurately mainly because of the spatial resolution of the extracellular recording techniques used, the difficulty in obtaining electrode penetrations perfectly aligned with cell columns, and the fact that cells in laminae above and below the thalamocortical target zone are relatively unresponsive under general anesthesia. Abeles and Goldstein (1970) carefully approached this problem by recording from isolated neurons in AI of unanesthetized cats and correcting the data for the orientation of the electrode path with respect to that of the radial cell columns. They found that, almost without exception, narrowly tuned units in the same or close radial cell chains have nearly equal CFs. Although CFs may be more difficult to judge in broadly tuned cells, the response range of frequencies of these neurons overlapped that of neighboring narrowly tuned cells in the same radial penetration. Cell columns representing the same CF are, when projected upon the brain surface, arrayed along what is referred to as an *isofrequency line* (when viewed through the depth of the cortex, the array of elements having the same CF occupies a ribbon of tissue). The presence of isofrequency lines (or ribbons) is the consequence of the projection of points from a one-dimensional linear array (the cochlear partition) to a three-dimensional structure (the cortex). By systematically determining the characteristic frequencies of neurons and neuron clusters in multiple closely spaced microelectrode penetrations, complete CF maps of AI and several surrounding fields have been obtained in a variety of mammalian species as described previously. The isofrequency lines on the cat AI (Fig. 10A) run in a dorsoventral direction; those lines representing high frequencies are found rostrally in the field, those representing lower frequencies are located progressively more caudally. Thus, the isofrequency lines in cat AI, and in surrounding fields, run roughly orthogonal to the high-to-low-frequency gradient of the field, a conclusion reached by Woolsey and Walzl in their evoked potential work more

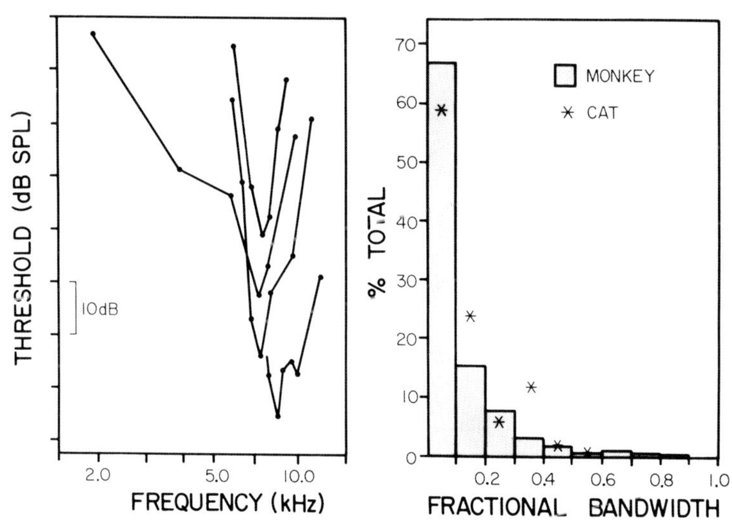

Figure 8. (Left) Threshold tuning curves derived from four AI neurons within a single electrode penetration oriented nearly perpendicular to the cortical surface. The curves are shifted along the ordinate for clarity. From Phillips and Irvine (1981a). (Right) Variation in CF of neurons within a single perpendicular penetration into AI of cat and monkey. Fractional bandwidth = (highest CF − lowest CF)/lowest CF in the penetration. Cat data from Phillips and Irvine (1981a), monkey data from Brugge and Merzenich (1973b).

than four decades ago. A similar situation obtains for tonotopically organized fields in other species studies so far.

In 1964 a controversy arose over the question of tonotopic organization in cat AI. At that time Evans and Whitfield (1964) reported that in unanesthetized cats the positions of neurons in what they believed to be area AI were poorly correlated with the neuron's CFs, suggesting that the organization of units in this field was not a tonotopic one. Part of the problem arose from the fact that tonotopy at the level of the cortex has traditionally been linked to a central "place" mechanism for frequency discrimination (see, e.g., Poliak, 1932). As frequency discrimination is not abolished in cats after large auditory cortical lesions, it was not difficult to accept the possibility that tonotopy in this region was blurred or nonexistent. Moreover, it does seem to be the case that general anesthesia reduced substantially the rich and varied afferent input to many cells, thereby rendering them unresponsive to sound or responsive mainly to the more tightly coupled input from the MGB. Third, the CF map of AI is not related in any systematic way to the positions of the sulci of the region. Hence, pooling of CF data from different animals and using sulci as landmarks for the construction of frequency maps can easily blur the organization that might exist there.

5.2. Binaural Sensitivity, Computational Cortical Maps, and the Cortical Representation of Sound in Space

It was discovered several years ago that AI neurons having similar binaural properties tend to form vertical columns through the cortical gray matter and that these columns aggregate into clusters or bands (Imig and Adrian, 1977; Middlebrooks *et al.*, 1980). Detailed microelectrode mapping studies of AI, similar to those that revealed the tonotopic organization of this field, were carried out to determine the distribution of these so-called *binaural interaction bands.*

Some variations may exist in the binaural interactions exhibited by neurons isolated in a single vertical electrode penetration into AI, at least under a limited set of binaural conditions (Phillips and Irvine, 1979). It is also the case that, qualitatively, the binaural response properties of the great majority of neurons and neuron clusters may be assigned to one of two major classes that are easily distinguished from one another simply by visual inspection on the oscilloscope screen of the neural discharge in response to dichotic tonal stimulation. This latter approach was taken by Imig and Adrian (1977) who classified about two-thirds of the neuronal cluster responses as being *summation* responses, i.e., the response to simultaneous binaural stimulation was greater than the response to stimulation of either ear alone. These neurons are generally of the kind referred to as EE (excitatory/excitatory) in the similar study of Middlebrooks *et al.* (1980). *Suppression* responses occurred when the neurons were excited less by binaural stimulation than by stimulation of the more effective monaural stimulus. These are also referred to as EI (excitatory/inhibitory) responses. Other types of responses are recorded but they make up a small proportion of the total recorded.

An electrode penetrating the cortex normal to the pial surface encounters successively units with either suppressive or summative properties (Fig. 9A,C).

Systematic mapping of the cortical surface of AI with closely spaced electrode penetrations revealed aggregations of binaural columns, some of which were elongated into bands running nearly orthogonal to the isofrequency contours (Fig. 9B).

Binaural sensitivity, unlike frequency sensitivity, is not a property represented on the basilar membrane. It arises as the result of the interaction of excitatory and inhibitory inputs impinging on target neurons located, presum-

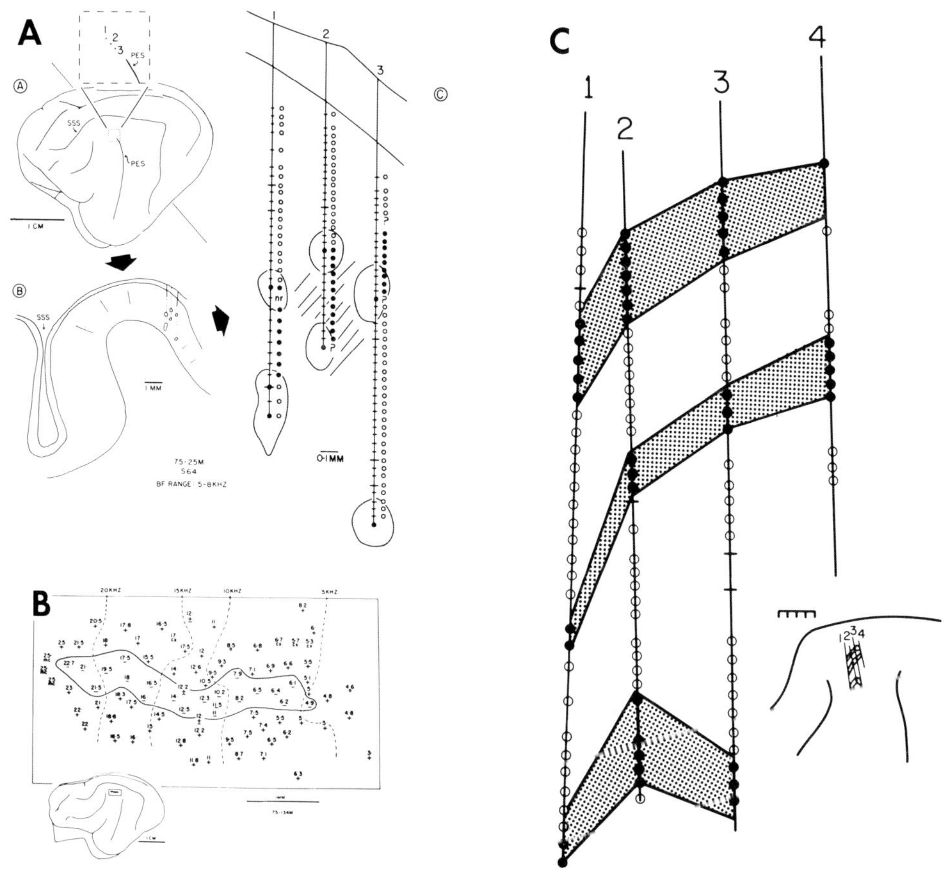

Figure 9. (A) Diagram showing the entry points of three microelectrode penetrations into AI of the cat and the distribution of binaural classes within each penetration. Open circles designate neurons or neuronal clusters classified as generating *summation* (EE) responses; closed circles designate suppression (EI) responses. (B) Surface view of a tonotopic and binaural map. Solid line surrounds a region containing neurons exhibiting only suppression (−) responses. Summation responses (+) are found surrounding this area. Numbers are CFs of neurons or neuron clusters in the penetration at that cortical site. (A, B) redrawn from Imig and Adrian (1977). (C) Distribution of binaural classes of neurons or neuron clusters obtained from four tangentially oriented electrode penetrations through AI of the cat. Open and closed circles represent EE and EI responses, respectively. Shaded areas connect EI zones to form the hypothesized binaural bands running somewhat rostrocaudally across AI. These bands probably correspond to the suppression regions outlined by Imig and Adrian as shown in (B). (C) Redrawn from Middlebrooks *et al.* (1980).

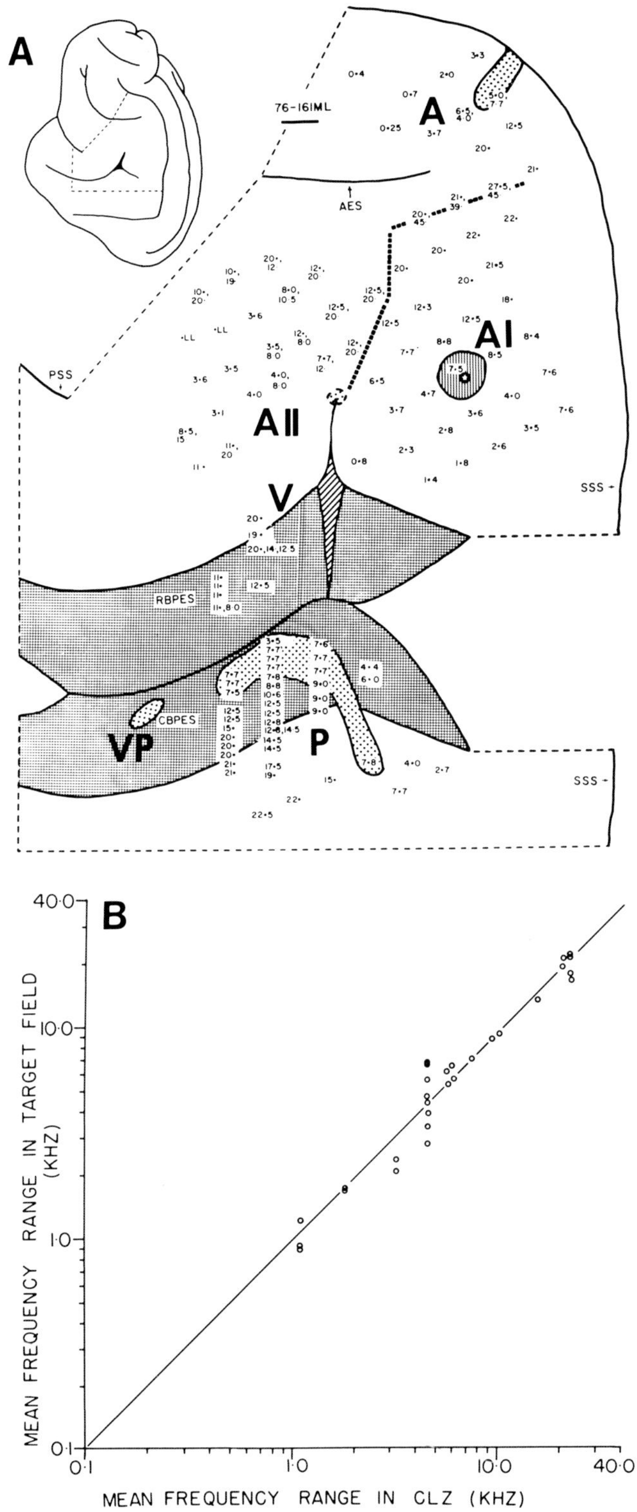
A
76-161ML
AES
PSS
AI
AII
V
RBPES
CBPES
VP
P
SSS
B
MEAN FREQUENCY RANGE IN TARGET FIELD (KHZ)
MEAN FREQUENCY RANGE IN CLZ (KHZ)
0·1
1·0
10·0
40·0

ably, in the lower auditory brain stem. Thus, whereas tonotopy in the auditory cortex is a map of the receptor surface projected topographically upon successive auditory relay nuclei, the binaural map is, in large part, a computational one the elements of which are formed by neuronal interactions.

6. Relationships between the Functional Organizations and Connectivity Patterns of Auditory Cortical Fields

Auditory cortical fields are linked to numerous subcortical areas via axons of thalamocortical, corticothalamic, corticotectal, corticostriatal, and corticopontine neurons and to each other via a network of corticocortical fibers. A variety of tract-tracing techniques have been employed to study these connections including those of lesion-degeneration and of axonally transported radioactive amino acids or horseradish peroxidase. Figure 11A shows the widespread distribution in the auditory fields around AI of terminal labeling of axons that had transported tritiated proline from the small injection site in AI. The earlier studies of auditory cortical connectivity, in both cat and monkey (Diamond *et al.*, 1968a,b, 1969; Pandya *et al.*, 1969; Heath and Jones, 1971; Kawamura, 1973; Pandya and Sanides, 1973; Sousa-Pinto, 1973; Paula-Barbosa *et al.*, 1975; Winer *et al.*, 1977; Niimi and Matsuoka, 1979), did not use electrophysiological mapping along with tract-tracing to determine the cortical field boundaries or the functional maps within the fields. More recently, several laboratories have combined microelectrode mapping of auditory cortex with anatomical pathway tracing in the same animal. This is a powerful combination of techniques that allows cortical input–output projection patterns to be related directly to the boundaries and functional organizations of cortical fields. Combined microelectrode mapping and tracer studies of auditory cortical connections with the thalamus (Andersen *et al.*, 1980a; Merzenich *et al.*, 1982), midbrain (Andersen *et al.*, 1980b,c), basal ganglia (Reale and Imig, 1983), and neighboring cortical fields (Imig and Brugge, 1978; Imig and Reale, 1980, 1981b) have revealed two outstanding features of these projection systems. First, the cortical auditory system is highly segregated cochleotopically; neurons with similar CFs are strongly connected, often in reciprocal fashion, whereas neurons with dissimilar CFs tend to remain relatively isolated from one another. Second, the projections are both highly divergent and highly convergent.

Figure 10. (A) Map of CFs of neurons or neuron clusters in auditory cortex of the cat. Small inset shows lateral view of the cat brain with the rostral end pointed upward and the ventral part to the left. The dashed lines outline the mapped area shown in the exploded view. Numbers are CFs recorded from neurons in penetrations at those points on the brain. Shaded areas represent the depths of the posterior ectosylvian sulcus which has been opened up in this drawing to show portions of the fields normally buried there. Area in AI circled is the site of a small injection of tritiated proline. Stippled areas in A, P, and VP are the sites of terminal labeling that resulted from the injection. (B) Plot of the CF at the site of a tritiated proline injection (CLZ) as a function of the CF at the target site where axonal terminal labeling was detected. From Imig and Reale (1980).

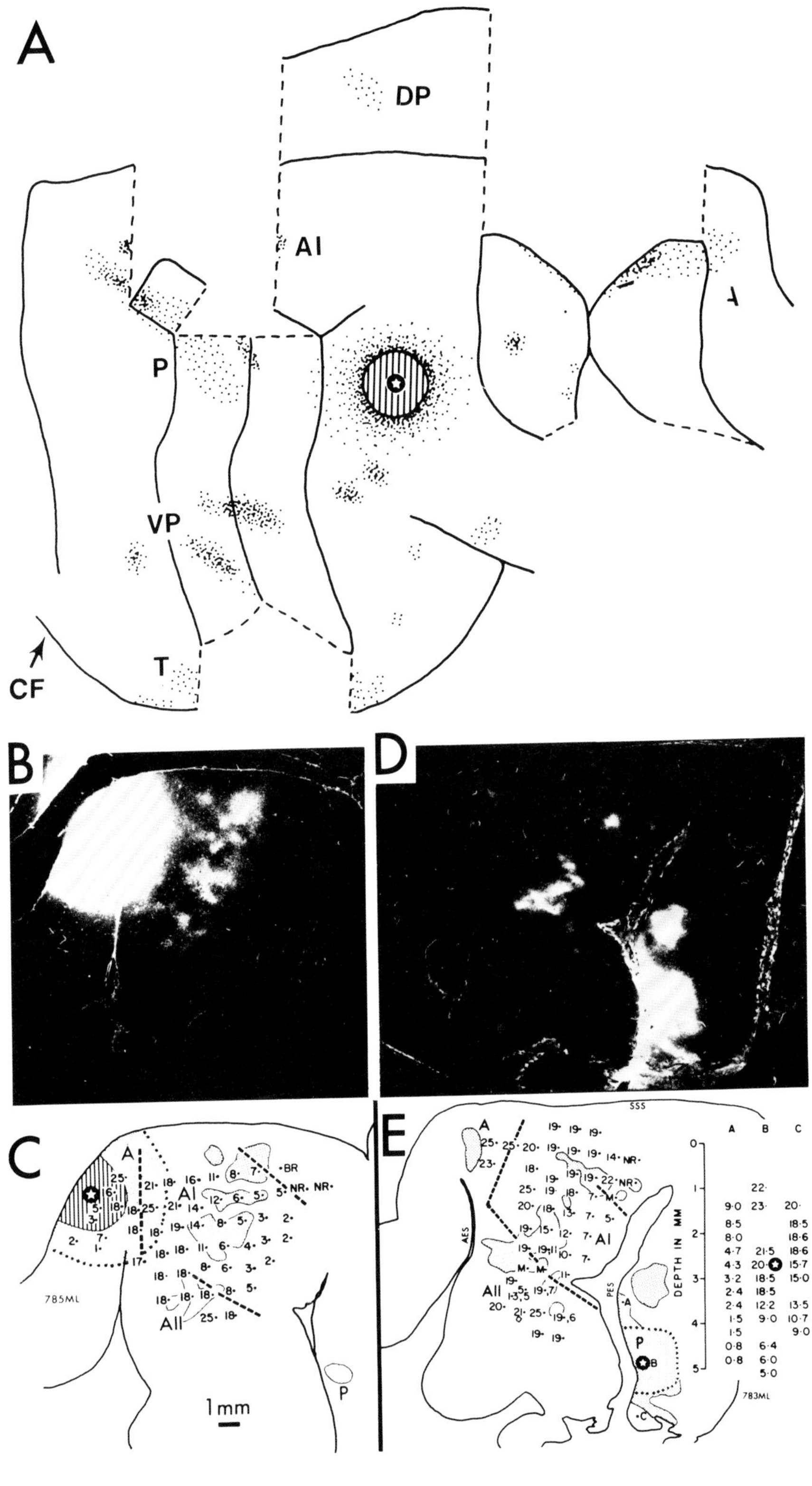
A
DP
AI
P
VP
T
CF
B
D
C
AII
785ML
1mm
E
SSS
AES
PES
DEPTH IN MM
783ML

6.1. Relationships to Tonotopic Organization

6.1.1. Corticocortical Connections

Evidence in the cat for a topographically organized system of auditory corticocortical connections was first provided by the experiments of Downman *et al.* (1960) in which electrical stimulation of a locus in one auditory field evoked a response in a restricted locus of another. Many of the details of this and other auditory corticocortical pathways were later worked out in studies in which single-unit mapping was combined with autoradiographic and HRP tract-tracing. Examples of results from several of these studies (Imig and Brugge, 1978; Imig and Reale, 1980, 1981b) are shown in Figs. 10–13. In the experiment illustrated in Fig. 10A, large parts of fields AI, A, P, and AII were mapped electrophysiologically and within AI a single small injection of tritiated proline was placed on the 7.5-kHz isofrequency contour. The transported label was later localized to the 7.5-kHz place in A and in P. The results of this experiment illustrate the two features of the auditory corticocortical connections mentioned previously. First, the projection is a topographic one; AI neurons project upon cells in both A and P having the same CF. Second, the projection is a divergent one; a single small injection at a known CF locus in AI produces a terminal labeling pattern that extends along the isofrequency contours in the target fields. Figure 11 illustrates the results of the reverse experiment and another feature of the corticocortical connection patterns. Here small injections of tritiated proline were made into mapped areas of A (Fig. 11B,C) or P (Fig. 11D,E). In both cases, the resultant projection patterns in AI are patchy ones, but within the appropriate CF representation. The uneven distribution of corticocortical connections is most dramatically shown in Fig. 12, which is taken from experiments in which large injections of tritiated proline or HRP were injected into AI of the opposite hemisphere.

The experiments illustrated here are typical of many others in which injections were made in AI, A, P, and VP at many different CF loci. The overall picture is that ipsilaterally fields AI, A, P, and VP appear topographically and reciprocally connected with one another. In addition, each of these fields projects to multiple areas within the peripheral auditory belt. Furthermore, each of the four tonotopically organized areas is reciprocally and topographically connected with the homotopic field on the opposite hemisphere. Figure 10B summarizes the results of many experiments of the kind described. In this figure the abscissa represents the CF at the center of the injection site and the ordinate the centers of the labeled terminal fields that resulted from the injection. The points on the plot cluster around the diagonal line indicating the high correlation between the CFs at the injection and target sites.

Figure 11. Drawing of the ectosylvian region of the cat, unfolded to show the banks of the sulci within which are found portions of auditory fields. The circle and star mark the center of a small injection of tritiated proline. Stippling indicates the terminal anterograde labeling in the ipsilateral fields to which AI projects. (B, C) Patchy labeling in AI after injection of tritiated proline into A. (D, E) Patchy labeling in AI after injection of tritiated proline into P. In both cases the transported label appears at the same CF region as that of the injection site. Numbers in (C) and (E) are CFs of neurons or neuron clusters.

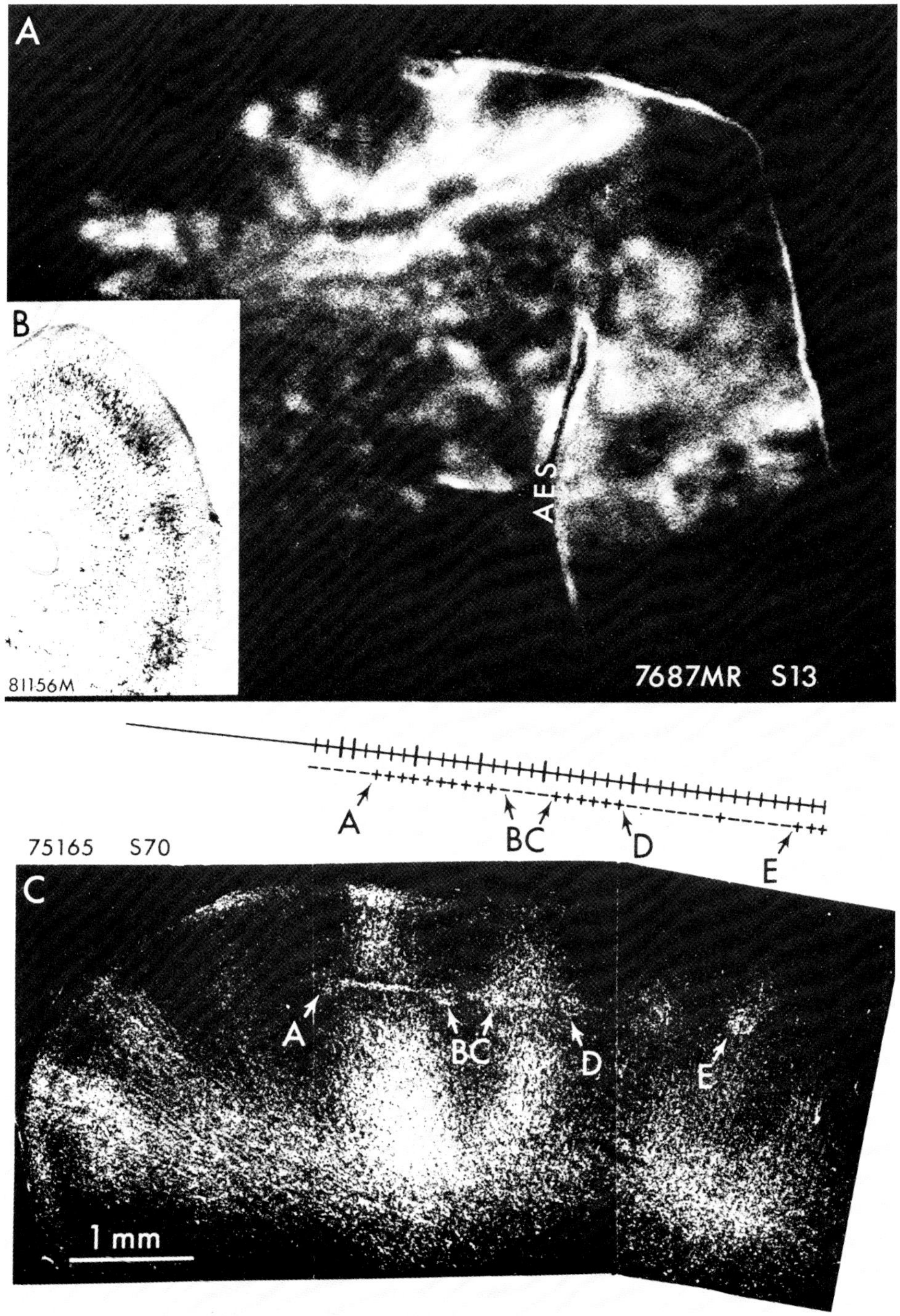

Figure 12. (A) Darkfield photomicrograph of a tangential section through the middle layers of the ectosylvian area of cat brain showing in an autoradiograph the distribution of terminals of the corpus callosum that were labeled as the result of a large injection of tritiated proline into AI of the opposite hemisphere. The suprasylvian sulcus is above, and anterior ectosylvian sulcus (AES) to the lower

The organization of the corticocortical connections in the monkey is far less well understood than that in the cat. Only the projections of AI and the field rostral to it (field R) have been studied in any detail using both electrophysiological and anatomical tracing methods. According to the results of autoradiographic tracer studies in the owl monkey (FitzPatrick and Imig, 1980, 1982), AI projects ipsilaterally to the five known cortical auditory fields and contralaterally to three fields, including AI. Field R is reciprocally connected to AI and projects to at least three other fields. Contralaterally, field R projects to the same fields as does AI. Like the projections in the cat, those in the monkey auditory cortex are also patchy in nature. In the rhesus monkey, auditory cortex has been shown by Seltzer and Pandya (1978) to be connected in a cascading fashion to the temporal pole and perhaps to a wider segment of the superior temporal gyrus (see Pandya and Yeterian, this volume).

6.1.2. Thalamocortical and Corticothalamic Connections

Andersen *et al.* (1980a) mapped the thalamocortical and corticothalamic connections between the MGB and three auditory fields—AI, A, and AII—using both anterograde and retrograde tracing methods. Microelectrode recording guided the cortical placement of the tracers and later was used to correlate the anatomical and CF maps. Restricted loci in AI and A were found to receive coextensive inputs from folded sheets of neurons passing through the ventral division of the MGB, from columns of cells through the deep dorsal nucleus of the dorsal division of the MGB, from the medial division of the MGB, and from cells in the lateral division of the posterior group. Many neurons of the ventral division that project upon field AI also send an axon collateral to field A (Imig and Reale, 1981a; Imig *et al.*, 1981). The main differences between the thalamic connections to fields AI and A are that the input to field AI is stronger from the ventral division whereas that to field A is stronger from the posterior group (Morel and Imig, 1984).

Fields P and VP receive their thalamic input from a different group of thalamic nuclei although the ventral nucleus, deep dorsal nucleus, and medial division of the MGB are also included in this grouping (FitzPatrick *et al.*, 1977; Imig and Reale, 1981a; Imig and Morel, 1983; Morel and Imig, 1984). The projection from the ventral division of the MGB is relatively weak to fields P and VP as compared to AI whereas the projections from nuclei surrounding

right. (B) Photomicrograph of a transverse section through AI of the cat showing the pattern of retrograde uptake of HRP following a large HRP injection into AI of the opposite hemisphere. (C) Darkfield photomicrograph through AI of cat showing the labeling of callosal terminals and the relationship of this pattern to the distribution of binaural neurons and neuron clusters recorded in an electrode penetration through that area. In the autoradiographs of both (A) and (C) the white dots are silver grains indicating relatively heavy callosal innervation. Arrows (A–E) point to microlesions produced during the experiment at points where the binaural interactions changed. The reconstructed penetration above shows the clustering of *summation* (EE) responses within areas of relatively heavy callosal innervation and *suppression* (EI) responses in regions of relatively sparse callosal innervation. From Imig and Brugge (1978).

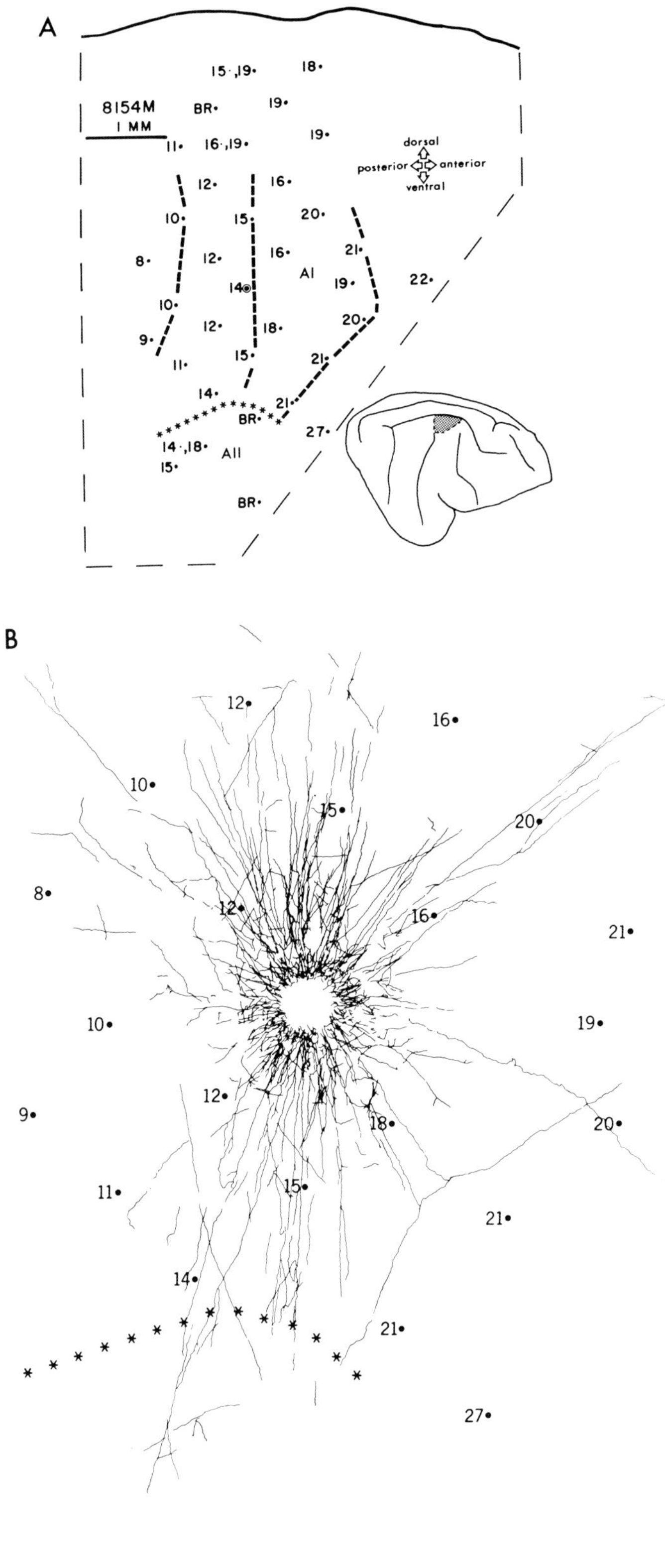
A
8154M
1 MM
dorsal
posterior
anterior
ventral
15·,19·
18·
BR·
19·
19·
11·
16·,19·
12·
16·
10·
15·
20·
8·
12·
16·
21·
AI
14
19·
22·
10·
12·
18·
20·
9·
15·
11·
21·
14·
21·
BR·
27·
14·,18·
AII
15·
BR·
B
12·
16·
10·
15·
20·
8·
12·
16·
21·
10·
19·
9·
12·
18·
20·
11·
15·
21·
14·
21·
27·

the ventral division along its dorsal, caudal, and ventral borders are relatively strong. Some neurons in the ventral nucleus, however, have been shown to project via axon collaterals to both fields P and AI and fields P and A (Imig and Reale, 1981a; Imig *et al.*, 1981).

Area AII receives its thalamic input from a different group of MGB subdivisions including the ventrolateral nucleus and the caudal part of the dorsal nucleus. It also receives from the medial nucleus, as does A and AI. These connectivity patterns are reciprocal in nature.

Imig and Morel (1984) have reconstructed the topography of the entire array of MGB neurons which give rise to the frequency-specific projections destined for fields A, AI, P, and VP. The array is composed of cells located in the ventral and medial divisions of the MGB as well as in the lateral posterior complex. In the ventral nucleus of the ventral division, the precise geometry of an array was found to be dependent on the specific target frequency representation in the cortex and was composed of both planar (sheetlike) and concentric components. This topography is consistent with previous descriptions of the distributions of thalamocortical cells projecting to fields A and AI (Colwell, 1977; Andersen *et al.*, 1980a; Merzenich *et al.*, 1982) and with the tonotopic organization of the ventral nucleus (Imig and Morel, 1985). Taking these data together, Imig and Morel (1984) propose that a single tonotopic representation exists within the ventral nucleus and can be modeled by three-dimensional arrangements of planar and concentric components.

A single cortical locus in either field AI or A having a restricted frequency representation receives input from neurons located in several separate cell aggregates of the ventral division of the MGB (Colwell, 1977; Andersen *et al.*, 1980a; Middlebrooks and Zook, 1983). Together these aggregates of thalamocortical projecting neurons are confined to sheetlike or slablike sectors of the ventral nucleus for a cortical locus representing frequencies near the middle of the acoustic spectrum. This geometry corresponds, in turn, to an MGB isofrequency lamina (Morel, 1980; Imig and Morel, 1985). Similarly, a single cortical locus in either field AI or A projects in a reciprocal fashion to these separate targets in the MGB isofrequency plane (Andersen *et al.*, 1980a). Middlebrooks and Zook (1983) have shown that two different cortical loci located along the same isofrequency strip in field AI receive input from many of the same aggregates of ventral nucleus cells distributed within an MGB isofrequency lamina. These observations are consistent with the findings of Dickson and Gerstein (1974) who recorded simultaneously from neighboring auditory cortical cells

Figure 13. Relationship between the CF map and intrinsic connectivity patterns in field AI of the cat. (A) Reconstructed CF map displayed on the surface of the right hemisphere. Stippling in the inset brain drawing shows the area in which the map was reconstructed. Recording sites are located at decimal points and CFs are in kHz. Thick dashed lines indicate isofrequency contours. Asterisks indicate the border between field AI and AII. BR indicates a response over a broad frequency range. The site of the iontophoretic injection of HRP on the 14-kHz isofrequency contour is shown by the bullseye. (B) Camera lucida tracing of all HRP-labeled processes located within a tangentially cut (90 μm) tissue section showing the orientation of the processes with respect to the orientation of the isofrequency contours. From Reale *et al.* (1983).

and analyzed the spontaneous and sound-evoked spike trains using cross-correlation techniques. Their data suggested that nearly all of the cortical neurons they studied that were located within about 250 μm of one another, and some others that were separated by as much as 1500 μm, received a shared input from some distant source, the most likely one being the MGB. Thus, there appear to be both divergent and convergent connections between similar portions of the thalamic and cortical frequency representations. Later we take up the evidence that points to the relationships between this convergent and divergent pattern of thalamic inputs and the binaural map of AI.

Two functionally separate pathways through the MGB to the cortex have been postulated to account for the physiology and connectivity patterns of the multiple cortical fields described so far in the cat (Reale and Imig, 1980; Andersen *et al.*, 1980a; Imig and Morel, 1983; Schreiner and Cynader, 1984). One pathway preserves cochleotopy through the brain stem, thalamus, and cortex. It would include the central nucleus of the inferior colliculus (ICC) and the pars lateralis and pars ovoidae, deep dorsal nucleus, and medial division of the MGB. Areas AI, A, P, and VP make up the cortical component of this system. A second pathway to the cortex routes the information from neurons that have poor frequency selectivity (e.g., the pericentral nucleus of the inferior colliculus), and, therefore, reside in regions where cochleotopy is blurred. Area AII, and perhaps portions of fields P, VP, and the "temporal" auditory field are the cortical components of this system.

6.1.3. Intrinsic Connections of Field AI

Neurons within AI may communicate with one another over local circuits that are intrinsic to AI. The patterns of these connections appear related to the AI cochleotopic organization. Reale *et al.* (1983) deposited iontophoretically small quantities of HRP into AI at a known CF locus within a larger tonotopic map and later traced the courses of neuronal processes that transported the enzyme (Fig. 13). When viewed in a plane tangential to the cortical surface, labeled processes radiated out asymmetrically from the injection site over distances of several millimeters. The heaviest concentrations of labeled fibers were along an axis parallel with the isofrequency line in which the injection was made. In Golgi preparations, neurons in the middle ectosylvian cortex have been shown to have dendritic processes with preferred orientations (Glaser *et al.*, 1979), and while in such preparations no physiological maps were available for comparison, the dorsoventral orientation of the dendritic fields is consistent with the orientation of isofrequency lines in AI. Thus, primary field neurons having the same or similar CFs have the potential of being preferentially interconnected. By the same token, disparate CF representations are unlikely to communicate strongly with one another over this intrinsic fiber system. Recall that along the isofrequency dimensions of AI there are alternating patches of neurons having different binaural properties. Consequently, at any frequency all binaural classes could be activated. It remains to be seen the extent to which the intracortical fiber system of AI interconnects these binaural patches.

6.1.4. Corticotectal Connections

The results of the most recent autoradiographic tracer studies of auditory corticotectal connections (FitzPatrick and Imig, 1978; Andersen *et al.*, 1980b,c) are consistent with and extend the earlier observations made with lesion-degeneration techniques (Diamond *et al.*, 1969; Rockel and Jones, 1973). Like the corticocortical and corticothalamic projections, these projections are divergent, convergent, and topographic. In both cat and monkey, corticotectal projections from AI loci terminate as sheets of terminals in the dorsomedial sector of the ipsilateral ICC. The contralateral ICC also receives AI input but to a lesser extent. The sheets of terminals are oriented roughly parallel to the isofrequency contours and to the flattened disk-shaped dendritic fields of neurons in the ventral part of the ICC (see Rockel and Jones, 1973; Merzenich and Reid, 1974; FitzPatrick, 1975). In the cat ICC, AI axons do not appear to project beyond the dorsomedial area of large multipolar neurons, whereas in the owl monkey the corticotectal terminal sheets extend well into the laminated ventral part. The topographic projection patterns are consistent with the cochleotopic organizations of AI and ICC which, considered in light of the tectothalamic and thalamocortical connectivity patterns (Andersen *et al.*, 1980b), means that cochleotopy is maintained throughout the ICC–MGB–AI–ICC loop.

AI of the cat, but apparently not of the monkey, also projects to the pericentral nucleus. The available evidence suggests that the topography and divergence–convergence observed for the AI–ICC pathway obtains here as well. Field A of cat may project sparsely to the dorsomedial area of the ipsilateral ICC. If it does so, then the same CF loci in AI and A may converge upon the same ICC target cells. AII does not send fibers to ICC but does project strongly upon the pericentral nucleus ipsilaterally. Field R of the monkey also projects to the dorsomedial sector of the ipsilateral ICC and to the pericentral and external nuclei.

6.1.5. Corticostriate Connections

Using the mapping-tracing methods outlined previously, Reale and Imig (1983) determined in the cat the patterns of projection from the four tonotopically organized fields (AI, A, P, and VP) to the basal ganglia. This subcortical projection system is topographically organized with respect to the relative positions of these four cortical fields. The neurons representing the full high-to-low-frequency axis in fields A and AI were found to project to dorsal striatal targets of the caudate nucleus and the putamen but not to the more ventral lateral amygdaloid nucleus. The field P projections show a low-to-middle-frequency representation in the caudate connections, a complete tonotopic representation in the connections to the putamen, and a middle-to-high-frequency representation in the projection to the lateral amygdaloid nucleus. In the VP projection, the full tonal spectrum is represented ventrally in the putamen and lateral amygdaloid nucleus but no connection is seen dorsally with the caudate nucleus. Moreover, double-tracer experiments in AI and P showed that the projections are also topographically organized with respect to the cortical frequency representation; projections from the high-frequency representations are

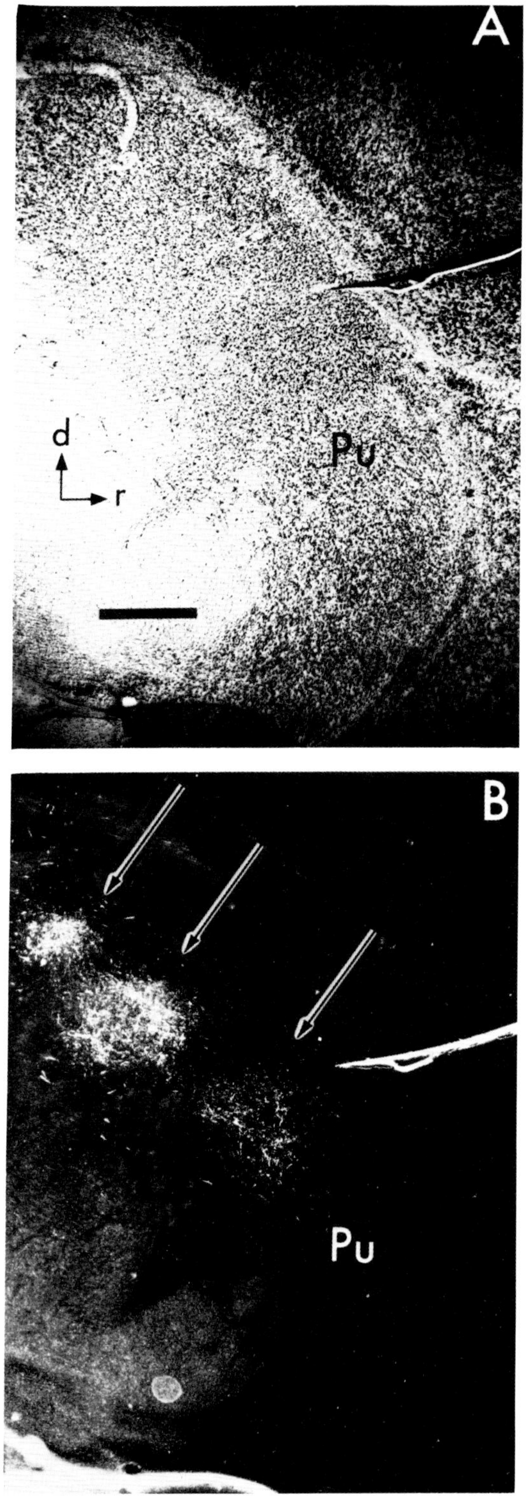

Figure 14. Typical configuration of labeled axonal terminals in the putamen (Pu) of the basal ganglia resulting from an injection of tritiated proline into an auditory cortical field. (A) Brightfield photograph of a Nissl-stained sagittally cut (50 μm) tissue section prepared for autoradiography. (B) Matching darkfield photograph of autoradiograph. Arrows point to divergent targets from a tritiated amino acid injection placed in the anterior (A) cortical field. Bar in (A) = 1 mm.

located more laterally in the striatum than are those from low-frequency areas. Like other auditory cortical projection systems, a restricted locus in each of the auditory fields sends axons to patches of striatal target neurons that are arrayed in a band (Fig. 14).

6.2. Relationships to Binaural Organization: Corticocortical and Corticothalamic Connections

Projections from AI of one hemisphere to field AI of the other via the corpus callosum have long been known to exhibit a complex topography (Ebner and Myers, 1965; Diamond *et al.*, 1968a; Pandya *et al.*, 1969; Karol and Pandya, 1971; FitzPatrick and Imig, 1980). Some of this complexity was resolved when Imig and Brugge (1978) combined electrophysiological and anatomical studies in the cat to show that the callosal projection pattern was closely related to the distribution of cells with different binaural properties (Fig. 12). Regions containing neurons that are excited by stimulation of one ear and inhibited by stimulation of the other make relatively sparse interhemispheric connections. From a previous discussion we saw that cells with these properties detect interaural intensity cues for sound localization and are usually sensitive to sound sources located in the opposite auditory hemifield. Cells that are excited by monaural stimulation of each ear, and which are often insensitive to interaural intensity differences, make relatively strong interhemispheric connections.

Fields A and P also project ipsilaterally upon AI in a complex fashion (Fig. 11). Imig and Reale (1981b) demonstrated that, like the callosal connections, the patchy ipsilateral projections are also related to binaural organization. In this case, however, target zones in AI receiving the most dense corticocortical projections are reversed from those of the callosal projection system.

The patchy and divergent–convergent patterns of thalamic inputs are also related to the binaural representation in AI (Middlebrooks and Zook, 1983). A cortical locus containing cells of a binaural class associated with strong interhemispheric connections and weak intrahemispheric connections derives its principal input from a topographically restricted and contiguous region of the ventral division of the MGB. By comparison, AI loci containing cells of the other binaural class (i.e., associated with weak interhemispheric and strong intrahemispheric connections) receive a differential input from three separate columns of medial geniculate cells which span the ventral division. Thus, it appears that the two major fiber systems that interconnect directly auditory cortical fields on the same and opposite hemisphere and link these fields with the thalamus are intimately related to representations of both frequency and binaurality.

ACKNOWLEDGMENTS. We wish to thank Jean Heinz and Carol Dizack for the artwork and Shirley Hunsaker for the photography. Supported by NIH Grants NS-12732 and HD-03352 and NSF Grant BNS-19893.

7. References

Abeles, M., and Goldstein, M. H., Jr., 1970, Functional architecture in cat primary auditory cortex: Columnar organization and organization according to depth, *J. Neurophysiol.* **33:**172–187.

Aitkin, L. M., Irvine, D. R. F., and Webster, W. R., 1984, Central neural mechanisms of hearing, in: *Handbook of Physiology,* Volume III (I. Darian-Smith, ed.), American Physiological Society, Washington, D.C., pp. 675–737.

Andersen, R. A., Knight, P. L., and Merzenich, M. M., 1980a, The thalamocortical and corticothalamic connections of AI, AII, and the anterior auditory field (AAF) in the cat: Evidence for two largely segregated systems of connections, *J. Comp. Neurol.* **194:**663–701.

Andersen, R. A., Roth, G. L., Aitkin, L. M., and Merzenich, M. M., 1980b, The efferent projections of the central nucleus and the pericentral nucleus of the inferior colliculus in the cat, *J. Comp. Neurol.* **194:**649–662.

Andersen, R. A., Snyder, R. L., and Merzenich, M. M., 1980c, The topographic organization of corticocollicular projections from physiologically defined loci in the AI, AII and anterior auditory cortical fields in the cat, *J. Comp. Neurol.* **191:**479–494.

Benson, D. A., and Teas, D. C., 1976, Single unit study of binaural interaction in the auditory cortex of the chinchilla, *Brain Res.* **103:**313–338.

Benson, D. A., Heinz, R. D., and Goldstein, M. H., Jr., 1979, Observations on unit activity in monkey auditory cortex and dorsolateral frontal cortex during a sound localization task, *Soc. Neurosci. Abstr.* **5:**16.

Brugge, J. F., 1975, Progress in neuroanatomy and neurophysiology of auditory cortex, in: *The Nervous System: Human Communication and Its Disorders* (D. B. Tower, ed.), Raven Press, New York, pp. 97–111.

Brugge, J. F., 1982, Auditory cortical areas in primates, in: *Cortical Sensory Organization,* Volume 3 (C. N. Woolsey, ed.), Humana Press, Clifton, N. J., pp. 59–70.

Brugge, J. F., and Imig, T. J., 1978, Some relationships of binaural response patterns of single neurons to cortical columns and interhemispheric connections of auditory area AI of cat cerebral cortex, in: *Evoked Electrical Activity in the Auditory Nervous System* (R. F. Naunton and C. Fernandez, eds.), Academic Press, New York, pp. 487–504.

Brugge, J. F., and Merzenich, M. M., 1973a, Patterns of activity of single neurons of the auditory cortex in monkey, in: *Basic Mechanisms in Hearing* (A. R. Moller, ed.), Academic Press, New York, pp. 745–772.

Brugge, J. F., and Merzenich, M. M., 1973b, Responses of neurons in auditory cortex of the macaque monkey to monaural and binaural stimulation, *J. Neurophysiol.* **36:**1138–1158.

Brugge, J. F., Dubrovsky, N. A., Aitkin, L. M., and Anderson, D. J., 1969, Sensitivity of single neurons in auditory cortex of cat to binaural tonal stimulation; effects of varying interaural time and intensity, *J. Neurophysiol.* **32:**1005–1024.

Brugge, J. F., Anderson, D. J., and Aitkin, L. M., 1970, Responses of neurons in the dorsal nucleus of the lateral lemniscus of cat to binaural tonal stimulation, *J. Neurophysiol.* **33:**441–458.

Calford, M. B., and Pettigrew, J. D., 1984, Frequency dependence of directional amplification of the cat's pinna, *Hearing Res.* **14:**13–19.

Calford, M. B., Webster, W. R., and Semple, M. M., 1983, Measurement of frequency selectivity of single neurons in the central auditory pathway, *Hearing Res.* **11:**395–401.

Casseday, J. H., Harting, J. K., and Diamond, I. T., 1976, Auditory pathways to the cortex in *Tupaia glis, J. Comp. Neurol.* **166:**303–340.

Colwell, S. A., 1977, Corticothalamic projections from physiologically defined loci within auditory cortex in the cat: Reciprocal structure in the medial geniculate body, Ph.D. thesis, University of California, San Francisco.

Creutzfeldt, O., Hellweg, F. C., and Schreiner, C., 1980, The thalamocortical transformation of responses to complex auditory stimuli, *Exp. Brain Res.* **39:**87–104.

De Ribaupierre, F., and Goldstein, M. H., Jr., 1972, Cortical coding of repetitive acoustic pulses, *Brain Res.* **48:**205–225.

Diamond, I. T., Jones, E. G., and Powell, T. P. S., 1968a, Interhemispheric fiber connections of the auditory cortex of the cat, *Brain Res.* **11:**177–193.

Diamond, I. T., Jones, E. G., and Powell, T. P. S., 1968b, The association connections of the auditory cortex of the cat, *Brain Res.* **11:**560–579.

Diamond, I. T., Jones, E. G., and Powell, T. P. S., 1969, The projection of auditory cortex upon the diencephalon and brain stem in the cat, *Brain Res.* **15:**305–340.

Dickson, J. W., and Gerstein, G. L., 1974, Interactions between neurons in auditory cortex of the cat, *J. Neurophysiol.* **37:**1239–1261.

Downman, C. B. B., Woolsey, C. N., and Lende, R. A., 1960, Auditory areas I, II and Ep: Cochlear representation, afferent paths and interconnections, *Bull. John Hopkins Hosp.* **106:** 127–142.

Ebner, F. F., and Myers, R. E., 1965, Distribution of corpus callosum and anterior commissure in cat and raccoon, *J. Comp. Neurol.* **124:**353–366.

Eisenman, L. M., 1974, Neural encoding of sound location: An electrophysiological study of auditory cortex (AI) of the cat using free-field stimuli, *Brain Res.* **75:**203–214.

Evans, E. F., 1968, Cortical representation, in: *Hearing Mechanisms in Vertebrates* (A. V. S. de Reuck and J. Knight, eds.), Little, Brown, Boston, pp. 272–295.

Evans, E. F., and Whitfield, I. C., 1964, Classification of unit responses in the auditory cortex of the unanesthetized and unrestrained cat, *J. Physiol (London)* **171:**470–493.

FitzPatrick, K. A., 1975, Cellular architecture and topographic organization of the inferior colliculus of the squirrel monkey, *J. Comp. Neurol.* **164:**185–208.

FitzPatrick, K. A., and Imig, T. J., 1978, Projections of auditory cortex upon the thalamus and midbrain in the owl monkey, *J. Comp. Neurol.* **177:**537–566.

FitzPatrick, K. A., and Imig, T. J., 1980, Auditory cortico-cortical connections in the owl monkey, *J. Comp. Neurol.* **192:**589–610.

FitzPatrick, K. A., and Imig, T. J., 1982, Organization of auditory connections: The primate auditory cortex, in: *Cortical Sensory Organization*, Volume 3 (C. N. Woolsey, ed.), Humana Press, Clifton, N.J., pp. 71–109.

FitzPatrick, K. A., Imig, T. J., and Reale, R. A., 1977, Thalamic projections to the posterior auditory field in the cat, *Soc. Neurosci. Abstr.* **3:**6.

Galaburda, A., and Sanides, F., 1980, Cytoarchitectonic organization of the human auditory cortex, *J. Comp. Neurol.* **190:**597–610.

Gates, G. R., and Aitkin, L. M., 1982, Auditory cortex in the marsupial possum *Trichosurus vulpecula*, *Hearing Res.* **7:**1–11.

Glaser, E. M., Van der Loos, H., and Gissler, M., 1979, Tangential orientation and spatial order in dendrites of cat auditory cortex: A computer microscope study of Golgi-impregnated material, *Exp. Brain Res.* **36:**411–431.

Glass, I., and Wollberg, Z., 1983, Responses of cells in the auditory cortex of awake squirrel monkeys to normal and reversed species-specific vocalizations, *Hearing Res.* **9:**27–33.

Goldberg, J. M., and Brown, P. B., 1969, Response of binaural neurons of dog superior olivary complex to dichotic tonal stimuli: Some physiological mechanisms of sound localization, *J. Neurophysiol.* **32:**613–636.

Goldberg, J. M., Diamond, I. T., and Neff, W. D., 1957, Auditory discrimination after ablation of temporal and insular cortex in cat, *Fed. Proc.* **16:**47.

Goldstein, M. H., Jr., 1968, Single unit studies of cortical coding of simple acoustic stimuli, in: *Physiological and Biochemical Aspects of Nervous Integration* (F. D. Carlson, ed.), Prentice–Hall, Englewood Cliffs, N.J., pp. 131–151.

Goldstein, M. H., Jr., and Abeles, M., 1975, Single unit activity of the auditory cortex, in: *Handbook of Sensory Physiology*, Volume 5, Part 2 (W. D. Keidel and W. D. Neff), Springer-Verlag, Berlin, pp. 199–218.

Goldstein, M. H., Jr., Hall, J. L., and Butterfield, B. O., 1968, Single-unit activity in the primary auditory cortex of unanesthetized cats, *J. Acoust. Soc. Am.* **43:**444–455.

Graybiel, A. M., 1973, The thalamo-cortical projection of the so-called posterior nuclear group: A study with anterograde degeneration methods in the cat, *Brain Res.* **49:**229–244.

Graybiel, A. M., 1974, Studies on the anatomical organization of posterior association cortex, in: *The Neurosciences: Third Study Program* (F. O. Schmitt and F. G. Worden, eds.), MIT Press, Cambridge, Mass., pp. 205–214.

Green, D. M., 1976, *An Introduction to Hearing*, Erlbaum, Hillsdale, N.J.

Hall, J. L., and Goldstein, M. H., Jr., 1968, Representation of binaural stimuli by single units in primary auditory cortex of unanesthetized cats, *J. Acoust. Soc. Am.* **43:**456–461.

Heath, C. J., and Jones, E. G., 1971, The anatomical organization of the suprasylvian gyrus of the cat, *Ergeb. Anat. Entwicklungsgesch.* **45:**1–64.

Hellweg, F. C., Koch, R., and Vollrath, M., 1977, Representation of the cochlea in the neocortex of guinea pigs, *Exp. Brain Res.* **29:**467–474.

Hind, J. E., Rose, J. E., Davies, P. W., Woolsey, C. N., Benjamin, R. M., Welker, W. I., and Thompson, R. F., 1960, Unit activity in the auditory cortex, in: *Neural Mechanisms of the Auditory and Vestibular Systems* (G. L. Rasmussen and W. Windle, eds.), Thomas, Springfield, Ill., pp. 201–210.

Imig, T. J., and Adrian, H. O., 1977, Binaural columns in the primary field (AI) of cat auditory cortex, *Brain Res.* **138:**241–257.

Imig, T. J., and Brugge, J. F., 1978, Sources and terminations of callosal axons related to binaural and frequency maps in primary auditory cortex of the cat, *J. Comp. Neurol.* **182:**637–660.

Imig, T. J., and Morel, A., 1983, Organization of the thalamocortical auditory system in the cat, *Annu. Rev. Neurosci.* **6:**95–120.

Imig, T. J., and Morel, A., 1984, Topographic and cytoarchitectonic organization of thalamic neurons related to their targets in low-, middle-, and high-frequency representations in cat auditory cortex, *J. Comp. Neurol.* **227:**511–539.

Imig, T. J., and Morel, A., 1985, Tonotopic organization in ventral nucleus of medial geniculate body in the cat, *J. Neurophysiol.* **53:**309–340.

Imig, T. J., and Reale, R. A., 1980, Patterns of cortico-cortical connections related to tonotopic maps in cat auditory cortex, *J. Comp. Neurol.* **192:**293–332.

Imig, T. J., and Reale, R. A., 1981a, Medial geniculate projections to auditory cortical fields A, AI, and P in the cat, *Soc. Neurosci. Abstr.* **7:**230.

Imig, T. J., and Reale, R. A., 1981b, Ipsilateral cortico-cortical projections related to binaural columns in cat primary auditory cortex, *J. Comp. Neurol.* **203:**1–14.

Imig, T. J., Ruggero, M. A., Kitzes, L. M., Javel, E., and Brugge, J. F., 1977, Organization of auditory cortex in the owl monkey (*Aotus trivirgatus*), *J. Comp. Neurol.* **171:**111–128.

Imig, T. J., Morel, A., and Reale, R. A., 1981, Organization of thalamic neurons projecting to auditory cortical fields in the cat, *International Seminar of Neuroscience*, Algiers, Algeria, pp. 62–63.

Irvine, D. R. F., 1980, Acoustic properties of neurons in posteromedial thalamus of cat, *J. Neurophysiol.* **43:**395–408.

Irvine, D. R. F., and Huebner, H., 1979, Acoustic response characteristics of neurons in non-specific areas of cat cerebral cortex, *J. Neurophysiol.* **42:**107–122.

Irvine, D. R. F., and Phillips, D. P., 1982, Polysensory "association" areas of the cerebral cortex: Organization of acoustic inputs in the cat, in: *Cortical Sensory Organization*, Volume 3 (C. N. Woolsey, ed.), Humana Press, Clifton, N.J., pp. 111–156.

Jenkins, W. M., and Merzenich, M. M., 1984, Role of cat primary auditory cortex for sound–localization behavior, *J. Neurophysiol.* **52:**819–847.

Kaas, J. H., Hall, W. C., and Diamond, I. T., 1972, Visual cortex of the grey squirrel (*Sciureus carolinensis*): Architectonic subdivisions and connections from visual thalamus, *J. Comp. Neurol.* **145:**273–306.

Karol, E. A., and Pandya, D. N., 1971, The distribution of the corpus callosum in the rhesus monkey, *Brain* **94:**471–486.

Kawamura, K., 1973, Corticocortical fiber connections of the cat cerebrum. I. The temporal region, *Brain Res.* **51:**1–21.

Khanna, S. M., and Leonard, D. G. B., 1982, Basilar membrane tuning in the cat cochlea, *Science* **215:**305–306.

Kitzes, L. M., Wrege, K. S., and Cassady, J. M., 1980, Patterns of responses of cortical cells to binaural stimulation, *J. Comp. Neurol.* **192:**455–472.

Knight, P. L., 1977, Representation of the cochlea within the anterior auditory field (AAF) of the cat, *Brain Res.* **130:**447–467.

Knudsen, E. I., and Konishi, M., 1978, A neural map of auditory space in the owl, *Science* **200:**795–797.

Kraus, N., and Disterhoft, J. F., 1982, Response plasticity of single neurons in rabbit auditory association cortex during tone-signalled learning, *Brain Res.* **246:**205–215.

Kuwada, S., and Yin, T. C., 1983, Binaural interaction in low-frequency neurons in inferior colliculus of the cat. I. Effects of long interaural delays, intensity, and repetition rate on interaural delay function, *J. Neurophysiol.* **50:**981–999.

McMullen, N. T., and Glaser, E. M., 1982, Tonotopic organization of rabbit auditory cortex, *Exp. Brain Res.* **75:**208–220.

Manley, J. A., and Müller-Preuss, P., 1978, Response variability in mammalian auditory cortex-objection to feature detection, *Fed. Proc.* **37:**2355–2359.

Masterton, R. B., and Imig, T. J., 1984, Neural mechanisms for sound localization, *Annu. Rev. Physiol.* **46**:275–287.

Merzenich, M. M., and Brugge, J. F., 1973, Representation of the cochlear partition on the superior temporal plane of the macaque monkey, *Brain Res.* **50**:275–296.

Merzenich, M. M., and Reid, M. D., 1974, Representation of the cochlea within the inferior colliculus of the cat, *Brain Res.* **77**:397–415.

Merzenich, M. M., Knight, P. L., and Roth, G. L., 1975, Representation of cochlea within primary auditory cortex in the cat, *J. Neurophysiol.* **38**:231–249.

Merzenich, M. M., Kaas, J. H., and Roth, G. L., 1976, Auditory cortex in the grey squirrel: Tonotopic organization and architectonic fields, *J. Comp. Neurol.* **166**:387–401.

Merzenich, M. M., Colwell, S. A., and Andersen, R. A., 1982, Auditory forebrain organization: Thalamocortical and corticothalamic connections in the cat, in: *Cortical Sensory Organization* (C. N. Woolsey, ed.), Humana Press, Clifton, N. J., pp. 43–57.

Middlebrooks, J. C., and Pettigrew, J. D., 1981, Functional classes of neurons in primary auditory cortex of the cat distinguished by sensitivity to sound location, *J. Neurosci.* **1**:107–120.

Middlebrooks, J. C., and Zook, J. M., 1983, Intrinsic organization of the cat's medial geniculate body identified by projections to binaural response-specific bands in the primary auditory cortex, *J. Neurosci.* **3**:203–224.

Middlebrooks, J. C., Dykes, R. W., and Merzenich, M. M., 1980, Binaural response-specific bands in primary auditory cortex (AI) of the cat: Topographical organization orthogonal to isofrequency contours, *Brain Res.* **181**:31–48.

Moore, D. R., and Irvine, D. R. F., 1979, A developmental study of the sound pressure transformation by the head of the cat, *Acta Oto-Laryngol.* **87**:434–440.

Morel, A., 1980, *Codage des sons dans le corps genouille median du chat: Evaluation de l'organisation tonotopique de ses differents noyaux*, Druckt Verlag, Zurich.

Morel, A., and Imig, T. J., 1984, Neurons in the tonotopic thalamus of the cat are topographically organized with respect to their target fields in auditory cortex, *Soc. Neurosci. Abstr.* **10**:244.

Newman, J. D., 1978, Perception of sounds used in species-specific communication: The auditory cortex and beyond, *J. Med. Primatol.* **7**:98–105.

Newman, J. D., and Symmes, D., 1979, Feature detection by single units in squirrel monkey auditory cortex, *Exp. Brain Res. Suppl.* **2**:140–145.

Newman, J. D., and Wollberg, Z., 1973, Responses of single neurons in the auditory cortex of squirrel monkeys to variants of a single cell type, *Exp. Neurol.* **40**:821–824.

Niimi, K., and Matsuoka, M., 1979, Thalamocortical organization of the auditory system in the cat studied by retrograde axonal transport of horseradish peroxidase, *Adv. Anat. Embryol. Cell Biol.* **57**:1–56.

Oliver, D. L., and Hall, W. C., 1978, The medial geniculate body of the tree shrew, *Tupaia glis*. II. Connections with the neocortex, *J. Comp. Neurol.* **182**:459–493.

Oonishi, S., and Katsuki, Y., 1965, Functional organization and integrative mechanism on the auditory cortex of the cat, *Jpn. J. Physiol.* **15**:342–365.

Orman, S. S., and Phillips, D. P., 1984, Binaural interactions of single neurons in posterior field of cat auditory cortex, *J. Neurophysiol.* **51**:1028–1039.

Pandya, D. N., and Sanides, F., 1973, Architectonic parcellation of the temporal operculum in rhesus monkey and its projection pattern, *Z. Anat. Entwicklungsgesch.* **139**:127–161.

Pandya, D. N., Hallet, M., and Mukherjee, S. K., 1969, Intra- and interhemispheric connections of the neocortical auditory system in the rhesus monkey, *Brain Res.* **14**:49–65.

Paula Barbosa, M. M., Feyo, P. B., and Sousa-Pinto, A., 1975, The association connexions of the suprasylvian fringe (SF) and other areas of the cat auditory cortex, *Exp. Brain Res.* **23**: 535–554.

Phillips, D. P., and Brugge, J. F., 1985, Progress in neurophysiology of sound localization, *Annu. Rev. Psychol.* **36**:245–274.

Phillips, D. P., and Irvine, D. R. F., 1979, Methodological considerations in mapping auditory cortex: Binaural columns in AI of cat, *Brain Res.* **169**:342–346.

Phillips, D. P., and Irvine, D. R. F., 1981a, Responses of single neurons in physiologically defined primary auditory cortex (AI) of the cat: Frequency tuning and responses to intensity, *J. Neurophysiol.* **45**:48–58.

Phillips, D. P., and Irvine, D. R. F., 1981b, Responses of single neurons in physiologically defined area AI of cat cerebral cortex: Sensitivity to interaural intensity, *Hearing Res.* **4**:299–307.

Phillips, D. P., and Irvine, D. R. F., 1982, Properties of single neurons in the anterior auditory field (AAF) of cat cerebral cortex, *Brain Res.* **248:**237–244.

Phillips, D. P., and Irvine, D. R. F., 1983, Some features of the binaural input to single neurons in physiologically defined area AI of cat cerebral cortex, *J. Neurophysiol.* **49:**383–395.

Phillips, D. P., and Orman, S. S., 1984, Responses of single neurons in posterior field of cat auditory cortex, *J. Neurophysiol.* **51:**1028–1039.

Phillips, D. P., Calford, M. B., Pettigrew, J. D., Aitkin, L. M., and Semple, M. N., 1982, Directionality of sound pressure transformation at the cat's pinna, *Hearing Res.* **8:**13–28.

Phillips, D. P., Orman, S. S., Musicant, A. D., Wilson, G. F., and Huang, C.-M., 1985, Primary auditory cortex in the cat: Classes of neurons distinguished by their responses to tones and noise, *Hearing Res.* (in press).

Poliak, S., 1932, Origin, course, termination, and internal organization of the auditory radiation, in: *The Main Afferent Fiber Systems of the Cerebral Cortex in Primates* (H. M. Evans and I. M. Thompson, eds.), University of California Press, Berkeley, pp. 81–104.

Reale, R. A., and Imig, T. J., 1980, Tonotopic organization in auditory cortex of the cat, *J. Comp. Neurol.* **192:**265–291.

Reale, R. A., and Imig, T. J., 1983, Auditory cortical field projections to the basal ganglia of the cat, *Neuroscience* **8:**67–86.

Reale, R. A., Brugge, J. F., and Feng, J. Z., 1983, Geometry and orientation of neuronal processes in cat primary auditory cortex (AI) related to characteristic-frequency maps, *Proc. Natl. Acad. Sci. USA* **80:**5449–5453.

Rhode, W. S., 1978, Some observations on cochlear mechanics, *J. Acoust. Soc. Am.* **64:**158–176.

Rockel, A. J., and Jones, E. G., 1973, The neuronal organization of the inferior colliculus of the adult cat. I. The central nucleus, *J. Comp. Neurol.* **147:**11–60.

Rose, J. E., 1949, The cellular structure of the auditory region of the cat, *J. Comp. Neurol.* **19:**409–439.

Rose, J. E., and Woolsey, C. N., 1949, The relations of thalamic connections, cellular structure and evocable electrical activity in the auditory region of the cat, *J. Comp. Neurol.* **91:**441.

Rose, J. E., and Woolsey, C. N., 1958, Cortical connections and functional organization of thalamic auditory system in the cat, in: *Biological and Biochemical Bases of Behavior* (H. F. Harlow and C. N. Woolsey, eds.), University of Wisconsin Press, Madison, pp. 127–150.

Rose, J. E., Gross, N. B., Geisler, C. D., and Hind, J. E., 1966, Some neural mechanisms in the inferior colliculus of the cat which may be relevant to localization of a sound source, *J. Neurophysiol.* **29:**288–314.

Schreiner, C. E., and Cynader, M. S., 1984, Basic functional organization of second auditory cortical field (AII) of the cat, *J. Neurophysiol.* **51:**1284–1305.

Seldon, H. L., 1981, Structure of human auditory cortex. I. Cytoarchitectonics and dendritic distributions, *Brain Res.* **229:**277–294.

Sellick, P. M., Patuzzi, R., and Johnstone, B. M., 1982, Measurement of basilar membrane motion in the guinea pig using the Mossbauer technique, *J. Acoust. Soc. Am.* **72:**131–141.

Seltzer, B., and Pandya, D. N., 1978, Afferent cortical connections and architectonics of the superior temporal sulcus and surrounding cortex in the rhesus monkey, *Brain Res.* **149:**1–24.

Sousa-Pinto, A., 1973, The structure of the first auditory cortex (AI) of the cat. I. Light microscopic observations on its structure, *Arch. Ital. Biol.* **111:**112–137.

Sovijarvi, A. R. A., and Hyvarinen, J., 1974, Auditory cortical neurons in the cat sensitive to the direction of sound source movement, *Brain Res.* **73:**455–471.

Steinschneider, M., Arezzo, J., and Vaughan, H. G. J., 1982, Speech evoked activity in the auditory radiations and cortex of the awake monkey, *Brain Res.* **252:**353–365.

Suga, N., 1977, Amplitude spectrum representation in the Doppler-shifted-CF processing area of the auditory cortex of the mustache bat, *Science* **196:**64–67.

Symmes, D., 1974, Discrimination of intermittent noise by macaques following lesions of the temporal lobe, *Exp. Neurol.* **16:**201–214.

Thompson, R. F., and Sindberg, R. M., 1960, Auditory response fields in association and motor cortex of cat, *J. Neurophysiol.* **23:**87–105.

Tunturi, A. R., 1960, Anatomy and physiology of the auditory cortex, in: *Neural Mechanisms of the Auditory and Vestibular Systems* (G. L. Rasmussen and W. F. Windle, eds.), Thomas, Springfield, Ill., pp. 181–200.

von Economo, C., 1929, *The Cytoarchitectonics of the Human Cerebral Cortex,* Oxford University Press, London.

Winer, J. A., Diamond, I. T., and Raczkowski, D., 1977, Subdivisions of the auditory cortex of the cat: The retrograde transport of horseradish peroxidase to the medial geniculate body and posterior thalamic nuclei, *J. Comp. Neurol.* **176:**387–418.

Wollberg, Z., and Newman, J. D., 1972, Auditory cortex of squirrel monkey: Response patterns of single cells to species specific vocalizations, *Science* **175:**212–214.

Woolsey, C. N., 1960, Organization of auditory cortical system: A review and a synthesis, in: *Neural Mechanisms of the Auditory and Vestibular Systems* (G. Rasmussen and W. Windle, eds.), Thomas, Springfield, Ill., pp. 165–180.

Woolsey, C. N., 1961, Organization of cortical auditory system, in: *Sensory Communication* (W. A. Rosenblith, ed.), MIT Press, Cambridge, Mass., pp. 235–257.

Woolsey, C. N., 1971, Tonotopic organization of the auditory cortex, in: *Physiology of the Auditory System* (M. B. Sachs, ed.), National Educational Consultants, Baltimore, pp. 271–282.

Woolsey, C. N., and Walzl, E. M., 1942, Topical projection of nerve fibers from local regions of the cochlea to the cerebral cortex of the cat, *Bull. Johns Hopkins Hosp.* **71:**315–344.

Woolsey, C. N., and Walzl, E. M., 1982, Cortical auditory area of *Macaca mulatta* and its relation to the second somatic sensory area (Sm II): Determination by electrical excitation of auditory nerve fibers in the spiral osseous lamina and by click stimulation, in: *Cortical Sensory Organization,* Volume 3 (C. N. Woolsey, ed.), Humana Press, Clifton, N.J., pp. 231–256.

Yin, T. C. T., and Kuwada, S., 1983a, Binaural interaction in low-frequency neurons in inferior colliculus of the cat. II. Effects of changing rate and direction of interaural phase, *J. Neurophysiol.* **50:**1000–1019.

Yin, T. C. T., and Kuwada, S., 1983b, Binaural interaction in low-frequency neurons in inferior colliculus of the cat. III. Effects of changing frequency, *J. Neurophysiol.* **50:**1020–1042.

Yin, T. C. T., and Kuwada, S., 1984, Neuronal mechanisms of binaural interaction, in: *Dynamic Aspects of Neocortical Function* (G. M. Edelman, W. M. Cowan, and W. E. Gall, eds.), Wiley, New York, pp. 263–313.

7

The Anatomy of Speech Perception
Human Auditory Cortex

H. LEE SELDON

1. Introductory Remarks

Language comprehension is an everyday phenomenon which we all accept, yet nobody knows what it means in terms of brain activity. This review will not resolve the matter. It purports only to mention the methods which are used in approaching the problem and to summarize the results to date. Since I try to synthesize clinical, histological, and neurolinguistic findings, as well as relevant results from animal experiments, the treatment of each individual field of research must of necessity be very compact. Due to my own training and inclinations, the emphasis is on the anatomy and physiology of language "centers" in the brain, and, as the title indicates, mainly on the sensory language regions. In these respects this review differs from other recent reviews (Caramazza and Berndt, 1978; Zurif and Blumstein, 1978; Saffran, 1982) which were written from the neurolinguistic point of view and make a more "high-level" attack on the problem of language. In the speculative part of the chapter I shall attempt to interpret neurolinguistic findings in anatomical/physiological terms. The models proposed there are not intended to be final explanations of cortical speech

---

H. LEE SELDON • Physiology Institute, Free University of Berlin, D-1000 Berlin 33, Federal Republic of Germany.

processing but ideas which will be modified by future research. Finally, I shall point out the obvious gaps in our current knowledge—for future investigation.

Due to the frequency of use of the expressions "left hemisphere" and "right hemisphere," they will be abbreviated LH and RH, respectively.

If I have made any mistakes in the interpretation of published findings, I beg to be corrected.

2. How and Where Do We Locate Speech Perception in the Brain?

In order to introduce some terminology immediately, let me state here that we shall be talking mostly about three areas in the cortex of the left temporal lobe of right-handed people (Fig. 1). Area TC, using the nomenclature of von Economo and Koskinas (1925), is generally recognized as the primary auditory cortex. It lies on the first transverse gyrus. Area TA1, covering the posterior half of the superior temporal gyrus, is the "secondary" auditory region and comprises part or all of the so-called Wernicke speech region. Area TB covers most of the planum temporale and has an as yet undeciphered auditory function. These are the classical auditory regions and are the topic of our discussion. If we start to examine those cortical areas which respond to auditory signals, we rapidly discover that large expanses of the cerebral cortex are involved.

Fig. 1

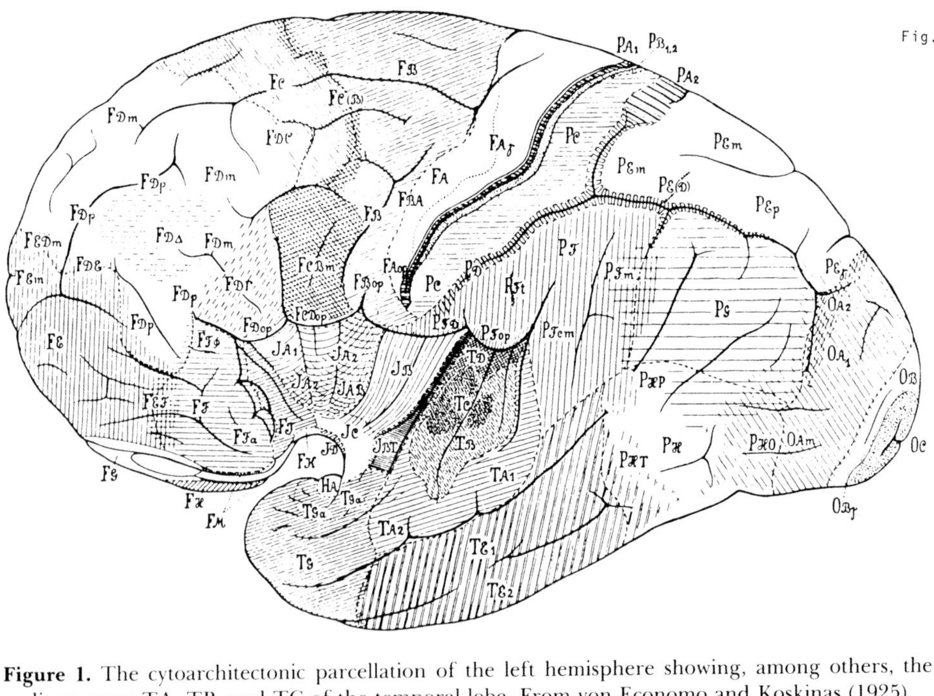

Figure 1. The cytoarchitectonic parcellation of the left hemisphere showing, among others, the auditory areas TA, TB, and TC of the temporal lobe. From von Economo and Koskinas (1925).

2.1. Morphological Methods

Since the classical papers of Broca (1865) and Wernicke (1874) the basic method for "localizing" language functions has remained the same—to correlate specific language deficits with specific brain lesions. The techniques used and the conclusions reached have varied. The earliest findings were interpreted as arguments for a strict localization of language functions. Later, von Monakow (1914) made an extensive review of autopsy findings, categorizing lesions as those which fit well with the classical aphasic syndromes and those which fit less well. For example, he disputed the claim that lesions of the classical Broca area were solely responsible for motor aphasia, pointing out that often the "main lesion" lay in the white matter beneath the Broca area and that lesions even of the basal ganglia could cause a motor aphasia syndrome. He and later Goldstein (1927) warned against a too strict localization. Luria (1966), in his treatise on higher cortical functions, also did not attempt a more precise language localization than his predecessors.

More recent attempts at morphological localization of language functions have concentrated on lesion analysis with computer tomography (CT), for instance by Naeser and her co-workers (1978, 1981a,b, 1982), the Damasios and their group (1979, 1980, 1982), Kertesz *et al.* (1979), Mazzochi and Vignolo (1979), and Blunk *et al.* (1981). Although the linguistic analyses of the language deficits (aphasia tests) are different, the CT evaluation is similar in all cases. The identified lesions are plotted onto "standard" brain sections, and the lesioned points are "summed up" for all patients with a given syndrome. The advantages of the CT method are its relatively high spatial resolution (modern CT scanners have a 256 × 256 point matrix for each 2- to 8-mm-thick "brain slice") and its capacity for quantification of tissue density. These allow a good localization of brain structures and a semirigorous definition of a "lesion." The results to date have been consistent with the classical descriptions. Speech perception disturbances are localized in the classical Wernicke area; other varieties of aphasia are assigned to other brain regions. Some research groups have also found effects of subcortical (basal ganglia) lesions on language perception and production, confirming the reports of Marie (1908, cited by von Monakow), von Monakow (1914), and Goldstein (1927).

The lesion analysis methods have some disadvantages (Caplan, 1981). Seen morphologically, they are susceptible to "false negative" and "false positive" results. The false negatives are due to the inability of the CT scan to detect reduced metabolism or activity in structurally intact regions. Considering the degree of connectivity among brain centers, a structural lesion in one can easily affect the activity in others, which may in turn be the real cause of an observed speech defect. False positive findings may arise from the nature of lesions themselves, which are mostly of vascular origin. Infarcts do not affect only regions directly involved in language, but also all those regions distal to the infarct in the arterial net. An example of this may be the reported involvement of the white matter subcortical to Broca's region in motor aphasia (von Monakow, 1914; Blunk *et al.*, 1981). Axons running through here may convey language-related information, but it is unlikely that they synthesize it. However, they are supplied by the same arterioles that supply the Broca area and anterior insula (Szikla *et al.*, 1977) and thus are likely to be involved in any infarct affecting the latter areas.

2.2. Functional Methods

The localization methods based on brain function vary from the very coarse to possibly very exact ones. They differ from lesion analysis in that they can be used to localize brain activity during language perception and production in "normal" subjects.

The Amytal injection technique (Wada and Rasmussen, 1960) temporarily anesthetizes one hemisphere, allowing the clinician to determine whether language capacity has thereby been eliminated or not. A more exact localization than the hemisphere is not possible. Since the intracarotid injections are also technically complicated, the method is rarely used.

A few attempts at localization have been made using radioisotopes since some radionuclides are preferably accumulated at infarct or tumor sites (Benson and Patten, 1967; Kertesz *et al.*, 1977). The results correspond to those from CT analysis, but the resolution of the radioactivity scan is somewhat poorer.

A major tool in the hemispheric localization of speech perception has been the dichotic listening test developed by Kimura (1967). In this test, differing acoustic stimuli are simultaneously applied to both ears, and the subject reports on what he or she has heard. The right-ear advantage observed for speech stimuli leads to the inference of LH localization for language, due to the stronger coupling of each ear with the contralateral hemisphere. Although the technique does not allow a more exact localization, it does offer the great advantage of enabling one to dissect speech stimuli into their component parts and to test the parts. This has been done, for example, by Studdert-Kennedy and Shankweiler (1970), Cutting (1974), and Blumstein *et al.* (1977b). With dichotic listening tests it has been inferred that vowel sounds, which last 100–300 msec, are perceived as a continuous spectrum and equally well in both hemispheres, whereas consonants (40–60 msec) are preferably processed in the dominant, language hemisphere. Interestingly, von Monakow (1914) also mentioned that aphasic patients have greater difficulties with consonant than with vowel perception. Furthermore, consonants are perceived categorically, i.e., consonant-like sounds are automatically assigned to some "recognized" stimulus even if the actual stimulus is on the border between two consonants in terms of its acoustic features. The same is true for vowel-like stimuli which are truncated to consonant length. This tells us that the LH auditory areas are processing very short stimuli, which the corresponding RH areas apparently cannot do. The LH also seems to distinguish speech from nonspeech stimuli. The right-ear advantage is, however, very weak for many phonemes during dichotic presentation, implying some RH participation in speech processing (see review by Hécaen, 1978).

Similar tests in macaque monkeys uncovered a right-ear advantage for processing species-specific vocalizations but no lateralization for "foreign" calls (Petersen *et al.*, 1978).

A refinement of the dichotic listening test is the use of recorded acoustic evoked responses (AER) to trace the relative activity of the hemispheres or cortical lobes. The techniques have been reviewed by Springer and Deutsch (1981) and Galambos and Hillyard (1981). Early results were summarized by Desmedt (1977) and Thatcher (1977). Since then, a small plethora of studies has appeared.

Such recordings are in principle an excellent means of following brain activity but have two major drawbacks. The spatial resolution has been very poor, being limited to cortical lobes. To acquire sufficient data for a statistical analysis and to achieve the necessary signal-to-noise ratio, numerous stimulus repetitions are necessary, which in turn can themselves influence the brain's reaction. This handicap has been overcome to some extent by using random stimulus sequences and random intervals between stimuli.

There are newer evaluation methods which greatly increase the spatial resolution of the AER, but which have not yet been systematically applied to speech-related analyses. These include BEAM [brain electrical activity mapping (Duffy, 1982)] and the dipole localization method (Wood, 1982). BEAM has been successfully used to localize abnormal brain activity in cases of dyslexia.

The spatial resolution may also be greatly increased with the use of magnetic encephalography (MEG), with which magnetic fields, which are much weaker and more locally bound than the electric ones, are detected. MEG is still being developed and has a number of technical problems to surmount. For example, a recording unit can support only a single, helium-supercooled detector. The problems of stimulus repetition and averaging which beset the AER method also apply to the MEG. The MEG method has been applied to the auditory cortex, and the first results on the tonotopic organization of the primary cortex are consistent with previous findings (Romani *et al.*, 1982).

Some knowledge of language localization in the brain has been derived during or after invasive brain operations. Penfield and his co-workers made extensive series of electrical stimuli on the cortex during operations and recorded the patients' responses (Penfield and Pernot, 1963). Their localization of auditory perception was to the accepted superior temporal region, yet they described occasional very complex auditory "experiences" upon stimulation of this region, hinting at the extreme complexity of auditory memory. Also along this line, Ojemann and his co-workers have included stimulation of subcortical structures such as the pulvinar and ventrolateral thalamus (summarized by Ojemann, 1976). They found that stimuli there interrupt speech, which, considering the central position of the thalamus in sensory and motor systems, is perhaps not surprising. [Brown (1975) postulated that the pulvinar may mediate between the sensory and motor speech regions, but contended that pulvinar lesions have no effect on aphasia.] Ojemann and Mateer (1979) stimulated cortical points during short tests of speech perception and production. They could discern disjoint areas for phoneme identification, verbal memory, and even syntax. In a recent summary, Ojemann (1983) described an attention system centered around the thalamus, a perisylvian system for sequencing, and around that a short-term memory system. He did, however, admit that the localization of functions is variable with IQ and with the native language of the subject. Although such work is restricted to isolated cases, the method does afford the most direct access to the human brain and will certainly continue to be used in the future.

Another line of work has concerned the analysis of speech in split-brain patients (Gazzaniga and Sperry, 1967). Besides confirming the location of the speech centers in the LH, this research has concentrated on the language capabilities of the RH (Nebes, 1978; Sidtis *et al.*, 1981). Due to the small number and great individual variability of the patients studied, these capabilities have been quite varied. The results cannot be applied ipso facto to "normal" brain function, since the subjects of these analyses are somewhat extraordinary persons.

The two methods which currently best register local cerebral activity clinically are the regional cerebral blood flow (rCBF or ^{133}Xe) method and positron emission tomography (PET). Only the ^{133}Xe technique has been extensively used for language localization until now. It is clinically easy to use and is based on the direct and substantiated correlation between local brain activity and local blood flow. ^{133}Xe is injected into a carotid artery or intravenously or is inhaled and is distributed through the brain according to the local blood volume. An array of up to 254 collimated detectors over one hemisphere registers the spatial distribution of radioactivity. The results using this method indicate generally an unexpectedly broad participation of brain centers in language perception and production. During verbal stimulation, Nishizawa *et al.* (1982) found an overall increase of 10% in LH blood flow and further local increases in the superior temporal, prefrontal, and orbitofrontal regions and frontal eye fields. Interestingly, during nonverbal auditory discrimination (between meaningless pairs of tone rhythms) the situation is reversed—parts of the RH, including some temporal, parietal, and frontal regions, show higher activity than the corresponding left regions (Roland *et al.*, 1981). The task of learning words (presented visually) leads to a marked increase in prefrontal brain activity (Maximilian *et al.*, 1978). The simple task of producing words (counting 1–20) turns out to involve activation of the face area of the primary motor and somatosensory cortex, the superior temporal region, and the supplementary motor region—all of these in both hemispheres (Larsen *et al.*, 1978). rCBF studies have also implicated prefrontal regions in language processing (Ingvar, 1982). With these indications of extensive brain participation in language processing, one must keep in mind, though, that the rCBF does not distinguish excitatory from inhibitory activity.

PET is another possible method for localization of metabolic or activity changes in the brain. For this method, short-half-life, positron-emitting radioisotopes are inhaled directly or incorporated into a metabolic substrate such as deoxyglucose, which is injected intravenously. Neurons take up the molecules in amounts proportional to their energy requirements, i.e., proportional to their activity (Frackowiak *et al.*, 1980). After nuclear decay, the emitted positron wanders a few millimeters, depending on its energy, before colliding with an electron to yield two equal and opposite γ-rays. Since the γ-rays do not arise at the source of the radioactive decay, there is a theoretical and practical limit on the resolution of this method. From the spatial distribution of the detected rays, their source density can be reconstructed and converted into metabolic terms according to the model of Sokoloff *et al.* (1977). Using PET and verbal stimuli, Mazziota *et al.* (1982) demonstrated increased activity bilaterally in the transverse gyri and posterior temporal lobe, as well as diffuse LH changes. With nonverbal stimuli there was a bilateral increase in the parietotemporal regions and a diffuse frontotemporal increase, more noticeable in the RH. In a series of aphasic patients,

Metter *et al.* (1983) found metabolic depression in the thalamus and caudate nucleus, regardless of the location of the brain lesions. Since the only symptom shared by all patients was a severe verbal memory defect, Metter *et al.* inferred a correlation between verbal memory and thalamic and caudate nucleus activity.

3. What Do the Sensory Auditory Regions Look Like?

3.1. Histological Methods

Parallel to the clinical attempts at localizing language comprehension, there have been, over the past century, a few histological studies aimed at discerning and distinguishing the cortical areas involved. Let us first look briefly at the history of the studies before going into detail on the microscopic structure of the auditory regions.

Flechsig (1896), following his theory that primary sensory projections are the first to be myelinated, described the transverse or Heschl gyri as the primary auditory cortex. Campbell (1905) described the myelin picture in more detail. Brodmann (1909), in his survey of cortical cytoarchitectonics, included a brief description of this region, dividing it into area 41, the primary auditory cortex, area 22, the secondary or speech cortex, and area 42, a mixture. But the definitive cytoarchitectonic studies remained to be done by von Economo and Koskinas (1925) and von Economo and Horn (1930). They also remarked on hemispheric differences in the macroscopic structure and extent (borders) of the three areas involved, TC, TA, and TB, corresponding respectively to Brodmann's 41, 22, and 42. These differences have been elaborated more recently by Geschwind and Levitsky (1968), Witelson and Pallie (1973), Galaburda *et al.* (1978a,b), and Galaburda and Sanides (1980). For a more detailed view of neuronal structure, other stains such as Golgi's silver impregnation are necessary. Ramón y Cajal (1902), the great master of silver impregnation, used this technique to exhaustively describe the dendritic structure of cortical auditory neurons and of afferent plexuses. His booklet remains the definitive work on the types and distribution of human auditory cortical neurons, on their dendritic trees and axonal projections. His only omission was to clearly delineate the area he studied—he described it as the "superior temporal gyrus." Although Conel (1939–1967) included Golgi studies (as well as Nissl ones) of this region in his survey of postnatal cortical development, his impregnations yielded considerably less detail than Ramón y Cajal's. Recently, Seldon (1981a,b, 1982) has described hemispheric differences in the dendritic trees, especially of pyramidal neurons, between corresponding areas. Actually, the Nissl and Golgi methods must be combined to get an idea of the overlap and interconnectivity between dendritic and axonal fields in a cortical area, and this will be the basis of the discussion in Section 3.3.

The histological examination of the human auditory cortex is hindered by the lack of applicability of most of the elegant methods used in animal experiments. There is correspondingly much more guesswork involved, at least at present.

3.2. General Neuron Structure from Golgi Studies

A few remarks on the structure of auditory cortical neurons are in order before we delve into the details of the auditory areas. Since the structures themselves have been thoroughly depicted by Ramón y Cajal (1902), we can restrict ourselves more to the implications of different patterns. We should view the neocortex as a columnarly organized, laminated plate in which information, represented by the electrical activity of cortical afferents and intracortical axons, is spatially and temporally distributed. The important point is that we must view the cortical neurons as units which make a spatial and temporal sampling of the information available. The dendrites (mostly) of a neuron sample this activity in a way depending of course on their geometry and synapse distribution (Szentágothai, 1978, 1979). For example, a synapse near the cell soma will have a greater influence on the outcome than one at the distal end of a dendritic tree (Rall, 1964). However, since nothing is currently known of the synapse distribution of human auditory cortical neurons, we must restrict our considerations to their geometry.

At this point we have to make an aside about the basic statement underlying this view of cortical information processing, namely that a neuron samples axonal activity within its dendritic field. The question is, how does the sampling occur? The extreme of one view would be that a dendrite synapses randomly with any axon within reach. Thus, the neuronal response would reflect the level of activity in the axons within the dendritic space. The random model would explain physiological precision on a mass-action basis. (An example of "random" connections is that pyramidal cells within the arborization of cortical afferents receive direct afferent synapses whereas other pyramidal cells do not.) The opposite extreme would propose exact specificity in synapses, for example, that a neuron is driven by a single axon and is oblivious to the activity of all others in its dendritic space. Proponents of the high-specificity model would point to specific connections between types of cells or among cortical laminae as an explanation of precise physiological neuronal responses. We make a very conservative assumption when we say that a neuron samples activity in its dendritic space, i.e., we say nothing about the specificity of sampling. [For recent discussions of this problem in animal sensory cortices, see Davis and Sterling (1979) and Hendry and Jones (1983).] The problem of specificity appears later in estimates of convergence and divergence.

On the basis of their dendritic trees we can describe several types (and subtypes) of human auditory neurons and comment on their ways of information sampling (Ramón y Cajal, 1902; Seldon, 1981a). For our purposes a rough classification according to dendritic geometry suffices. We would not yet benefit from a detailed morphological differentiation, since too little is known about specific neuronal connections to allow us to interpret minor anatomical differences.

Pyramidal cells are found throughout laminae II–VI, with the highest densities in II–III and V. In sagittal sections the basal dendrites of pyramidal neurons sometimes show flat branchings within a single cortical lamina, especially in layers IV and V. Other pyramidal cells have more vertically oriented basal dendrites (Fig. 2). Still others show a peculiar organization into columns, with the horizontal primary dendrites splitting into clearly vertical branches at some distance from the soma, apparently sampling disjoint columns of information. In the tangential plane the primary basal dendrites often exit from the soma in a cross pattern (Fig. 3). In some cases, especially in the middle layers, the higher-order branches continue in roughly the same orientation, but this is not the general rule. The tangential area occupied by the basal dendrites may be round (in the upper layers) or slightly elliptical (in the lower layers). The radius varies within each layer, yet is often smaller (75–100 μm) in layer II and increases in lower layers up to 200–250 μm. The spatial density of dendrites also varies markedly. The apical dendrites have not been so well studied. Ramón y Cajal noted that pyramidal cells in all layers send apical dendrites up to layer I. Tangentially, they do not expand so far as the basal dendrites, except for cells in layer II, in which the radius of the apical branches in layer I is 100–300 μm. The axon arborizations of the pyramidal cells, as well as of all other neurons in the human auditory cortex, have been as yet poorly quantified as regards their vertical and tangential dimensions. The axons show the classical structure, descending to the white matter while giving off recurrent collaterals. Ramón y Cajal (1902) described a few pyramidal cells whose axons "dissolve completely into arced collaterals." According to recent findings in tree shrews, the recurrent collaterals may contact not the immediate surroundings, but rather specific, disjoint groups of neurons (Rockland *et al.*, 1982). (Tree shrews admittedly belong to Tupaiidae, the lowest family of primates, so these findings remain to be confirmed in higher-order primates.)

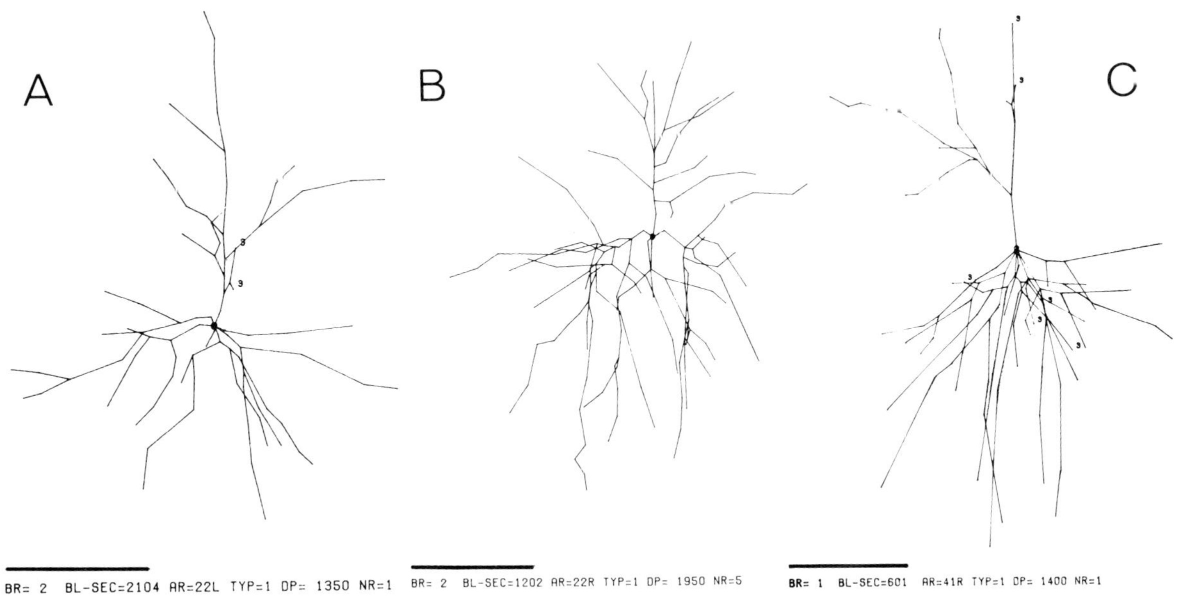

Figure 2. Sagittal view of pyramidal neurons showing various distributions of basal dendrites. BR = brain number; BL-SEC = block and section number; AR = area, using Brodmann's nomenclature; TYP = neuron type; DP = depth under the pial surface, NR = serial number; "3" at the end of a dendrite signifies a dendrite projecting out of the section. Bars = 100 μm. (B) from Seldon (1981a) with permission.

Smooth stellate cells were described by Ramón y Cajal (1902) in layers II–VI although his definition of smooth stellates was equivocal. He classified stellate cells according to their axon arborizations rather than their spine content. Smooth stellates, generally acknowledged as inhibitory neurons [in cat and monkey (LeVay, 1973; see also Peters and Saint Marie, Volume 1 of this treatise)], show poorly organized dendritic trees (Fig. 4). They have a vertical (radial) extent up to 450

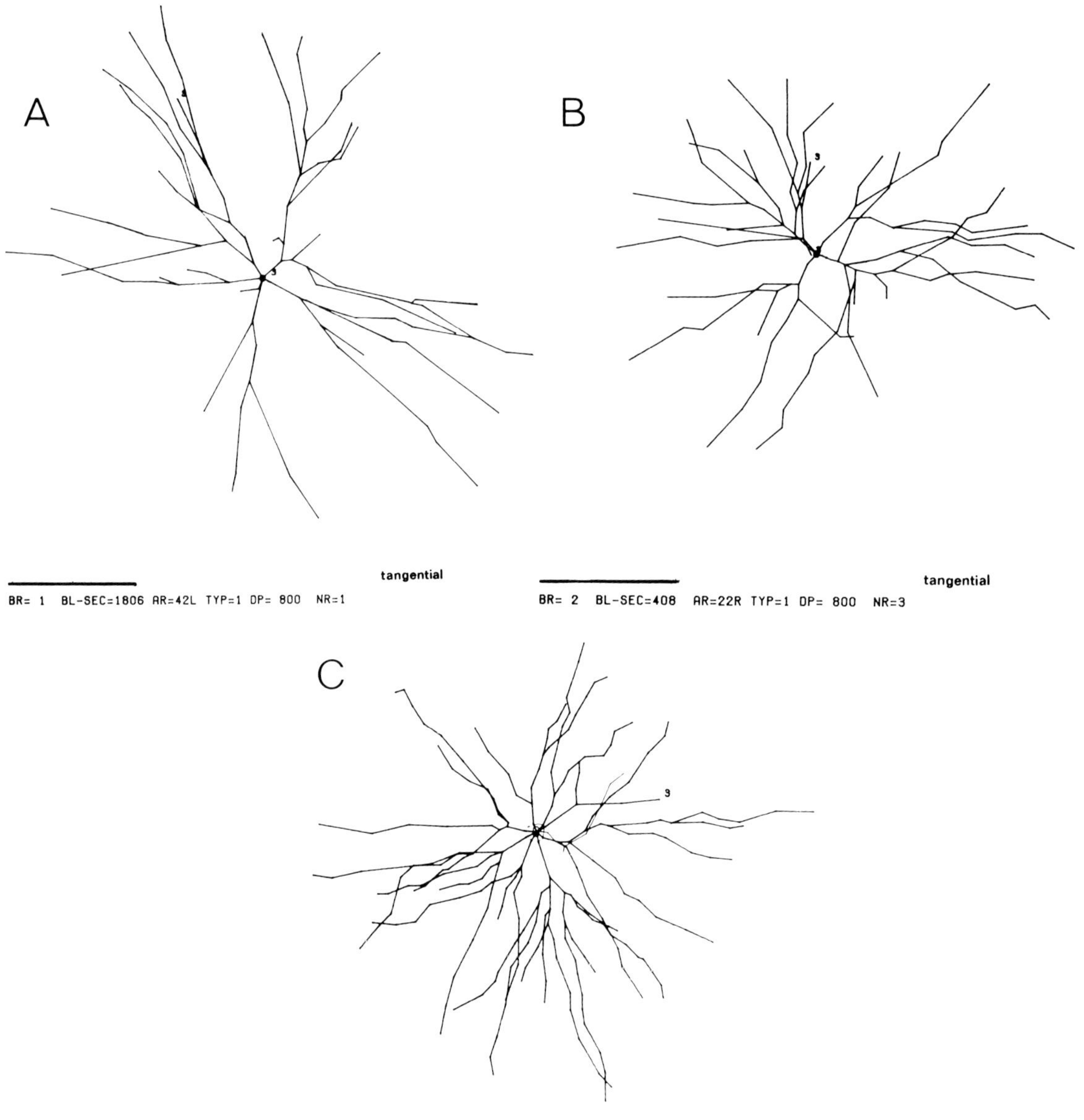

Figure 3. Tangential view of basal dendrites of pyramidal neurons. "3" in (B) is a branch of the apical dendrite, which projects toward the reader. (B) from Seldon (1981a) with permission. Labels as in Fig. 2.

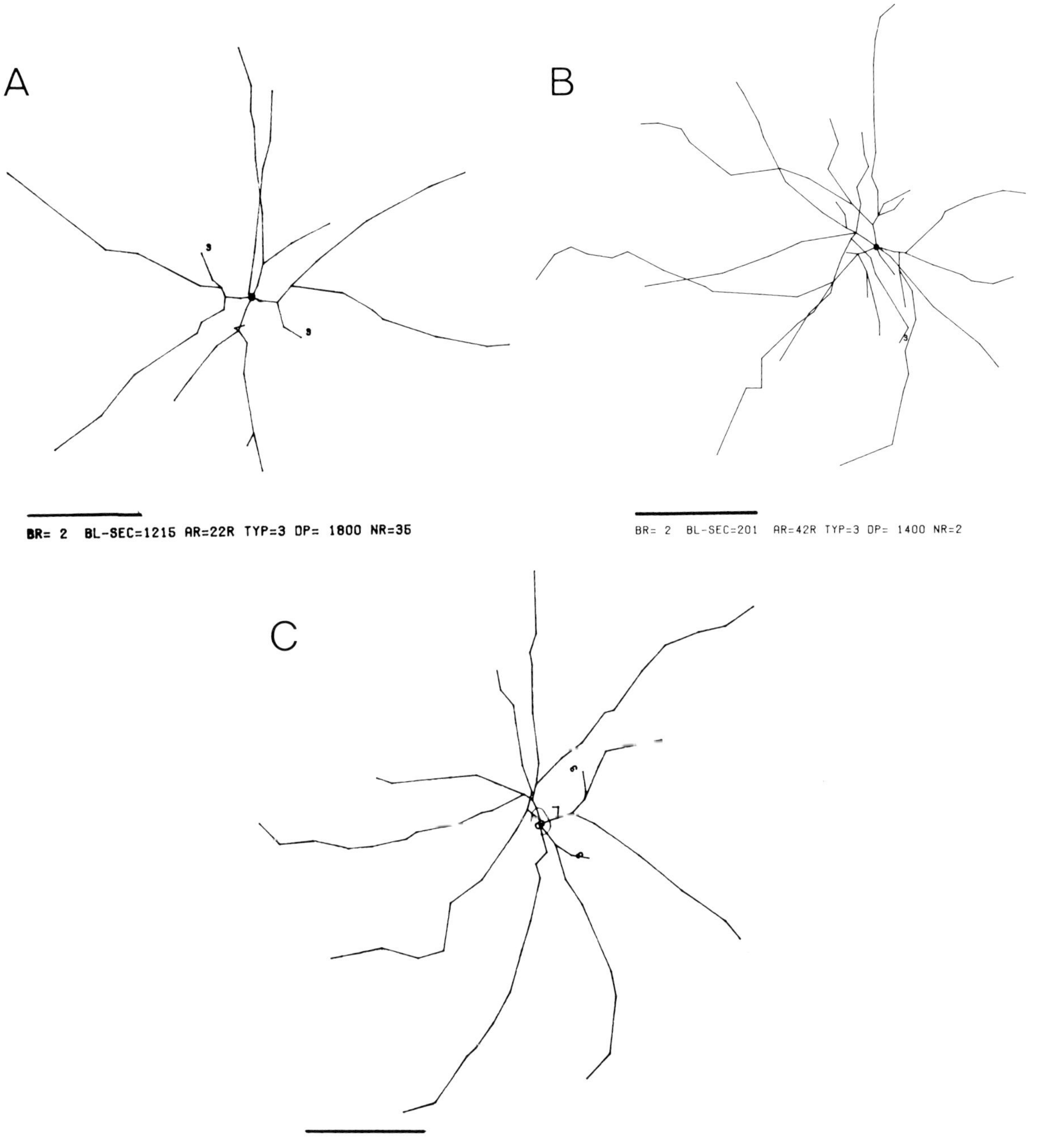

Figure 4. Smooth stellate cells, emphasizing the diffuse, low-density dendritic trees in sagittal views. (B) from Seldon (1981a) with permission. Labels as in Fig. 2.

μm and a tangential one up to 400 μm. Their most remarkable characteristic is the long, unbranched dendrites extending several hundred micrometers, often splayed in all directions but sometimes showing a more vertical organization. In the tangential plane the dendritic trees are often elliptical. The spatial density of dendrites of smooth stellate cells is low, implying that their information sampling is less precisely localized than, for example, pyramidal neurons. The axon initially ascends but then, according to Ramón y Cajal, splits into horizontal branches or sometimes yields a thick, localized network within or near to the dendritic field. Finally, there are a few exotic neurons whose branching patterns resemble those of smooth stellates, but which have tremendous perikarya and thick dendrites.

Double-tufted cells are found in all cortical layers except I. In contrast to pyramidal neurons and smooth stellates, double-tufted cells ("cellules à bouquet double dendritique" of Ramón y Cajal (see Somogyi and Cowey, Volume 1 of this treatise) have a rather strict vertical organization with almost all the dendrites grouped in one or a few narrow vertical columns (Fig. 5). Apart from the usual variations in size, Ramón y Cajal found that spiny dendrites were associated with larger double-tufted neurons, whereas small and medium ones had varicose dendrites. One particular variety, prominent in layer IV, has noticeably smooth vertical dendrites with especially large intervals between them. The vertical range of double-tufted cells may reach 400 μm, yet the diameter in the tangential plane may be as small as 50 μm. In cases in which the dendrites form two columns, these columns are separated by about 50 μm, implying a sampling from disjoint

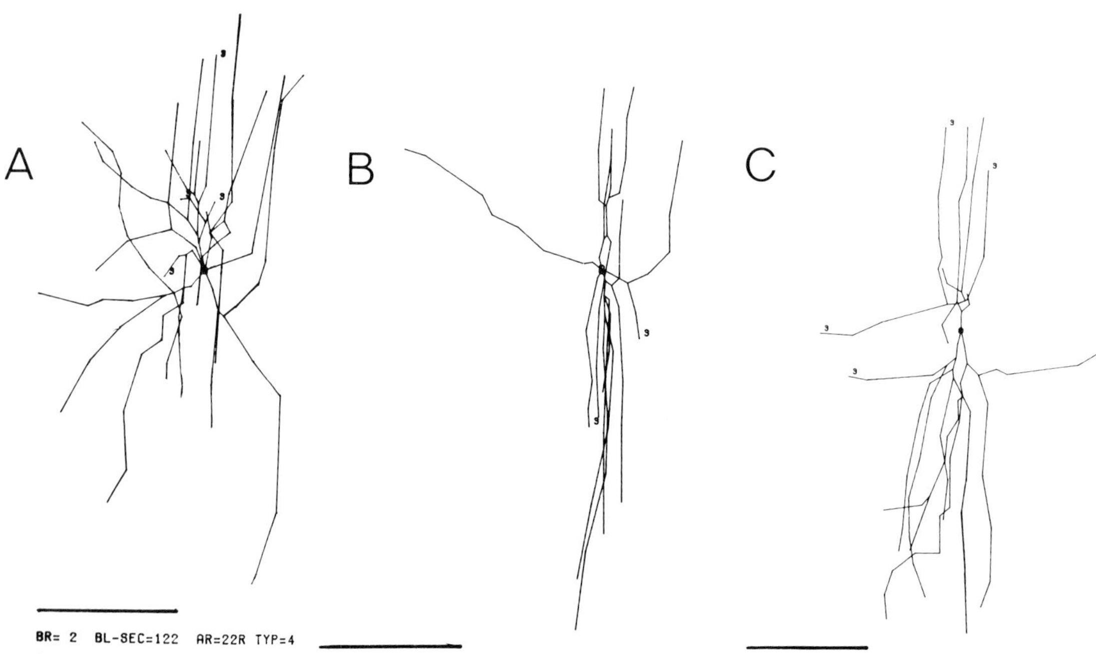

Figure 5. Double-tufted cells showing various widths of the cylindrical dendritic trees, also showing the occasional horizontal secondary or tertiary dendrite. (C) from Seldon (1981a) with permission. Labels as in Fig. 2.

information pools. A puzzling feature of these neurons is the occasional long, unbranched, horizontal tertiary dendrite. Since they are apparently interested in only a small part of the tangential information distribution, these cells are good candidates to coordinate several laminae within their narrow columns. On the output side, the axon arborizations fall into four general categories (Ramón y Cajal, 1902). The most obvious is a dense arborization within a narrow vertical cylinder either within or immediately neighboring the dendritic field. Another variety occupies a similar space but consists of only a few ascending and descending branches. A third type of axon arises from a superior primary dendrite and ascends sometimes as far as layer I (from IV or V), giving off some tangential collaterals. The last type, associated with the spiny double-tufted cells, descends and yields ascending collaterals within the dendritic field. As far as Ramón y Cajal could determine, the axons of double-tufted cells contact the apical dendrite and soma of pyramidal neurons, probably making excitatory synapses at least with the former (see also Szentágothai, 1973). In any case, the information collected by a double-tufted cell is distributed largely in the vertical direction.

Fusiform cells, especially prominent in layer VI, form a long cylinder with the axis parallel to afferent (and efferent) fibers (Fig. 6). Their tangential width is 150–200 μm, intermediate for cortical neurons, and their role is unknown. The axon is long and ascends, yielding horizontal collaterals at least in layer VI.

"Specific giant cells" were so labeled by Ramón y Cajal (1902) because their processes showed a unique organization and stretched through a volume of 1 mm^3. He emphasized that their existence in human cortex is peculiar to the auditory areas. They are found in all layers except I and are particularly numerous in layers V–VI. They have smooth dendrites; the primary ones course horizontally and give off vertically ascending (and sometimes descending) secondary branches at intervals of 50–100 μm (Fig. 7). Below the level of the cell soma there are often also oblique dendrites, but the organization in discrete columns is still obvious. The axon descends obliquely and tortuously and has many horizontal and some vertical recurrent collaterals. The axons of only deep giant cells were seen to enter the white matter (Ramón y Cajal, 1902). Ramón y Cajal postulated a role in "acoustic memory" for these cells.

Concluding the list are the spiny stellate cells, mentioned last because they have been hard to find in my material. They appear in layers II–VI but are most obvious in IV and rare in IIIC. Ramón y Cajal described various sizes and shapes of these, with diffuse or bushy, winding dendrites, spatially more restricted than the pyramidal neurons. The branching pattern may be fairly dense, and their radii vary from 100 to 200 μm in the middle cortical layers. The axons are as variable as the dendritic patterns. Some descend and dissolve into recurrent collaterals; others build dense plexuses. In any case the arborization seems to be confined tangentially either to the same extent as the dendritic tree or to a localized space nearby.

There are in addition numerous variants and types of seldomly appearing neurons, which are excellently summarized by Ramón y Cajal (1902).

Since the axon arborizations of cell types described by Ramón y Cajal (1902) are quite similar to those described in the visual cortex, the pattern of interlaminar activity flow is also probably similar to that inferred for the visual cortex by Lund *et al.* (1979).

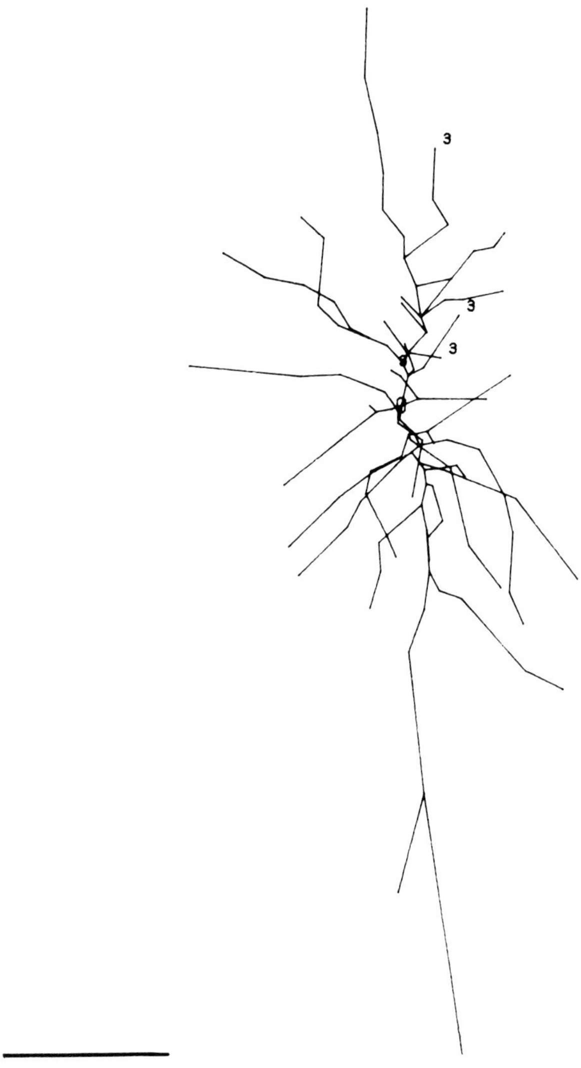

Figure 6. A fusiform cell in layer VI, sagittal view. From Seldon (1981a) with permission. Labels as in Fig. 2.

Figure 7. Ramón y Cajal's specific giant cells, indicating the division of the dendritic tree into several columns. (A) from Seldon (1981a) with permission. (C) is a reproduction of Ramón y Cajal's (1902) prime example of this type of neuron. For (A) and (B), labels as in Fig. 2. For (C), A, lamina II; B, lamina IIIA–B; C, lamina IIIC; D, lamina IV; a, axon.

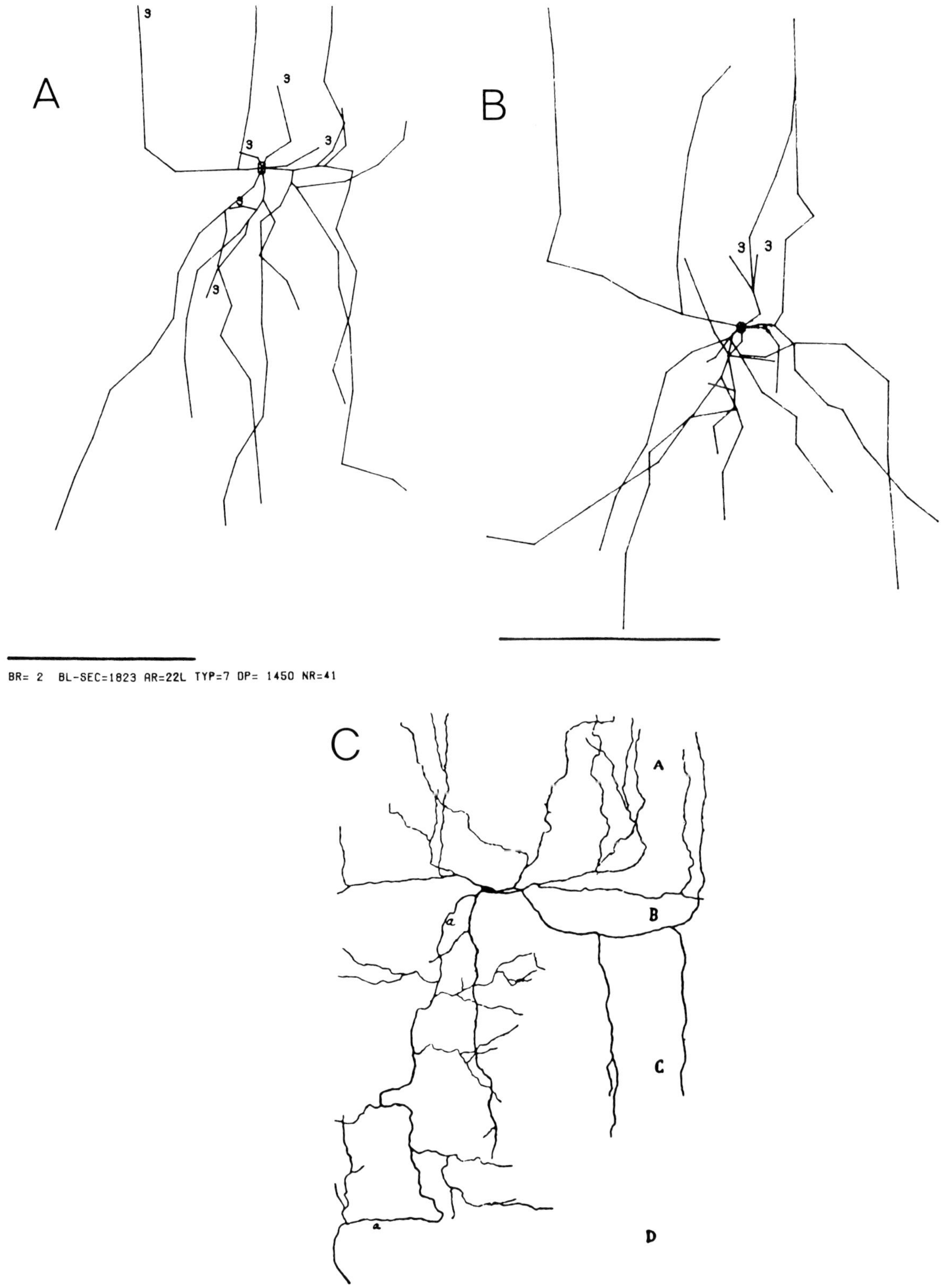
A
B
BR= 2 BL-SEC=1823 AR=22L TYP=7 DP= 1450 NR=41
C
A
B
C
D
a
a

3.3. Cytoarchitectonics, Dendritic and Axonal Organization, and Hemispheric Asymmetry

Macroscopic hemispheric asymmetries in the sylvian fissure have been found anatomically in humans (von Economo and Koskinas, 1925; Geschwind and Levitsky, 1968) and in higher primates (Yeni-Komshian and Benson, 1976). Chimpanzees show less asymmetry than humans, and rhesus monkeys have none. We should look, though, at the individual areas.

3.3.1. Area TC, Area Supratemporalis Granulosa

Let us take the primary cortex, area TC, first. This is a small (3–4 cm^2) area located on the first transverse gyrus. Its size and location are variable, and it is sometimes completely replaced by a mixed area TBC (von Economo and Koskinas, 1925). It shows the typical laminar structure of primary sensory cortices with a thick, well-defined layer IV (Fig. 8). The relative thicknesses of the laminae (in %) are: I 10%, II 10%, III 29%, IV 16%, V 18%, VI 17% with a total thickness of 2.7–2.9 mm (von Economo and Koskinas, 1925). The most particular feature of this area is the presence of "raindrop" clusters of neuron somata (von Economo and Koskinas, 1925), short vertical groupings distributed among less regular formations throughout all layers. Smith and Moskowitz (1979), on the basis of electron microscopic studies, proposed electrotonic coupling among somata in these groups, at least in layer IV—this increases the possibility of functional coupling. The fact that the individual "raindrops" are rather short indicates a less strict columnar functional organization.

von Economo and Horn (1930) described numerous hemispheric variations in area TC. The orientation of the first Heschl gyrus varies from transverse to anterolateral. Its length is greater on the left side (in 8 of 10 cases). Area TC, which occupies much of this gyrus, can be parcellated into up to 11 cytoarchitectonic subtypes. Type 1 is the "most granular," occupies a central position (the "middle plaque") in the area, and usually has a greater extent in the LH. The central position leads one to suspect that this subarea corresponds roughly to the representation of a central frequency range, perhaps the human voice range. The hemispheric size differences are interesting since the total area of TC is practically the same on both sides. In addition, Seldon (1981a) found that the tangential center–center distance between "raindrops" is significantly greater in the left, "speech" hemisphere (46.1 μm vs. 39.7 μm in the RH, Tables I and II). The cytoarchitectonic heterogeneity in TC could lead us to infer many possible variations in response properties even at this "primary" level.

Hemispheric differences in the dendritic trees further support this idea. Although the absolute lengths of dendritic segments are roughly equal (as shown by the *t* test and analysis of variance), the dendrites in the left area TC have distinctly greater tangential components than their counterparts on the right (Seldon, 1982). This is true of pyramidal and nonpyramidal cells alike, whereas RH pyramidal cells have significantly greater vertical components. For purposes of comparison of dendritic overlap among neighboring columns, we need a measure of the tangential input area of a neuron or column. We cannot use the maximum measured tangential radius, since this is subject to freak fluctuations. Neither can we use the average weighted tangential component (segment length

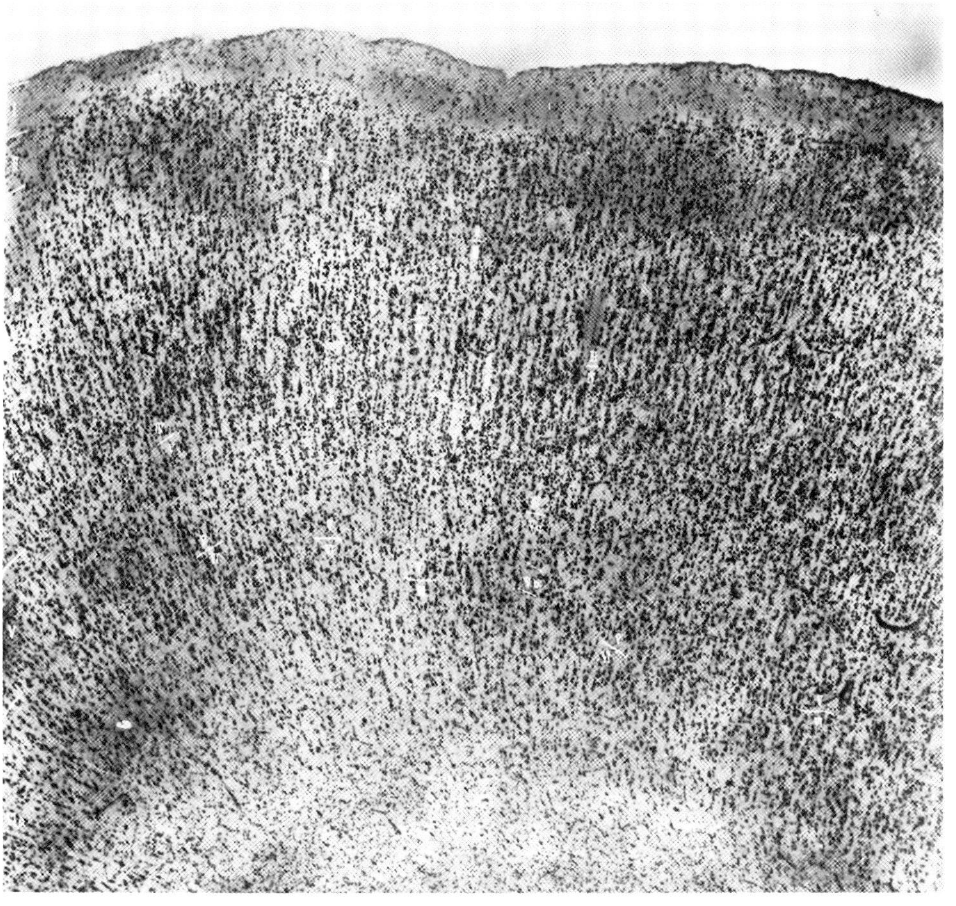

Figure 8. Nissl-stained section from area TC, taken from the atlas of von Economo and Koskinas (1925). Note the "raindrop" formation as compared with the more emphatic columnar structure of areas TA1 and TB (Figs. 10 and 11). 32×.

Table I. Width of Cell-Somata Columns and Intervals between Columns (Nissl Stain)[a]

	TA		TB		TC	
	Column	Interval	Column	Interval	Column	Interval
Right						
No. of measurements	91	100	71	70	34	33
Average width (μm)	21.2	17.9	18.4	18.3	20.7	19.0
Left						
No. of measurements	47	47	71	71	39	39
Average width (μm)	25.4	21.3	23.8	21.0	24.0	22.1
t value	4.45	4.38	7.61	3.27	3.07	1.42
Probability	<0.001	<0.001	<0.001	<0.001	≃0.002	<0.16

[a] From Seldon (1981a) with permission.

Table II. Ratios of the Weighted Tangential Dendritic Extents to the Center–Center Distances between Cell-Somata Columns[a]

	Left hemisphere			Right hemisphere		
Area	C–C distance	Tangential extent	Ratio	C–C distance	Tangential extent	Ratio
TA	46.7 μm	56 μm	1.20	39.1 μm	53 μm	1.36
TB	44.8	54	1.21	36.7	57	1.55
TC	46.1	54	1.17	39.7	51	1.28

[a] From Seldon (1982) with permission.

times segment's center of mass), since it is heavily influenced by vertical dendrites, e.g., the apical one. As a compromise measure of the area of cortex from which a neuron collects its major input, Seldon (1982) defined the "tangential extent" as a radius:

$$\frac{\sum \text{(Segment length)} \times \text{(Segment's maximum tangential distance from soma)}}{\sum \text{(Segment length)}}$$

The average tangential extent is clearly greater in the LH (Table II); however, not so much that it can compensate for the greater center–center distance between "raindrops." This in turn means that the dendritic overlap between neighboring units is actually less on the left in spite of the greater absolute tangential extent. The absolute volume of an input region for a unit on the left is greater than on the right, but the input regions on the left are more disjoint. Thus, neurons in the LH can at least theoretically react more specifically than those on the right (in congruence with Ramón y Cajal's idea that dendritic spread and specificity are inversely related). Since pyramidal and nonpyramidal neurons alike show lesser overlap in the LH, clearly the geometric proportions between the two types of neurons are roughly constant within each hemisphere.

Very little work has been done on the axon picture in area TC. Vogt and Vogt (1919) found poorly organized radial myelin bundles corresponding to the relatively weak organization of cell-somata columns. They also described a band of myelinated fibers in layer IIIB and a generally dense myelin infrastructure in the area. Ramón y Cajal (1902) described many horizontal afferents in layer III and a thick horizontal plexus in layer IV, where he also found a high concentration of nonpyramidal cells with ascending axons. Seldon (1981b) found in area TC, as in TA and TB, hints of a macroscopic banding pattern of axons (Fig. 9). The bands are about 350 μm wide in sagittal sections. In the sagittal plane they are divided into two or three layers, the lowest one being in lamina V and the upper one(s) being in laminae III–IV. When reconstructed over several sections and then viewed from the pial surface, the bands appear as irregular, interdigitating stripes. Within the bands the proportion of horizontal axons is significantly greater than between them. The origin of these bands is still a matter of speculation, and analogies with data from animal experiments allow for either a thalamic or a cortical origin (Conel, 1939–1967; Jones and

Burton, 1976; Imig and Brugge, 1978; Middlebrooks *et al.*, 1980). In their width and pattern they are similar to the ocular dominance columns described by LeVay *et al.* (1975) in the monkey visual cortex.

Since we are trying to elucidate the information-processing properties of the cortex, it is also interesting to look at the convergence and divergence between cortical afferents and the neuronal columns. In this we are hampered by a lack of data on (1) the density of afferents, (2) inhomogeneity in the afferent density, (3) the tangential arborization patterns of afferents, and (4) hemispheric differences. Thus, any estimates we make now are subject to future revision. To point (1), the density estimates range from 16,000 afferents/mm^2 (derived from Conel, 1939–1967) to 25,000 in cat visual cortex (Sholl, 1956) to 35,000. The last figure was obtained by taking an average axon diameter of 3 μm (with myelin) (Ferster and LeVay, 1978), which yields a cross-sectional area of 7 μm^2 and thus 140,000 axons/mm^2. If a fourth of these are assumed to be afferents (Sholl, 1956), we get 35,000. For afferent density let us assume the average of 25,000/mm^2. To point (2), numerous recent studies have demonstrated banding patterns of subcortical afferents in animals (e.g., LeVay *et al.*, 1975) and human visual cortex (Hitchcock and Hickey, 1980) and columnar organization of corticocortical afferents (e.g., Imig and Reale, 1981). As mentioned above, Seldon (1981b) found a banding pattern of axons in the human auditory areas. Whether the overall afferent density is homogeneous, in spite of this, has not been studied. We must

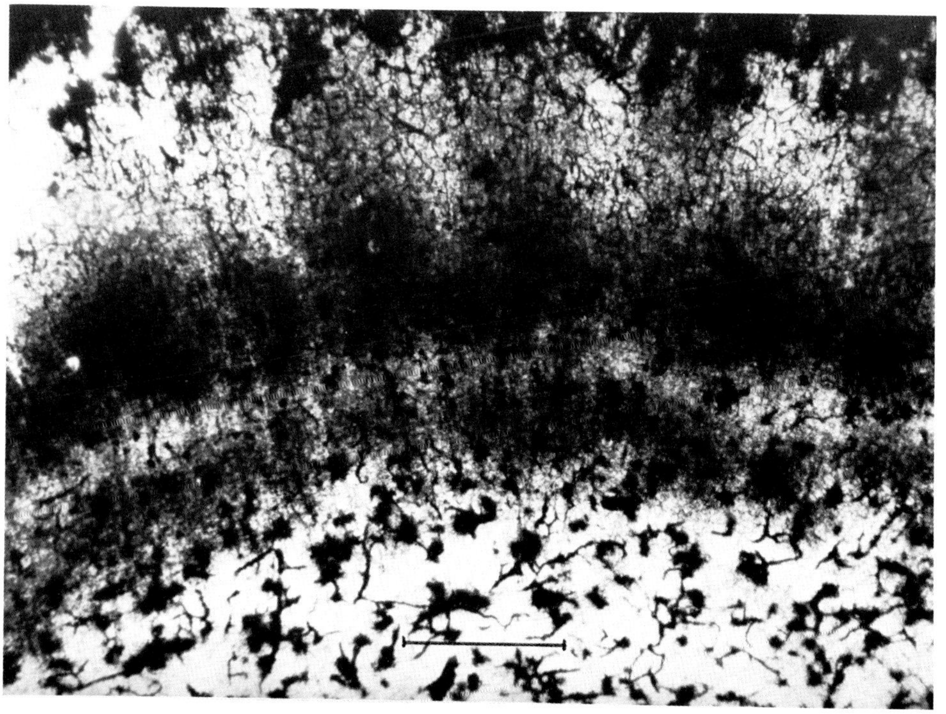

Figure 9. Low-power view of the axon distribution in area TC, showing the "macrocolumns." Sagittal view. Bar = 500 μm. From Seldon (1981b) with permission.

accept homogeneity as a first (and simple) approximation but must point out that when data on banding patterns are more definitive, then convergence/divergence estimates will have to be revised. To point (3), there are no data. Pictures (Szentágothai, 1973; Ferster and LeVay, 1978) indicate a diameter for an individual terminal arborization of about 400 μm (range 150–680) in cat visual cortex, and estimates from the banding pattern of axons in human auditory cortex (Seldon, 1981b) range from 350 to 500 μm, depending on the area. We shall have to accept a reasonable compromise (Table III). About hemispheric differences we know nothing, but the differing dendritic orientations encourage us to infer the existence of variations in afferent distributions.

To calculate the convergence we need the number of axons whose arborizations can (note emphasis) touch the dendritic field, i.e., the number which lie within the dendritic plus the axonal radius in the tangential plane. For the dendritic radius we take not the average but rather the maximum weighted tangential extent (Seldon, 1982) of neuronal columns. The results, meant to be order-of-magnitude approximations, are given in Table III. Convergence in the left area TC is estimated to be higher than on the right, due probably to the more tangential orientation of the LH dendrites. Here we return to the problem of specificity. The estimates in Table III were made assuming random connections, i.e., that a neuron can be contacted by every afferent within its dendritic space. Thus, the estimates represent maximum values which would decrease with increasing connection specificity. However, it must be emphasized that we are not estimating the input to a single cell, but rather to a column of functionally related cells whose summated input is certainly greater than that of an individual member.

To calculate the divergence we need the number of columns which can be contacted by one axon. An additional problem here is subcortical axonal branching which yields several terminal arborizations. Let us assume a simple bifurcation and that the terminals contact disjoint sets of neuronal columns. [See, for example, Gilbert and Wiesel's (1979) HRP study of neurons and afferents in cat visual cortex.] For this calculation we also need the spatial density of the columns, taken from Seldon (1981a). The results are also given in Table III. According to this estimate the divergence from afferents to cortical columns is slightly

Table III. Estimated Convergence and Divergence between Afferents and Neuronal Columns

	Left area			Right area		
	TA	TB	TC	TA	TB	TC
Max. tangential extent	133	115	128	131	157	92
Estimated tangential radius of afferent	215	240	175	215	240	175
Column density/mm^2	1,120	1,140	1,060	1,490	1,780	1,440
No. of afferents contacting one column	10,300	9,900	7,200	9,400	12,400	5,600
No. of columns contacted by one afferent	860	900	600	1,120	1,760	640

greater in the RH, due perhaps to the higher packing density of the neuronal columns there.

3.3.2. Area TA, Area Temporalis Superior

The second area of interest is TA, which occupies the superior temporal gyrus including the upper bank of the superior temporal sulcus. The sensory speech region is usually maintained to include the posterior half of this area, although the anterior part also plays an as yet undefined role in auditory processing. von Economo and Koskinas distinguished the two halves slightly, labeling them TA1 (posterior) and TA2. Galaburda and Sanides (1980) call the posterior half Tpt and set its posterior border in the parietal lobe. Parallel to the greater macroscopic area of the LH auditory regions mentioned above, in cytoarchitectonic studies Galaburda *et al.* (1978b) and Galaburda and Sanides (1980) have found that the area TA1 left has a greater extent than that on the right. The left/right ratio of extents ranges from about 1 to over 7. Galaburda *et al.* (1978b) differ from von Economo and Horn on the borders of area TA1, but this may be due to individual variations in the two different brain samples or to deviating architectonic parcellations (Galaburda and Sanides, 1980). The former contend that it covers most of the excess planum temporale in the LH, whereas the latter set its upper border at the lateral edge of the planum.

The relative thicknesses of the laminae are: I 11%, II 7%, III 36%, IV 9%, V 19%, and VI 18% with a total thickness of 2.1–2.8 mm (von Economo and Koskinas, 1925). The major cytoarchitectonic features of this area are clear columns of cell somata ranging through practically the entire cortex (layers VI–II)—von Economo and Koskinas called them "organ pipes," also noting the similarity to the secondary parietal cortex (Fig. 10). The columns are separated by soma-free neuropil and are thickest in layer IV. In other layers the pyramidal cell bodies are literally stacked on top of one another, thus leading to the grouping of apical dendrites observed also by Ramón y Cajal (1902). The cell bodies themselves are often larger than in the primary auditory cortex, but the center–center distances between columns are almost the same as in the ipsilateral area TC (46.7 μm left, 39.1 μm right, Tables I and II). This of course means that the spacing in the LH is again significantly greater than on the right.

The structural similarities to the parietal cortex lead us to infer a similar functional organization. In area 7 Hyvärinen and his co-workers (1974, 1981) have described a distinctly columnar functional architecture, whereby between neighboring columns they found often closely related electrophysiological behavior but sometimes discrete jumps in response properties.

To this organization comes the added difference in columnar spacing between right and left areas TA1, allowing different functional connectivities. The dendritic orientations are also distinctly different between the two areas. Although the absolute segment lengths are equivalent, the tangential component of those on the left is significantly greater than those on the right, and vice versa for the vertical components. Together with the wider column spacing on the left, this implies that neurons of an LH column are filling their neighborhood with more horizontal dendrites than the corresponding columns on the right, where the more vertical tendencies are probably related to the greater proximity of neighboring columns and their dendrites. These organizational tendencies are shared by pyramidal and nonpyramidal cells alike. As in the areas TC, we find here again that the ratio of the tangential extents to the center–center intercolumn distance is less in the LH than in the RH (1.20 vs. 1.36, Table II), implying that the cells of LH columns share their input region (information) with fewer neighboring columns. The segregation of neighboring columns is better, at least concerning the dendritic overlap. This sharper columnar parcellation allows, at least theoretically, for more independent responses.

Another conclusion which we might draw from the relative sizes and separations of the columns is that the convergence/divergence relationships show hemispheric differences. Using the same sorts of estimates described above for Table III, we guess that LH columns receive more cortical afferents than do RH ones, which also speaks for a greater specificity and variability of action.

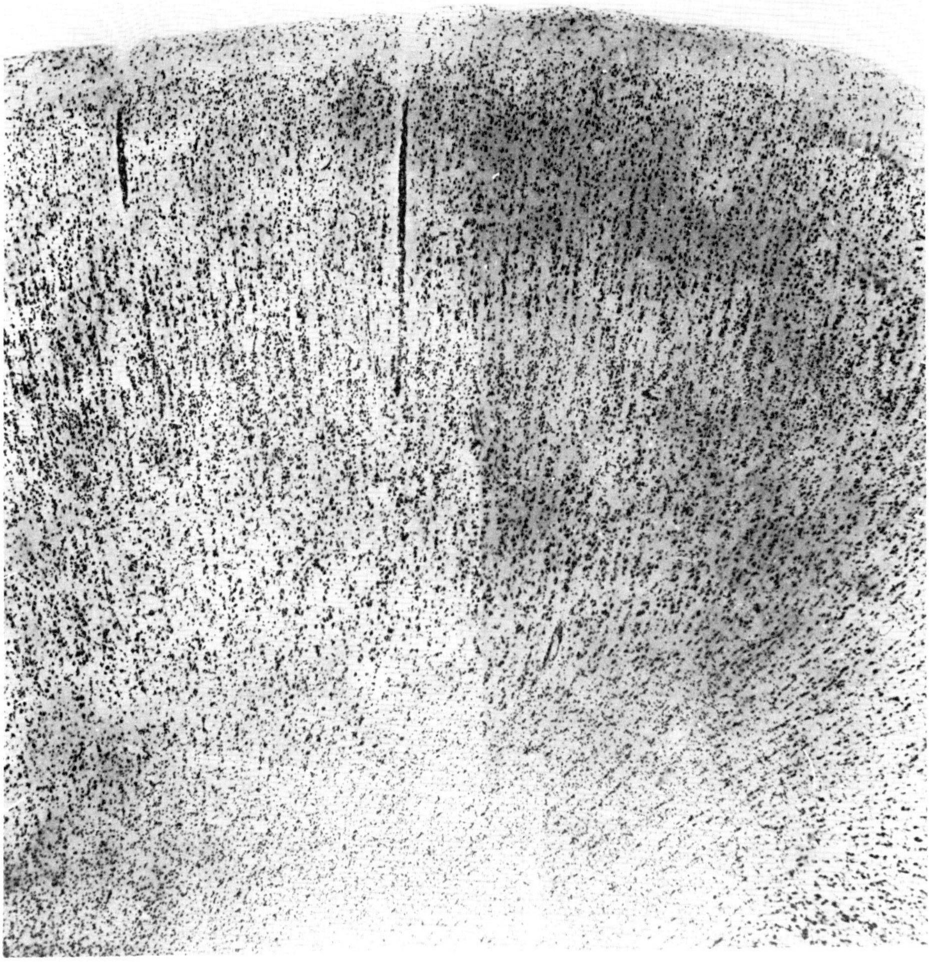

Figure 10. Nissl-stained section from area TA1, taken from the atlas of von Economo and Koskinas (1925). Note the "organ-pipe" structure, the alignment of somata in columns separated by neuropil. 32×.

Likewise, the afferent-to-column divergence is apparently less on the left (Table III).

As in area TC, the axon distribution in TA is characterized by vertical bands or slabs (Seldon, 1981b). In the tangential plane the slabs are roughly 500 μm wide, are separated by 100-μm strips, and have an irregular course. In the radial plane we find a thin band in layer VI and a broad one stretching from layer V up to IIIA and narrowing at the top. Within the bands there is a slight emphasis on radial orientation of axons, but there are no such clear preferential orientations as in area TC.

3.3.3. Area TB, Area Supratemporalis Simplex

In the rhesus monkey, von Bonin and Bailey (1947) found only rudiments of area TB, but later Pandya and Sanides (1973) and Galaburda and Pandya (1983) discerned a cytoarchitectonically similar area (paAlt) lateral to the primary cortex (see Pandya and Yeterian, this volume). In the chimpanzee it appears more as an integral area, a strip lateral and posterior to TC (Bailey *et al.*, 1950). In humans it lies lateral and posterior to area TC on the transverse gyrus (gyri) and extends back to cover most of the planum temporale, according to von Economo and Horn (1930), but in contrast to Galaburda *et al.* (1978b), who restrict it to a portion of the planum (see above). Although the studies are discrepant regarding the phylogenetic appearance and development of TB, there is general agreement that it is the rapid growth of this area which lengthens the caudal end of the sylvian fissure in the LH (see Sanides, 1975). von Economo and Horn (1930) mention that the planum temporale appears more often in the LH, whereas the RH more often has a second transverse gyrus, which, however, is partially occupied by area TC. Geschwind and Levitsky (1968) showed that in adults the area of the left planum temporale was usually greater than that on the right, which was confirmed by Wada *et al.* (1975). Witelson and Pallie (1973) found this asymmetry in neonates, and Chi *et al.* (1977) described it in fetuses of 31 weeks. The effect is that the left area TB is markedly larger than the right one. The laminar structure is that of a secondary cortex, with the relative thicknesses being: I 9%, II 8%, III 34%, IV 16%, V 15%, and VI 18% with a total thickness of 2.7–3.1 mm (von Economo and Koskinas, 1925). The cytoarchitectonic structure of TB resembles that of TA1, but the "organ-pipe" formations of cell somata are even more obvious, extending again from layer VI to layer II (Fig. 11). The columns are also amplified by occasional giant pyramidal cells in layer IIIC—the landmarks of area TB (also called "magnocellularis" by von Economo and Koskinas). The center–center distance between the columns here is, in each hemisphere, slightly less than in the neighboring area TA1, but the great hemispheric asymmetry is present here too. The columns in the LH are significantly more widely spaced than those on the right (44.8 μm left, 36.7 μm right, Table II). Looking at the dendritic trees, we find much more irregularity than in the other auditory areas. In the RH the average absolute dendritic segment lengths are longer than on the left. In addition, they show both greater tangential and greater vertical components on the right. However, if we examine the neuron types individually, we find that pyramidal cells on the left have a greater vertical and lesser tangential expansion than those on the right. Unfortunately (for clarity), the opposite is true of nonpyramidal cell dendrites. So the pyramidal–nonpyramidal cell interaction apparently has a different form here than in areas TA1 and TC. Looking at the average tangential extent we see that it is greater on the right (57 μm vs. 54 μm, Table II), in contrast to areas TA1 and TC but in congruence with the tangential expansions of the dendritic segments in TB. In addition, the greater tangential extent leads, as in areas TA1 and TC, to a greater overlap among dendrites of neighboring columns on the right—the hemispheric difference here is the most extreme of all three areas. It is rather remarkable that the ratio of tangential extents to center–center distances between columns is constant in the LH (1.17–1.21) and greater and much more variable in the RH (1.28–1.55). This leads us to believe that the

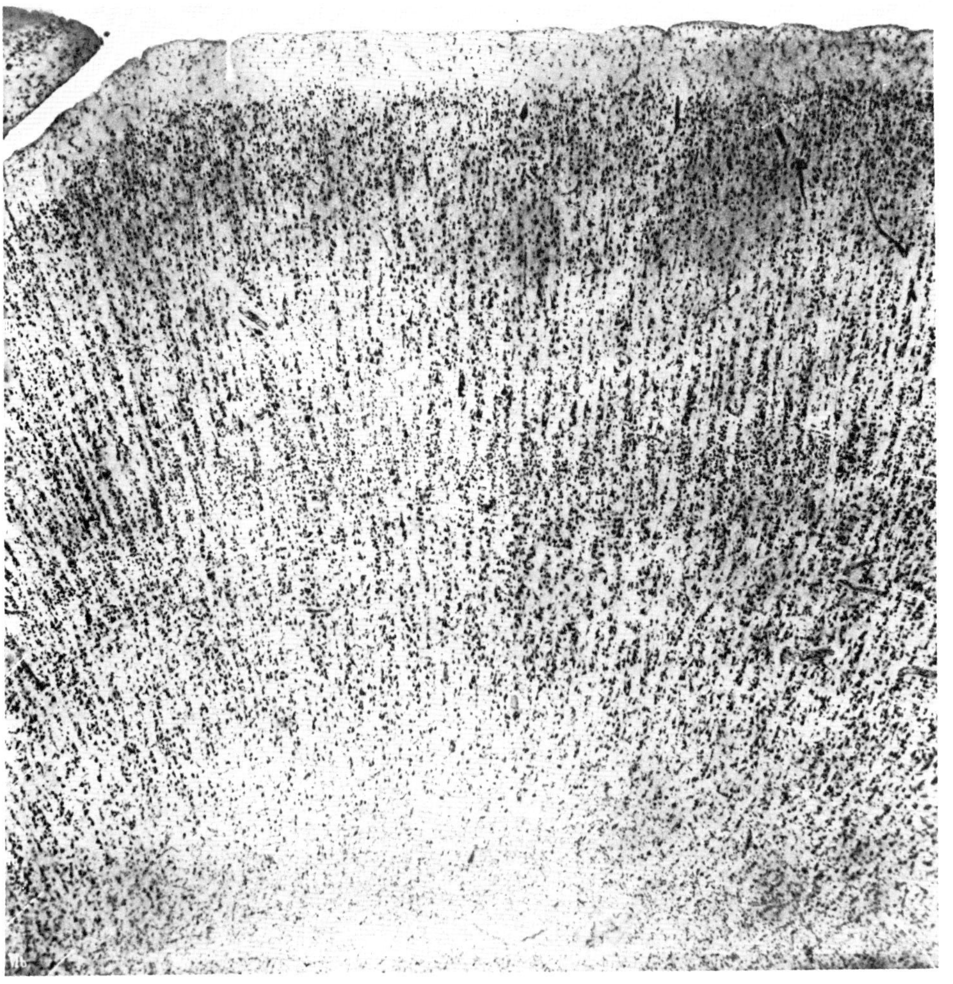

Figure 11. Nissl-stained section from area TB, taken from the atlas of von Economo and Koskinas (1925). Here the organ-pipe formation is more emphatic than in TA1. 32×.

organization of all auditory areas is somewhat more diffuse on the right, again in keeping with the higher specificity involved in "speech perception" on the left.

The implication for convergence of afferents is that here RH columns receive more input (Table III), but the significance of this clearly discrepant finding is not apparent. The phylogenetically recent expansion of TB may be a cause or explanation for its seeming structural inconsistencies. The divergence from afferents to columns also appears to be greater in the right area TB, due to the closer column packing and in congruence with the results for TA and TC.

In TB we again find a banding pattern among the axons, with dense stripes in layers III–IV and V–VI (Seldon, 1981b). In the tangential plane they are about 550 μm wide and separated by 100-μm intervals. As in TA there is no significant preferential orientation for the axons—the gridlike pattern found in TC yields to a more even distribution of orientations here.

4. What Else Is Connected to the Auditory Cortex?

The purpose of this section is to impress upon the reader the multitude and complexity of influences on our auditory cortex. The problem of elucidating them is not simplified by the fact that almost all the work has been/has to be done in animal experiments. [From human brains, von Monakow (1914) compiled a—in hindsight certainly incomplete—list of connections of the motor speech centers.] The tracing methods have all been exhaustively reviewed in several books, e.g., Nauta and Ebbeson (1970), Cowan and Cuénod (1975), and Heimer and Robards (1981), so I need not mention details. The first thorough studies were done using degeneration methods, whereby the pattern of necrotic efferent projections from lesion sites can be demonstrated with selective silver stains. Using these methods, Pandya and Kuypers (1969) and Jones and Powell (1970) examined corticocortical connections generally; Pandya *et al.* (1969) looked at the connections of the auditory cortex in particular (see Pandya and Yeterian, this volume); and Mesulam (1979) traced selected human cortical connections. Newer tracing methods have confirmed and elaborated the connections found previously. Anterograde projections are well delineated using radioactively labeled amino acids or fucose. For tracing in the retrograde direction, HRP has been a valuable tool. The resolution, exactness, and clarity of the newer methods enable detection of very sparse projections, and with them a veritable wealth of connectivity data has accumulated.

A summary of the results of these methods for the sensory speech areas is given in Fig. 12. Individual projections will be described below. They have almost all been described in cats and/or monkeys and thus remain to be definitively confirmed in humans. The connections listed are, unless otherwise stated, all ipsilateral (i.e., LH) connections, and area TB has been largely omitted since there are no reliable data available for it. The most important point, which becomes obvious at a glance at the figures, is that large regions of the brain are connected to the sensory speech area within very few processing stages. This implies, very generally, that (1) speech perception can be influenced by many things, and (2) the influence can be felt very quickly. This confirms what we know from everyday experience.

The primary input pathway, from the ear, goes through the cochlear nucleus, the trapezoid body, the superior olive, the inferior colliculus, and the medial geniculate nucleus (MGN) before arriving at the cortex. Harrison (1978) has written an excellent review of the brain-stem portion of this system, so it will only be summarized here. Harrison describes a well-known principal and two poorly understood accessory ascending systems. The principal pathway goes

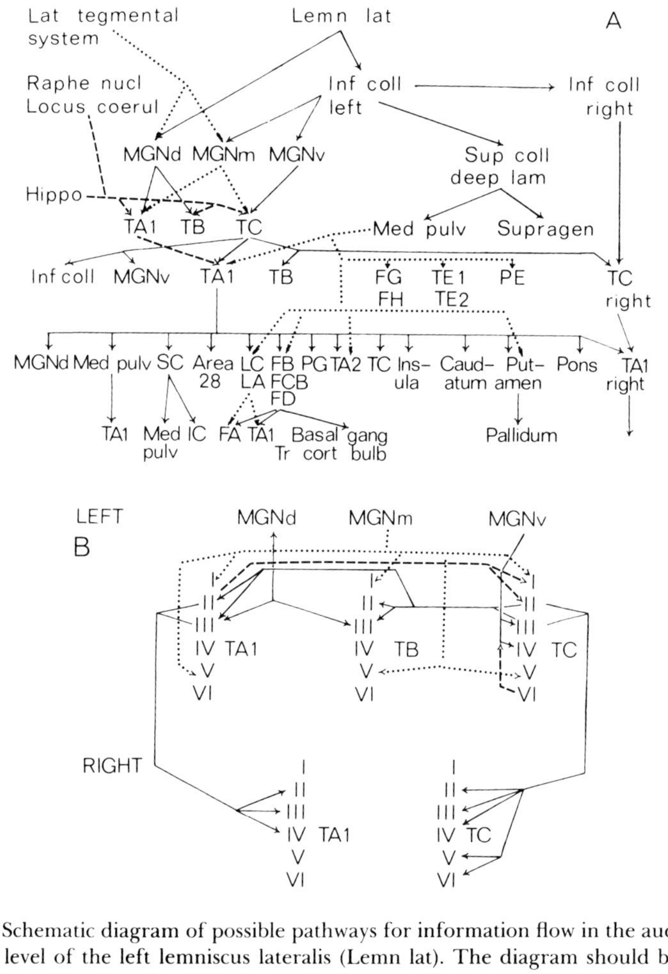

Figure 12. (A) Schematic diagram of possible pathways for information flow in the auditory system, starting at the level of the left lemniscus lateralis (Lemn lat). The diagram should be read slowly from the top down. For the sake of clarity, projections are shown as solid, dashed, or dotted lines. Many corticocortical feedback loops have been omitted, as have been most connections of area TB (since they are uncertain). With the exception of area 28, the cortical areas are designated with the nomenclature of von Economo and Koskinas. Inf coll (IC), inferior colliculus; Sup coll (SC), superior colliculus; MGN, medial geniculate nucleus (d, dorsal; v, ventral; m, magnocellular); Supragen, nucleus suprageniculatus; Med pulv, medial pulvinar; Tr cort bulb, tractus corticobulbaris. (B) Details of projections among the laminae of the auditory cortical areas in both hemispheres.

through the ventral part of the cochlear nucleus, the medial nucleus of the trapezoid body, the lateral superior olive, and the central nucleus of the contralateral inferior colliculus. All levels above the cochlear nucleus receive bilateral innervation, and it is the bilateral comparison in the superior olive and inferior colliculus which is probably the basis for spatial localization of auditory stimuli. The system is tonotopic at all levels, but individual neurons do not respond just to a very narrow frequency band. Neuronal responses are of the sustained type up to the inferior colliculus, where the neurons respond to various combinations of stimulus parameters. The two accessory systems may include the dorsal cochlear nucleus, the dorsal acoustic striae, the striae of Held, the pericentral nucleus of the inferior colliculus, the dorsal part and magnocellular nucleus of the MGN, and the suprageniculate nucleus. The systems are also connected to the lateral tegmental system and the superior colliculus, and the projections within them are more diffuse than in the principal one (Kudo and Niimi, 1980). Many neurons in the deep layers of the superior colliculus do respond to noise stimuli (Wise and Irvine, 1983). Most of these auditory neurons receive excitatory input from the contralateral ear but are also influenced positively or negatively by ipsilateral input. The function of the accessory systems is unclear, though they may be associated with the descending, feedback and control system and apparently can also play a role in sound localization. Going a step higher, Pandya and Sanides (1973), Jones and Burton (1976), and Burton and Jones (1976) showed that the MGN projects to both primary and secondary auditory cortices.

Since each ear yields bilateral input to the ascending system, and since cortical columns showing binaural summation or suppression are segregated (see Section 5), the question arises as to the segregation or integration of binaural input in the ascending system. Roth *et al.* (1978) and Semple and Aitkin (1979) have described classes of neurons with different binaural properties in the inferior colliculus. Recently, Middlebrooks and Zook (1983), using multiple retrograde tracers in cats, demonstrated that the regions of MGNv inputting to cortical summation columns are distinct from those projecting to suppression columns. Thus, the binaural interaction properties represent a second dimension (after frequency) in the higher auditory centers.

The primary and secondary cortices show reciprocal anatomical projections (Pandya and Sanides, 1973; FitzPatrick and Imig, 1980; Galaburda and Pandya, 1983), and neuronographic tracing in the chimpanzee indicated a sequential projection TC → TB → TA (Bailey *et al.*, 1950). So the secondary cortex receives two versions of the primary afferent input, one directly from the MGN and the other via the primary cortex, and is thus simultaneously involved in more than one stage of auditory processing (Fig. 12).

The left primary and secondary auditory cortices are reciprocally connected via homotopical projections to their counterparts in the RH (Myers, 1967; Pandya and Sanides, 1973). FitzPatrick and Imig (1980) also described heterotopical projections. Viewed sagittally, the callosal afferents build vertical columns extending through most of the cortical laminae (Ramón y Cajal, 1902; Majorossy and Bodoky, 1977). Viewed in a plane parallel to the cortical surface, the afferents are probably arranged in irregular stripes (e.g., Imig and Brugge, 1978). Within the bands of callosal connections are, at least in the primary auditory cortex, neurons which sum bilateral input (Imig and Brugge, 1978). Just what kind of signals cross the corpus callosum is still unknown. For the primary auditory cortex they seem to be derived from the tonotopic map and carry frequency-dependent information (Imig and Adrian, 1977; Middlebrooks *et al.*, 1980). Interwoven with the stripes of callosal projections may be similar bands of ipsilateral corticocortical projections (Imig and Reale, 1981). In the bands of ipsilateral projections are neurons in which input from the contralateral ear usually suppresses that from the ipsilateral one. Finally, Seldon (1981b) did discover a banding pattern of afferent axons in Golgi-impregnated sections of human auditory cortex, but the source of the axons was unknown.

Rostral projections of the auditory areas lie within the temporal lobe or go via the arcuate fasciculus. The rostral superior temporal gyrus (area TA2) receives some auditory input (Penfield and Pernot, 1963), apparently from area TA1. There are anatomical projections going further from area TA2 to the temporal pole, area TG (Jones and Powell, 1970), and both areas TA2 and TG project to the amygdala (Turner *et al.*, 1980). Turner *et al.* interpreted these latter projections as the connecting link between acoustic stimuli and emotional responses. A further elaboration of temporal rostral projections was recently made by Galaburda and Pandya (1983). They divided the superior temporal lobe (of rhesus monkeys) into a matrix of three rostral–caudal rows of four cytoarchitectonic areas each. The areas in each row are connected by a series of two-way rostral–caudal projections. Each area is also connected to its nearest neighbors in the other rows. The primary auditory cortex occupies place three in the middle row; its rostral projections lie in the middle row or go via the neighboring areas in the medial or lateral rows. The secondary auditory areas paAlt (TB?) and Tpt (TA1) lie in the lateral row, so their rostral projections are in that row or go indirectly via the middle or even medial row. A preliminary indication of the role of anterior temporal areas was given by Sherwin and Efron (1980), who found that sequence order judgments for pairs of tones were impaired after anterior temporal lobectomy. Also impaired is the ability to follow a selected signal among many simultaneous ones (the "cocktail party" effect) in the contralateral ear (Efron *et al.*, 1983). A remarkable exception is the human voice, which is easily identified even after such a lobectomy.

The physically most extensive and macroscopically most obvious projection from the sensory language region is the arcuate fasciculus, which leads to area 44 (FCB or Broca's area). In spite of its importance, little is known about the fasciculus, such as the number of axons it contains or whether it is a one-way or reciprocal projection. Analogy with other corticocortical projections and functional considerations such as speech comprehension deficits in patients with motor aphasia (Blumstein *et al.*, 1977a) imply that the fasciculus contains two-way connections. According to some of the localization studies mentioned earlier, its interruption may lead to conduction aphasia, a form in which the patients cannot repeat what is said to them, but can produce meaningful speech themselves. However, an anatomical–physiological explanation for this syndrome has not yet been presented.

The motor speech cortex is itself by nature a secondary motor cortex, more or less an extension of area 6 (FC). von Economo and Koskinas (1925) labeled it FCB due to its similarity to other secondary motor areas. As such an area, it probably also has the same input sources as the others, namely the primary and secondary somatosensory cortices (areas 1, 2, 3, and 5) as well as anterior thalamic nuclei (von Monakow, 1914; Scheibel and Scheibel, 1966; Pandya and Kuypers, 1969; Jones and Powell, 1970). Indirect confirmation of the somatosensory input comes as mentioned earlier from Ojemann and his co-workers (see Ojemann, 1976) who found that stimulation of the ventrolateral thalamus interrupts speech production. Another putative source of input to the motor speech area is the prefrontal cortex (von Monakow, 1914), which according to Fuster *et al.* (1982) is involved in temporal integration of sensory information (see Fuster, this volume). By this roundabout way the motor speech cortex and perhaps even the sensory language cortex can be influenced by multimodal sensory stimuli.

Occasionally, some question has been raised as to whether area 44 is the true or only motor speech cortex (von Monakow, 1914; Blunk *et al.*, 1981). Some findings implicate the anterior insula instead of area 44, others show a wide cortical participation in speech production. The problem may be one of definition, depending on whether area 44 is designated as "the motor speech cortex" or as one of several areas necessary for speech production. In the latter case the existence of a hierarchy among the various areas has yet to be demonstrated.

In cats Imig and Reale (1980) found a projection from AII to the insular cortex. In humans such a projection could be a subset of the arcuate fasciculus, and von Monakow (1914) indicated its existence on the basis of degeneration studies.

Associated with the arcuate fasciculus and perhaps contributing to it is area 40 (PF), the supramarginal gyrus. Lesions in this area have been associated with conduction aphasia (Green and Howes, 1977; Damasio and Damasio, 1980), although such lesions almost always appear in combination with those of the underlying white matter, including the arcuate fasciculus. The role of the supramarginal gyrus in the brain is not understood. Its position at the corner between the temporal and parietal lobes indicates that it may be a multimodal integration area, and electrophysiological recordings (Leinonen *et al.*, 1980) seem to substantiate this. Neurons here respond to spatially localized auditory signals and/or to contralateral or bilateral somatosensory cues. Cytoarchitectonically, area PF shows the "organ-pipe" organization of neuron somata typical of areas TA, TB, and the parietal cortex, all secondary sensory integration centers.

Posterior to the supramarginal gyrus is the angular gyrus, often implicated in reading disturbances and thus indirectly in speech processing. Von Monakow's (1914) observations on degenerating neuron processes led him to infer a direct connection between the angular and the superior temporal gyrus.

One of the highest-order connections of the posterior superior temporal gyrus is with areas in the prefrontal cortex (FB, FC, FD) (Pandya and Kuypers, 1969; Jones and Powell, 1970). The functions of these areas are poorly understood. Interestingly, Luria (1966) postulated that the prefrontal cortex generally regulates motor behavior with linguistic control. As mentioned above, Fuster *et al.* (1982; see also Fuster, this volume) assign it the role of temporally integrating multimodal sensory information. Roland (1982) corroborated the connections with measurements of regional cerebral blood flow. With auditory stimuli he elicited significantly increased blood flow in the frontal operculum and supplementary motor area bilaterally and part of the left prefrontal cortex (in addition to bilateral temporal, medial parietal, and Rolandic areas). Finally, depressed frontal activity is associated with language deficits resembling sensory or transcortical sensory aphasia (Appell *et al.*, 1982).

The superior temporal cortex is also connected reciprocally with the limbic cortex, area LA, probably via the extreme capsule (Müller-Preuss *et al.*, 1980). Stimulation in the anterior cingulate gyrus produces initiation of vocalization as well as a predominantly inhibitory effect on the sensory auditory cortex. Similarly, during self-produced phonation, neurons in the auditory cortex are inhibited (Müller-Preuss and Ploog, 1981).

Votaw (1960) placed lesions in Ammon's horn and subsequently found degenerating fibers in, among others, all three temporal gyri. He also mentioned that stimulation of Ammon's horn leads to face and body movements, apparently mediated through the temporal lobe.

Among subcortical connections of the auditory cortex, the one to the medial pulvinar has received the most attention. It has been shown to be topographical (and probably also reciprocal) (Simpson, 1952; Locke, 1960; Siqueira, 1965; Mauguiere and Baleydier, 1978). Pulvinar lesions may or may not lead to aphasia (Streitfeld, 1980). The pulvinar is a complex of several nuclei which are involved in various stages of visual and probably auditory processing (Brown, 1975). The inferior and lateral subdivisions receive input from the retina and superior colliculus and are also reciprocally and topographically connected with both primary and secondary visual cortices (Benevento and Davis, 1977; Ogren and Hendrickson, 1979). On the other hand, the story on the medial pulvinar is unclear, and the reports are conflicting. Locke (1960) described a projection from the posterior medial pulvinar to the anterior superior temporal cortex, and from the medial medial pulvinar to the posterior superior temporal cortex. Trojanowski and Jacobsen (1976) confirmed and elaborated this, finding projections to Brodmann areas 8a, 45, 46, 11, 12, 20, 21, 22 and the thalamic reticular nucleus. The ventral medial pulvinar is connected to the superior temporal gyrus, the caudal part to the middle temporal gyrus. Burton and Jones (1976) described projections to areas around the cortical auditory regions. Mauguiere and Baleydier (1978) found a connection to visuo-somatic area 7, and Benevento and Davis (1977) saw one to the prestriate areas. Finally, DeVito (1978) described connections to the primary and secondary somatosensory areas, and Niimi *et al.* (1978) found some to the limbic cortex. All the cortical connections of the medial pulvinar seem to be reciprocal and topographic. Subcortical inputs include the basal ganglia and the deep layers of the superior colliculus. In a review article, Streitfeld (1980) contended that the evidence indicates a purely auditory role for the medial pulvinar, but the findings listed above cast some doubt on this conclusion. The problems are whether there is a strict functional separation among the pulvinar nuclei, and whether the functional role of the medial pulvinar is invariable and species-independent. In any case, the connections enable the pulvinar to play an integrative or regulatory role in one or more of these sensory systems, separately or together.

The auditory areas also have several other subcortical projections. Some of these belong to the descending auditory system (Forbes and Moskowitz, 1977; Andersen *et al.*, 1980). There is a tonotopic projection from area TC to the ventral part of the MGN and bilaterally to the central nucleus of the inferior colliculus. The secondary areas project back to the dorsal MGN and the pericentral collicular nucleus. Other subcortical projections belong to the motor or other nervous systems (Whitlock and Nauta, 1956; Forbes and Moskowitz, 1977; Van Hoesen *et al.*, 1981). The secondary auditory cortex sends fibers to the dorsal putamen, the deep layers of the superior colliculus, the lateral tegmentum, the pons, and the spinal trigeminal nucleus.

A perusal of the connections mentioned above shows that the sensory speech region, area TA1, can be involved in several loops: TA1 → MGNd → TA1; TA1 → medial pulvinar →TA1; TA1 → deep layers of superior colliculus → medial pulvinar and inferior colliculus → → TA1; TA1 → TC → TA1; TA1 → frontal cortex → TA1; TA1 left → TA1 right → TA1 left.

This list of connections of the sensory speech regions is not exhaustive. Moore and Bloom (1979) described a general neocortical noradrenergic projection from the locus coeruleus. The afferent fibers build a dense network in layer I and have extensive radial collaterals in layers II, V, and VI. The locus coeruleus also innervates several nuclei of the ascending auditory projection. In rats Lidov *et al.* (1980) found a dense serotoninergic innervation in all layers of the entire neocortex, apparently originating in the dorsal and median raphé. The future may show that the sensory auditory areas, as well as other neocortical areas, have afferents from dopaminergic systems. Connections with numerous unspecific thalamic nuclei will certainly also be found.

In short, from the point of view of information processing, it seems that speech input is processed in a shallow (front-to-back) yet very wide and profusely cross-connected system.

5. How Does It Behave Physiologically?

Due to a gross lack of data, this section is very short, and some of the results are borrowed from animal experiments.

In the primary cortex the latency for binaural stimulation with a click is 50 msec, and the latency for monaural stimulation is 8 msec shorter for the contralateral ear (Romani *et al.*, 1982).

The primary cortex contains a tonotopic map in which the relative position of a frequency projection is proportional to the logarithm of the frequency (Romani *et al.*, 1982). Deep tones lie lateral, high ones medial. Thus, on a logarithmic scale, frequencies in the human voice range (250–4000 Hz) would lie roughly in the middle, corresponding to the position of the cytoarchitectonic subtype 1 region. Detailed mapping has not been done in humans, so we have to infer a lot from animal experiments. In unanesthetized monkeys, Pfingst and O'Connor (1981) found a less strict tonotopic organization, with several neurons showing exceptional behavior. Also, the majority (three-fourths) of the neurons showed nonmonotonic intensity functions with several "best intensities," implying that they are not only involved in intensity and/or frequency processing. Pfingst *et al.* (1977) showed that, with exceptions, the intensity of the responses to auditory stimuli was correlated with the monkey's degree of alertness.

In secondary auditory areas of monkeys, Pfingst and O'Connor (1981) found a tonotopic organization in some cases but only broadly tuned cells with no obvious organization in other cases.

As mentioned earlier, the mode of binaural interaction represents another dimension in the electrophysiological behavior of higher auditory neurons. In the auditory cortex this has been most thoroughly studied by Imig and his associates (Imig and Brugge, 1978; Imig and Reale, 1980, 1981) in cats. Neurons showing binaural summation are localized to cortical regions receiving callosal afferents. In the other regions, ipsilateral corticocortical connections are maintained, and the neurons show contralateral suppression.

Cellular responses to speech stimuli have been recorded in monkey primary auditory cortex by Steinschneider *et al.* (1982) (Fig. 13). Previous work had shown that monkeys can discriminate syllables composed of stop consonants (/b,p,d,t,g,k/) and vowels, so Steinschneider *et al.* used the synthetic syllables /ba/, /da/, /ta/ as stimuli. Some cortical neurons differentiated /da/ and /ba/ from /ta/, responding with one or two short bursts, respectively, reflecting a sensitivity to voice-onset-time. Other neurons differentiated /da/ and /ta/ (alveolar syllables) from /ba/ (bilabial syllable). Still others reflected the amplitude of the stimuli and could

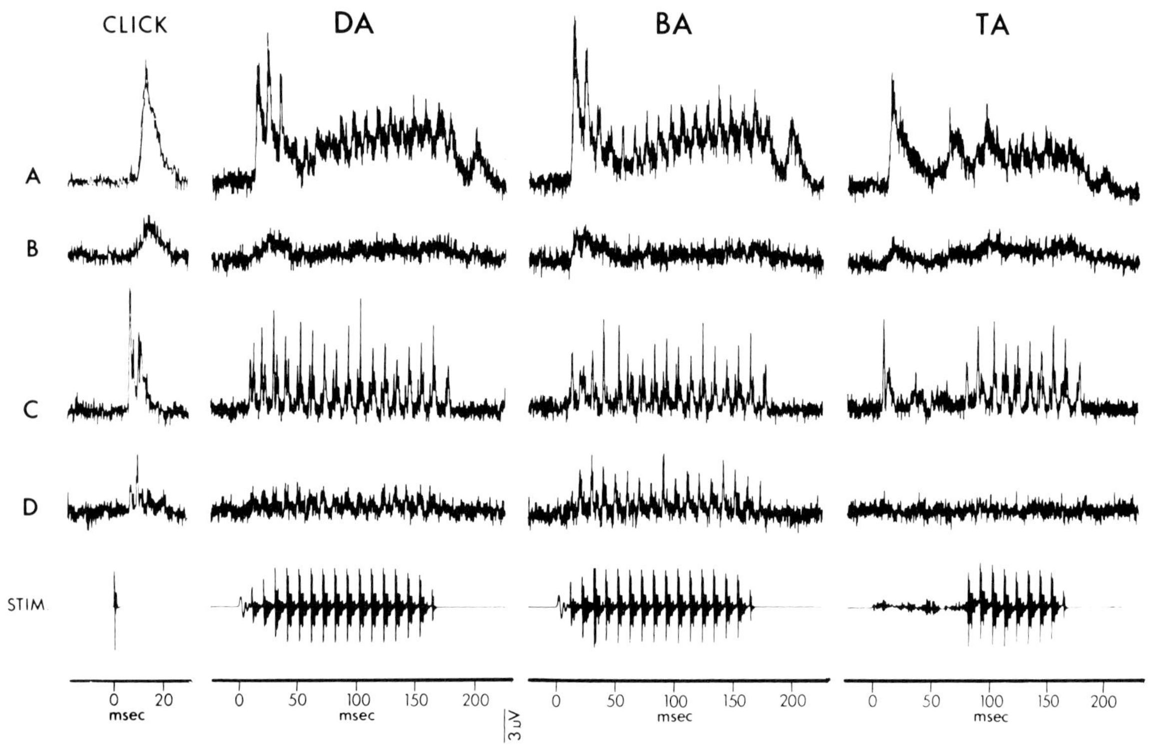

Figure 13. Multiple unit activity at progressively deeper sites (A–D) in the posterior primary auditory cortex of a monkey. A and D were recorded simultaneously, as were B and C. A is a cortical neuron in lamina III, B is one in upper lamina IV. C is an afferent axon in lower lamina IV, D is one in the subcortical white matter. A shows a summated response to all three syllables /da,ba,ta/; C shows phase-locking. The responses in A and C reflect voice-onset-time differences (compare /ta/ with the others). D responds more strongly to /ba/ than to the others, thus differentiating according to place of articulation. Note that the responses to the vowel /a/ are influenced by the preceding consonant. From Steinschneider *et al.* (1982) with permission.

thus hypothetically distinguish voiced from unvoiced consonants. About 20% showed phase-locking to the fundamental frequency of the syllable. The amplitude of the phase-locked responses increased during formant transition, serving as a measure of the latter, which could distinguish stop consonants from semivowels (see Section 6.3). [Corroborating evidence comes from a study of cat primary auditory cortex, in which Whitfield and Evans (1965) found neurons which responded most vigorously to frequency changes (see also Whitfield, this volume).] Many cortical units showed simultaneous discrimination of more than one parameter, reacting differently to each stimulus syllable.

Interestingly, the response to the vowel part of the stimulus was influenced by the preceding consonant, implying a summation effect. This also corroborates some neurolinguistic studies (Blumstein and Stevens, 1980).

The most surprising point of the study was that, contrary to expectations, the differentiated responses could be found even more easily in the cortical afferents than in the cortical neurons. The major differences were that 20% of the cortical neurons showed sustained responses (vs. 3% of the afferents), and only 20% were phase-locked to the stimulus (vs. 44% of the afferents).

A further differentiation in response properties has been described by Vaadia *et al.* (1982). In a discrimination reversal task, monkeys were presented with two stimuli and had to choose alternatingly between two possible reactions. Four-fifths of the auditory cortical neurons recorded during the task responded the same to both decisions, implying that they are purely sensory in nature. However, 9–17% of the neurons showed a response component dependent on the stimulus and another dependent on the required reaction. The remarkable points are that the associating units are not clustered (e.g., in columns) and that they are evenly distributed throughout both primary and secondary cortices. These findings contrast with Hyvärinen's (1981) in the parietal cortex and with the indication by Smith and Moskowitz (1979) of electrotonic junctions between auditory cortical cells (with the implication of functional coupling).

Lability in responses at the single-cell level was recorded in monkeys by Glass and Wollberg (1979). They used species-specific calls as stimuli and found that the great majority of neurons in the secondary auditory cortex responded to at least one call (but showed more selectivity than in the primary cortex), and that the selectivity for calls and/or the response to individual calls changed in the course of only a few hours. As in earlier reports of lability in primary and secondary cortical neuron responses (Pfingst *et al.*, 1977; Manley and Müller-Preuss, 1978), the cause or source of the instability was not determined.

Penfield's stimulation experiments on human auditory cortex gave a rough indication of functional characteristics of the neurons there (Penfield and Pernot, 1963). Stimulation in the left superior gyrus led to vague impressions of "voices far away" or a "radio," in any case involving phonetic perception. On the other hand, stimulation in the right superior gyrus yielded reports of "music," "singing," or "an instrument," even in a case in which speech was localized to the RH. Audiovisual experiential responses were elicited bilaterally from the lateral superior temporal gyrus and from the anterior supratemporal plane (area TA2). These areas notably overlap with the sensory speech region but not with the transverse gyri or the planum temporale, i.e., they coincide with area TA as defined by von Economo and Koskinas. Mixed experiences were also elicited at the posterior end of the left superior temporal gyrus. Responses were definitely labile—stimulation at the same place two or more times, even with pauses, led to different responses or extinction.

Molfese and Schmidt (1983) stimulated college students with various consonant–vowel combinations and recorded the auditory evoked responses (AERs) from temporal and parietal positions. Using factor analysis of the AERs averaged over each hemisphere, they found a factor which reflected LH changes to consonants independent of the vowels and which had a maximum at 450 msec, i.e., after completion of the syllable. Another factor was related to the consonants, had a peak latency of 170 msec, but showed no lateralization. [The significance of the latencies is unclear, since latencies recorded with MEG are only 50 msec (Romani *et al.*, 1982).] Hemispheric differences were variable among consonants (Molfese, 1978, 1980b), and Molfese (1980a) contended that phonetic differentiation based on frequency discrimination occurs in the LH, whereas voice-onset-time (VOT) discrimination occurs more in the RH. (The latter contention contrasts with the results of dichotic-listening tests mentioned in Section 2.) As far as vowels are concerned, Molfese and Erwin (1981) found no hemispheric differences in the AER after stimulation with vowel sounds. They did, however, say that at no single recording site were all vowels distinguished and implied that the task of differentiating vowels is distributed through temporal and parietal regions. Davis and Wada (1977) looked at the power spectrum of the AER, finding that the range 3–9 Hz contains 65–75% of the total power. Stimulation with speech leads to a high correlation among activities at LH electrodes, but not to a clearly greater amplitude (which contradicts the rCBF studies).

6. What Are the Sensory Auditory Areas Doing?

It is of course a delicate matter when an anatomist ventures into the field of neurolinguistics, but a synthesis of the two fields must start somewhere. Also, a review of psycholinguistics for the uninitiated anatomist would go beyond the scope of this review, so the interested reader is referred to Chapter 5 of Clark and Clark (1977) for an explanation of linguistic terms and details. I must emphasize that the conclusions reached here are speculative and will hopefully be proven, revised, or refuted in the coming years.

6.1. Summary of Neurolinguistic Findings

Before we try to reconstruct language processing in the brain, we should briefly review pertinent findings that require an explanation. Since psycholinguistic research to date indicates a role for the auditory areas in the phonological processing of speech signals, the emphasis in this summary will be on this aspect. For sake of completeness, and because it is not always possible to disentangle phonological from lexical or maybe even syntactical processing, reports on the latter areas are also included. The following summary is, however, only meant to be an indexed reference list for the discussion. Since the precision of localization varies considerably in research reports (from hemispheres to individual gyri), we have to restrict our interpretation to the level of the original work.

A. Speech perception generally
   1. From Studdert-Kennedy (1976):
      1. The temporal spectrum of the signal is important.
      2. We need a short-term auditory memory (about 100 msec).
      3. Vowels are distinguished by their formants.
      4. Vowel differentiation is continuous.
      5. The perception of consonants occurs in phoneme-length units since the consonants are so short; vowels are rests between consonants, allowing time for identification of the latter.
      6. VOT boundary is in the range 20–50 msec, also for nonspeech stimuli, and it varies with language environment and adaptation.
      7. Processing of pitch and phonemic information is independent and parallel and occurs in the left temporal and central brain regions.
   2. From Blumstein *et al.* (1977a,b):
      1. Vowels are recognized on acoustic parameters.
      2. Dichotic differentiation of vowels is not influenced by the number of their shared phonetic characteristics, implying that vowels are not phonetically recoded.
      3. Consonants are encoded.
      4. There is categorical perception and a right-ear advantage for consonants and truncated vowels.
   3. From Bond (1976):
      1. The primary perception unit appears to be a syllable or phrase, and individual sounds are identified within this context.

B. The left hemisphere
   1. General
      1. Overall rCBF increases during listening to words (Nishizawa *et al.*, 1982).
      2. LH lexical information includes visual, phonetic, and semantic characteristics of words and is thus more complete than in the RH (Leiber, 1982) where more purely visual information is stored.
      3. There is no overall activity increase during automatic speech (repeating numbers) (Larsen *et al.*, 1978).
   2. Posterior regions
      1. With lesions here, patients retain superordinate lexical categories but confuse or ignore distinguishing features; they do not separate human from nonhuman categories; they show broadened lexical classes; functional (use) association is also disrupted (Caramazza and Berndt, 1978; Zurif and Blumstein, 1978); with lesions, patients are insensitive to linguistic antonyms (Gardner *et al.*, 1978).
      2. With lesions, patients make phonemic and semantic substitutions in speech, resulting in paragrammatic speech or jargon; however, syntax is apparently retained (Caramazza and Berndt, 1978).

3. With lesions, patients make more consonant than vowel errors; substitution errors involving a single phonetic feature (voicing, place of articulation, manner of articulation) are most common (Blumstein, 1981).
4. With lesions, patients show poorer processing of phonological kana than ideographic kanji symbols (Sasanuma, 1975).
5. During auditory word recall, regional blood flow increases here (Maximilian *et al.,* 1978).

3. Anterior regions
   1. With lesions, patients understand concrete nouns and associations better than abstract ones (Caramazza and Berndt, 1978); patients make errors in the hierarchical organization of nouns; lexical classes are more functional than taxonomic (Caramazza and Berndt, 1978; Zurif and Blumstein, 1978); patients can group nouns according to their "best fit together," but cannot do this for function words (which they understand poorly) (Caramazza and Berndt).
   2. Sentence production is simplified, grammatically impoverished, and telegramlike; syntax is disturbed (Caramazza and Berndt; Zurif and Blumstein); patients cannot integrate lexical items into certain syntactical frames.
   3. With lesions, patients preserve time–space–quantity perception (Zurif and Blumstein).
4. Posterior upper temporal region
   1. The rCBF increases 29% during binaural stimulation (Nishizawa *et al.,* 1982); it increases during monaural stimulation (Maximilian, 1982) and during automatic speech (Larsen *et al.,* 1978).
   2. The rCBF here is reduced in sensory aphasia (Larsen *et al.,* 1978).
   3. Categorical perception of consonants is localized here (Studdert-Kennedy, 1976); the discrimination of VOT may take place here; it is a prerequisite for correct labeling of voiced/voiceless consonants but not for language comprehension (Blumstein *et al.,* 1977a,b); in sensory aphasia the perception of consonants is disturbed more than that of vowels (von Monakow, 1914); vowel discrimination shows no categories.
   4. The local EEG varies according to whether the subject is distinguishing pitch or phonetic classes, implying independent processing of these two characteristics; the EEG does not vary between pitch and formant transition identification (Studdert-Kennedy, 1976; Wood, 1977).
   5. With lesions here, auditory sequence perception is also disturbed (von Monakow, 1914).
   6. Naming is only disturbed sometimes during electrical stimulation here (Ojemann and Whitaker, 1978).
5. Posterior middle temporal region
   1. Naming is disturbed during stimulation (Ojemann and Whitaker, 1978).

6. Broca region
   1. There is no rCBF increase while listening to words (Nishizawa *et al.*, 1982).
   2. This region is not always involved in cases of motor aphasia (von Monakow, 1914; Larsen *et al.*, 1978; Blunk *et al.*, 1981).
7. Face area of Rolandic and premotor regions
   1. The rCBF increases during automatic speech and is reduced in cases of motor aphasia (Larsen *et al.*, 1978).
   2. Naming is disturbed during stimulation (Ojemann and Whitaker, 1978).
8. Insula
   2. Naming is always disturbed during stimulation of the posterior part and only occasionally during stimulation of the anterior part (Ojemann and Whitaker, 1978).
   2. Lesions here correlate with motor aphasia (Marie, cited in von Monakow, 1914; Blunk *et al.*, 1981).
9. Supplementary motor region
   1. The rCBF increases during automatic speech (Larsen *et al.*, 1978).
10. Regions surrounding the Wernicke–Broca complex
    1. Lesions here lead to poorer processing of ideographic kanji than of phonological kana symbols, implying a form of transcortical aphasia (Sasanuma, 1975).

C. The right hemisphere
1. General
   1. rCBF increases generally while listening to words (Nishizawa *et al.*, 1982) and during automatic speech (Larsen *et al.*, 1978).
   2. The RH recognizes principally concrete words (Marin *et al.*, 1979).
   3. Words are stored as visual images of themselves or their referents (Leiber, 1982); ideographic kanji words are processed better than in the LH in contrast to phonological kana words (Sasanuma *et al.*, 1977).
   4. "Emotional" words are stored better than in the LH (Wogan and Kaplan, 1981; Landis *et al.*, 1982).
   5. This hemisphere helps process linear, spatial syllogisms (A is taller than B—who is smaller?); it shows a preference for synonyms and is insensitive to nonlinguistic antonyms (Gardner *et al.*, 1978).
   6. Damage in the RH leads to dysarthria, word production defects, sentence anagram deficits, and visual perception defects (Hécaen, 1978; Cavalli *et al.*, 1982).
   7. The RH is more proficient than the LH in recognition and production of pitch (Sidtis, 1984).
   8. The RH has a deficient auditory short-term memory (e.g., for phrases) compared to the LH (Marin *et al.*, 1979).
2. Frontotemporal regions
   1. The rCBF increases during left-ear verbal stimulation (Maximilian, 1982).

3. Posterior upper temporal region
   1. The rCBF increases 18% locally while listening to words (Nishizawa *et al.*, 1982).
   2. There is no difference in the local EEG while the subject is distinguishing pitch or phonetic classes, or pitch or initial formant transition (Studdert-Kennedy, 1976).
   3. The rCBF increases locally during discrimination of nonverbal sounds (Roland *et al.*, 1981) and during automatic speech (Larsen *et al.*, 1978).
4. Rolandic face area and supplementary motor regions
   1. Activity increases during automatic speech (Penfield and Pernot, 1963; Larsen *et al.*, 1978).
5. Inferior posterior frontal region
   1. The rCBF increases during discrimination of nonverbal auditory stimuli (Roland *et al.*, 1981).

With our current knowledge it is of course impossible to explain all of the phenomena summarized, yet many of them do fit into a reasonable scheme. In the discussion we should first mention the problem of parallel versus sequential processing of speech signals generally. Then we shall discuss various aspects of phonological processing: consonants versus vowels, lateralization, the time segments involved in processing (unit of perception), and pitch versus phoneme discrimination. Finally, we shall quite briefly mention the higher-level problem of lexical organization and localization, and the role of the RH.

6.2. Parallel or Sequential Processing?

The debate in the psycholinguistic literature centers around the question of which perception processes run sequentially and which in parallel. The answer will clearly be a combination of the two: sequential processing since the input is sequential, and parallel processing due to the anatomy of the system and to the obvious time limitations.

Marslen-Wilson and Tyler (1980) tested subjects for recognition of a target word, given the word, a rhyming word, or the taxonomic category of the word as a cue. The target word was embedded in a normal sentence, a syntactically correct but nonsense sentence, or a scrambled "sentence." Judging from the speed with which subjects recognized the target word (and pressed a button), Marslen-Wilson and Tyler concluded that recognition had occurred after only 200 msec of presentation in a normal sentence, i.e., before the presentation of the 367-msec word was completed. Even if only the category was given as a cue, recognition was completed by the time word presentation was finished. Supporting evidence for this processing time comes from Molfese (1980a,b), Romani *et al.* (1982), and Rugg (1983)—they recorded the AER after presentation of speech sounds or words and found peaks around 50, 100, and 200 msec, from which we infer that the cortex processes the stimuli for 150–200 msec. Rugg (1983) concluded that the word/nonword decision is completed in less than 200 msec. Marslen-Wilson and Tyler interpreted their (own) findings to mean that word recognition is an interactive process involving parallel lexical, syntactic, and even interpretive searches which successively narrow the range of possible choices immediately upon reception of each cue. Their model is compatible with Leiber's (1982) experiments indicating that LH word storage is based on several parameters (B.1.2). One inherent difficulty in Marslen-Wilson and Tyler's tests is that the subjects were given cues, i.e., they were expecting a particular word or kind of word. Thus, the role of "attention" in facilitating word recognition cannot be ruled out. When the target word was embedded in a scrambled sentence, reaction times increased by 90–150 msec, which is thus the amount of time saved by parallel syntactic and interpretive searches possible with a normal sentence. The final conclusion is that word recognition can (note emphasis) be a result of interactive, parallel searches.

Another argument for parallel processing derives from the work of Blumstein *et al.* (1977a) who showed that in aphasic patients, discrimination of VOT (voiced/voiceless consonants) is not a prerequisite for speech comprehension. In normal subjects, Warren (1976) found that "complete" phonetic sequences were perceived even in cases in which a phonetic unit was missing, i.e., the missing unit was filled-in—this phenomenon has come to be known as the phonemic restoration effect. In either case the brain is clearly capable of "perceiving" or guessing a correct message from an incomplete signal.

Caramazza *et al.* (1983) discriminated three types of word processing: acoustic processing (including consonant transitions), phonemic discrimination, and phoneme and word identification. They presented a patient whose only lack was in the third type of processing—this showed that some types are distinct, but not whether they are sequential or parallel.

Since processing at the level of word identification is apparently a parallel analysis of word components, what remains for sequential processing is at most the phonetic analysis (i.e., of acoustic features), the decision of whether a signal is speech or nonspeech, and perhaps phoneme identification.

6.3. Components of Phoneme Processing

This section is devoted to those aspects of speech processing which are presumably located in the auditory cortex.

Phoneme processing apparently consists of two sequential stages: (1) discrimination of the parts on acoustic parameters, and (2) phonological identification for meaning. Vowels are discriminated largely on their formant structure (A.1.3), but consonants are discriminated on several phonetic features. These include (a) voicing (VOT, when the vocal cords start to vibrate—e.g., /p/ vs. /b/), (b) place of articulation (bilabial—/b/, labiodental—/f/, dental—/th/, alveolar—/d/, palatal—/sh/, velar—/k/, or glottal—/h/), and (c) manner of articulation (stops—abrupt opening of the vocal tract—/p/, fricatives—constriction of the vocal tract—/f/, affricates—stop plus fricative—/ch/, nasals—closure of the mouth and opening of the nasal passages—/m/, and laterals and semivowels—from different tongue shapes—/l/, /w/). The place of articulation is reflected acoustically in the duration of formant transitions. Discrimination is, however, just part of the speech processing, and many aphasic patients can distinguish phonemes without being able to identify them (Blumstein, 1981).

6.3.1. Consonant and Vowel Processing

Speech stimuli probably, as expected, excite a set of acoustic afferents related to the formant structure of the stimuli (Voigt *et al.*, 1982). The bilateral structure of the ascending auditory system implies that acoustic input from one ear is distributed to both auditory cortices (B.4.1, C.3.1), yet the input reaches the contralateral hemisphere sooner. The difference between vowel and consonant processing is apparently in the length of the signal, since vowels truncated to 50 msec are processed like consonants.

The border between acoustic cues, e.g., between voiced and voiceless consonants, arises somewhere in the ascending system and is also valid for nonspeech stimuli (A.1.6). One population of cortical afferents carries indiscriminately voiced cues, and another one voiceless ones (Steinschneider *et al.*, 1982). Adaptation experiments corroborate this. By habituating the human auditory system with one category and thereby exhausting the response, it can be shown by testing with borderline VOTs that the border is shifted toward the adapted response A.1.6, Clark and Clark, 1977). For example, by adapting with the voiced consonant /d/, signals with borderline VOTs which were previously perceived as /b/ are then perceived as /p/. Similar adaptation results can be obtained for other acoustic cues.

Since single auditory thalamocortical axons already discriminate speech stimuli on one or more acoustic parameters (Steinschneider *et al.*, 1982), the question remains as to the role of the auditory cortex. As mentioned above, neurolinguistic and evoked-response studies have indicated that the auditory areas in both hemispheres do acoustic processing, i.e., can distinguish signals according to their acoustic parameters. What these studies may have tested is the cortical afferents and/or the first cortical processing stage. Since both the afferents and the primary cortical neurons recorded by Steinschneider *et al.* (1982) responded to acoustic parameters rather than to individual phonemes, they cannot be doing the next stages (speech/nonspeech decision; phoneme identification). The cortical neurons do often show three response characteristics not associated with the afferents, and which may give a clue as to their function: (1) the "on" response at the beginning of a signal (the first 50 msec) is much stronger and lasts longer (correlating with the AER peak at 50 msec); (2) the phase-locked response is superimposed on an integrated response (activity increases for the duration of the syllable); (3) there is an "off" response at the end of the signal. If phoneme identification occurs in the primary cortex, it must be represented by an activity pattern in apparently disjoint sets of neurons. If it is equivalent to activity in particular single units or columns, these apparently must be in the secondary auditory areas. This is where the hemispheric lateralization could take place (Wood, 1977).

Besides Steinschneider *et al.* (1982), other groups have found cortical auditory cells which could contribute to phoneme processing. In cats Whitfield and

Evans (1965) found neurons which were sensitive only to rapid frequency modulations, and which could thus help discriminate frequency transitions in consonants. More complex feature-detecting neurons were found in the superior temporal cortex of monkeys by Wollberg and Newman (1972). They described a class of cells which have high spontaneous activity and which show complex responses to many different species-specific calls. Another class has low spontaneous activity and a simple response to only one particular call. The question remains as to the morphological identification or position in the cortical circuitry of such cells. Also, as yet no synthetic signals have been used to determine which characteristics of the calls are related to the response.

The discrimination on acoustic parameters which is present in the afferents has to be refined to discrimination of phonemes at the cortical level. Apparently, the first step in phoneme identification is categorical encoding based on the first 50 msec of a signal. Categorical perception implies that different signals excite disjoint subpopulations of neurons and that there is a lack of perceptual units for borderline signals. The amplified "on" responses of primary cortical neurons mentioned above likely play a role in defining a categorical response. For such short signals, the 8-msec head start of input from the contralateral ear is a factor in determining which hemisphere will do the processing. The temporal advantage should help in those bands of cortex in which contralateral input suppresses ipsilateral input (see Section 6.3.2). The categorization is most pronounced and best studied for consonants differing only in VOT. Since the VOT boundary between voiced and voiceless consonants lies between +20 and +40 msec, there can be about twice as many cortical convergence steps for the first formant in voiceless consonants as in voiced ones. The additional convergence steps must suffice to define a new category of responding neurons. Since truncated vowel sounds are also perceived in disjoint sets, the reaction of the auditory cortex to the first 50 msec of any speech stimulus is to assign it to a category. Such short stimuli are processed to varying degrees better in the LH (A.1.5, A.2.3–4), but there is as yet no evidence indicating an anatomical separation between consonants and vowel analysis within the LH.

The anatomical evidence fits with a model of LH phoneme identification. Discrimination of acoustic cues could be found in cortical afferents and neurons in both hemispheres, but final identification and retention is reserved for the LH. The greater extent of areas TC, TB, and TA1 in the LH could indicate relatively increased consonant processing associated with phoneme identification. According to our estimates above, the left auditory areas are better suited for specific responses due to the wider neuron column spacing and lesser dendritic overlap among columns. Since the dendrites of LH columns interact with more afferents than their RH counterparts, due to the greater tangential extents (at least in TA and TC), the columns should be more responsive to converging afferent signals and, which is very important, more capable of establishing the early categorical response. Activity of cells in a column can be synchronized in the excitatory or inhibitory sense by narrow (double-tufted) cells. Lateral inhibition is likely mediated by smooth stellates, whose dendritic trees indicate a less precise data sampling and whose axons have long horizontal expansions. The role of Ramón y Cajal's specific giant cells in defining columnar activity is unclear.

Although the dendritic tree indicates a clearly columnar information sampling, the axon shows little obvious organization, except that it lies completely below the soma. It is also presently hard to elucidate the role of pyramidal neurons in delineating responsive regions.

The role of the longer stimulus for vowels is a question for further speculation. Both cortices can process vowel stimuli purely by frequency analysis (A.1.3–4, A.2.1–2) due to the long signal duration (A.1.2), and phonetic features (grave, tense, compact) play no role in distinguishing dichotically presented vowels (Blumstein *et al.*, 1977b). The processing "centers" for various vowels may be spatially distinct within each hemisphere (Molfese and Erwin, 1981). The integrative response of cortical cells mentioned above modifies the initial reaction to the first 50 msec of a speech stimulus, i.e., probably modifies the categorical nature of the early response and the distribution of responding cells. Steinschneider *et al.* (1982) remarked that the cortical neuronal response to a vowel is influenced by the preceding consonant. The question remains as to what happens to the stimulus after the first 50 msec. Interhemispheric transfer is possible but apparently not necessary since split-brain patients do not have sensory aphasia. A role of interhemispheric summation in vowel perception would imply activity in columns receiving callosal connections (Imig and Brugge, 1978; Imig and Reale, 1980, 1981; see also Section 6.3.2)—recordings directly from the corpus callosum would certainly help elucidate its role. Such binaural cortical columns are also unilaterally excitable, although less so with reduced binaural summation (i.e., only via the ascending auditory system). In connection with vowel perception we could also see a role for pyramidal cells in distributing the signal to other cortical areas or laterally to disjoint columns in the same area (Rockland *et al.*, 1982).

Crucial experiments on neuronal responses and the topography of consonant and vowel perception will probably continue to be made with animals. The work of Steinschneider *et al.* (1982) should be expanded. Considerable progress could be made by combining electrophysiological recordings with tracing in monkeys trained to differentiate speech signals. Extensive injections of labeled leucine/proline in the RH could anterogradely delineate the callosal projection in the LH, and current-source-density analysis (Mitzdorf and Singer, 1978) in the tangential plane could relate patterns of activity flow to these projections. On the anatomical side, what must be done is to better define the inhomogeneities of afferent distributions in human auditory cortex, hopefully relating them to animal models.

Ojemann and Mateer (1979) implicated other cortical areas in phoneme identification. They tested it by presenting one of /p/, /b/, /t/, /d/, /k/, /g/ in the syllable /a_ma/ during neurosurgical operations. Identification was disturbed by simultaneous electrical stimulation not only in the superior temporal gyrus but also in the supramarginal region and sometimes in an area anterior to Broca's region. Stimulation in all three areas also interrupted sequences of facial movements, leading Ojemann and Mateer to support a theory that motor imitation (at least mentally) is an integral part of phoneme identification. They could not test whether all three areas represented the same or different stages in phoneme identification.

6.3.2. Right-Ear Advantage

The 8-msec shorter latency for right-ear stimuli (Romani *et al.*, 1982) correlates with the observed right-ear advantage (REA) for some linguistic stimuli (A.2.4, B.4.1, C.3.1). In bands of primary auditory cortex which receive only ipsilateral connections from other auditory areas, ipsilateral input (e.g., from left ear) is suppressed by contralateral input (e.g., from right ear) (Imig and Reale, 1981). (There is also a class of ipsilateral connections which is not systematically related to the binaural map.) The majority of primary auditory cortical neurons show some binaural interaction and respond more strongly when interaural stimulus intensity or temporal differences favor the contralateral ear (Phillips and Gates, 1982). Phillips and Gates interpret this and clinical reports to mean that the auditory cortex directs its attention mainly to the contralateral "sound space." In the case of simultaneous stimuli to both ears, the stimulus duration becomes important. The shorter a stimulus, the more the processing must be done in each hemisphere individually due to the longer latencies involved in binaural summation via the callosum. In the analysis of longer stimuli (vowels), binaural summation, which occurs in regions supporting callosal connections, can play a greater role. This model does imply some spatial localization of perception—vowels would be preferentially processed bilaterally in cortical regions with callosal connections.

Recently, Dwyer *et al.* (1982) have presented evidence that only the abrupt onset of consonants is responsible for the REA. They tested consonant–vowel syllables at durations of 300, 40, and 20 msec, and found that the magnitude of the REA for consonant perception was the same at all three durations, thus excluding duration as a sole determinant. This corroborates the finding that the REA is greater for stop consonants than for liquids (Cutting, 1974) and implies that the LH auditory areas are superior to the RH in the categorization of abrupt-onset signals. In apparent contradiction to other findings [A.1.3–4, A.2.4, Godfrey (1974)], Dwyer *et al.* (1982) found an REA for vowels in the 300-msec signal. The difference is probably due to the consonant–vowel stimuli they used (instead of pure vowels), corroborating the contention by Steinschnedier *et al.* (1982) that the cortical neuronal response to vowels is influenced by the preceding consonant.

Whether the contralateral-ear advantage is bilaterally symmetrical has recently been questioned by Sidtis (1982), who found that half of his right-handed subjects did not show both an REA for speech and an LEA for complex pitch stimuli.

In the future the REA could be better differentiated and quantified by combining brain electrical activity mapping (see Section 2.2) with monaural and dichotic speech-related stimuli.

6.3.3. Unit of Perception

That temporal summation is important for comprehension is obvious (A.1.1, A.1.5, A.3.1); the summation seems to occur at several levels. There is a short-term summation around 100–250 msec (A.1.2), which corresponds to vowel or syllable length and, interestingly, to the gestalt fusion frequency (Foster, 1971). (The gestalt fusion frequency is 4–6 Hz and is the flicker frequency at which a picture appears to be "continual," i.e., at which the cortical retention bridges the gap between actual appearances of the picture.) This time segment is necessary and sufficient for detailed frequency analysis of vowels and for encoding and retention of consonants (A.1.3–5, A.2.1–2). It also corresponds to the time between the first and last (third) major peaks in the cortical AER (Molfese, 1980b), to the time required by subjects to recognize monosyllabic words (Marslen-Wilson and Tyler, 1980), and roughly to the interval between "on" and "off" responses of cortical neurons (see above).

Another way to measure this perception unit would be an adaptation of the gestalt-fusion-frequency method. By presenting subjects with different syllables alternating at various frequencies, around 5 Hz we should find a transition between perception of distinct syllables and perception of a mixed signal.

Other temporal segments in speech comprehension are longer and must be related to further, more widespread cortical and subcortical summation (Fig. 12; see also Desmedt, 1979). In the AER there are as yet unexplained variable late peaks (latency at least 300 msec) which may be related to longer time units in perception, as well as to input selection (attention) or even motor responses. There is a certain redundancy in the long-term temporal summation (perhaps due to parallel or converging paths) which can lead to auditory illusions and to the phonemic restoration phenomenon.

6.3.4. Pitch versus Phoneme Discrimination

EEG recordings indicate that pitch and phoneme discrimination in the LH are not identical processes, whereas they seem to be identical in the RH (A.1.7, B.4.4, C.3.2), implying that the RH does not perform phoneme identification. Pitch and formant-transitions seem to be discriminated the same way in each hemisphere. Wood (1977) tested this by presenting /bæ/ and /gæ/, each at two different fundamental frequencies. In temporal and central AER recordings, he found hemispheric differences related to /bæ/ and /gæ/ but not to the fundamental frequencies. Almost all of the studies on phoneme discrimination have taken great trouble to eliminate the pitch variable, often by the use of synthetic stimuli with a fixed fundamental frequency.

In both hemispheres pitch must be discriminated in structures with a tonotopic frequency representation. This is the case in the entire principal ascending system and in the primary auditory cortex. No hemispheric differences in frequency maps have yet been reported. The tonotopic order breaks down only in secondary or higher regions (Imig and Reale, 1980).

6.4. Hierarchy in Word Grouping

A spatial division in the recognition of vowels was indicated by Molfese and Erwin (1981) as mentioned above. Spatial separation in the recognition of nouns versus verbs beyond the boundaries of the auditory cortex was indicated by Brown and Lehmann (1979). They stimulated with homophone words whose noun or verb meaning was derived from the context. In the AER they found different topographical locations for maximum and minimum field values dependent on the word category but independent of the language (English or Swiss-German).

Differential noun processing in the two hemispheres, inferred from variations in processing strategies, is based on the visual form of words and their phonetic and semantic characteristics, including their emotional content (B.1.2, C.1.3–4). It also indicates that the noun lexicon is not purely holistic but may show a hierarchy or taxonomic grouping. In aphasia, especially sensory aphasia, the taxonomic grouping of nouns is defective (B.2.1, B.3.1). This can occur by disturbance in the storage of the nouns and/or disturbance in the access to the nouns. Distortion or destruction of auditory perception of course disrupts the assignment of auditory stimuli to categories. An example of this is the failure of some aphasics to differentiate and label VOTs. The breakdown in noun hierarchy could be the same phenomenon at a higher level. Whitehouse *et al.* (1978) found that sensory aphasics could not differentiate schematic pictures into categories (glass, cup, bowl) and concluded that the word meaning or the organization of the lexicon was disrupted. This would imply that the lexicon is at least partly congruent with the sensory speech area. Stimulation in the posterior temporal, insular, and lower motor regions leads to disturbances in naming (B.4.6, B.5.1, B.7.2, B.8.1), but the disturbance can equally likely be in the lexicon or in its input/output channels.

The possibility that auditory comprehension deficits are not so much due to disturbances in semantic organization as to deficient access to semantic information received support from Caramazza *et al.* (1983), who presented a case study of a patient who has difficulty in perception and production of phonological sequences but whose lexicon is quite intact (when tested by visual recognition). The patient can identify nonspeech sounds and can discriminate (but not identify) syllables. A CT scan indicated roughly an infarct of areas TA1 and PF. However, the neurolinguistic tests could not exclude defects in the output as an additional explanation for the patient's performance. In any case, the debate about defect access versus disturbed organization may itself be of a semantic nature, since the organization of a file partially determines the access to it.

Further progress in determining whether lexical classes have distinct topographical locations in the brain could come from extended rCBF or PET experiments on a very homogeneous group of subjects. Using single word categories (e.g., nouns or verbs) as stimuli, hot spots in scans could give a clue to localization. Slight increases in activity may only become apparent after correlating measurements at corresponding anatomical points in many scans. A negative result would, however, indicate either the absence of a topographic map or considerable individual variability.

An interesting and relevant anatomical study would be the demonstration of inhomogeneities in the projections from TA1 to other cortical and subcortical sites.

Lesion analysis indicates that the lexicon and syntax can be independently disrupted and thus do not overlap completely (B.2.1–2, B.3.2). The neuroanatomical basis for the latter may be closely related to the circuitry for motor speech production (Marin *et al.*, 1979).

6.5. Right Hemisphere

Molfese (1980a) found that the RH does detect VOT boundaries, and other studies (C.1.1, C.2.1, C.3.1) showed increased RH blood flow while listening to words, indicating RH participation in speech perception. Current discussions about the linguistic capabilities and role of the RH center around (1) its relation to visual processing [C.1.2–3, C.1.5–6, Levy-Agresti and Sperry (1968)], (2) its relation to "emotional" words (C.1.4), which may actually be a subset of its holistic processing capabilities, (3) its lack of ability to process syntactical sequences (C.1.8), and (4) its role in processing melody (Penfield and Pernot, 1963; C.1.7).

Springer and Deutsch (1981) gave a good discussion of the visual processing, which need not be repeated here.

The preference of emotional words for the RH cannot be easily explained. A possibility is that emotional words such as those tested by Landis *et al.* (1982) are more easily visually imageable than abstract, nonemotional words. The RH could participate in the accompanying visual processing. Another possibility is that emotional words evoke a more general, i.e., holistic reaction than do abstract words. Kutas and Hillyard (1982) found that concluding a sentence with a semantically anomalous word led to a large negative peak (N400) in the AER over the RH. However, they could not tell whether the effect was due to language processing or perhaps to an emotional reaction to a "false sentence." The RH advantage for other abstract sounds, e.g., in differentiating nonverbal stimuli, must derive from other mechanisms. Without wanting to overdo a comparison, we could point out that the anatomical structure of RH auditory areas corresponds to a more "holistic processing" than the LH, i.e., the dendritic overlap among cortical columns is greater, allowing for the possibility of more joint (synchronous) responses (Seldon, 1982).

Although studies on hemispherectomized or commissurotomized patients imply that the RH has little or no ability to generate phonological sequences alone (Marin *et al.*, 1979), CBF studies indicate RH participation in speech production (C.1.1, C.1.6, C.3.3, C.4.1).

The role of the RH in pitch analysis has been discussed above.

7. Summary and Points for Future Thought

Starting from the top, we have seen that our auditory cortex, especially that of the left temporal lobe, is involved in the perception, acoustic analysis, and phonological identification of parts of speech. What differentiates the LH auditory regions from the RH ones is that only the former are capable of retaining, encoding, and identifying very short (around 50 msec) signals such as appear in some consonants. It is probably this ability which enables the LH to process speech. At the level of electrophysiology we have seen that cortical afferents and primary cortical neurons differentiate speech signals on acoustic parameters but do not unequivocally identify individual phonemes. As yet no hemispheric differences have been reported from intracortical recordings. Recordings of acoustic-evoked potentials in people have shown a lateralization to the LH for processing of some consonants but not for vowels. The peaks in such recordings correspond in part with response peaks of individual neurons and in part with reaction times in recognition tests. At the anatomical level we have seen that the primary auditory areas are the same size in both hemispheres, but that the secondary areas are clearly larger in the LH. There is a strongly columnar organization in the secondary areas TB and TA1 and less so in the primary area TC. The most granular part of area TC corresponds roughly to the tonotopic projection of the frequency range in human speech. Neurons of the LH generally have a larger tangential dendritic extent, implying a more intense interaction with afferents. However, the center–center distance between columns in the LH is so much greater than in the RH that the relative dendritic overlap among neighboring columns is actually less in the LH, implying a better segregation of their input and more independence in their responses.

The most glaring gap in our knowledge about our auditory cortex is at present the electrophysiology, which is also the most inaccessible to research. There is a limit to the extrapolations we can make from animal models. The localization (or lack thereof) of neurons reacting to speech stimuli remains a mystery, as does their response to frequency or intensity functions. The most direct solution to this problem would be a refinement of Penfield's and Ojemann's work; the most likely solution would be refined measurement and analysis of the AER. In either case the use of defined, reproducible, synthetic signals will be necessary to differentiate the roles of various populations of auditory neurons.

As far as the anatomy is concerned, three main problems await solution. The connections of our auditory cortex, described above from animal experiments, remain to be confirmed. Within the cortex itself, the density and spatial distribution of individual afferents and of axons of the various neuron types remains poorly known. Yet, this, as well as the dendritic spread, determines the interconnectivity of neuronal columns. For lack of data we have assumed, for the estimates presented here, similar densities and distributions in both hemispheres. This assumption will probably have to be revised. The last and most prodigious problem is that of the specificity of intracortical connections. The answer would certainly reveal in detail the functional circuitry of the auditory areas and confirm or refute the current working assumptions.

It is hard to point to a single missing item in the neurolinguistic research, since this has emphasized determinations of laterality for speech functions or separation of semantic, lexical, and syntactical functions. From the purely linguistic point of view it is difficult to approach the local functional topography and circuitry of the auditory cortex, and it is likely that substantial progress will be made only in connection with AER analysis as mentioned above.

ACKNOWLEDGMENTS. I wish to express my deep appreciation to Drs. M. Davies, R. Eckhorn, A. M. Galaburda, B. Grover, O.-J. Grüsser, and S. Kelter for critical reading of the manuscript during several stages of its development. I also wish to thank Dr. P. H. Schreier for the use of his word processor.

8. References

CHAPTER 7

Andersen, R. A., Snyder, R. L., and Merzenich, M. M., 1980, The topographic organization of corticocollicular projections from physiologically identified loci in the AI, AII, and anterior cortical fields of the cat, *J. Comp. Neurol.* **191:**479–494.

Appell, J., Kertesz, A., and Fisman, M., 1982, A study of language functioning in Alzheimer patients, *Brain Lang.* **17:**73–91.

Bailey, P., von Bonin, G., and McCulloch, W. S., 1950, *The Isocortex of the Chimpanzee,* University of Illinois Press, Urbana.

Benevento, L. A., and Davis, B., 1977, Topographical projections of the prestriate cortex to the pulvinar nuclei in the macaque monkey: An autoradiographic study, *Exp. Brain Res.* **30:**405–424.

Benson, D. F., and Patten, D. H., 1967, The use of radioactive isotopes in the localization of aphasia-producing lesions, *Cortex* **3:**258–271.

Blumstein, S. E., 1981, Phonological aspects of aphasia, in: *Acquired Aphasia* (M. T. Sarno, ed.), Academic Press, New York, pp. 129–155.

Blumstein, S. E., and Stevens, K. N., 1980, Perceptual invariance and onset spectra for stop consonants in different vowel environments, *J. Acoust. Soc. Am.* **67:**648–662.

Blumstein, S. E., Cooper, W. E., Zurif, E. B., and Caramazza, A., 1977a, The perception and production of voice-onset time in aphasia, *Neuropsychology* **15:**371–383.

Blumstein, S. E., Tarttir, V. C., Michel, D., Hirsch, B., and Letter, E., 1977b, The role of distinctive features in the dichotic perception of vowels, *Brain Lang.* **4:**508–520.

Blunk, R., DeBleser, R., Willmes, K., and Zeumer, H., 1981, A refined method to relate morphological and functional aspects of aphasia, *Eur. Neurol.* **20:**69–79.

Bond, Z. S., 1976, On the specification of input units in speech perception, *Brain Lang.* **3:**72–87.

Broca, P., 1865, Remarques sur le siège de la faculté du langage articulé, *Bull. Soc. Anthrop.* **6:**377–393.

Brodmann, K., 1909, *Vergleichende Lokalisationslehre der Grosshirnrinde in ihren Prinzipien dargestellt auf Grund des Zellenbaues,* Barth, Leipzig.

Brown, J. W., 1975, On the neural organization of language: Thalamic and cortical relationships, *Brain Lang.* **2:**18–30.

Brown, W. S., and Lehmann, D., 1979, Verb and noun meaning of homophone words activate different cortical generators: A topographical study of evoked potential fields, in: *Hearing Mechanisms and Speech* (O. Creutzfeldt, H. Schleich, and C. Schreiner, eds.) Springer-Verlag, Berlin, pp. 159–167.

Burton, H., and Jones, E. G., 1976, The posterior thalamic region and its cortical projection in New World and Old World monkeys, *J. Comp. Neurol.* **168:**249–302.

Campbell, A. W., 1905, *Histological Studies on the Localisation of Cerebral Function,* Cambridge University Press, Cambridge.

Caplan, D., 1981, On the cerebral localization of linguistic functions: Logical and empirical issues surrounding deficit analysis and functional localization, *Brain Lang.* **14:**120–137.

Caramazza, A., and Berndt, R. S., 1978, Semantic and syntactic processing in aphasia: A review of the literature, *Psychol. Bull.* **85:**898–918.

Caramazza, A., Berndt, R. S., and Basili, A. G., 1983, The selective impairment of phonological processing: A case study, *Brain Lang.* **18:**128–173.

Cavalli, M., DeRenzi, E., Faglioni, P., and Vitale, A., 1982, Impairment of right brain-damaged patients on a linguistic cognitive task, *Cortex* **17:**545–556.

Chi, J. G., Dooling, E. C., and Gilles, F. H., 1977, Gyral development of the human brain, *Ann. Neurol.* **1:**86–93.

Clark, H. H., and Clark, E. V., 1977, *Psychology and Language,* Harcourt/Brace/Jovanich, New York.

Conel, J., 1939–1967, *The Postnatal Development of the Human Cerebral Cortex,* Volumes 1–6, Harvard University Press, Cambridge, Mass.

Cowan, W. M., and Cuénod, M., 1975, *The Use of Axonal Transport for Studies of Neuronal Connectivity,* Elsevier, Amsterdam.

Cutting, J. E., 1974, Two left hemisphere mechanisms in speech perception, *Percept. Psychophys.* **16:**601–612.

Damasio, A. R., Damasio, H., Rizzo, M., Varney, N., and Gersh, F., 1982, Aphasia with nonhemorrhagic lesions in the basal ganglia and internal capsule, *Arch. Neurol.* **39:**15–20.

Damasio, H., and Damasio, A. R., 1980, The anatomical basis of conduction aphasia, *Brain* **103:**337–350.

Damasio, H., Damasio, A. R., Hamsher, K., and Varney, N., 1979, CT scan correlates of aphasia and allied disorders, *Neurology* **29:**572.

Davis, A. E., and Wada, J. A., 1977, Spectral analysis of evoked potential asymmetries related to speech dominance, in: *Language and Hemispheric Specialization in Man: Cerebral ERP's* (J. E. Desmedt, ed.), Karger, Basel, pp. 127–139.

Davis, T. L., and Sterling, P., 1979, Microcircuitry of cat visual cortex: Classification of neurons in layer IV of area 17, and identification of the patterns of lateral geniculate input, *J. Comp. Neurol.* **188:**599–628.

Desmedt, J. E. (ed.), 1977, *Language and Hemispheric Specialization in Man: Cerebral ERP's,* Karger, Basel.

Desmedt, J. E., 1979, *Cognitive Components in Cerebral Event-Related Potentials and Selective Attention,* Karger, Basel.

DeVito, J. L., 1978, A horseradish peroxidase-autoradiographic study of parieto-pulvinar connections in *Saimiri sciureus, Exp. Brain Res.* **32:**581–590.

Duffy, F. H., 1982, Topographic display of evoked potentials: Clinical applications of brain electrical activity mapping (BEAM), *Ann. N.Y. Acad. Sci.* **388:**183–196.

Dwyer, J., Blumstein, S. E., and Ryalls, J., 1982, The role of duration and rapid temporal processing on the lateral perception of consonants and vowels, *Brain Lang.* **17:**272–286.

Efron, R., Crandall, P. H., Koss, B., Divenyi, P. L., and Yund, E. W., 1983, Central auditory processing. III. The "cocktail party" effect and anterior temporal lobectomy, *Brain Lang.* **19:**254–263.

Ferster, D., and LeVay, S., 1978, The axonal arborizations of lateral geniculate neurons in the striate cortex of the cat, *J. Comp. Neurol.* **182:**923–944.

FitzPatrick, K. A., and Imig, T. J., 1980, Auditory cortico-cortical connections in the owl monkey, *J. Comp. Neurol.* **192:**589–610.

Flechsig, P., 1896, *Die Lokalisation der geistigen Vorgaenge,* Veit, Leipzig.

Forbes, B. F., and Moskowitz, N., 1977, Cortico-cortical connection of the superior temporal gyrus in the squirrel monkey, *Brain Res.* **136:**547–552.

Foster, D. H., 1971, The response of the human visual system to moving spatially-periodic patterns: Further analysis, *Vis. Res.* **11:**57–81.

Frackowiak, R. S. J., Lenzi, G.-L., Jones, T., and Heather, J. D., 1980, Quantitative measurement of regional cerebral blood flow and oxygen metabolism in man using 15-O and positron emission tomography: Theory, procedure, and normal values, *J. Comput. Assist. Tomogr.* **4:** 727–736.

Fuster, J. M., Bauer, R. H., and Jervey, J. P., 1982, Cellular discharge in the dorsolateral prefrontal cortex of the monkey in cognitive tasks, *Exp. Neurol.* **77:**679–694.

Galaburda, A. M., and Pandya, D. N., 1983, The intrinsic architectonic and connectional organization of the superior temporal region of the rhesus monkey, *J. Comp. Neurol.* **221:**169–184.

Galaburda, A. M., and Sanides, F., 1980, Cytoarchitectonic organization of the human auditory cortex, *J. Comp. Neurol.* **190:**597–610

Galaburda, A. M., LeMay, M., Kemper, T. L., and Geschwind, N., 1978a, Right–left asymmetries in the brain, *Science* **199:**852–856.

Galaburda, A. M., Sanides, F., and Geschwind, N., 1978b, Human brain: Cytoarchitectonic left–right asymmetries in the temporal speech region, *Arch. Neurol.* **35:**812–817.

Galambos, R., and Hillyard, S. A., 1981, Electrophysiological approaches to human cognitive processing, *Neurosci. Res. Progr. Bull.* **20:**145–265.

Gardner, H., Silverman, J., Wapner, W., and Zurif, E., 1978, The appreciation of antonymic contrasts in aphasia, *Brain Lang.* **6:**301–317.

Gazzaniga, M. S., and Sperry, R. W., 1967, Language after section of the cerebral commissures, *Brain* **90:**131–148.

Geschwind, N., and Levitsky, W., 1968, Human brain: Left–right asymmetries in temporal speech region, *Science* **161:**186–187.

Gilbert, C. D., and Wiesel, T. N., 1979, Morphology and intracortical projections of functionally characterized neurones in the cat visual cortex, *Nature* **280:**120–125.

Glass, I., and Wollberg, Z., 1979, Lability in the responses of cells in the auditory cortex of squirrel monkeys to species-specific vocalizations, *Exp. Brain Res.* **34:**489–498.

Godfrey, J. J., 1974, Perceptual difficulty and the right ear advantage for vowels, *Brain Lang.* **1**:323–335.

Goldstein, K., 1927, Die Lokalisation in der Grosshirnrinde, in: *Handbuch der Normalen und Pathologischen Physiologie,* Volume X, Springer, Berlin, pp. 600–795.

Green, E., and Howes, D. H., 1977, The nature of conduction aphasia: A study of anatomic and clinical features of underlying mechanisms, in: *Studies in Neurolinguistics,* Volume 3 (H. Whitaker, ed.), Academic Press, New York, pp. 123–156.

Harrison, J. M., 1978, Functional properties of the auditory system of the brain stem, in: *Handbook of Behavioral Neurobiology,* Volume 1 (R. B. Masterton, ed.), Plenum Press, New York, pp. 409–458.

Hécaen, H., 1978, Right hemisphere contribution to language functions, in: *Cerebral Correlates of Conscious Experience* (P. A. Buser and A. Rogeul-Buser, eds.), North-Holland, Amsterdam, pp. 199–214.

Heimer, L., and Robards, M. J. (eds.), 1981, *Neuroanatomical Tract-Tracing Methods,* Plenum Press, New York.

Hendry, S. H., and Jones, E. G., 1983, The organization of pyramidal and non-pyramidal cell dendrites in relation to thalamic afferent terminations in the monkey somatic sensory cortex, *J. Neurocytol.* **12**:277–298.

Hitchcock, P. F., and Hickey, T. L., 1980, Ocular dominance columns: Evidence for their presence in humans, *Brain Res.* **182**:176–179.

Hyvärinen, J., 1981, Regional distribution of functions in parietal association area 7 of the monkey, *Brain Res.* **206**:287–304.

Hyvärinen, J., and Poranen, A., 1974, Function of the parietal associative area 7 as revealed from cellular discharges in alert monkeys, *Brain* **97**:673–692.

Imig, T. J., and Adrian, H. O., 1977, Binaural columns in the primary field (A1) of cat auditory cortex, *Brain Res.* **138**:241–258.

Imig, T. J., and Brugge, J. F., 1978, Sources and terminations of callosal axons related to binaural and frequency maps in primary auditory cortex of the cat, *J. Comp. Neurol.* **182**:637–661.

Imig, T. J., and Reale, R. A., 1980, Patterns of corticocortical connections related to tonotopic maps in cat auditory cortex, *J. Comp. Neurol.* **192**:293–332.

Imig, T. J., and Reale, R. A., 1981, Ipsilateral corticocortical projections related to binaural columns in cat primary auditory cortex, *J. Comp. Neurol.* **203**:1–14.

Ingvar, D. H., 1982, Frontal lobe activity in normal and abnormal mentation, *Neurosci. Lett. Suppl.* **10**:14.

Jones, E. G., and Burton, H., 1976, Areal differences in the laminar distribution of thalamic afferents in cortical fields of the insular, parietal and temporal regions of primates, *J. Comp. Neurol.* **168**:197–248.

Jones, E. G., and Powell, T. P. S., 1970, An anatomical study of converging sensory pathways within the cerebral cortex of the monkey, *Brain* **93**:793–820.

Kertesz, A., Lesk, D., and McCabe, P., 1977, Isotope localization of infarcts in aphasia, *Arch. Neurol.* **34**:590–601.

Kertesz, A., Harlock, W., and Coates, R., 1979, Computer tomographic localization, lesion size, and prognosis in aphasia and nonverbal impairment, *Brain Lang.* **8**:34–50.

Kimura, D., 1967, Functional asymmetry of the brain in dichotic listening, *Cortex* **3**:163–178.

Kudo, M., and Niimi, K., 1980, Ascending projections of the inferior colliculus in the cat: An autoradiographic study, *J. Comp. Neurol.* **191**:545–556.

Kutas, M., and Hillyard, S. A., 1982, The lateral distribution of event-related potentials during sentence processing, *Neuropsychologia* **20**:579–590.

Landis, T., Graves, R., and Goodglass, H., 1982, Aphasic reading and writing—Possible evidence for right hemisphere participation, *Cortex* **18**:105–112.

Larsen, B., Skinhøj, E., and Lassen, N., 1978, Variations in regional cortical blood flow in the right and left hemispheres during automatic speech, *Brain* **101**:193–210.

Leiber, L., 1982, Interhemispheric effects in short-term recognition memory for single words, *Cortex* **18**:113–124.

Leinonen, L., Hyvärinen, J., and Sovijarvi, A. R. A., 1980, Functional properties of neurons in the temporo-parietal association cortex of awake monkey, *Exp. Brain Res.* **39**:203–216.

LeVay, S., 1973, Synaptic patterns in the visual cortex of the cat and monkey: Electron microscopy of Golgi preparations, *J. Comp. Neurol.* **150**:53–86.

LeVay, S., Hubel, D. H., and Wiesel, T. N., 1975, The pattern of ocular dominance columns in macaque visual cortex revealed by a reduced silver stain, *J. Comp. Neurol.* **159:**559–575.

Levy-Agresti, J., and Sperry, R., 1968, Differential perceptual capacities in major and minor hemispheres, *Proc. Natl. Acad. Sci. USA* **61:**1151.

Lidov, H. G. W., Grzanna, R., and Molliver, M. E., 1980, The serotonin innervation of the cerebral cortex in the rat—An immunohistochemical analysis, *Neuroscience* **5:**207–227.

Locke, S., 1960, The projection of the medial pulvinar of the macaque, *J. Comp. Neurol.* **115:**155–167.

Lund, J. S., Henry, G. H., MacQueen, C. L., and Harvey, A. R., 1979, Anatomical organization of the primary visual cortex (area 17) of the cat: A comparison with area 17 of the macaque monkey, *J. Comp. Neurol.* **184:**599–618.

Luria, A. R., 1966, *Higher Cortical Functions,* Tavistock Publications, London.

Majorossy, K., and Bodoky, M., 1977, Neuronal structure and intrinsic connections of the auditory cortex, *Acta Anat.* **99:**292.

Manley, J. A., and Müller-Preuss, P., 1978, Response variability of auditory cortex cells in the squirrel monkey to constant acoustic stimuli, *Exp. Brain Res.* **32:**171–180.

Marin, O. S., Schwartz, M. F., and Saffran, E. M., 1979, Origins and distributions of language functions in the brain, in: *Handbook of Behavioral Neurobiology,* Volume 2 (M. S. Gazzaniga, ed.), Plenum Press, New York, pp. 192–213.

Marslen-Wilson, W., and Tyler, L. K., 1980, The temporal structure of spoken language understanding, *Cognition* **8:**1–71.

Mauguiere, F., and Baleydier, C., 1978, Topographical organization of medial pulvinar neurons sending fibres to Brodmann's areas 7, 21, and 22 in the monkey, *Exp. Brain Res.* **31:**605–607.

Maximilian, V. A., 1982, Cortical blood flow asymmetries during monaural verbal stimulation, *Brain Lang.* **15:**1–11.

Maximilian, V. A., Prohovnik, I., Risberg, J., and Hakansson, K., 1978, Regional blood flow changes in the left cerebral hemisphere during word pair learning and recall, *Brain Lang.* **6:**22–31.

Mazziota, J. C., Phelps, M. E., Carson, R. E., and Kuhl, D. E., 1982, Tomographic mapping of human cerebral metabolism: Auditory stimulation, *Neurology* **32:**921–937.

Mazzochi, F., and Vignolo, L. A., 1979, Localisation of lesions in aphasia: Clinical–CT scan correlations in stroke patients, *Cortex* **15:**627–654.

Mesulam, M.-M., 1979, Tracing neural connections of human brain with selective silver impregnation, *Arch. Neurol.* **36:**814–818.

Metter, E. J., Riege, W. H., Hanson, W. R., Kuhl, D. E., Phelps, M. E., Squire, L. R., Wasterlain, C. D., and Benson, D. F., 1983, Comparison of metabolic rates, language, and memory in subcortical aphasias, *Brain Lang.* **19:**33–47.

Middlebrooks, J. C., and Zook, J. M., 1983, Intrinsic organization of the cat's medial geniculate body identified by projections to binaural response-specific bands in the primary auditory cortex, *J. Neurosci.* **3:**203–224.

Middlebrooks, J. C., Dykes, R. W., and Merzenich, M. M., 1980, Binaural response-specific bands in primary auditory cortex (A1) of the cat: Topographical organization orthogonal to isofrequency contours, *Brain Res.* **181:**31–48.

Mitzdorf, U., and Singer, W., 1978, Prominent excitatory pathways in the cat visual cortex (A 17 and A 18): A current source density analysis of electrically evoked potentials, *Exp. Brain Res.* 33:371–394.

Molfese, D. L., 1978, Left and right hemisphere involvement in speech perception: Electrophysiological correlates, *Percept. Psychophys.* **23:**237–243.

Molfese, D. L., 1980a, Hemispheric specialization for temporal information: Implications for the perception of voicing cues during speech perception, *Brain Lang.* **11:**285–299.

Molfese, D. L., 1980b, The phoneme and the engram: Electrophysiological evidence for the acoustic invariant in stop consonants, *Brain Lang.* **9:**372–376.

Molfese, D. L., and Erwin, R. J., 1981, Intrahemispheric differentiation of vowels: Principal component analysis of auditory evoked responses to computer-synthesized vowel sounds, *Brain Lang.* **13:**333–344.

Molfese, D. L., and Schmidt, A., 1983, An auditory evoked potential study of consonant perception in different vowel environments, *Brain Lang.* **18:**57–70.

Moore, R. Y., and Bloom, F. E., 1979, Central catecholamine neuron systems: Anatomy and physiology of the norepinephrine and epinephrine systems, *Annu. Rev. Neurosci.* **2:**113–168.

Müller-Preuss, P., and Ploog, D., 1981, Inhibition of auditory cortical neurons during phonation, *Brain Res.* **215:**61–76.

Müller-Preuss, P., Newman, J. D., and Jurgens, U., 1980, Anatomical and physiological evidence for a relationship between the "cingular" vocalization area and the auditory cortex in the squirrel monkey, *Brain Res.* **202:**307–316.

Myers, R. E., 1967, Cerebral connectionism and brain function, in: *Brain Mechanisms Underlying Speech and Language* (G. N. Millikan and F. L. Darley, eds.), Grune & Stratton, New York, pp. 61–69.

Naeser, M. A., and Hayward, R. W., 1978, Lesion localization in aphasia with cranial computed tomography and the Boston Diagnostic Aphasia Exam, *Neurology* **28:**545–551.

Naeser, M. A., Hayward, R. W., Laughlin, S. A., and Zatz, L. M., 1981a, Quantitative CT scan studies in aphasia. I. Infarct size and CT numbers, *Brain Lang.* **12:**140–163.

Naeser, M. A., Hayward, R. W., Laughlin, S. A., Becker, J. M., Jernigan, T. L., and Zatz, L. M., 1981b, Quantitative CT scan studies in aphasia. II. Comparison of the right and left hemispheres, *Brain Lang.* **12:**165–189.

Naeser, M. A., Alexander, M. P., Helm-Estabrooks, N., Levine, H., Laughlin, S. A., and Geschwind, N., 1982, Aphasia with predominantly subcortical lesion sites, *Arch. Neurol.* **39:**2–14.

Nauta, W. J. H., and Ebbeson, S. O. E., 1970, *Contemporary Research Methods in Neuroanatomy*, Springer-Verlag, Berlin.

Nebes, R. D., 1978, Direct examination of cognitive function in the right and left hemispheres, in: *Asymmetrical Function of the Brain* (M. Kinsbourne, ed.), Cambridge University Press, London, pp. 99–137.

Niimi, K., Niimi, M., and Okada, Y., 1978, Thalamic afferents to the limbic cortex in the cat studied with the method of retrograde axonal transport of horseradish peroxidase, *Brain Res.* **145:**225–238.

Nishizawa, Y., Olsen, T. S., Larsen, B., and Lassen, N., 1982, Left–right cortical asymmetries of regional cerebral blood flow during listening to words, *J. Neurophysiol.* **48:**458–466.

Ogren, M. P., and Hendrickson, A. E., 1979, The morphology and distribution of striate cortex terminals in the inferior and lateral subdivisions of the macaca monkey pulvinar, *J. Comp. Neurol.* **188:**179–200.

Ojemann, G. A., 1976, Subcortical language mechanisms, in: *Studies in Neurolinguistics,* Volume 1 (H. Whitaker, ed.), Academic Press, New York, pp. 103–138.

Ojemann, G. A., 1983, The intrahemispheric organization of human language, derived with electrical stimulation techniques, *Trends Neurosci.* **6:**184–189.

Ojemann, G. A., and Mateer, C., 1979, Human language cortex: Localization of memory, syntax and sequential motor-phoneme identification systems, *Science* **205:**1401–1403.

Ojemann, G. A., and Whitaker, H., 1978, Language localization and variability, *Brain Lang.* **6:**239–260.

Pandya, D. N., and Kuypers, H. G. J. M., 1969, Cortico-cortical connections in the rhesus monkey, *Brain Res.* **13:**13–36.

Pandya, D. N., and Sanides, F., 1973, Architectonic parcellation of the temporal operculum in rhesus monkeys and its projection pattern, *Z. Anat. Entwicklungsgesch.* **139:**127–161.

Pandya, D. N., Hallett, M., and Mukherjee, S. K., 1969, Intra- and interhemispheric connections of neocortical auditory system in rhesus monkey, *Brain Res.* **14:**49–65.

Penfield, W., and Pernot, P., 1963, The brain's record of auditory and visual experience: A final summary and discussion, *Brain* **86:**595–696.

Petersen, M. R., Beecher, M. D., Zoloth, S. R., Moody, D. B., and Stebbins, W. C., 1978, Neural lateralization of species-specific vocalizations by Japanese macaques (*Macaca fuscata*), *Science* **202:**324–327.

Pfingst, B. E., and O'Connor, T. A., 1981, Characteristics of neurons in auditory cortex of monkeys performing a simple auditory task, *J. Neurophysiol.* **45:**16–34.

Pfingst, B. E., O'Connor, T. A., and Miller, J. M., 1977, Response plasticity of neurons in auditory cortex of the rhesus monkey, *Exp. Brain Res.* **29:**393–404.

Phillips, D. P., and Gates, G. R., 1982, Representation of the two ears in the auditory cortex: A re-examination, *Int. J. Neurosci.* **16:**41–46.

Rall, W., 1964, Theoretical significance of dendritic trees for neuronal input–output relations, in: *Neural Theory and Modeling* (R. F. Reiss, ed.), Stanford University Press, Stanford, Calif., pp. 73–97.

Ramón y Cajal, S., 1902, *Studien ueber die Hirnrinde des Menschen,* No. 3, Barth, Leipzig.

Rockland, K. S., Lund, J. S., and Humphrey, A. L., 1982, Anatomical banding of intrinsic connections in striate cortex of tree shrews (*Tupaia glis*), *J. Comp. Neurol.* **209:**41–58.

Roland, P. E., 1982, Cortical regulation of selective attention in man: A regional cerebral blood flow study, *J. Neurophysiol.* **48:**1059–1078.

Roland, P. E., Skinhøj, E., and Lassen, N., 1981, Focal activations of human cerebral cortex during auditory discrimination, *J. Neurophysiol.* **45:**1139–1151.

Romani, G. L., Williamson, S. J., Kaufman, L., and Brenner, D., 1982, Characterization of the human auditory cortex by the neuromagnetic method, *Exp. Brain Res.* **47:**381–393.

Roth, G. L., Aitkin, L. M., Andersen, R. A., and Merzenich, M. M., 1978, Some features of the spatial organization of the central nucleus of the inferior colliculus of the cat, *J. Comp. Neurol.* **182:**661–680.

Rugg, M. D., 1983, Further study of the electrophysiological correlates of lexical decision, *Brain Lang.* **19:**142–152.

Saffran, E. M., 1982, Neuropsychological approaches to the study of language, *Br. J. Psychol.* **73:**317–337.

Sanides, F., 1975, Comparative neurology of the temporal lobe in primates including man with reference to speech, *Brain Lang.* **2:**396–419.

Sasanuma, S., 1975, Kana and kanji processing in Japanese aphasics, *Brain Lang.* **2:**369–383.

Sasanuma, S., Itoh, M., Mori, K., and Kobayashi, Y., 1977, Tachistoscopic recognition of kana and kanji words, *Neuropsychologia* **15:**547–554.

Scheibel, M. E., and Scheibel, A. B., 1966, The organization of the ventral anterior nucleus of the thalamus: A Golgi study, *Brain Res.* **1:**250–268.

Seldon, H. L., 1981a, Structure of human auditory cortex. I. Cytoarchitectonics and dendritic distributions, *Brain Res.* **229:**277–294.

Seldon, H. L., 1981b, Structure of human auditory cortex. II. Axon distributions and morphological correlates of speech perception, *Brain Res.* **229:**295–310.

Seldon, H. L., 1982, Structure of human auditory cortex. III. Statistical analysis of dendritic trees, *Brain Res.* **249:**211–221.

Semple, M. N., and Aitkin, L. M., 1979, Representation of sound frequency and laterality by units in central nucleus of cat inferior colliculus, *J. Neurophysiol.* **42:**1626–1639.

Sherwin, I., and Efron, R., 1980, Temporal ordering deficits following anterior temporal lobectomy, *Brain Lang.* **11:**195–203.

Sholl, D. A., 1956, *The Organization of the Cerebral Cortex*, Methuen, London.

Sidtis, J. J., 1982, Predicting brain organization from dichotic listening performance: Cortical and subcortical functional asymmetries contribute to perceptual asymmetries, *Brain Lang.* **17:**287–300.

Sidtis, J. J., 1984, Music, pitch perception, and the mechanisms of cortical hearing, in: *Handbook of Cognitive Neuroscience* (M. Gazzaniga, ed.), Plenum Press, New York, pp. 91–114.

Sidtis, J. J., Volpe, B. T., Wilson, D. H., Rayport, M., and Gazzaniga, M. S., 1981, Variability in right hemisphere language function after callosal section: Evidence for a continuum of generative capacity, *J. Neurosci.* **1:**323–331.

Simpson, D. A., 1952, The projection of the pulvinar to the temporal lobe, *J. Anat.* **86:**20–28.

Siqueira, E. B., 1965, The temporo pulvinar connections in the rhesus monkey, *Arch. Neurol.* **13:**321–330.

Smith, D. E., and Moskowitz, N., 1979, Ultrastructure of layer IV of the primary auditory cortex of the squirrel monkey, *Neuroscience* **4:**349–360.

Sokoloff, L., Reivich, M., Kennedy, C., Des Rosiers, M. H., Patlak, C. S., Pettigrew, K. D., Sakurada, O., and Shinohara, M., 1977, The (14-C) deoxyglucose method for the measurement of local cerebral glucose utilization: Theory, procedure and normal values in the conscious and anesthetized albino rat, *J. Neurochem.* **28:**897–916.

Springer, S., and Deutsch, G., 1981, *Left Brain, Right Brain*, Freeman, San Francisco.

Steinschneider, M., Arezzo, J., and Vaughan, H. G., 1982, Speech evoked activity in the auditory radiations and cortex of the awake monkey, *Brain Res.* **252:**353–366.

Streitfeld, B. D., 1980, The fiber connections of the temporal lobe with emphasis on the rhesus monkey, *Int. J. Neurosci.* **11:**51–71.

Studdert-Kennedy, M., 1976, Speech perception, in: *Contemporary Issues in Experimental Phonetics* (N. J. Lass, ed.), Academic Press, New York, pp. 243–293.

Studdert-Kennedy, M., and Shankweiler, D., 1970, Hemispheric specialization for speech perception, *J. Acoust. Soc. Am.* **48:**579–594.

Szentágothai, J., 1973, Synaptology of the visual cortex, in: *Handbook of Sensory Physiology*, Volume VII/3B (R. Jung, ed.), Springer, Berlin, pp. 267–324.

Szentágothai, J., 1978, The neuron network of the cerebral cortex: A functional interpretation, *Proc. R. Soc. London Ser. B.* **201:**219–248.

Szentágothai, J., 1979, Local neuron circuits of the neocortex, in: *The Neurosciences: Fourth Study Program* (F. O. Schmitt and F. G. Warden, eds.), MIT Press, Cambridge, Mass., pp. 399–418.

Szikla, G., Bouvier, A., Hori, T., and Petrov, V., 1977, *Angiography of the Human Brain Cortex*, Springer-Verlag, Berlin.

Thatcher, R. W., 1977, Evoked-potential correlates of hemispheric lateralization during semantic information processing, in: *Lateralization in the Nervous System* (S. Harnard, R. W. Doty, L. Goldstein, J. Jaynes, and G. Krauthamer, eds.), Academic Press, New York, pp. 429–448.

Trojanowski, J. Q., and Jacobsen, S., 1976, Areal and laminar distribution of some pulvinar cortical efferents in rhesus monkey, *J. Comp. Neurol.* **169:**371–392.

Turner, B. H., Mishkin, M., and Knapp, M., 1980, Organization of the amygdalopetal projections from modality-specific cortical association areas in the monkey, *J. Comp. Neurol.* **191:**515–543.

Vaadia, E., Gottlieb, Y., and Abeles, M., 1982, Single-unit activity related to sensorimotor association in auditory cortex of a monkey, *J. Neurophysiol.* **48:**1201–1213.

Van Hoesen, G. W., Yeterian, E. H., and Lavizzo-Mourey, R., 1981, Widespread corticostriate projections from temporal cortex of the rhesus monkey, *J. Comp. Neurol.* **199:**205–220.

Vogt, C., and Vogt, O., 1919, Allgemeine Ergebnisse unserer Hirnforschung, *J. Psychol. Neurol.* **25:**279–462.

Voigt, H. F., Sachs, M. B., and Young, E. D., 1982, Representation of whispered vowels in discharge patterns of auditory-nerve fibers, *Hearing Res.* **8:**49–58.

von Bonin, G., and Bailey, P., 1947, *The Neocortex of Macaca Mulatta*, University of Illinois Press, Urbana.

von Economo, C., and Horn, L., 1930, Ueber Windungsrelief, Masse und Rindenarchitektonik der Supratemporalflaeche, ihre individuellen und ihre Seitenunterschiede, *Z. Gesamte Neurol. Psychiatr.* **130:**678–757.

von Economo, C., and Koskinas, G. N., 1925, *Die Cytoarchitektonik der Hirnrinde des erwachsenen Menschen*, Springer, Berlin.

von Monakow, C., 1914, *Die Lokalisation im Grosshirn*, Bergmann, Wiesbaden.

Votaw, C. L., 1960, Certain functional and anatomical relations of the cornu ammonis of the macaque monkey. II. Anatomical relations, *J. Comp. Neurol.* **114:**283–293.

Wada, J., and Rasmussen, T., 1960, Intracarotid injection of sodium amytal for the lateralization of cerebral speech dominance: Experimental and clinical observations, *J. Neurosurg.* **17:**266–282.

Wada, J., Clarke, R., and Hamm, A., 1975, Cerebral hemispheric asymmetry in humans, *Arch. Neurol.* **32:**239–246.

Warren, R. M., 1976, Auditory illusions and perceptual processes, in: *Contemporary Issues in Experimental Phonetics* (N. J. Lass, ed.), Academic Press, New York, pp. 389–417.

Wernicke, C., 1874, *Der aphasische Symptomenkomplex*, Cohn & Weigert, Breslau.

Whitehouse, P., Caramazza, A., and Zurif, E., 1978, Naming in aphasia: Interacting effects of form and function, *Brain Lang.* **6:**63–74.

Whitfield, I. C., and Evans, E. F., 1965, Responses of auditory neurons to stimuli of changing frequency, *J. Neurophysiol.* **28:**655–672.

Whitlock, D. G., and Nauta, W. J. H., 1956, Subcortical projections from the temporal neocortex in *Macaca mulatta*, *J. Comp. Neurol.* **106:**183–212.

Wise, L. Z., and Irvine, D. R. F., 1983, Auditory response properties of neurons in deep layers of cat superior colliculus, *J. Neurophysiol.* **49:**674–685.

Witelson, S. F., and Pallie, W., 1973, Left-hemisphere specialization for language in the human newborn: Neuroanatomical evidence of asymmetry, *Brain* **96:**641–646.

Wogan, M., and Kaplan, C. D., 1981, Lateralized differences associated with semantically varied auditory material, *Int. J. Neurosci.* **14:**163–168.

Wollberg, Z., and Newman, J. D., 1972, Auditory cortex of squirrel monkey: Response patterns of single cells to species-specific vocalizations, *Science* **175:**212–214.

Wood, C. C., 1977, Average evoked potentials and phonetic processing in speech perception, in: *Language and Hemispheric Specialization in Man: Cerebral ERP's* (J. E. Desmedt, ed.), Karger, Basel, pp. 73–86.

Wood, C. C., 1982, Application of dipole localization methods to source identification of human evoked potentials, *Ann. N.Y. Acad. Sci.* **388:**139–155.

Yeni-Komshian, G. H., and Benson, D. A., 1976, Anatomical study of cerebral asymmetry in the temporal lobe of humans, chimpanzees, and rhesus monkeys, *Science* **192:**387–389.

Zurif, E. B., and Blumstein, S. E., 1978, Language and the brain, in: *Linguistic Theory and Psychological Reality* (M. Halle, J. Bresnan, and G. A. Miller, eds.), MIT Press, Cambridge, Mass., pp. 229–246.

8

The Role of Auditory Cortex in Behavior

I. C. WHITFIELD

1. The Auditory Pathway and Behavior

Prior to the research expansion that followed the Second World War, little work seems to have been done on the relationship of the parts of the central auditory pathway to either internal or external behavior. Neuroanatomy had delineated the structures involved and included the cytoarchitectonic studies that at least approximately located or delimited the auditory cortex.

Of course, many auditory behavioral studies were made on intact organisms, especially in the 1950s when electronic methods of generating and controlling sound stimuli became available, and an extensive literature accumulated. Responses to signal direction, to changes of frequency and intensity, and such phenomena as masking were studied (Stevens and Davis, 1938). However, insofar as attempts were being made to relate these phenomena to physiological mechanism, that was being done almost entirely in terms of the peripheral receptor mechanism. Natural and artificial lesions, those important experimental tools, were studied in the cochlea, but naturally occurring central lesions were able to throw little light on the problems. The reason, of course, is that such lesions are inevitably widespread and involve other structures, damage to which has behavioral effects that tend to overshadow any auditory effects.

I. C. WHITFIELD • Neurocommunications Research Unit, University of Birmingham, Birmingham B15 2TJ, England. *Present address:* Department of Electronic and Electrical Engineering, University of Birmingham, Birmingham B15 2TT, England.

Knowledge of the electrical activity of the auditory system starts in the 1930s with the famous work of Wever and Bray on the cochlear microphonic (1930) and the detection of neural activity by Saul and Davis in 1932. However, it was not until 1939 that Bremer made the first electrical recording of auditory responses in the cat cortex (and incidentally failed to find any spatial distribution of stimulus frequency). We might perhaps date interest in auditory cortical behavior *per se* from that point. Bremer's results were followed by a report from McCulloch *et al.* (1942) that in the anesthetized monkey (Bremer's cats had been unanesthetized) a tonotopic arrangement of response to stimulus frequency was discernible on the cortical surface. The other key paper was that of Woolsey and Walzl (1942), who showed that electrical stimulation of restricted bundles of auditory nerve fibers at the modiolus produced localized evoked responses at the cortex that were arranged in an orderly cochleotopic manner. The stage was set for the great tonotopicity argument that continues mildly to this day and has very much obscured the elucidation of the real behavioral role of the auditory cortex.

The idea that the auditory cortex plays a role in frequency discrimination probably goes back to the "resonance" versus "telephone" theories of the late 19th century (Helmholtz, 1863; Rutherford, 1886). The telephone or periodicity theory of pitch required analysis by the "higher centers" and where better than the highest center, the auditory cortex? The subsequent demonstration by Butler *et al.* (1957) that removal of the whole auditory cortex need not have any effect whatsoever on the discrimination of pure tones did little to dampen the argument, so firmly entrenched had the idea become.

I have emphasized the issue at this point because it is extremely important to distinguish between behavior of the organism—which we shall be discussing here—and behavior of the individual components that make up the nervous system. Unquestionably, there are individual neurons in the auditory cortex that respond to a specific band of frequencies and do not respond to frequencies outside that band. We are not, on that account, entitled to assume that those neurons are the seat of the behavior we call frequency discrimination. It is, of course, the latter type of behavior we shall be concerned with in this chapter.

One problem is to discover what types of behavior are associated with the presence and intactness of an auditory cortex. The other will be to try to relate that behavior to the behavior of the neural networks present in the cortex, and ultimately to that of individual neurons. However, it will be fatal to assume without evidence that the behavior of the organism can be equated with the sum total of the recordable behaviors of the individual neurons.

2. The Auditory Cortex

Having discussed what we shall mean by behavior, it is equally important to consider what we mean by cortex. The classic experiments of Lashley carried out over the period 1926 to 1948 (see Beach, 1960) seemed to show that "the cortex" was to a large extent unnecessary for those behavioral responses involving sensory discriminations, or at least that the function of any particular part is highly unspecific. Much work since then has been devoted to trying to ascertain what the cortex is *for* (see Whitfield, 1979a). However, implicit in Lashley's experiments was the equation of "cortex" with "neocortex". The dichotomy of cortical (meaning neocortical) and subcortical in discriminatory behavior has been largely followed ever since. This is strange, because the role of the paleocortex in behavior has been known for at least 40 years (Papez, 1937; MacLean, 1949) and evidence has been accumulating of direct sensory projections to this cortical division (Cuénod *et al.*, 1965). The striatum, although strictly it is subcortical, has in some classes (birds, reptiles) a behavioral role that overlaps that of the allocortex. In looking at the role of a sensory cortex in behavior, we certainly ought not to confine ourselves to neocortex, and might usefully extend the discussion to embrace the whole telencephalon. However, just as it is difficult to associate specific losses with damage to specific areas of neocortex, it is still more difficult to do so with other systems. There does not seem to be any such structure as auditory paleocortex or even sensory paleocortex. The reasons for this will emerge when we consider functional cortical organization. One of the results is that little work has been done to test behaviorally for modality-specific deficits in non-neocortical telencephalic structures.

We might be able to approach the problem via anatomy, but here again there is little information except in primitive forms. Ebner (1969) reported that lesions in the medial geniculate body (GM) of the opossum revealed the presence of dense terminals in putamen, globus pallidus, and lateral amygdaloid nucleus, and LeVay and Sherk (1981) used HRP injection to demonstrate a pathway from GM to putamen in the cat. Diamond and Casseday (personal communication) have found projections from GM to amygdala and possibly to putamen in the bat, using the HRP technique. It seems likely that such projections are to be found in all mammals and thus a large new area of study for auditory behavioral mechanisms is opened up.

3. The Cortex and Discrimination

The reason the neocortex has loomed so large in the study of modality-specific behavior, such as vision and audition, is not that it has been regarded as the sole seat of behavioral response patterns, but rather because experimental research in this field has been concerned with limits of discrimination, and the neocortex has been regarded, probably wrongly, as the seat of the "highest" discriminatory powers. The question asked has been, in effect, can we tell A from B♭, rather than can we tell a march from a polka, or a shout of joy from a cry of pain, and that has probably been the wrong question.

Neither can we avoid the problem of trying to confine ourselves to those forms of behavior known to be impaired by neocortical damage alone, for there are specific differences. Animals as different as the monkey (Heffner and Masterton, 1975), the cat (Neff and Diamond, 1958) and the ferret (Kavanagh and Kelly, 1983) have their concept of the location of a sound in space destroyed by extensive removal of auditory cortex, yet the rat (Kelly, 1980) is apparently unaffected by this operation.

4. Frequency Discrimination

The work of Bremer, and of Woolsey and Walzl, referred to earlier, raised questions about the role of the auditory cortex in the discrimination of frequency. Allen (1945) and Meyer and Woolsey (1967) investigated the behavioral effects of cortical ablation on frequency discrimination in the dog and the cat, respectively. The results they obtained were different, but so were the stimuli used. Allen used an "absolute" discrimination test (Fig. 1a), in which the dogs were required to recognize one or other of two tones presented alone. This might be described as a crude "absolute pitch" experiment. Allen found that the dogs could not relearn this problem after bilateral ablation of the auditory cortex. On the other hand, Meyer and Woolsey found that their cats could perform as well after bilateral removal of AI, AII, Ep as they could normally. However, those workers used the stimulus paradigm of Fig. 1b, where a direct comparison of the two tones is possible in the avoidance situation. Thus, the differences are stimulus-based differences rather than species differences. Thompson (1959, 1960) found that it required over 1000 trials for a normal cat to learn the Fig. 1a discrimination, while 600 were sufficient for the type b discrimination and only 120 trials were required for type c. Thompson also found that the type c paradigm was the one that gave the smallest discriminable frequency-differences, the DL at 1 kHz having a value of $\Delta f/f$ equal to about 0.02. Butler *et al.* (1957) found that the performance of cats with the whole of AI, AII, Ep, and SII removed bilaterally is quite equal to that of normal animals, when this last

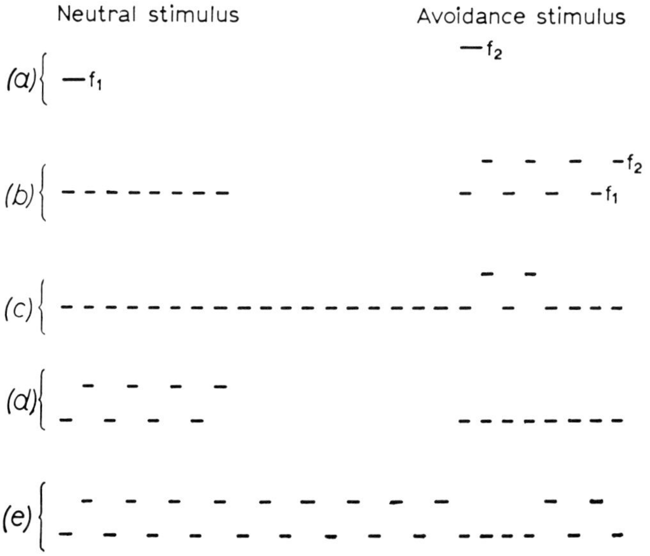

Figure 1. Some types of stimulus pattern used in frequency discrimination experiments. Except in (a) the bars represent time pulses 1 sec in duration separated by silent intervals of 1 sec. (a) requires an absolute judgment, whereas in (b) and (c) both tones are present in the avoidance situation. Note that (b) and (c) introduce a "new" tone into the avoidance stimulus, while (d) and (e) do not. From Whitfield (1967).

stimulus paradigm is employed. Taken together, these findings suggest that frequency discrimination *per se* is not mediated by auditory neocortex even when that area is widely drawn. One cannot even claim that discrimination is "refined" by that cortex, since the most accurate discrimination possible can be carried out without it.

Whereas stimuli a–c in Fig. 1 involve an additional stimulus in the "warning" situation, those depicted in d and e involve the cessation of part of the "safe" signal. Another possible paradigm (not shown here) is to exchange the tone pattern $f_1f_2f_1$ for the pattern $f_2f_1f_2$. Unlike the additional-signal situations, all these latter patterns require the presence of at least some part of the auditory cortex to enable a response to be learned (Diamond *et al.*, 1962). The cortex seems to be necessary, not to make the discrimination, but to assess the nature of the problem and thereby attach the stimulus to the response.

It is perhaps not sufficiently appreciated that when an animal is placed in an experimental situation it has no idea at all what it is being called upon to do. One can explain to a human subject "when the tone sounds, press the lever", but an animal does not even know it is involved in an auditory experiment. There is (or used to be) a children's game called "I hand you the scissors crossed", in which one child hands the other a pair of scissors a number of times saying sometimes "I hand you the scissors crossed" and sometimes "I hand you the scissors uncrossed". The recipient is invited to notice the difference in the actions accompanying the two statements and reproduce these actions. It can take quite a long time for the recipient to "tumble" to the solution because attention is drawn away from the key point, which is that the crossing refers to the position of the performer's legs and not the relation between scissors and hands.

An additional stimulus may well capture an animal's attention in a way that the cessation of a stimulus does not, for the stimulus being withdrawn will be only one among a great number of stimuli it is already experiencing; thus, the concept necessary for the solution is more rapidly reached in the first situation than in the second. It is a matter of common observation that difficult problems take longer to solve, but that once solved, the final performance is not necessarily worse than that on more quickly learned problems.

5. The Site of Relearning

It is a frequent finding that when the auditory neocortex is ablated the previous training is lost "but can be relearned". It has generally been assumed that this implies that the discrimination was previously carried out by the part ablated, but is now being carried out elsewhere. Since the discrimination is often as good as before, that seems *prima facie* to be unlikely, unless there is a great deal of normally unused capacity. What we shall here argue is that it is the *concept of the problem* that is destroyed by neocortical ablation, the actual discrimination remaining intact. It is a new understanding of the problem that has to be learned after the ablation.

Some old experiments of Raab and Ades (1946) on intensity discrimination illustrate this point. Cats were trained on an intensity discrimination at 1000 Hz (where the value of the DL was about 2 dB). After bilateral ablation of the whole auditory cortex the training was lost, but the animals could be retrained to the same level of discrimination as before. If the tectal auditory pathway to the forebrain were then destroyed, retraining was again necessary and possible, but the level of discrimination achieved was much poorer (10–12 dB). Now the interesting point is that when the auditory cortex was left intact, but the primary pathway from the inferior colliculus was destroyed by ablating the latter, there was no loss of discrimination, even temporarily—no relearning was necessary.

In 1967 I rather unsuccessfully attempted an explanation of these superficially puzzling results in terms of possible descending pathways, but in the light of more recent knowledge of the forebrain it seems that they are readily explicable in terms of the thesis here advanced, namely that the discriminatory process and its coupling to the response are mediated via the sensory-paleocortical or sensory-striatal systems, while the neocortex normally serves only to identify the problem. That this should be so is not surprising. Many species with poorly developed neocortices (birds, for example) are quite capable of making discriminations equal to those of higher mammals. It must be remembered too that the training methods normally used—either shock avoidance or food reward—involve affective responses such as we associate with the paleocortical (limbic) system.

Having involved the paleocortex, it would be as unrewarding to try to locate the "site of the discrimination" in it as it has been in the classical auditory cortex. Strictly speaking, discrimination has already taken place at the periphery; if the total receptor array has not responded differentially to two stimulus situations, no subsequent operations on the neural data will separate them. The information may, however, be transformed. It is now well established, for example, that information about auditory intensity and frequency become concentrated in "edges", the points separating active and inactive regions in the neural array (Whitfield, 1956, 1967). However, when behaviorally we talk of discrimination, it is usually implied that the discrimination manifests itself as a differential response. Thus we are interested in the complete series of transformations that link the sensory input to the motor output and constitute the behavior.

6. Recognition Systems

These input/output transformations are of two general types (Fig. 2), although intermediate forms are possible. The first we might describe as hierarchical, and involves the extraction and combination of features of the input. These feature detectors have commonly been equated with cortical neurons and the idea is that they build up more and more complex combinations of simple input features in successive neural layers. The theoretical basis of such classifying systems was treated by Uttley (1954), and neural feature detectors were first described in the visual system by Hubel and Wiesel (1962). Auditory equivalents have been described by Whitfield and Evans (1965), Feher and Whitfield (1966) and Swarbrick and Whitfield (1972). Many other quite complex feature responses have been described by various authors.

There are obvious defects in this hierarchical hypothesis. The first is theoretical. The sort of combinatorial tree that has been described can grow indefinitely. Do we stop at a unit that recognizes an apple?, an apple tree?, an orchard?—a chord? or a symphony?

The second is practical. It now seems that complex and hypercomplex feature units in the visual cortex do not always receive their inputs from simple units, as was originally supposed, but that each type may receive separate inputs in parallel (Stone and Dreher, 1973).

Furthermore, we must be very careful in building theories on the supposedly specific feature-extracting properties of particular neurons. Experiments with gratings (Campbell and Robson, 1968) indicate that those neurons in visual cortex previously regarded as detecting such features as orientation and extension may equally be regarded as responding to elements of a Fourier transform of the input pattern (Andrews and Pollen, 1979). The important word here is *equally*. Once we admit that neurons may extract two (or more) quite different classes of feature, the whole basis of the hierarchical hypothesis becomes suspect.

A scheme that avoids the theoretical objections inherent in a hierarchical system of recognition—but not all the practical difficulties of feature detection—was proposed by Uttley (1975) and is based on the arrangement of Fig. 2B. Uttley's problem was to design a fairly simple blind reading aid. The method employed was to make use of some 15 or 20 key features of the printed alphabet, such as free ends, corners, T-junctions, left-facing curves, right-facing curves, etc. These were matched with an equal number of the features of spoken phonemes, such as the frequencies of the first, second and third formants, presence or absence of fricatives, etc. Each single optical feature was connected to a single acoustic output feature, so that (say) the presence of a T-junction would produce a fricative in the output sound. Uttley discusses in his paper ways to construct

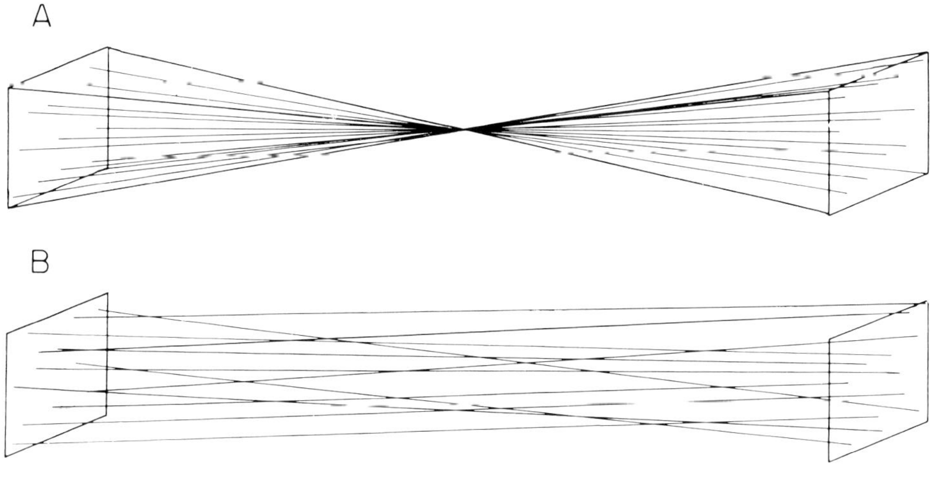

Figure 2. Diagrammatic representations of a classical (hierarchical) recognition system (A) in which the sensory inputs are progressively categorized, ultimately reaching a recognition point from which motor activity fans out to the peripheral effectors, and a parallel system (B) in which individual features of the input are connected to individual features of the output. There is no recognition involved. From Whitfield (1984).

the connective matrices to ensure the maximum discriminability. The resulting machine when presented with the letters A, B, C, etc. produces a unique phonemic sound (say shoo, gah, doh) in response to each.

Although this system employs features, these features are all of the same level of complexity (which need not be very high). The need to construct more complex features out of less complex ones is removed, and with it the problem of "where to stop". The disadvantage is that the difficulty of constructing the appropriate matrices rises very steeply as the number of features, and hence the order of the matrix, is increased. It is to be noted, however, that when the visual stimulus "B" gives rise to the response "gah", there is nowhere in the system that we can identify as "recognizing" the letter B. This is an extremely important property since it relieves us of the necessity, always implicitly present in recognition systems, for the presence of Adrian's (1947) hobgoblin—the little man who has to sit somewhere up top and eventually do the recognizing.

7. Discrimination and Transfer

Physiologists and anatomists have drawn attention to three aspects of the structure and function of auditory (and other sensory) neocortex. Firstly, many cortical neurons are selective in their response to stimuli, a property giving rise in the limit to feature detection (see Whitfield, 1982). Secondly, there are several subareas within the auditory area each of which receives input from outside the area. Thirdly, the input to at least one of the subareas is highly spatially organized.

Given these undisputed facts, we might be forgiven for supposing that lesions confined to small regions of the whole auditory area would give rise to behavioral deficits that differed according to where the lesion was placed. For example, the medial geniculate body is tonotopically organized, with the higher-frequency fibers projecting to the cortex anteriorly and the low-frequency fibers posteriorly. This arrangement can be demonstrated electrophysiologically in the anesthetized (but not in the unanesthetized) animal. Yet limited lesions of AI do not produce behavioral deficits confined to a particular frequency region. Indeed, as we have seen, removal of the *whole* auditory cortex does not affect frequency discrimination at all.

Likewise, there are units in the auditory cortex that respond only when the sound source occupies a particular region of space in relation to the animal's head (Evans, 1968), and units that are sensitive to time disparity of arrival time of a sound at the two ears. Yet removal of one auditory cortex does not destroy behavioral localization of a sound source on one side of the head, or indeed affect it at all. Animals lacking all auditory cortex can too be trained to distinguish a right-ear-before-left-ear click pair from a left-ear-before-right-ear click pair (Masterton and Diamond, 1964). Whitfield and Evans (1965) demonstrated the presence in cat auditory cortex of units responding rather specifically to direction of frequency change, yet as Kelly and Whitfield showed (1971), extensive bilateral ablation of auditory cortex by no means completely destroys the animal's ability to distinguish a rising from a falling pitch.

The fact is that information concerning all these primary discriminations—intensity, frequency, direction of frequency change, time of arrival—is already present in usable form at the brainstem level, and appears to be passed directly to the midbrain/paleocortical system concerned with the affective behavior patterns that constitute the demanded responses.

We ask ourselves again, therefore, what is the cortex *for?* What sort of deficit should we be seeking?

The key is to be found in the experiment of Masterton and Diamond (1964) just referred to. A normal cat is trained to distinguish between a train of clicks presented to the left ear and a similar train presented to the right ear. If it is then presented with trains of clicks to both ears (Fig. 3) but with a time disparity between the click pairs, it will treat right-before-left just as it treated the single train on the right, and left-before-right as it treated a single sound on the left. It *transfers* immediately from the one problem to the other. What Masterton and Diamond discovered was that when they tried the experiment with cats without auditory cortex *there was no transfer.* Both problems could still be learned, but had to be learned separately; they had no apparent relationship. This was a most important observation because it suggests that the function of the sensory cortex is not the detection of differences (discrimination) but the detection of similarities among stimuli. The idea of behavioral "transfer" is indeed precisely that—the

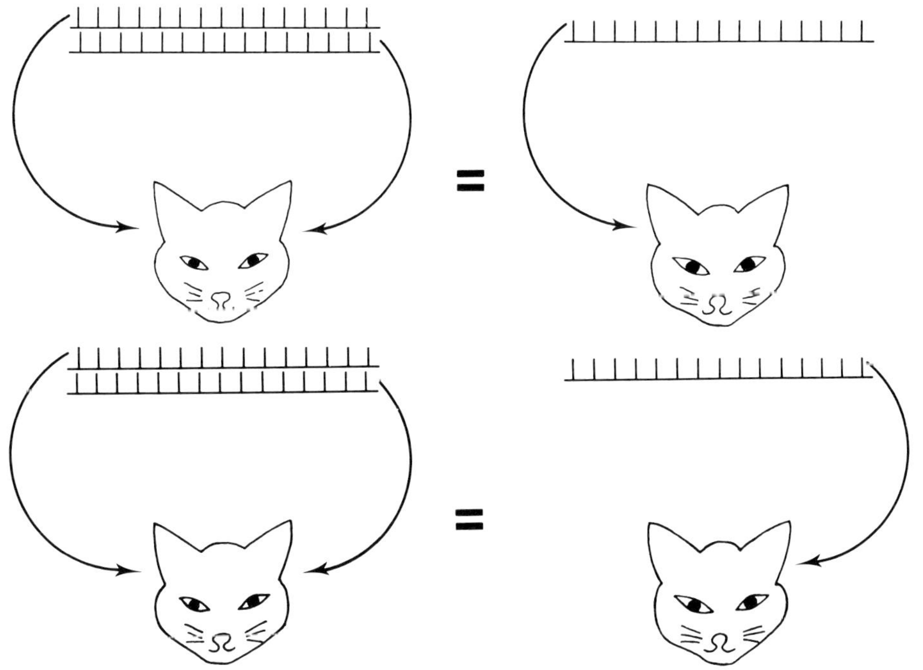

Figure 3. Trains of clicks presented dichotically to the two ears. If the right-ear click of each pair precedes the left-ear click by around 100 μsec, the effect is equivalent in the normal animal to a train of clicks in the right ear alone. Conversely, if left precedes right, the effect is that of a train of clicks in the left ear alone. These equivalences are lost in the decorticate animal. From Whitfield (1979a).

organism says "these situations, though different, are essentially the same; let us try the same response."

One of the areas in which this similarity detection is most important is in the recognition of objects. As an animal moves, an observed object presents an ever-changing stream of stimuli to the sense organs, yet the object remains an unchanging entity. I have argued elsewhere (Whitfield, 1984) that the world of separate objects-on-a-ground—the world of trees and chairs and people—is as real (or unreal) as is the description of the world by atomic physics as consisting of vast empty spaces punctuated at great intervals by rather indeterminate packets of energy; the fact that we *do* see the world as a set of objects is an important clue to how the mammalian nervous system processes information.

8. Constancies and Externalization

In the auditory realm the position of a sound source in space has many of the properties of an "object"—indeed we often talk of a "sound image". In general, for example, the position of a sound source does not appear to move relative to the external world, as we move our heads, irrespective of the fact that the sensory input is changing all the time. The sound image is like a chair or a tree in this respect, but unlike most physical objects, it has the experimental advantage of having only a single variable attribute, that of position. It has therefore proved very useful in the investigation of the role of the sensory neocortex in behavior.

Unlike frequency and intensity discrimination, sound localization appears to be severely affected by cortical ablation. In 1958 Neff and Diamond carried out a series of studies in which cats were required to go to a food-reward box 6 feet away, behind which a buzzer had been sounded. Normal cats could discriminate angles of about 5°, while those with bilateral lesions of AI, AII and Ep (Fig. 4A) could do no better than 40° at the same criterion level. For larger lesions involving additionally SII, I-T and SS, the loss was virtually complete (Neff, 1961).

We saw in the case of frequency discrimination (p. 332) that the effect of

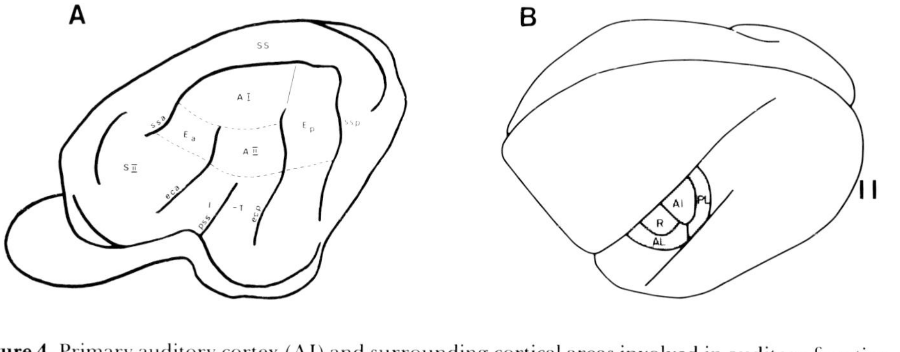

Figure 4. Primary auditory cortex (AI) and surrounding cortical areas involved in auditory function. (A) Cat; (B) owl monkey. (A) from Whitfield (1967); (B) after Merzenich (1982).

cortical ablation depended somewhat on the task, and especially on how difficult it was to discover what the task was. In the sphere of localization also, we find that the result is task-dependent. If we change the task so that the animal has to discriminate between two positions of a source and respond not by approaching the source but by, say, pressing a lever or withholding a lick response, then correct performance is possible in the absence of auditory cortex. This has been shown by Ravizza and Masterton (1972) in the case of the opossum (lick response) and by Heffner and Masterton (1975) in the monkey. Indeed, the latter authors were able to show in the same individual that monkeys without auditory cortex could signal their (successful) discrimination by lever pressing, but were unable to approach the source of the sound for a food reward. Thus the animal can tell that the two signals are different and which is which, but lacks the concept of out-there-ness, the existence of the source in an external world.

9. Sound Images and the Precedence Effect

This conceptualization of the source as an external object can be illustrated even better by a study of the precedence effect (Wallach *et al.*, 1949). In the "free field" experiments quoted above, it is a *sine qua non* that there should be no reflections or echoes to confuse the experimental situation. However, in "real life" many sounds are generated in enclosed spaces such as a room or hall. Under these conditions, a sound may be reflected many times before it falls below audible threshold, and so the listener hears the same sound arriving from many different directions; yet all the sound appears to come from the original source, even if that source be actually weaker than some of the echoes. It is not that the later-arriving signals are suppressed; the use of the phenomenon in sound-reinforcement equipment demonstrates that these later signals actually

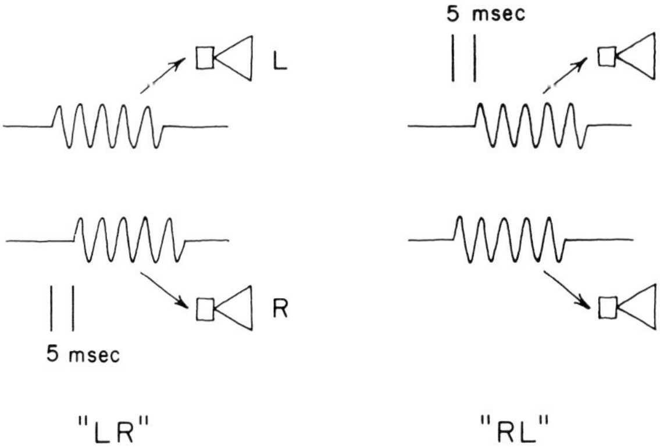

Figure 5. Stimuli used to study the effect of cortical lesions in the precedence effect. Each speaker delivers a train of 1-kHz pulses, each 25 msec long, at about 5/sec. The members of one train are displaced by 5 msec with respect to the corresponding members of the other train. Normally, all the sound appears to come from the speaker carrying the leading pulse-train. From Whitfield (1979a).

add to intelligibility. The auditory system, it seems, combines all these repetitions into the concept of a single source located in the position of the primary signal. This image formation takes place in, or at least requires the presence of, auditory cortex, for it is disrupted by cortical removal. Furthermore, unlike the localization of a single free-field source, it can be disrupted unilaterally by removal of auditory cortex on one side only (Whitfield *et al.*, 1972).

The precedence effect can be mimicked by having two loudspeakers fed by tone pulses so arranged that the start of the pulse to one speaker precedes that to the other by (say) 5 msec (Fig. 5). If left precedes right (LR), all the sound appears to come from the left speaker; if right precedes left (RL), all the sound appears to come from the right speaker. (Note that this is a different situation from that of Fig. 3 where disparate click trains were delivered to separate ears. Here both signals reach both ears.) Cats were trained in a Y-maze to approach the goal box from behind which the sound appeared to come. They will transfer to this problem quite readily after training to two real single sources behind the goal boxes. Indeed, the two types of signal can be intermingled without upsetting performance (Fig. 6A).

If now the left auditory cortex be removed, the animal treats the single left and right signals as before, and also responds correctly to LR (what might be called the "virtual left"). However, the response to RL, the "virtual right", is disrupted and falls to around chance. A similar disruption of the response to LR occurs when the right auditory cortex is removed.

It will be remembered that although bilateral cortical ablation disrupted the animal's ability to approach a sound source, it did not destroy the discrimination as such. We have observed an analogous phenomenon with the unilateral lesions. Although the majority of cats behave in the way described above, a few (about one in eight of those tested) were able to relearn the problem and overcome the

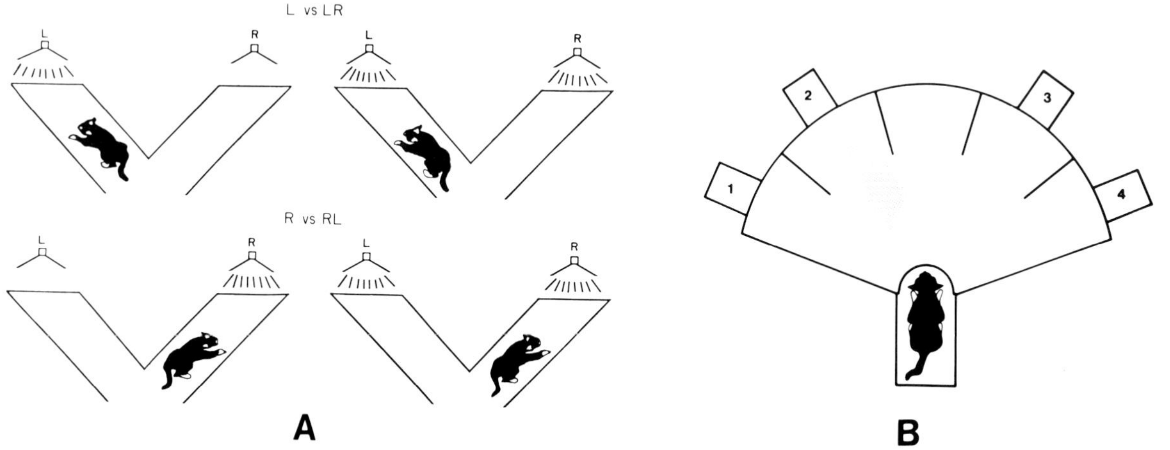

Figure 6. Localization apparatus for the precedence-effect problem. In (A) the animal having made a left/right choice between the two stimuli is constrained by the wire tunnel to reach the goal box. In (B) this constraint is absent; the animal goes to the correct side, but not necessarily to the correct box.

deficit. The difference did not seem to be in any way related to the lesion. This proved puzzling until it was shown to be a property of the test situation. Some of these same "successful" cats were tested in an apparatus that, instead of having the two tunnels of the Y-maze, had four goal boxes situated across an open space, two to the left of the start cage and two to the right (Fig. 6B). It was now found (Whitfield, 1979b) that a cat with, say, a left lesion could go correctly to each of the boxes on the left of the start box. When asked to go to one of the boxes on the right, however, it usually got the *side* correct but distributed its responses randomly between the two boxes on that side. Presumably, some of the cats were able to utilize the difference in neural pattern provided by the LR and RL signals to judge the side, but not the precise position, of the source. In the Y-maze that was sufficient, since having selected the correct arm, left or right, the tunnel automatically guided them to the goal box.

10. The Constancy of Pitch

During the discussion of localization it has emerged that animals are able to learn to discriminate between sound sources in different positions by utilizing the differences in brainstem neural patterns, patterns of which the temporal and spectral aspects have been thoroughly studied. We noted too (p. 337) that while a single attribute of the signal can be used as stimulus, there is no transfer to other attributes; each has to be learned, as it were, in isolation. We can observe an analogous situation with frequency and pitch.

Because we have no words analogous to red or green to describe pitch, other than the rather vague descriptions "high" or "low", we tend to describe pitch in terms of frequency, as "so many hertz". This has in turn led to a good deal of confusion about pitch mechanisms (Whitfield, 1980a).

Although, in certain circumstances, the pitch associated with a given single frequency can vary slightly (Zurmühl, 1930; Stevens, 1935), there is, in the case of single tones, a reasonable one-to-one correspondence between pitch and frequency. With complex tones the case is very different; many different harmonic combinations can have the same pitch, and such combinations need not include the fundamental at all. Thus, the complexes 700, 800, 900, 1000 Hz; 600, 700, 1100, 1200 Hz; 100, 400, 500, 700 Hz will all have the same pitch, a pitch that can be matched by a single tone of 100 Hz.

Periodicity theories (Schouten, 1940, 1970) have proved unsatisfactory in explaining this "missing fundamental" effect, and in 1970 I put forward a "best fit" theory to account for this and related phenomena (Whitfield, 1970). The brainstem auditory mechanisms can be regarded as resolving a complex tone into its component frequencies each of which occupies a separate block of fibers in the total array (Whitfield, 1967). Thus, the complete set of harmonics, 1st, 2nd, 3rd . . . . *n*th, will produce a neural spectrum like that shown diagrammatically in Fig. 7a. Any sufficient subset (usually three or more components) of this spectrum—which may (b) or may not (c) contain the fundamental—is recognized as belonging, and so has the pitch of (a). Note, however, that if only the even harmonics are present, then these belong to a new series with a pitch one octave higher.

This theory was named the "best fit" theory because of the way it accounts for the pitch of anharmonic series. The harmonic combination 1200, 1400, 1600 Hz has a difference frequency of 200 Hz and a pitch that can be matched by a single tone of 200 Hz. The combination 1300, 1500, 1700 Hz also has a difference frequency of 200 Hz, but does not have the 200-Hz pitch. The pitch is in fact ambiguous, and can be matched by tones of either 214 or 187 Hz approximately. It will be noted that the 6th, 7th and 8th harmonics of 214 are 1284, 1498 and 1712, respectively, while the 7th, 8th, and 9th harmonics of 187 are 1309, 1496 and 1683. 1500 Hz is actually the 7th harmonic of 214. 3 Hz and the 8th harmonic of 187.5Hz. Thus, it seems the system adopts the nearest true harmonic series as a measure of the pitch. In some instances it is possible to detect more than two pitches, and generally speaking a whole spectrum of pitches exists, though since the worse the fit the weaker the pitch, the outlying members are rapidly swamped.

On the basis of the foregoing theory, pitch exhibits constancy across a range of input patterns, just as we saw that an object had constancy across the range of inputs it might generate, or a sound source preserved its position in space as the observer moved. We might therefore expect that pitch would be disrupted by damage to the auditory cortex, and that proves to be the case. In saying earlier that frequency discrimination is unimpaired by cortical ablation, we mean just that; frequency can be discriminated but not pitch. In the case of single tones it is impossible to separate the two, but in the case of complex tones separation is readily achieved.

Heffner and Whitfield (1976) showed by the following experiment that cats perceive the missing fundamental in much the same way as humans. Four pairs of complex tone signals were prepared (Table I). The left-hand triad in each pair was succeeded by the right-hand triad, as shown (dah-dah–dah-dah–), and was presented to the animal in a lick-suppression apparatus. The signal A is a rising-pitch signal, and it is to be noted that the actual frequency components also increase in frequency or stay the same. Likewise, B is a falling-pitch signal in which the frequency components also fall or stay the same. By contrast the

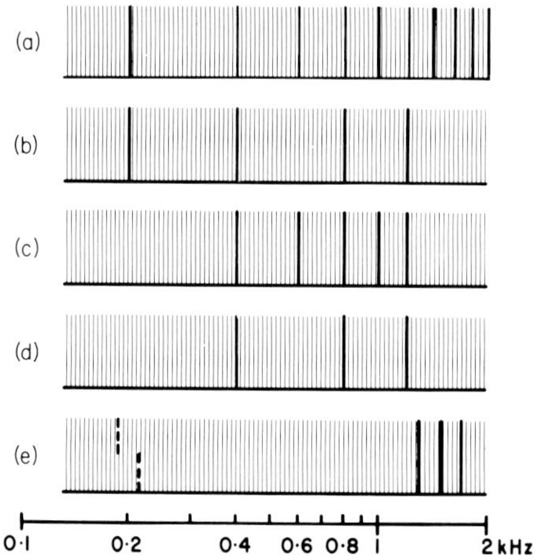

Figure 7. (a) Neural response pattern along the auditory array for a set of harmonics based on a fundamental frequency of 200 Hz. All signals containing identifiable parts of this pattern (b) will be assigned the same pitch, even if the fundamental component is missing (c). However, if the odd harmonics are systematically absent (d), then the best fit is with a tone an octave higher and the higher pitch is assigned. The anharmonic triad in (e) fits equally well a harmonic series based on 187 Hz or on 214 Hz, and the pitch is ambiguous. After Whitfield (1970).

signal C, which also has rising frequencies, has a falling "pitch", while D has falling frequencies and a rising pitch. In other words, the pitch and frequency move in opposite directions. The cats were first of all trained on the pair AB, one forming the "safe" and the other the warning signal. When the animals reached criterion they were presented with C and D as test stimuli. It was found that they treated C like B and D like A, showing that it is the pitch shift and not the frequency shift that is significant.

In further experiments (Whitfield, 1980b), auditory cortex was removed bilaterally. Animals previously trained on the AB discrimination lost that training but could be retrained to make the discrimination as accurately as before. However, when now presented with the CD pair they no longer made the transfer; the pitch equivalence was lost. It is important to note that they do not recognize the frequency change either; they can, however, be retrained to it. Indeed, the decorticate cats can be trained to discriminate any pair of these signals, but each pair requires separate training, just as separate training was required for the elements of the pulse-train signals in the Masterton and Diamond experiment (p. 337).

11. Levels of Conceptualization

We noted earlier (p. 334) that once the problem of discriminating two intensities had been learned, the input to the auditory neocortex could be severed without loss of the learned response, and it was argued that the role of the cortex lay in "identifying the problem". That, however, is clearly a very crude statement of the conceptual roles of the auditory cortex. Obviously, there must be a large number of conceptual levels disrupted in an involved manner by any one of the comparatively crude ablations that experimenters have to date been able to carry out. I believe, however, that we can now identify at least three. These may be illustrated by an experiment carried out by Whitfield *et al.* (1972), known as the "one-versus-two" experiment.

Cats were required to choose the left arm of a Y-maze when a single train of tone pips was sounded on the left (L) and to choose the right arm when identical trains of tone pips sounded on both sides (L/R). The repetition rates of the two trains were not locked together, so that there was a slight drift or jitter. Thus, the precedence effect did not operate, and the double stimulus was

Table I. Pitch versus Frequency Responsiveness[a]

A	1600–1800–2000 → 1600–1850–2100 ↑
B	1600–1800–2000 → 1600–1750–1900 ↓
C	1600–1800–2000 → 1700–1850–2000 ↓
D	1600–1800–2000 → 1500–1750–2000 ↑

[a] The signal consists of a triad of frequencies (Hz); the left-hand triad is succeeded by the right-hand triad to form a signal pair with a rising or falling pitch (vertical arrows). In A and B, the frequencies and the pitch move in the same direction; in C and D, they move in opposite directions.

quite neutral as far as sidedness was concerned. Indeed, the cats could just as readily be trained to go right to a single right stimulus and left to the double stimulus as the other way round.

After training to criterion (>95% success) on the L vs. L/R problem, the animals had the left auditory cortex removed. The success rate fell to about 70%, the errors being made about equally to the two stimuli L and L/R. If, on the other hand, animals were not trained until *after* the lesion had been made, a different picture emerged. The success rate was lower (57%), but more importantly, the performance was asymmetric, many more mistakes being made for the L/R signals that the L signals; the success rate for L was in fact enhanced above the 70% rate for pre-ablation training, while that for L/R was correspondingly depressed below the mean.

This result can be explained by the following hypothesis (see Whitfield, 1979a). Suppose that after ablation the simple L signal can still be attached to the response "go left", but that the complex L/R cannot be attached to "go right". The latter will be run at random. If there is no change in the original degree of discrimination (70%), then out of 100 presentations of L, 70 will be correctly run, but since the 30% "wrong" identifications are run at random, a proportion of these will be run to the left, giving an apparent, but spurious, success. The result is that the overall success rate on L is raised above 70%. Conversely, of those stimuli correctly identified as L/R, only around half will be run to the correct side, thus enhancing the asymmetry. The theoretical mean rate on the basis of pure randomness of the L/R response would be 60%, but with many more successes to L than to L/R.

Now it is to be noted that in the experiment described on p. 340, in which the signals L, R, LR, RL were intermixed, removal of the left auditory cortex disrupted only the response to RL. *The responses to the other three were unchanged.* In the experiment just described (where the lesion was the same) the performances on both L and L/R were equally impaired, even though if the single left L were opposed to a single right R, the response to L was perfectly normal. In the first case, the cortical ablation evidently disrupted the concept that RL represents a sound "out there" on the right. In the second stimulus situation, a different concept appears to be disrupted, namely that it is the difference between L and L/R that is critical for the cat to solve its food reward problem. This uncertainty produces a symmetrically degraded performance. Note that this result is unaffected by whether the learning takes place before or after cortical removal.

The third conceptual category might be termed "attachment of the response". This is an important, and apparently distinct part of a behavioral pattern. A cat can readily be taught to choose the left arm of a Y-maze in response to a sound on the left and the right arm in response to a sound on the right, and this pattern can even be reversed. However, a normal cat seems quite unable to form the concept "choose the left arm when you hear a rising tone and the right arm when you hear a falling tone". Nevertheless, it can readily learn "continue licking milk when you hear a rising tone, stop licking when you hear a falling tone".

In the experiment described, removal of the left auditory cortex resulted in the animal being unable to attach the signal L/R to the response "choose the right arm", although the lesion did not apparently affect the ability to attach the signal L to the response "choose the left arm." The interesting feature is that once this concept had been formed, the actual attachment was apparently made elsewhere, and the cortex could then be dispensed with.

Thus, we can recognize three conceptual classes disrupted by major lesions of the auditory cortex:

a. Object formation, e.g., the concept of the position of sound sources in space.
b. The significance or otherwise of particular stimulus situations
c. Decisions about the appropriate responses to be attached to (a) or (b)

12. Multiple Representation in the Auditory Cortex

In discussing the effects of cortical ablation, reference has already been made to various cortical areas, AI, AII, Ep, etc. Since the early studies of Woolsey and Walzl, it has been recognized that there is more than one "representation" of the periphery in the auditory cortex and, indeed, the same is true of sensory cortex in general. By multiple representation is meant that a given point in the periphery projects to several different cortical points, as determined either anatomically or from electrical responsiveness (see Brugge and Reale, this volume). The number of such areas that can be delineated will depend very much on the experimental methods, the criteria used, and the species studied. For the auditory system it is of the order of three or four (Fig. 4). What do these representations represent? So far there has been no satisfactory explanation. It is true that "the more you take out the worse the deficit". Thus (p. 338), sound localization was greatly impaired by removal of AI, AII, and Ep, and completely abolished by the additional removal of SII, I-T, and SS. However, we are certainly not in a position to assign a particular type of deficit to a particular area. This may be in part due to the fact that experimental situations that would distinguish between conceptual categories such as those outlined in (a)–(c) above have simply not been tried in the presence of limited lesions, or at all. The categories proposed are not necessarily the right ones if our purpose is to assign distinct functions to particular cortical areas, nor indeed may the limits of the presently described cortical areas be functionally correct. It is possible to trace the organized build-up of cortical structures from sensory cortical columns, through large subunits to the still larger labeled cortical areas, although the homologies between these areas in different modalities must be treated with caution (see Whitfield, 1984, p. 281ff for discussion). However, if we are to produce a systematic explanation of the apparent repetitions, we clearly have to reconcile this structural system with the behavioral deficits, if any, to be associated with limited lesions.

What mainly seems to emerge from the totality of experimentation in auditory neocortex is that we have been incorrect in examining it in terms of discrimination. The losses that ensue when it is damaged are those concerned with similarities rather than differences, so that the function of auditory cortex and probably all sensory cortex is to group together those sets of diverse stimuli that have a common origin. It cannot be too strongly emphasized that this is entirely different from classification—indeed diametrically opposed to it. Classification on the basis of the presence or absence of a particular feature or property assists discrimination. *Recognition* requires an understanding that diverse sets of stimuli, which may have no feature in common, "belong to" the same external source (the "object"). Objects and concepts are, however, man-made—or at least nervous-system-made. A stereo sound-image does not have any existence in the absence of a suitable auditory system, and the same might be argued, in more general terms, about a chair. A traveling α-particle would give a very different description from that given by a cat. We ask ourselves therefore: Why does a cat see objects while an α-particle does not?

13. Transforms

We are so accustomed to listening to sounds in terms of their harmonic structure that we often overlook the peculiarity involved in doing so. Sound waves, even periodic sound waves, do not "contain" harmonics. Fourier analysis, which can represent such waves in harmonic terms, is a quite sophisticated mathematical technique originally devised for the study of heat flow. The fact that we hear in harmonic terms seems to be a reflection of the fact that the peripheral auditory system carries out something closely approximating to a Fourier transformation on the incident sound wave. The Fourier transform has the effect of changing the variable from time to frequency, and this change is reflected in the so-called "place theory" of hearing. The point is illustrated by considering the sensations produced by two acoustic square waves differing only in that one has unity mark/space ratio while the other has a ratio of, say, 6 : 1. On switching from the first to the second, the listener is not at all aware that there has been a change of mark/space ratio, but very much aware of the entry of the octave (the second harmonic). Examination of the Fourier transforms shows that whereas the second waveform contains a large second term, that term is zero in the first. One could correspondingly demonstrate the response or lack of response of the appropriate units in the auditory pathway as the switch is made.

This characteristic may well be quite general. We hear sound in terms of harmonics because of the nature of the auditory brainstem transform. It is suggested that we see or hear the external world as "objects on a ground" not because objects exist as such, but because that is how the cortical transform naturally represents them; no cortex, no object. If we can discover such a transform, we shall have a clue both to the structural organization of the cortex and the kinds of behavioral deficits to be expected from circumscribed lesions.

14. Series

The Fourier transform is represented by a series of terms (the harmonic series), and knowledge of such series enables us to predict the sort of deficit that omission of part of the series will produce. It is an important property of many such (convergent) series that omission of some of the terms produces a general degradation, rather than a total loss. For example, we may store values of sin θ in a table of numbers; destruction of part of the table will result in the total loss of some values and total retention of others. But we can also store values of sin θ as a series:

$$\sin \theta = \theta - \frac{\theta^3}{3!} + \frac{\theta^5}{5!} - \frac{\theta^7}{7!} + \cdots$$

Here, destruction of one or more terms will progressively impair the accuracy for all values of θ, but no particular value will be entirely lost. This corresponds to the situation repeatedly described with respect to behavioral loss when cortical lesions are made. It seems that the known patterns of interconnection of the substructures of sensory cortex would be well-adapted to carrying out a series of transformations of this type, to yield levels of varying degrees of conceptual elaboration (Whitfield, 1984).

15. Summary

Auditory neocortex appears not to be essential for simple discriminations involving, for example, frequency, intensity, or position. Such discriminations can be carried out by other telencephalic structures, on the substrate of patterns existing in the sensory pathway.

Cortical lesions affect transfer of training, presumably by destroying the ability to detect *similarities* between signals. This is in turn related to the destruction of the ability to form concepts concerning constancies, such as the constancy of the position of a sound source when the listener moves, or the constancy of pitch as the harmonic composition changes.

Other classes of concept that have been shown to be affected by cortical lesions are the ability to understand the nature of the problem, and the attachment of a particular response to the discriminatory output.

It is suggested that the various "representations" of the periphery at the cortex constitute different conceptual levels and that the neural transform at each of these levels closely determines the way the concept is formed. Some understanding of these transforms will be essential to an understanding of the sort of behavioral deficit to be expected from circumscribed cortical lesions.

16. References

Adrian, E. D., 1947, *The Physical Background of Perception,* Oxford University Press (Clarendon), London.

Allen, W. F., 1945, Effect of destroying three localized cerebral areas for sound on correct differential responses of the dog's foreleg, *Am. J. Physiol.* **144:**415–428.

Andrews, B. W., and Pollen, D. A., 1979, Relationship between spatial frequency selectivity and receptive field profile of simple cells, *J. Physiol. (London)* **287:**163–176.

Beach, F. A. (ed.), 1960, *The Neuropsychology of Lashley,* McGraw–Hill, New York.

Bremer, F., 1939, Analyse oscillographique des reponses sensorielles de l'aire acoustique corticale chez le chat, *C.R. Soc. Biol.* **130:**257–261.

Butler, R. A., Diamond, I. T., and Neff, W. D., 1957, Role of auditory cortex in discrimination of changes of frequency, *J. Neurophysiol.* **20:**108–120.

Campbell, F. W., and Robson, J. G., 1968, Application of Fourier analysis to the visibility of gratings, *J. Physiol. (London)* **197:**551–566.

Cuénod, M., Casey, K. L., and MacLean, P. D., 1965, Unit analysis of visual input to posterior limbic cortex. I. Photic stimulation, *J. Neurophysiol.* **28:**1101–1117.

Diamond, I. T., Goldberg, J. M., and Neff, W. D., 1962, Tonal discrimination after ablation of auditory cortex, *J. Neurophysiol.* **25:**223–235.

Ebner, F. F., 1969, A comparison of primitive forebrain organization in metatherian and eutherian mammals, *Ann. N.Y. Acad. Sci.* **167:**241–257.

Evans, E. F., 1968, Cortical representation, in: *Hearing Mechanisms in Vertebrates* (A. V. S. de Reuck and J. Knight, eds.), Churchill, London, pp. 272–287.

Feher, O., and Whitfield, I. C., 1966, Auditory cortical units which respond to complex tonal stimuli, *J. Physiol. (London)* **182:**39P.

Heffner, H. H., and Masterton, R. B., 1975, Contribution of auditory cortex to sound localization in the monkey, *J. Neurophysiol.* **38:**1340–1358.

Heffner, H. H., and Whitfield, I. C., 1976, Perception of the missing fundamental by cats. *J. Acoust. Soc. Am.* **59:**915–919.

Helmholtz, H. L. F., 1863, Die Lehre von den Tonempfindungen als physiologische Grundlage für die Theorie der Musik, English translation of 3rd ed. by A. J. Ellis, *On the Sensations of Tone,* Longmans Green, London, 1875.

Hubel, D. H., and Wiesel, T. N., 1962, Receptive fields, binocular interaction and functional architecture in the cat's visual cortex. *J. Physiol. (London)* **160:**106–154.

Kavanagh, G., and Kelly, J. B., 1983, The effects of auditory cortical lesions on seven choice sound localizations by ferrets, *Soc. Neurosci. Abstr.* **9:**956.

Kelly, J. B., 1980, Effects of auditory cortical lesions on sound localization by the rat, *J. Neurophysiol.* **44:**1161–1174.

Kelly, J. B., and Whitfield, I. C., 1971, Effect of auditory cortical lesions on discriminations of rising and falling frequency-modulated tones, *J. Neurophysiol.* **34:**802–816.

LeVay, S., and Sherk, H., 1981, The visual claustrum of the cat. I. Structure and connections, *J. Neurosci.* **1:**956–980.

McCulloch, W. S., Garol, H. W., Bailey, P., and von Bonin, G., 1942, The functional organization of the temporal lobe, *Anat. Rec.* **82:**430.

MacLean, P. D., 1949, Psychosomatic disease and the 'visceral brain': Recent developments bearing on the Papez theory of emotion, *Psychosom. Med.* **11:**338–353.

Masterton, R. B., and Diamond, I. T., 1964, Effects of auditory cortical ablation on discrimination of small binaural time differences, *J. Neurophysiol.* **27:**15–36.

Merzenich, M. M., 1982, Organization of primate sensory forebrain structures: A new perspective, in: *New Perspectives in Cerebral Localization* (R. A. Thompson and J. R. Green, eds.), Raven Press, New York, pp. 47–62.

Meyer, D. R., and Woolsey, C. N., 1957, Effects of localized cortical destruction on auditory discriminative conditioning in the cat, *J. Neurophysiol.* **15:**149–162.

Neff, W. D., 1961, Neural mechanisms of auditory discrimination, in: *Sensory Communication* (W. A. Rosenblith, ed.), Wiley, New York, pp. 259–278.

Neff, W. D., and Diamond, I. T., 1958, The neural basis of auditory discrimination, in: *Biological and Biochemical Bases of Behavior* (H. F. Harlow and C. N. Woolsey, eds.), University of Wisconsin Press, Madison, pp. 101–126.

Papez, J. W., 1937, A proposed mechanism of emotion, *Arch. Neurol. Psychiatry* **38:**725–743.

Raab, D. H., and Ades, H. W., 1946, Cortical and midbrain mediation of a conditioned discrimination of acoustic intensities, *Am. J. Psychol.* **59:**59–83.

Ravizza, R. S., and Masterton, R. B., 1972, Contribution of neocortex to sound localization in opossum (*Didelphis virginiana*), *J. Neurophysiol.* **35:**344–356.

Rutherford, W., 1886, A new theory of hearing, *J. Anat. Physiol.* **21:**166–168.

Saul, L. J., and Davis, H., 1932, Action currents in the central nervous system. I. Action currents of the auditory tracts, *Arch. Neurol. Psychiatry* **28:**1104–1116.

Schouten, J. F., 1940, The perception of pitch, *Philips Tech. Rev.* **5:**286–294.

Schouten, J. F., 1970, The residue revisited, in: *Frequency Analysis and Periodicity Detection in Hearing* (R. Plomp and G. F. Smoorenburg, eds.), Sijthoff, Leiden, pp. 41–54.

Stevens, S. S., 1935, The relation of pitch to intensity, *J. Acoust. Soc. Am.* **6:**150–154.

Stevens, S. S., and Davis, H., 1938, *Hearing: Its Psychology and Physiology*, Wiley, New York.

Stone, J., and Dreher, B., 1973, Projection of X and Y cells of the cat's lateral geniculate nucleus to areas 17 and 18 of visual cortex, *J. Neurophysiol.* **36:**551–567.

Swarbrick, L., and Whitfield, I. C., 1972, Auditory cortical units selectively responsive to stimulus 'shape,' *J. Physiol. (London)* **224:**68–69P.

Thompson, R. F., 1959, The effect of training procedure upon auditory frequency discrimination in the cat, *J. Comp. Physiol. Psychol.* **52:**186–190.

Thompson, R. F., 1960, Function of the auditory cortex of cat in frequency discrimination, *J. Neurophysiol.* **23:**321–334.

Uttley, A. M., 1954, The classification of signals in the nervous system, *Electroencephalogr. Clin. Neurophysiol.* **6:**479–494.

Uttley, A. M., 1975, The conversion of printed characters into speechlike sound for a blind reading aid, *Med. Biol. Eng.* **1975:**285–293.

Wallach, H., Newman, E. B., and Rosenzweig, M. R., 1949, The precedence effect in sound localization, *Am. J. Psychol.* **62:**315–316.

Wever, E. G., and Bray, C. W., 1930, Action currents in the auditory nerve in response to acoustical stimulation, *Proc. Natl. Acad. Sci. USA* **16:**344–350.

Whitfield, I. C., 1956, Electrophysiology of the central auditory pathway, *Br. Med. Bull.* **12:**105–109.

Whitfield, I C., 1967, *The Auditory Pathway,* Arnold, London.

Whitfield, I. C., 1970, Neural integration and pitch perception, in: *Excitatory Synaptic Mechanisms* (P. Andersen and J. K. Jensen, eds.), Universitets forlaget, Oslo, pp. 277–285.

Whitfield, I. C., 1979a, The object of the sensory cortex, *Brain Behav. Evol.* **16:**129–154.

Whitfield, I. C., 1979b, Auditory cortical lesions and the precedence effect in a four-choice situation, *J. Physiol. (London)* **289:**81P.

Whitfield, I. C., 1980a, The relation between pitch and frequency in complex tones, in: *Psychophysical, Physiological and Behavioural Studies in Hearing* (G. van den Brink and F. A. Bilsen, eds.), Delft University Press, pp. 361–366.

Whitfield, I. C., 1980b, Auditory cortex and the pitch of complex tones, *J. Acoust. Soc. Am.* **67:**644–647.

Whitfield, I. C., 1982, Coding in the auditory cortex, in: *Contributions to Sensory Physiology* (W. D. Neff, ed.), Volume 6, Academic Press, New York, pp. 159–178.

Whitfield, I. C., 1984, *Neurocommunications: An Introduction,* Wiley, New York.

Whitfield, I. C., and Evans, E. F., 1965, Responses of auditory cortical neurones to stimuli of changing frequency, *J. Neurophysiol.* **28:**655–672.

Whitfield, I. C., Cranford, J., Ravizza, R., and Diamond, I. T., 1972, Effects of unilateral ablation of auditory cortex in cat on complex sound localization, *J. Neurophysiol.* **35:**718–731.

Woolsey, C. N., and Walzl, E. M., 1942, Topical projection of nerve fibers from local regions of the cochlea to the cerebral cortex of the cat, *Johns Hopkins Hosp. Bull.* **71:**315–344.

Zurmühl, G., 1930, Abhangigkeit der Tonhöhenempfindung von der Lautstärke und ihre Beziehung zur Helmholtzschen Resonanztheorie des Hörens, *Z. Sinnesphysiol.* **61:**40–86.

Index